Astrophysical Recipes (Second Edition)

The art of AMUSE

Online at: https://doi.org/10.1088/2514-3433/adfc3c

AMERICAN ASTRONOMICAL SOCIETY IOP | ebooks™

About the program:

AAS-IOP Astronomy ebooks is the official book program of the American Astronomical Society (AAS) and aims to share in depth the most fascinating areas of astronomy, astrophysics, solar physics and planetary science. The program includes publications in the following topics:

- GALAXIES AND COSMOLOGY
- INTERSTELLAR MATTER AND THE LOCAL UNIVERSE
- STARS AND STELLAR PHYSICS
- EDUCATION, OUTREACH, AND HERITAGE
- HIGH-ENERGY PHENOMENA AND FUNDAMENTAL PHYSICS
- THE SUN AND THE HELIOSPHERE
- THE SOLAR SYSTEM, EXOPLANETS, AND ASTROBIOLOGY
- LABORATORY ASTROPHYSICS, INSTRUMENTATION, SOFTWARE, AND DATA

Books in the program range in level from short introductory texts on fast-moving areas, graduate and upper-level undergraduate textbooks, research monographs, and practical handbooks.

For a complete list of published and forthcoming titles, please visit iopscience.org/books/aas.

About the American Astronomical Society

The American Astronomical Society (aas.org), established 1899, is the major organization of professional astronomers in North America. The membership (~7,000) also includes physicists, mathematicians, geologists, engineers, and others whose research interests lie within the broad spectrum of subjects now comprising the contemporary astronomical sciences. The mission of the Society is to enhance and share humanity's scientific understanding of the universe.

Editorial Advisory Board

Astrophysical Recipes (Second Edition)

The art of AMUSE

Simon Portegies Zwart
Sterrewacht Leiden, Leiden University, Einsteinweg 55, 2333 CC, Leiden, The Netherlands

Steve McMillan
Drexel University, Philadelphia, PA 19104, USA

Steven Rieder
Anton Pannekoek Instituut, Science Park 904, 1098 XH Amsterdam, The Netherlands

IOP Publishing, Bristol, UK

ISBN 978-0-7503-5331-1 (ebook)
ISBN 978-0-7503-5329-8 (print)
ISBN 978-0-7503-5332-8 (myPrint)
ISBN 978-0-7503-5330-4 (mobi)

DOI 10.1088/2514-3433/adfc3c

Version: 20260201

AAS–IOP Astronomy
ISSN 2514-3433 (online)
ISSN 2515-141X (print)

British Library Cataloguing-in-Publication Data: A catalogue record for this book is available from the British Library.

Published by IOP Publishing, wholly owned by The Institute of Physics, London

IOP Publishing, No.2 The Distillery, Glassfields, Avon Street, Bristol, BS2 0GR, UK

US Office: IOP Publishing, Inc., 190 North Independence Mall West, Suite 601, Philadelphia, PA 19106, USA

In memoriam

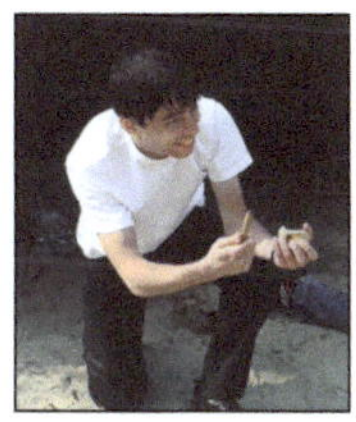

In memoriam Inti Pelupessy Gallo 1977–Leiden 2024

This photograph shows Inti Pelupessy at the Archeon Museum Park in the Netherlands, trying to make fire by rubbing together two wooden sticks. Inti was strongly involved from the start in the AMUSE project. He passed away in the summer of 2024, and is missed by all.

As a brilliant hydrodynamics expert (trained by the very best), Inti quickly became a key figure in the development of AMUSE, his swift mastery of this complex framework is a testament to his remarkable intellect. His pioneering work in this area resulted in important contributions to the field. Without his outstanding qualities as a researcher, his sharp thinking, his excellent understanding of the physics and his unsurpassable programming skills, AMUSE would simply not exist.

Possibly even more striking were his social skills. As you can see on the photograph, Inti was a natural educator, teaching on subjects ranging from procedural operations to thermodynamics, here demonstrated in the making of fire. Inti was reluctant to promote his own work, and always listened to the opinions of others before elaborating on his own ideas. Still, he was often one of the strongest minds in a collaboration.

Those of us fortunate enough to have known Inti will remember his deep insight, kindness, patience, and his dedication to science. His loss is felt deeply by us, and by all in his periphery but his legacy in both science and mentorship will live on.

Contents

Preface

This is the second, revised, edition of the academic textbook on the Astrophysics Multipurpose Software Environment, or AMUSE, for short (Portegies Zwart & McMillan, 2018). The AMUSE project started at the MODEST-5 conference in Hamilton, Ontario, in August 2004. Steve McMillan and Simon Portegies Zwart were sipping coffee, nibbling on cookies, and discussing the future of `starlab`.

`Starlab` began in the early 1990s as an alternative and competitor to Sverre Aarseth's NBODY series of codes. In collaboration with Piet Hut and Jun Makino, we wrote a modular package combining stellar dynamics and evolution using modern programming paradigms and data structures. The code was written in C++, an up-and-coming language at the time, which afforded many opportunities to go beyond the rigid arrays used in existing languages at the time.

The `starlab` package contained many novel features not found in comparable state-of-the-art codes even today. Most codes at the time, and certainly the leading NBODY series developed by Sverre Aarseth, were "monolithic" solvers, designed, written, and integrated for one particular purpose. A direct summation N-body code with a hard-coded realization of stellar evolution is an excellent example of a monolithic code. Because monolithic codes are written in a single programming language with simple data structures, they are generally relatively uncomplicated, straightforward to compile, and quite portable.

By contrast, `starlab` was a nonmonolithic code composed of two independent components; a gravitational N-body solver `kira` and the binary population synthesis code `SeBa`. Much of the internal coding in `starlab` was devoted to maintaining consistent communication between its constituent parts. The result was still quite tightly coupled, but it preserved the essential kernel of independent communicating parts. But, as we contemplated expanding our scope by including hydrodynamics, we realized that, although we had anticipated this possible expansion, we still did not really know how to do it without radically changing the basic data structures and rewriting the code. Hard-coding a hydro module into `starlab` would be counter to all our design choices and would be horribly complicated to code and maintain, but how else might we expand the code's functionality?

After more than a decade of reliable service and hundreds of scientific papers, `starlab`, we realized, faced complicated design choices for the future expansion of the package. Or, should we start all over again?

We had no clear solution to the problem, and we felt insufficiently confident to start all over until Sverre Aarseth and Peter Eggleton joined us. We started talking about the codes they had written. Sverre initiated and led the NBODY series of direct summation N-body solvers. His codes are highly optimized and always written in a historic FORTRAN dialect. His latest incarnation NBODY6, at that time, was the old-school counterpart of `starlab`. NBODY6 was faster than `starlab`, but (we think) was less elegantly structured, if one can speak of any formal structure. Peter is a pioneer in writing stellar structure and evolution codes.

The time Sverre spent writing N-body codes, Peter spent improving `EVTwin`, his stellar evolution code, also in vintage FORTRAN.

Sverre looked very happy to hear that we considered `starlab` past its prime. "NBODY6 is brand new" he told us, "very alive, and kicking bodies, yes." We thought that, in some way, we might benefit from the many decades of combined experience of these two grandfathers of computational astrophysics. Why redesign code? Why rewrite any software if we already have the right tools at hand? If only they would turn into useful tools with a proper interface. Could it be possible to incorporate one of these codes into the other? We asked Sverre how he would feel about this. He responded: "Wouldn't it be nice to have Peter's stellar evolution code as a part of my beautiful N-body code?" Peter responded: "But, Sverre, my dear friend, how splendid would it be to have your N-body code as part of my stellar evolution code."

That got us thinking. Maybe we should get away from "mine" and "thine," and operate instead in a collective framework, and we proposed: "What about if both your codes could work together?" While hardly ever in agreement, both men thought this would be an interesting idea, like the Borg (of whom we suspect neither had ever heard) assimilating the positive characteristics of everybody to become better as a whole. But they were even more sure that such an attempt would be impossible. You would have codes written in all sorts of languages. How would you deal with all the nasty communications and unit conversions worldwide? Also, the communication between codes will become a major bottleneck. No, better to have another cookie and dream about NBODY7.

By now, NBODY7 turned into NBODY6++; still in FORTRAN, though. But, in that conversation, we had the kernel needed to get started; together, we can rule the Galaxy. That evening, we had a quiet dinner and discussed the possibilities of a language-independent framework in which codes would operate as modules that carry out specific tasks. The communication between the various codes would have to go through some sort of interface, to be determined, allowing it to perform operations on the data, formatting it to the requirements of the other codes, and scheduling the execution of those codes. Maybe it was not so impossible after all.

The MODEST initiative, which gave rise to the Hamilton conference where this exchange took place, was a direct outgrowth of our experience with `starlab`. However, it eventually became clear that, even with the best intentions, we still had combined algorithms in a way that made it hard to make radical changes to the fundamentals.

This realization led Piet Hut and Mike Shara in 2001 to organize a conference at the American Museum of Natural History in New York, to bring together experts in stellar dynamics, stellar evolution, stellar hydrodynamics, computer science, and other relevant areas, to seek truly modular ways to combine codes to create a new computational framework. The meeting was subsequently named MODEST-1 (for Modeling Dense Stellar systems). Perhaps unsurprisingly, the early sample codes that emerged from the meeting looked very much like `starlab`, but MODEST (subsequently reacronymed as Modeling and Observing Dense Stellar systems) rapidly became an important forum for the discussion of all aspects of dynamical modeling and observations of dense stellar systems.

Early MODEST meetings focused more on astrophysics than computation. Still, following the MODEST-5 meeting in Hamilton, a rapid series of satellite meetings changed the software landscape. We organized a meeting in Lund, Sweden, in December 2005 and formulated a concrete plan for combining existing codes. The discussion continued a few months later in Amsterdam, where we organized another small workshop to discuss the possibilities. Piet Hut was there, along with Jun Makino, Peter Teuben, Pierro Spinnato, and Breanndán Ó Nualláin. Breanndán was a `Python` aficionado, and with his help, we had the first version of a rudimentary working code operational in a few days. We called this framework MUSE (for Multipurpose Software Environment).

The key idea of MUSE was that no code got to "drive" the simulation. Dynamical, stellar evolution, and other codes were invoked as modules, with suitably defined interfaces, but the top-level code was a `Python` script that created and coordinated the scientific study. The implementation involved heavy use of `Python` tools such as `f2py` and `swig`, and ran into numerous practical problems that need not concern us here, but the basic idea of a script-driven, democratic code framework was established.

After the fundamental ideas had been put into practice, and once some of their shortcomings were recognized, we realized that a bunch of astronomers would not be able to make quick progress on such a large software environment. We searched for funding to hire a dedicated software architect, Arjen van Elteren, who could help design our framework, and two computational astronomers, Inti Pelupessy and Nathan de Vries, to develop the first science cases with the new framework. The team turned out to be very effective. The framework is now called AMUSE (for Astrophysical Multipurpose Software Environment). Over the last decade, we worked hard with students to improve the package, broaden its scope, and widen the scientific applications. The most recent developments include incorporating codes for simulating exoplanetary atmospheres and interstellar chemistry, and the entire build-system was renewed. This manuscript explains some of the functionality and operations possible with the framework.

The main objective of this book, and the entire AMUSE framework, is to explore new science. To seek out new phenomena and new discoveries. To boldly go where no computer has gone before! This is the continuing mission of this exciting journey. The first edition of the AMUSE book, published in 2018, reflected the authors' roots in simulations of dense stellar systems, as well as the historical development of the field. In this new version, we sought a new collaborator and found him in Steven Rieder. He was closely connected to the AMUSE endeavor from the beginning; somewhat junior at the time of the initial development, he has developed to a true AMUSE master, and took care of most of the maintenance, documentation, updates, bug fixes, etc.; AMUSE's presence is strong in our academic family.[1]

This new book is divided into three parts. In Part I, we discuss the fundamentals of AMUSE and the working of the monolithic solvers; gravity, stellar evolution,

[1] Adopted from Luke Skywalker to Leia Organa in Star Wars: Return of the Jedi (1983).

hydrodynamics, and radiative transfer. In Part II, we discuss elementary coupling strategies adopted in AMUSE. Part III brings all AMUSE capabilities together and shows the reader how to design complicated multiscale and multiphysics experiments. In the last part, we also show how AMUSE can be adapted and applied to an even broader range of problems. We present complete AMUSE scripts for performing high-fidelity simulations and provide some examples outside the domain of astrophysics.

But above all, the goal of this book is to have fun. To enjoy computational astrophysics, and to explore the nonlinear couplings between different physical domains without having to worry too much about the details of energy conservation in a close three-body interaction or managing the arcana of high-dimensional differential equations. Decades of computational astrophysics research have gone into some of the modules incorporated into AMUSE. These are trustworthy codes and form an excellent basis for further study. Still, never trust a code that does not work, but trust a working code even less.

The AMUSE framework is still a work in progress. Even after more than a decade of coding, thinking, and writing, we suspect this book and the framework to contain errors and typos that neither we, our reviewers, nor our editors caught. For these reasons, we will maintain an evolving document in AMUSE that lists corrections to the published text and highlights developments in AMUSE that may interest readers, such as new algorithms, functionality, or insights into the many problems we address. It can be found in `AMUSE_updates.pdf`, which can be saved via

```
from amuse.examples import export
export("AMUSE_updates.pdf", savedir="./")
```

where `amuse.examples` is the examples package companion to this book (see Section 1.4). Incidentally, the "laptop" cited in the timing estimates for most figures is Simon's Lenovo P14s, 20-core 3rd Gen Intel Core i7-1370P with an NVIDIA RTX A500 GPU. Your timing may differ.

Simon Portegies Zwart
Steve McMillan
Steven Rieder

Reference

Portegies Zwart, S., & McMillan, S. 2018, *Astrophysical Recipes: The Art of AMUSE* (Bristol: IOP Publishing)

Acknowledgments

> I don't know half of you half as well as I should like; and I like less than half of you half as well as you deserve.
>
> —J.R.R. Tolkien

These words were uttered by Bilbo Baggins on 22 September 3001 in front of 144 invited guests to celebrate his eleventy-first birthday. With this speech, he wanted to say farewell to and acknowledge friends and family. We also would like to acknowledge a large number of people who contributed to this book, although we will not go so far as to insult half of them. We at least do not consider any of them at the same level as Otho and Lobelia.

AMUSE is really a community effort, and a large number of people have contributed. The Master's students at Leiden Observatory helped to eliminate bugs and typos during the various courses on computational astrophysics. We apologize to them for having used them as guinea pigs for the first version of this manuscript.[2] We particularly thank the many people who contributed by adding new code, developing or debugging scripts, or initiating new ideas through stimulating discussions. They include: Assilkan Adilkhan, Eric Andersson, Sabrina Appel, Jeroen Bédorf, Tjarda Boekholt, Anthony Brown, Maxwell Cai, Claude Cournoyer-Cloutier, Niels Drost, Will Farner, Michiko Fujii, Evghenii Gaburov, Joseph Glaser, Guilherme Goncalves, Adrian Hamers, Erwan Hochart, Shuo Huang, Piet Hut, Masaki Iwasawa, Lucie Jílkova, Nadia Kempers, Rony Keppens, Ralph Klessen, Rob Knop, Jun Makino, Tünde Meyer, Sean Lewis, Carmen Martínez-Barbosa, Mordecai-Mark Mac Low, Breanndán Ó'Nualláin, Jan-Pieter Paardekooper, Veronica Saz Ulibarrena, Alison Sills, Jelmer Stroo, Andrew Pellegrino, Brooke Polak, Silvia Toonen, Aaron Tran, Alessandro Trani, Gijs Vermariën, Serena Viti, Joshua Wall, Long Wang, Julia Wasala, Maite Wilhelm, Alfred Whitehead, and Thomas Wijnen.

Several of the nicest figures were made by Edwin van der Helm (Figure 8.5), Alex Rimoldi (Figure 6.4), Nora Lützgendorf (Figure 10.3), and Sabrina Appel (Figure 10.4).

We received illuminating private lectures on gravitational dynamics (Chapter 4) by Sverre Aarseth, on stellar evolution (Chapter 5) by Onno Pols and Peter Eggleton, on hydrodynamics (Chapter 6) by James Lombardi and Volker Springel, and on radiative transfer (Chapter 7) by Barbara Ercolano and Kees Dullemond.

During the development of AMUSE and the writing of this book, we made extensive use of many additional public facilities and packages, including `ADS`,

---

[2] Here, we refer to the Cavia porcellus, in honor of the 1901 Nobel prize in Medicine, awarded to E. Adolf von Behring, who used this cute furry rodent to develop a serum therapy against diphtheria.

`arXiv`, `CDS`, `Git`, `HDF5`, `IPython`, `Jupyter`, `LaTeX`, `Linux`, `MPI`, `Matplotlib`, `NumPy`, `OpenMP`, `Python`, `Seaborn`, `Wikipedia`, and probably many more we forgot to mention.

All three authors had the opportunity to spend an extensive amount of time on this book, for which we are grateful to our (former) host institutions: Leiden Observatory, Drexel University, Geneva Observatory, KU Leuven, and the Anton Pannekoek Institute for Astronomy. But, we are also grateful to the Center for Planetary Science in Kobe, the Lorentz Center in Leiden, Geneva Observatory, the University of Strasbourg, the Anton Pannekoek Institute, and KU Leuven for hosting our discussions. Parts of the development of AMUSE were financed by NOVA NWO (639.073.803 and 614.061.608), NASA (grants NNX07AG95G, NNX08AH15G), NSF (AST0708299), H2020 FET ComPat, the EU P7/08 CHARM project, and the NLeSc. We used copious amounts of supercomputer time on SURF/Snellius (2023.008, 2024.058, EINF-10097, EINF-11229, EINF-12718, EINF-13923) and LUMI (EINF-11368).

Without Arjen van Elteren, Inti Pelupessy, and Nathan de Vries, AMUSE would not exist, and we would never have been able to write this book. A special thank you also is due to the anonymous referee Douglas Heggie (of the first version), who identified many omissions and errors, which we hope we have corrected. We also thank our editor and staff at IoP, and in particular Robbie Trevellian, Leigh Jenkins, and Daniel Heatley, for their support of this writing project.

Our spouses and children earn considerable praise for their patience. They are very happy that this is over, and we are happy that they are still there.

Author biographies

Simon Portegies Zwart

Simon Portegies Zwart leads since 2009 an interdisciplinary research team on Computational Astrophysics at Leiden Observatory, focusing on understanding the universe through simulation while developing algorithms and tools to handle its complexity and multi-scale nature. He founded the IAU Commission C.1B on Computational Astrophysics, as well as the AMUSE project, and was honored by the IAU for naming an asteroid 12946 "portegieszwart".

Steve McMillan

Steve McMillan was Head of the Department of Physics at Drexel University until his retirement in 2024. He is now Professor Emeritus. His research interests lie in stellar dynamics and the few-body problem, computations of stellar systems, and the formation of stars and star clusters. These interests laid the foundation for the development of AMUSE, which is now central to this research.

Steven Rieder

Steven Rieder is a Scientific Research Software Engineer at the Anton Pannekoek Institute for Astronomy at the University of Amsterdam. His interests lie in the formation and evolution of dense stellar systems, the development of software that enables such research, and in science visualization.

Part I

Computational Astrophysics

Chapter 1

What is Computational Astrophysics?

> Informatica gaat net zo min over computers als astronomie over telescopen.
> Informatics is as little about computers as astronomy is about telescopes.
> —Edsger W. Dijkstra

Computational astrophysics has come a long way since the early days when astronomers began to use support devices for their calculations. In this chapter, we review some key developments in this field, embrace the relatively new concept of multiscale and multiphysics simulations, and introduce the `Python` programming environment that we will use for the rest of this book. We do this by presenting a series of examples which may look arcane now, but will (we hope) become clearer as we progress through the text.

1.1 Computational Astrophysics

Computational astrophysics bridges observation and theory. The improved quality of observations and the need for deeper understanding of the underlying physics have driven demand for higher resolution and increased physical realism in numerical simulations. With exponentially growing hardware performance, we expect that the importance of large-scale simulations will only increase, but the software must also improve to keep pace with observations and theory.

1.1.1 Origin of This Book

The idea for this book originated from writing the first version of the Astrophysical Multipurpose Software Environment (AMUSE) in 2009, which at the time we called the Multiphysics Software Environment (MUSE; Portegies Zwart et al. 2009). Since then, we have taught numerous courses in computational astrophysics based on our software, and supervised many students in research projects based on AMUSE.

doi:10.1088/2514-3433/adfc3cch1

The notes from these lectures and the example scripts written by our students and ourselves formed the first version of what we called the AMUSE Primer. Later, the idea gelled into a book, of which you are reading the second version right now.

Writing this book was fun. Writing the software was even more fun. Writing the example scripts was the hardest thing to do. The simplest looking codes often caused us most work. Our objective was to make these codes as short and crisp as possible. Making scripts educational without losing their main purpose was our primary challenge.

1.1.2 Hands on *is* Hands on

In our experience, the best way to learn is by doing. And the best way to start doing is to take an example that resembles your objective to some degree, and build on it. We therefore include a large number of examples in a separate repository within the AMUSE code distribution, with many code snippets incorporated into this book.

In practice, we prefer to work as follows:

- Start with a scientific question,
- Identify the minimal physics needed to address the question,
- Find an example script that resembles the target problem as closely as possible (or ask a large language model for help),
- Rewrite that script to make it solve our problem.

Making figures and animations can be time consuming but can also be very rewarding. Making a publishable figure can take as much as a day. In our examples, we adopt `matplotlib`[1] (Hunter 2007) as our graphics library. There are many alternatives to `matplotlib` for making animations or figures, but its principal advantage is its transparent interface with `Python`, making it easy to incorporate into AMUSE.

A simple script can be written in a few hours; a more complicated script may take days to get right. But then the long painstaking process from test script to production code starts. This generally entails a number of stages, as illustrated in Figure 1.1. The main steps include adjusting the script to one's needs, then refactoring the script, followed by profiling and optimization. We include the refactor step to stress the importance of preserving the readability of the source code, which so often leads to reproducible results (see Beck & Andres 2004 for more about refactoring).

1.1.3 What about the Math?

As Albert Einstein wrote on 1943 January 7 to nine-year-old Barbara Lee Wilson (Calaprice 2002):

> Do not worry about your difficulties in Mathematics. I can assure you mine are still greater.

[1] Matplotlib is well documented and publicly available as open source from http://matplotlib.org/api/pyplot_api.html

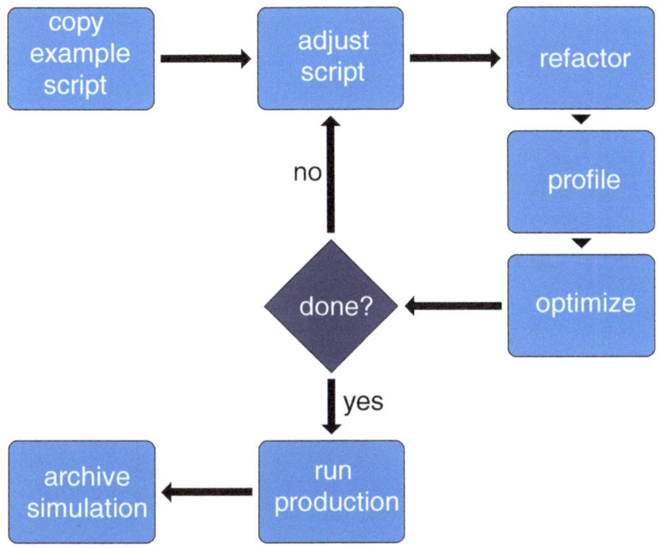

Figure 1.1. Workflow to design a production script.

A solid background in math is nice, and surely helpful, but you can probably read and understand most of this book (up to Part II) with only rudimentary mathematics and calculus. Similarly, a PhD or other strong background in astrophysics or computer science will certainly make some of the material easier to digest, but we have successfully used AMUSE and this text with students at all levels, from freshman undergraduate to Masters to PhD. We even have users with a high-school background.

1.1.4 Alternative Simulation Environments

There are no real alternatives to AMUSE, at least not if you desire to perform multiscale and multiphysics simulations. Here, we mention a few projects with similar objectives that most closely comply to the listed requirements for a multi-purpose simulation environment. These include `Cactus`, a multiscale modeling framework used in `muscle3`, `FDPS`, and `MMSF`. We will briefly discuss each of these frameworks.

`Cactus` (Goodale et al. 2003) is a general-purpose problem-solving environment for the physical sciences (Allen et al. 2001). It is modular and supports parallel computing across the network with a variety of hardware. `Cactus` was last updated (version 4.12) in May 2022.[2] `Cactus` offers several advantages over other frameworks. It uses a component-based architecture, where the core functionality is

[2] See http://www.cactuscode.org

provided by a central *"flesh"* and various *"thorns"* (modules) that can be independently developed, compiled, and loaded. The thorns can be parallelized, for multicore architectures and distributed computing environments. Its high level of abstraction and interoperability with other tools simplifies the development of complex simulations. Using `Cactus` can be very rewarding, but understanding the details is cumbersome, and it may require numerous recompilations to generate a working implementation.

`FDPS` stands for Framework for Developing Particle Simulators (Iwasawa et al. 2016).[3] It is built with much the same philosophy as AMUSE, but is much more fine-grained. If `FDPS` were some sort of Lego (in software terms) for simulating self-gravitating systems, then AMUSE would be more like Duplo: big blocks that allow you to build an interesting colorful building with only a little effort.

`FDPS` is a high-performance, open-source library, written in C++ and designed to simplify and accelerate the development of large-scale, parallel particle-based simulation codes—such as N-body or SPH simulations—on distributed-memory supercomputers (Iwasawa et al. 2016). `Petar` (Wang et al. 2020), the N-body code used in Torch (Section 10.2.1), is written using `FDPS`. It provides a good example of how code written in `FDPS` can be embedded in AMUSE.

`FDPS` is optimized for building fast, scalable, single-physics particle simulation codes from scratch, with a focus on parallel performance, whereas AMUSE is designed for integrating and orchestrating multiple existing codes within a high-level, user-friendly `Python` environment, prioritizing flexibility and multiphysics capabilities. Researchers should choose between them based on whether their priority is raw computational performance for custom particle simulations (`FDPS`) or flexible, multiphysics coupling and ease of scripting (AMUSE). As with `Cactus`, `FDPS` also requires recompilation after changes are made.

Finally, we briefly mention the Multiscale Modeling and Simulation Framework (MMSF, Borgdorff et al. 2013; Chopard et al. 2014), which is based on a similar philosophy as that underlying AMUSE. `MMSF` systematically organizes and couples models at different scales, providing a rigorous and flexible approach to simulating complex, multiscale systems. It is modular and is supported by formal languages and software tools for efficient implementation and execution. `MMSF` has been applied effectively in the `muscle3` environment (Veen & Hoekstra 2020). However, in contrast to AMUSE's structure of a driver script instructing a set of models to run and exchange information, it instead manages a set of distributed, mostly independent models which exchange messages via externally configurable connections, which block if a result they need is not yet available.

All three framework environments require recompilation before reuse, and are therefore not as suitable for experimentation as AMUSE. Each of them, however, is easier to tune to the appropriate hardware environment, but may lack native GPU support. Each package has its advantages and disadvantages. AMUSE stands out for its ease of use, breadth of included solvers, robust unit and data handling, and its

[3] See: https://github.com/FDPS/FDPS.

focus on reproducible, modular, and flexible multiphysics simulations. AMUSE and `FDPS` are geared towards astrophysics, whereas `MMSF` and `Cactus` target a broader range of applications.

1.1.5 Objective of This Book

This book is about how to perform simulations of astrophysical phenomena. These simulations have the goal of mimicking nature, or at least certain aspects of it, in order to better understand the Universe. By the end of this book, you will be able to perform many high-end simulations yourself. You will have learned how to set up initial conditions, how to design and develop a simulation environment to address your astrophysical question, how to perform the runs, and how to analyze the resulting data.

You can read this book in several ways. You can simply start at the beginning and work through to the end, or you may prefer to pick out the pieces you need in order to bootstrap your own research project. Either way, we recommend that you read the first couple of chapters to understand how AMUSE works and grasp some of its basic programming ideas.

1.1.6 What Is Missing from This Book

This book is far from exhaustive. Many topics, such as data mining, high-performance and distributed computing, programming paradigms for graphics processing units, operating systems, and the effects of limited numerical precision are entirely separate disciplines that we can only direct you to. AMUSE contains far more functionality than we can describe. We will only briefly address the algorithms underlying many of the phenomena discussed. Our discussion of the underlying physics is also necessarily limited. This is not an astrophysics text—rather, our goal is to present just enough astrophysics to allow you to harness the modules we describe.

Of the many astrophysical phenomena we could include, we have decided to limit our scope to gravitational dynamics, stellar evolution, hydrodynamics, radiative transfer, and chemistry. We neglect a lot of physics, including asteroseismology, astrobiology, planetary geology, cosmology, YORP,[4] and tidal dissipative effects. These are not unimportant, but the book is finite, and new modules addressing such areas are constantly being added to the AMUSE framework.

1.1.7 Outline of the Book

This book is divided into three parts. In Part I, Chapters 4 to 7, we discuss scripts designed to address single physical processes, for which only one numerical solver is required. A *minimal solver* is the simplest possible script needed to run a particular solver. In AMUSE, the low-level solvers—usually written in high-performance languages and often optimized for parallel architectures—are called *community codes*. When we include the AMUSE `Python` interface structure, we generally use

[4] The Yarkovsky–O'Keefe–Radzievskii–Paddack effect (Rubincam 2000) influences the spin of a minor body by reprocessing stellar radiation as thermal emission.

the more generic term *community module*. In practice, we often need additional modules to generate initial conditions, boundary values, or for postproduction analysis, but the main work in a minimal code is performed by a single community module. Minimal solvers run a particular code and stop after producing just enough information to gauge the quality of the solution. The solvers presented here should not only be viewed as examples to illustrate the AMUSE programming style and module interfaces but they can also be used as a basis for further work.

The next level of complexity is a *simple solver*, which is an extended version of a minimal solver. Simple solvers are meant to illustrate more complex and practical mono-physics applications. Within the AMUSE framework, we can run a single code as a standalone program. Since AMUSE contains multiple solvers for each physical domain it addresses, this allows for detailed and instructive comparison between them. The ease with which initial conditions can be generated, and with which the user can change solvers and analyze the results, makes running multiple solvers on the same problem a useful exercise. In this regard, the main advantage of using AMUSE is the unified way in which the input is parsed and delivered to, and results returned from, the solver. This enables the possibility of using exactly the same initial conditions for each of a set of numerical solvers, and analyzing the resulting data in a uniform way.

In Part III, Chapters 8 to 9, we discuss elementary and hierarchical code coupling strategies, followed by Chapter 10, in which we present a few example cases for multiscale and multiphysics simulations.

In the last part of the book, Part III, we dwell on expert knowledge and discuss the subtleties of multiscale and multiphysics simulations. In the last chapter (Chapter 10), we discuss possible expansions and other packages derived from AMUSE.

These chapters are followed by several appendices. In Appendix A, we further discuss the fundamental concepts of AMUSE and give some details about its internal operation. The generation of initial conditions is important for multiscale or multiphysics applications, but is often not explicitly considered, as we discuss in Chapter 3. How to add your own code to AMUSE is discussed in Appendix B. The last appendix includes a brief `Python` primer (Appendix C).

We have adopted this presentation order due to its steady increase in physical complexity and its systematic introduction of more and more sophisticated AMUSE capabilities.

1.2 A Brief History of Simulations in Astrophysics

The first recordings of theoretical astronomy date back to the Egyptian 18th dynastic (1550–1292 BCE) calculations by pharaoh Hatshepsut's architect Senmut; portions of the murals found in his tomb at Deir el-Bahri can be found at https://www.metmuseum.org (Dorman 2006). Figure 1.2 presents part of this mural. Such calculations enabled astronomers to recognize structure and describe patterns in the heavens, an aspect of astronomy that is still relevant today.

Although the Egyptian civilization did not show much academic interest in the heavens, this changed dramatically when the Babylonian civilization emerged.

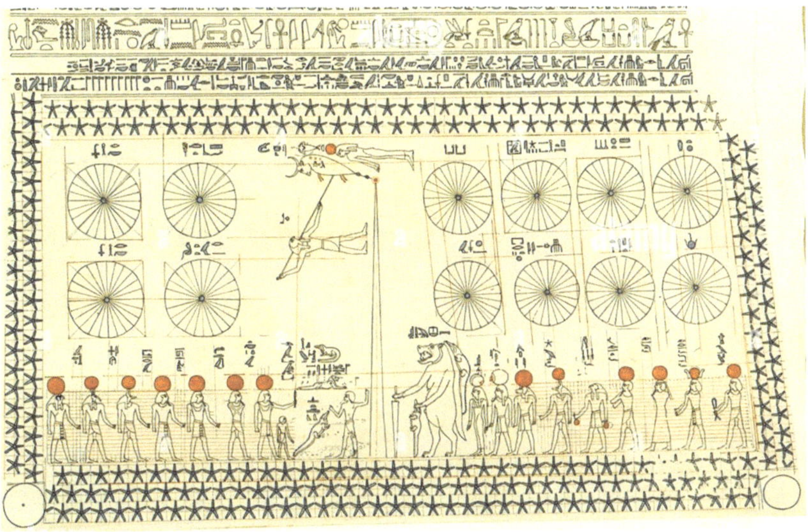

Figure 1.2. Reproduction of the lower part of the second tomb of Senenmut's stela, dated around 1479–1458 BCE (accession number MMT:48.105.52 (TT353), Figure 27 of Wilkinson 1979, GNU General Public License, Rogers Fund 1948). This unique mural shows, apart from all the Egyptian gods, the most important constellations and a much debated conjunction between the Sun and the Pleiades star cluster. The original photograph was presented in Dorman (2006, 1991). Credit: Rogers Fund, 1948.

Around 350 to 50 BCE, the Babylonians already showed a keen interest in mathematics and algorithms. This is demonstrated in several clay tablets with cuneiform writing, which shows a calculation of Jupiter's orbital integration on the sky (see Ossendrijver 2016). The complexity of this calculation, due to the epicycles the planet makes when its orbit is folded in with the Earth's and subsequently projected on the sky, requires a second order integration. The method depicted in Ossendrijver (2016) resembles a Leap-frog or Verlet integrator. Although we do not know who wrote the tablet, or who designed the algorithm, it is clear that the Babylonians were already aware of second order error-reducing algorithms (Ossendrijver 2016). It would be better to call this the Babylonian integrator.

Many centuries later, modern astronomy started with the discovery of the first refracting telescope, built by Hans Lipperhey in 1608, which had a collecting area of about 1 cm^2. Galileo Galilei reported the first astronomical applications in 1609, using a telescope of similar design. In the 400 years since that momentous first application of new technology, the collecting areas of the largest research telescopes have increased by a factor of about 10^6, becoming the 10 m class telescopes in widespread use today.

Digital computers were introduced in the late 1930s (Zuse 1967, 1993), and the first commercial programmable model, ENIAC, achieved in 1945 a computational speed equivalent to about 200 modern floating point (double precision) operations per second (flops). (We say "equivalent" because ENIAC was essentially a single precision machine.) Today, the fastest supercomputers achieve speeds exceeding 10^{18} (Exa) flops, a speedup of a factor of roughly 10^{15} in raw performance in just eight decades. Thus, while observing technology has grown impressively but relatively slowly over the past four centuries, computer scientists have witnessed a technological explosion in living memory.

The ongoing revolution in the availability of digital computers has led to an entirely new branch of research in which key facilities are not confined to high mountain tops in Hawaii or the Chilean Andes, but more often in the room next door. Astronomers quickly realized that they could use computers to archive, process, analyze, and mine the copious amounts of data taken by observing campaigns. However, arguably the biggest impact on the way astronomers pursue their scientific questions has been in numerical simulations of astrophysical phenomena.

Astronomical research lends itself naturally to computation because it is characterized by large ranges in temporal and spatial scales, complex and coupled nonlinear processes, and the extreme conditions found in interstellar space. With the appearance of digital computers, it became possible to study processes in the intergalactic vacuum, hot plasmas at stellar surfaces, billion-year timescale processes, and black-hole physics, none of which can be studied in Earth-based laboratories. The major current limitation to studying the Universe by means of computation lies in software, and in particular our limited understanding of algorithms that can solve multiscale, multiphysics problems. AMUSE is an attempt to address these issues.

Much help in generating AMUSE scripts can be expected from artificial intelligence (AI) via large language models (LLMs). One can simply ask an LLM to write you a script using AMUSE, and since these models are trained on the book's first version examples, it generally does a reasonable job. They tend to be a bit confused about channels and bridges though, which might indicate that our explanations are not always as easily apprehended by an AI.

1.2.1 The First Simulation Experiments

One of the first astrophysical simulations was carried out in 1941 by Erik Holmberg, a graduate student at Lund University who had access to a number of photocells. Using the fact that the intensity of radiation drops off as the inverse square of the distance, like the force of gravity, he devised an experiment to simulate two colliding galaxies. The initial setup for a single "galaxy" in Holmberg's experiment is presented in Figure 1.3. The progress of his experiment is shown in Figure 1.4. The arrows indicate the directions in which the galaxies move with respect to one another, and how they rotate. This experiment made Holmberg one of the founders of what is now an immense body of astronomical research in gravitational dynamics, some of it reviewed by Binney & Tremaine (1987).

Holmberg had no access to a computer, so all of his calculations were performed by hand. His article (Holmberg 1941) is interesting in many respects, but unclear on exactly how the forces (photon counts in his case) were used to update the positions of the particles (the light bulbs). Today, in a comparable paper, we might state that the integration was performed using (say) a fourth-order Hermite predictor-corrector scheme with direct force evaluation and zero softening (see Makino & Aarseth 1992). The choice of algorithm (in this case the Hermite scheme) is an

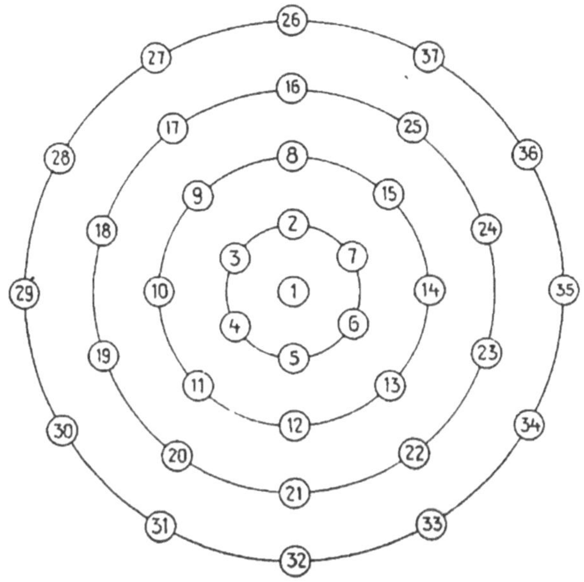

Figure 1.3. Copy of Figure 3 from Holmberg (1941), presenting the initial conditions for a single galaxy in Holmberg's simulation of two colliding galaxies. Reproduced from (Holmberg 1941).

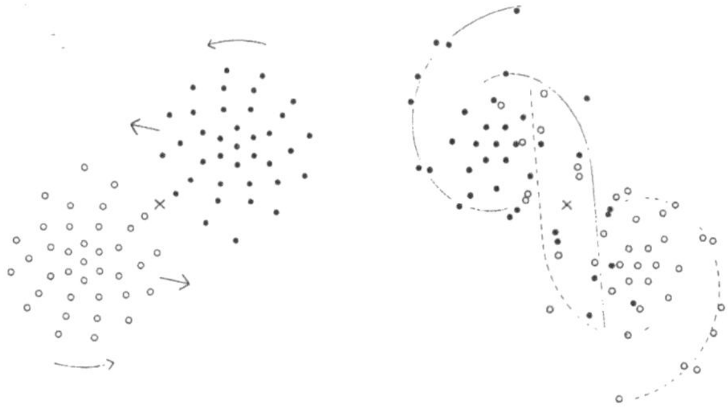

Figure 1.4. Copy of Figure 4(a) from Holmberg (1941), presenting two snapshots of the results from his calculations. Both snapshots were taken at the moment the two galaxies were separated by twice their radii, one before the interaction (left) and one after the first close approach (right). Reproduced from (Holmberg 1941).

important part of the solution. Although this level of detail might sound excessive, you will better appreciate such considerations by the time you have read Chapter 4.

1.3 Software Used in This Book

1.3.1 Motivation for a Homogeneous Software Environment

Developing high-quality simulation software is relatively straightforward when the physical scope of the problem is limited. The task becomes considerably more difficult once the application spans multiple physical domains, or if the dynamic range in temporal or spatial scales exceeds the precision of the hardware. Successful examples of complex multiphysics software in computational astrophysics are NBODY7 (Aarseth 1999, 2011), FLASH (Fryxell et al. 2000, 2010), and GADGET (Springel et al. 2001; Springel 2000), which have broad user bases and can be applied to a wide variety of problems. (Note that, where possible, we provide two references for each software implementation, one to the publication in which the package was first described, and one to the Astrophysics Source Code Library at http://ascl.net/ or ZENODO https://zenodo.org/.) Some of these codes address relatively narrow ranges of physics and are limited by the specific numerical algorithms or data topology adopted to solve the underlying equations, but others, such as FLASH, are highly modular in their internal organization, in some ways similar to AMUSE. Most of this software is hosted on US servers. In the near future, we expect to move all our activities to a less autocratic location.

Figure 1.5 presents typical temporal and spatial scales for a number of astronomical phenomena. The ellipses indicate specific areas of interest. The colors indicate physical domains: gravity (blue), stellar evolution (red), hydrodynamics (yellow), and radiative processes (green). Several domains are identified by name, such as the collapse of a giant molecular cloud (top right in yellow), and the range of habitable planet orbits (blue circle at lower left). Gravity and hydrodynamics span a broad range of temporal and spatial scales, whereas radiative processes and stellar evolution exhibit much narrower ranges. Note that both spatial and temporal scales span a range of about 14 orders of magnitude, from galactic orbits at one extreme to stellar mergers or intricate orbital dynamics at the other.

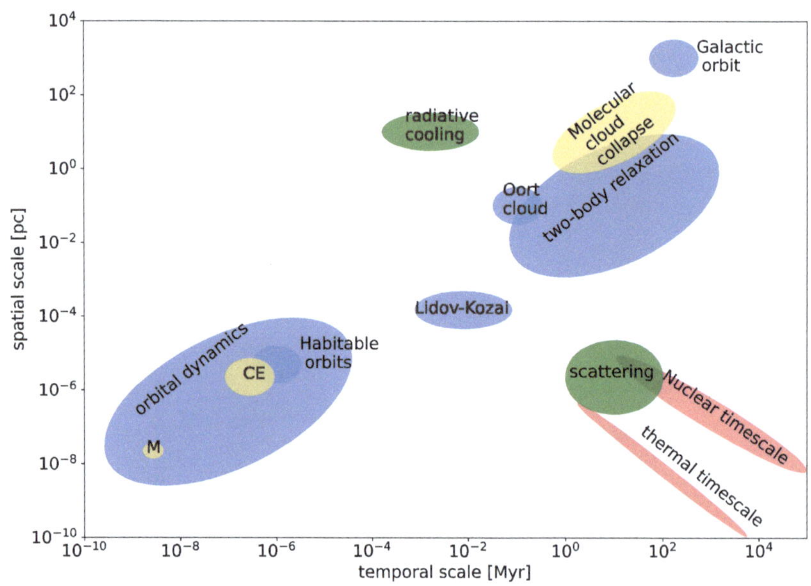

Figure 1.5. A number of typical scales, for gravitational dynamics (blue), stellar evolution (red), hydrodynamics (yellow), and radiative transfer (green), are indicated at several locations in the plane of temporal and spatial scales. Here, several small scales (mergers indicated by M) and common envelope evolutions (indicated by CE) are embedded within the orbital dynamics scale ellipse.

Many existing simulation codes provide excellent solvers within their specific physical domains, but astrophysical problems often require solvers for a wide variety of domains. For example, understanding the ejection of primordial gas from an embedded star cluster, born during a galaxy merger, requires simultaneous treatments of collisional stellar dynamics (the protostellar cluster), collisionless stellar dynamics (the galaxy merger), hydrodynamics (interstellar gas), stellar evolution (embedded protostars and stars), and radiative processes (feedback between stars and the interstellar gas). All these processes occur on comparable temporal and spatial scales.

Instead of designing new algorithms to solve all of these problems simultaneously, we adopt the philosophy that each of the physical domains can be solved separately, and can then be combined into a consistent solution at discrete intervals in time and space. Validation of such a simulation could subsequently be accomplished by recomputing the model with a different solver for each of the physical domains. In this way, we can study the convergence (or the absence thereof) of the solution with a sequence of numerical implementations. AMUSE (Portegies Zwart 2011; Portegies Zwart et al. 2013) allows a user to combine existing solvers to build new applications that can be combined again to study increasingly complex problems. This enables the growth of multiphysics and multiscale application software in a hierarchical fashion, and allows testing of each intermediate step as the complexity of the software increases.

AMUSE has a hierarchical component architecture that encapsulates dynamically shared libraries for simulating stellar evolution, gravitational dynamics, (magneto)hydrodynamics, radiative processes, and also has limited support for astrochemistry. Applications are able to use some or all of these resources simultaneously, and can exchange data in a unique and deterministic way. With this hybrid simulation environment, it becomes possible to study multiphysics processes operating on a broad range of length and timescales.

1.3.2 Choice of Programming Languages

We have adopted `Python` (van Rossum 1995a, b, c, d) as the implementation language for the AMUSE framework and high-level management functions, including bindings to the communication interface. The earlier version of this book used `Python` 2.7, but we have since migrated to `Python` 3 (Van Rossum & Drake 2009) (as of writing 3.8 or higher).[5] The choice of `Python` is motivated by its flexibility, broad acceptance in the scientific community, object-oriented design, and ability to allow rapid prototyping, which shortens the software development cycle and enables easy access to the community code in the community module, albeit at the cost of reduced performance (Portegies Zwart 2020). However, the entire AMUSE framework is organized in such a way that relatively little computer time is actually spent in the framework itself—most time is (or should be) spent in the community code(s). The overhead from `Python` compared to a compiled high-level language is typically $\lesssim 10\%$, and often much less (Pelupessy et al. 2013).

[5] `Python` version 3.13 was the most recent release while writing this.

As discussed further in Section B.1.1.2, our implementation of the communication layer in AMUSE uses the standard message-passing interface protocol (MPI; Gropp et al. 1996) and SmartSockets (Maassen & Bal 2007). Normally, MPI and SmartSockets are used for communication between compute nodes on parallel distributed-memory systems or within computational grids. However, in AMUSE they are used as communication channels between *all* processes, whether or not they reside on the same node as the control script, and whether or not they are themselves parallel.

AMUSE community codes are written in a variety of languages, reflecting the programming backgrounds and preferences of their authors. As a result, for each community code there are two interface functions. The one on the AMUSE side is written in `Python` and is called the proxy. The interface on the community code side is called the partner and is written in the native language of that code. These two interfaces communicate via one of the message-passing protocols just described. Thus, our only real restriction on supported languages is the requirement that the community code is written in a language with MPI or Smart Sockets bindings, such as C (Ritchie 1993b, a), C++, (Stroustrup 2013), Java (Gosling et al. 1996, 2014), and members of the FORTRAN family (Backus 1978).

We emphasize that, from the user's perspective, the details of the interface and communication protocols are invisible. These details are central to the AMUSE design, but most high-level users are blissfully unaware of them. To the user, a `Python` AMUSE script looks just like a traditional serial program.

1.3.3 Design of AMUSE

The AMUSE project goal stipulated three basic functionalities:

(1) A homogeneous, physically motivated interface for existing astronomical simulation codes;
(2) The incorporation of multiple community codes from four fundamental domains (stellar evolution, gravitational dynamics, hydrodynamics, and radiative transfer); and
(3) The ability to design new simulation experiments by combining one or more of the community codes in various ways.

In this way, researchers can concentrate on physically relevant details, such as the calling sequence for the various physical domains, without having to worry about the many technical details.

Each of the existing community codes in AMUSE is embedded in an interface layer that realizes communication between the AMUSE framework and a community code. One important advantage of this strict separation is that it explicitly prohibits short-cut communication between independent subroutines, greatly diminishing the opportunities to create "spaghetti code".[6] In addition, it allows us to manipulate data as it moves between the community code and the framework. One such data manipulation is the conversion of units, as discussed in Section 3.4.1. The

[6] In our opinions, the term "ramen code" would be more appropriate.

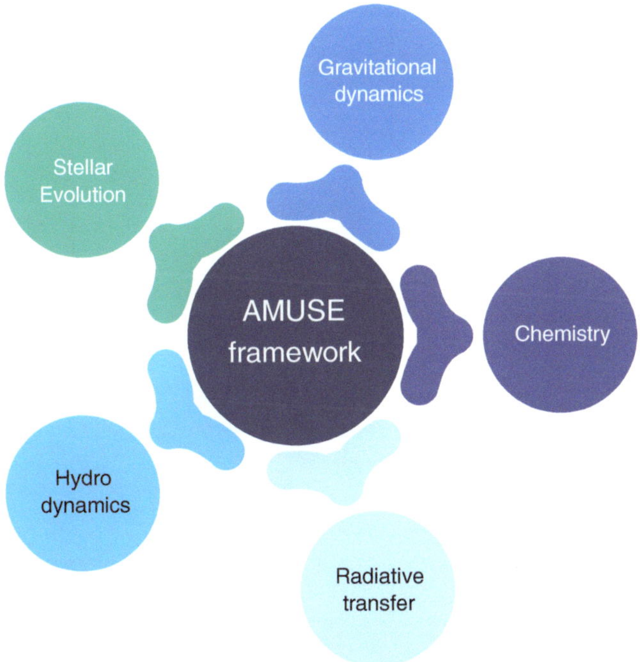

Figure 1.6. Multiscale and multiphysics codes are connected disciplines, as depicted here in the domain–application plane.

handling of community-code dependent parameters is largely managed by the framework, as is the acquisition and mining of simulation data. Figure 1.6 illustrates schematically the philosophy of AMUSE, with the central AMUSE framework surrounded by its four fundamental physical solvers.

New hybrid codes can be written in AMUSE by combining community codes using `Python`. An AMUSE `Python` script for solving a particular problem prescribes the order in which events are handled and specifies the numerical solver that should be used to resolve each event. Once initial conditions are generated and the simulation is completed, the resulting data can be stored or further analyzed using the tools available in AMUSE, or developed by the user. The technical problems associated with coupling codes, setting boundary conditions, specifying run-time parameters, and guaranteeing that the various elements communicate in the proper units are handled entirely by AMUSE. A thorough understanding of the limitations of each of the incorporated codes is important for assessing the quality of the simulation, but the burden of managing the technical details is not forced on the user.

We have demonstrated modularity and interchangeability of the AMUSE community codes for solving the basic physical processes described above. The user can select from a variety of gravitational N-body solvers (symplectic N-body codes, direct N-body codes, and tree codes) by changing a single word in the script. The same holds for stellar evolution codes (parameterized and Henyey codes), hydrodynamics solvers (smooth particle hydrodynamics and adaptive mesh refinement codes), and radiative transfer codes (Monte Carlo photon transport codes and tessellated photon flow solvers). The modularity, and how a complex cascade of numerical solvers can grow to an end product, is illustrated in Chapter 10 and in Part III. This cascade of jargon may sound like Jabberwocky (Carol 1871) right now, but we hope that much of it will eventually make sense by the time you reach the end of this book.

1.3.4 Terminology

The basic structure of the AMUSE framework is illustrated in Figure 1.7. The figure is subdivided into three blocks. The top block is mostly user written, the

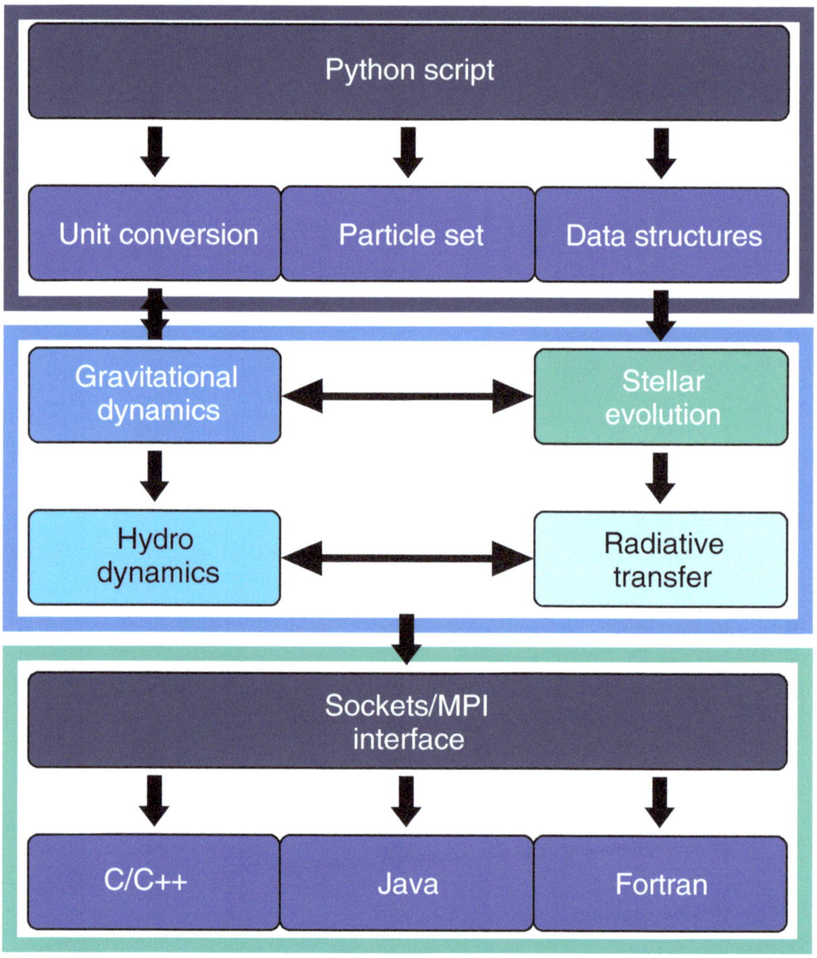

Figure 1.7. General structure of the AMUSE framework. The user writes the `Python` script (top layer). The lower layers in the first block are responsible for units and data structures. The framework in the second block represents the interfaces to the community modules (dark blue for gravitational dynamics, green for stellar evolution, turquoise blue for hydrodynamics, and radiative transfer in light blue to conform with the color coding of Figure 1.6), which connect via sockets or MPI (third block) to the community codes at bottom. Information flows downward to the community modules, as indicated by the arrows, as well as upward, back to the `Python` script.

middle block implements the interfaces onto the community codes, and the bottom block is the community modules and the actual community source code, which is connected to the rest of the framework via MPI or sockets.

The components can be described as follows:

- The `Python` *user script* (top layer in Figure 1.7), which addresses a specific physical problem or set of problems, in the form of system-provided or user-written `Python` commands that serve as the user interface onto the framework. This is the only part of a production script that needs to be written by the user. The coupling between community codes is implemented in this layer, with the help of support classes to convert units and data structures provided by AMUSE (see Section 2.3.1).
- The *community module*, comprising three key elements:
  - The *manager* (second layer in top block), which provides an object-oriented interface onto the communication layer via a suite of system-provided utility functions. This layer handles unit conversion, and also contains the state engine and the associated data repository, all of which are required to guarantee consistency of data across modules. This layer is generic—it is not specific to any particular problem or physical domain.
  - The *communication layer* (second block), which realizes the bidirectional communication between the manager and the community-code layer. This is implemented via a proxy and an associated partner, which together provide the connection to each community code.
  - The *community code layer* (third block), which contains the actual community codes and implements control and data management operations on them. Each piece of code in this layer is domain-specific, although the code may be designed to be very general within its particular physical domain.

1.4 Installing AMUSE

Before you continue reading, we advise you to install the AMUSE package. Here is a brief overview of the installation steps. We assume that you are working in a `Linux` or macOS environment. We do not recommend or support AMUSE under Windows, except within a `Linux` environment such as Windows subsystem for `Linux`[7] or Multipass.[8] For the following, we assume that you have a basic understanding of your operating system and the command line.

To install AMUSE, we strongly recommend creating a new virtual environment, either via `conda`[9] or via `Python` 3's `venv`.[10] AMUSE can be installed either as a package or from source in either. Other virtual environments could work equally well, but are not supported by us at the time of writing.

[7] https://en.wikipedia.org/wiki/Windows_Subsystem_for_Linux
[8] https://multipass.run
[9] https://conda-forge.org/download
[10] https://packaging.python.org/en/latest/guides/installing-using-pip-and-virtual-environments

1.4.1 Installing AMUSE in a Virtual Environment

Here, we will start from a `Python venv` or `conda` environment on an Ubuntu `Linux` laptop. This tutorial was written on a fresh Ubuntu 24.04.2 LTS (6.5.0-1027-oem) installation on a Lenovo P14s laptop. This is not a particularly high-end computer, but it should suffice for all example calculations in this book (if only because all the example calculations in this book are generated with this same machine). We would like to emphasize here that the revenues we receive for writing this book are by far insufficient to purchase such a laptop.

Open a terminal and copy the lines below to create an environment for your project (note that the leading characters only indicates the terminal prompt, and may be different on your computer. The instructions are written for `bash` or another Bourne shell.

```
%> git clone git@github.com:amusecode/amuse.git .
%> cd amuse
%> ./setup
```

This downloads the AMUSE source code and runs the AMUSE setup script. This script will detect the environment you are using and guide you through the installation process.

Since the details of the installation process may evolve over time, we do not print the full instructions here.

However, if you feel uncertain about what to do, try:

```
%> ./setup help
```

which will give you some options.

After following the instructions, the AMUSE framework should be built, and probably most of the packages and libraries.

In case something is broken, you can visit our `GitHub` issue tracker (https://github.com/amusecode/amuse/issues) and ask for help.

1.4.1.1 Testing Your AMUSE Installation

After a successful installation, we encourage you to run a standard set of tests. The general tests can be run quite similarly to installing using the setup routine

```
%> ./setup test amuse-framework
```

for testing the framework, or

```
%> ./setup test ph4
```

for testing an individual code, such as ph4 in this case.

1.4.2 Running the Examples

The examples in this book, and throughout the entire AMUSE framework, can be run by invoking the `Python examples` package

```
%> ./python -m amuse.examples.{example_name}
```

Here, `example_name` is the name of the example.

To make this work, one first has to install the amuse example package through pip,

```
%> pip install amuse-examples
```

And then run,

```
%> ./python -m amuse.examples.sun_venus_earth
```

which will run the example script `sun_venus_earth` and reproduce Figure 3.4.

1.4.3 Running AMUSE

By now, you should have built and tested AMUSE, as described in Section 1.4, and are probably wondering "What can AMUSE do for me?" This section will get you started. If you are new to `Python`, you will find the official `Python` documentation[11] a valuable resource. Most of the scripts in this book were rewritten with at least `Python` version 3.7 in mind, but much of the coding will remain up-to-date with the current version 3.12 and higher.

You can run AMUSE interactively from `Python` or `IPython`:

```
%> python
```

Once `Python` has started, you can load AMUSE by

```
from amuse.lab import *
```

This line is a lazy but very quick way to import most of the AMUSE package in a single call. However, we generally do not advise it because it pollutes the `Python` name space. Rather, we recommend that modules are imported from AMUSE individually, and in the examples that follow this is how we will present source code.

The code examples and snippets presented in the book are designed to be cut and pasted into your `Python` interpreter or your file editor. However, due to limitations

[11] https://docs.python.org/3

of online book formatting, whitespace may not be conserved when doing so. To facilitate using the code examples directly, we provide an `amuse-examples` package that contains the source code for all examples in this book as well as all the generated figures. Once installed, the examples can be imported in the following way (using the `ae` name space as shorthand for `amuse.examples`):

```
import amuse.examples as ae
```

The complete source code of an example (e.g., `gravity_minimal`) can be seen with the `ae.print_example` function:

```
ae.show("gravity_minimal")
```

If you are using AMUSE from a Jupyter notebook, you can copy the example directly to a new cell, ready for editing:

```
ae.to_cell("gravity_minimal")
```

To run an example with default settings, you can use the `run` function:

```
ae.run("gravity_minimal")
```

We hope that these examples and code snippets encourage you to experiment and initiate interesting research projects.

1.4.3.1 Running AMUSE without Network Access

Running AMUSE without a network connection is curiously not always supported by your favorite operating system. The underlying reason hides in the requirement for MPI to have a network connection. This can be solved by using the Smart Sockets connection channel, or by using a loopback network. The latter can be done with the following command line:

```
%> mpirun -n 1 --mca btl_tcp_if_include lo python -m amuse.examples.gravity_minimal
```

This is a rather complex command, but fundamentally it boils down to running the MPI command `mpirun` using a single core with the specific command for process binding for the tcp protocol on your local network. This so-called `MCA` command (-`mca`) can be invoked in mpirun with the arguments `btl_tcp_if_include` and an indicator for the local network (`lo`). This is a whole list of jargon, which may be hard to parse if you barely installed the AMUSE package, but maybe it shows that some of the things we do are not always as well supported by local operating systems as we would wish. There are quite a few other complex message-passing and scheduling issues if you desire to run AMUSE on larger machines, but since the architectures of such machines change within the mean timescale of a PhD (and on the timescale of a new release of this book), we will not dwell further on this topic.

References

Aarseth, S. J. 1999, PASP, 111, 1333

Aarseth, S. J. 2011, NBODY Codes: Numerical Simulations of Many-body (N-body) Gravitational Interactions, Astrophysics Source Code Library, ascl:1102.006

Allen, G., Benger, W., Dramlitsch, T., et al. 2001, Euro-Par 2001 Parallel Processing. Euro-Par 2001. Lecture Notes in Computer Science 2150, ed. R. Sakellariou, J. Gurd, L. Freeman, & J. Keane; Berlin: Springer) 817

Backus, J. 1978, The History of Fortran I, II, and III Vol. 25 (New York: ACM)

Beck, K., & Andres, C. 2004, Extreme Programming Explained: Embrace Change (2nd ed.; Reading, MA: Addison-Wesley)

Binney, J., & Tremaine, S. 1987, Galactic Dynamics (Princeton, NJ: Princeton Univ. Press)

Borgdorff, J., Falcone, J.-L., Lorenz, E., et al. 2013, JPDC, 73, 465

Calaprice, A. 2002, Dear Professor Einstein: Albert Einsteinas Letters to and from Children (New York: Prometheus)

Carol, L. 1871, Through the Looking-Glass, and What Alice Found There (London: Macmillan)

Chopard, B., Borgdorff, J., & Hoekstra, A. 2014, RSPTA, 372, 20130378

Dorman, F. 2006, Hatshepsut: From Queen to Pharaoh, ed. C. Roehrig (New Heaven, CT: Yale Univ. Press) 107

Dorman, P. F. 1991, The Tombs of Senenmut: The Architecture and Decoration of Tombs 71 and 353 (New York: Metropolitan Museum of Art)

Fryxell, B., Olson, K., Ricker, P., et al. 2000, ApJS, 131, 273

Fryxell, B., Olson, K., & Ricker, 2010, FLASH: Adaptive Mesh Hydrodynamics Code for Modeling Astrophysical Thermonuclear Flashes, Astrophysics Source Code Library, ascl:1010.082

Goodale, T., Allen, G., Lanfermann, G., et al. 2003, High Performance Computing for Computational Science: VECPAR 2002 2565, ed. J. M. L. M. Palma, A. A. Sousa, J. Dongarra, & V. Hernández; Berlin: Springer) 197

Gosling, J., Joy, B., & Steele, G. L. 1996, The Java Language Specification (1st ed.; Reading, MA: Addison-Wesley)

Gosling, J., Joy, B., Steele, G. L., Bracha, G., & Buckley, A. 2014, The Java Language Specification, Java SE 8 Edition (1st ed.; Reading, MA: Addison-Wesley)

Gropp, W., Lusk, E., Doss, N., & Skjellum, A. 1996, ParC, 22, 789

Holmberg, E. 1941, ApJ, 94, 385

Hunter, J. D. 2007, CSE, 9, 90

Iwasawa, M., Fukushige, T., Nitadori, D., & Makino, J. 2016, PASJ, 68, 54

Maassen, J., & Bal, H. E. 2007, HPDC '07: Proc. of the 6th Int. Symp. on High Performance Distributed Computing (New York: ACM) 1

Makino, J., & Aarseth, S. J. 1992, PASJ, 44, 141

Ossendrijver, M. 2016, Sci., 351, 482

Pelupessy, F. I., van Elteren, A., de Vries, N., et al. 2013, A&A, 557, A84

Portegies Zwart, S., McMillan, S. L. W., van Elteren, E., Pelupessy, I., & de Vries, N. 2013, CoPhC, 184, 456

Portegies Zwart, S. 2011, AMUSE: Astrophysical Multipurpose Software Environment, Astrophysics Source Code Library, ascl:1107.007

Portegies Zwart, S. 2020, NatAs, 4, 819

Portegies Zwart, S., McMillan, S., Harfst, S., et al. 2009, NewA, 14, 369

Ritchie, D. M. 1993a, SIGPLAN Not., 28, 201

Ritchie, D. M. 1993b, HOPL-IIACM: The Second SIGPLAN Conf. on History of Programming Languages (New York: ACM) 201

Rubincam, D. P. 2000, Icar, 148, 2

Springel, V. 2000, GADGET-2: A Code for Cosmological Simulations of Structure Formation, Astrophysics Source Code Library, ascl:0003.001

Springel, V., Yoshida, N., & White, S. D. M. 2001, NewA, 6, 79

Stroustrup, B. 2013, The C.. Programming Language (4th ed; New York: Addison-Wesley)

van Rossum, G. 1995a, Extending and Embedding the Python Interpreter CS-R9527, *CWI*

van Rossum, G. 1995b, Python Library Reference CS-R9524, *CWI*

van Rossum, G. 1995c, Python Reference Manual CS-R9525, *CWI*

van Rossum, G. 1995d, Python Tutorial CS-R9526, *CWI*

Van Rossum, G., & Drake, F. L. 2009, Python 3 Reference Manual (Scotts Valley, CA: CreateSpace)

Veen, L. E., & Hoekstra, A. G. 2020, CCS 2020: Computational Science – I ed. V. V. Krzhizhanovskaya, G. Závodszky,, M. H. Lees, et al. (Berlin: Springer) 425

Wang, L., Iwasawa, M., Nitadori, K., & Makino, J. 2020, MNRAS, 497, 536

Wilkinson, C. K. 1979, Egyptian Wall Paintings: The Metropolitan Museum's Collection of Facsimiles Vol. 36 (New York: The Metropolitan Museum of Art)

Zuse, K. 1967, Elektron. Datenverarb, 8, 336

Zuse, K. 1993, The Computer - My Life (Berlin: Springer)

Simon Portegies Zwart, Steve McMillan and Steven Rieder

Chapter 2

Fundamental Concepts

When I let go of what I am,
I become what I might be.

—Modern Taoist interpretation of a possible saying by Laozi, but no ancient Chinese version exists in the Tao Te Ching (Laozi 2009).

We live in an era of multimessenger astronomy, meaning that various instruments operating at multiple electromagnetic and non-electromagnetic wavelengths cooperate to collect data. In computational astrophysics, we have the concepts of multiscale and multiphysics modeling, which are the equivalents of multimessenger astronomy. Here, we discuss the fundamental aspects that enable multimessenger computational astrophysics. Fundamental parts include data handling, communication, unit conversion, and scaling.

2.1 Programming

The majority of code development that you will do as an apprentice AMUSE user will be in the `Python` programming language (van Rossum 1995a, b, c, d). In the previous version of this book, we adopted `Python` version 2.7, this version of the book is written with `Python` 3.7+ in mind (Van Rossum & Drake 2009).

2.1.1 Why `Python`

Different programming languages, such as Fortran (American National Standards Institute 1978), C (Kernighan & Ritchie 1988), and `Python`, have been the main languages that we have used. The first two authors moved to C++ (Stroustrup 2013) to write an extensive N-body simulation environment called Starlab (Hut et al. 2010).

doi:10.1088/2514-3433/adfc3cch2

Many of the modules in AMUSE are written in these languages. Although our preferences differ, and we may nowadays prefer C++, Julia (Bezanson et al. 2017), or Rust (S. Klabnik & C. Nichols 2023), the main AMUSE package is written in Python.

Python has become one of the most popular and versatile programming languages, it is easy to use, versatile, reasonably efficient, data focused, and it has a great and active development community. In addition, it makes working with unstructured text almost fun, whereas this has always been one of the older languages' weaker points.

Our choice of Python gives AMUSE users access to a powerful, flexible, and widely used language that opens doors to numerous career opportunities and enables them to address a wide range of programming challenges.

For clarity, we briefly review some of the most important style recommendations when you write or develop any scripts on your own. We present these small guidelines to ensure uniform formatting across the code you write. Homogeneous formatting reduces decision making so that you can focus on solving problems. It facilitates teamwork, and improves maintainability through adherence to best practices.

2.1.2 Style Guide

Here, we cover the most important aspects of the Python Enhancement Proposal (PEP 8, van Rossum et al. 2001). Some further style recommendations are available in Warsaw (2022), which we recommend reading. The Python style guide includes recommendations for indentation, imports, the use of white space, naming conventions, comments, docstrings, and basic syntax preferences. Keep in mind though that these are just guidelines. When your specific requirement needs you to abandon them, abandon them.

Resist the urge to put in more than one import per line. But group them in order of the standard Python libraries, third party libraries, and as last your local sources.

```
import os
import numpy as np
from amuse.units import units
```

The first two lines in this snippet should look familiar. The third line directly uses a specific module from the AMUSE framework. This is a module that allows us to convert units from one set to another. The unit module is quite handy and general, and AMUSE was the first astronomical code that incorporated a general unit module. We will go into some depth in explaining the unit module in Section 2.3.

Parameters in your script can be provided with units using the "pipe" operator; for example, a parameter mass can be provided with the mass of 1 solar mass by

```
from amuse.units import units
mass = 1 | units.MSun
```

Avoid the use of additional white space between braces, brackets, and parenthesis. Please write

```
sun_and_jupiter = Particles(name=["Sun", "Jupiter"], m=(1, 0.001) | units.MSun)
```

rather than this **bad example**:

```
SunAndJupiter = Particles ( name = [ "Sun", "Jupiter" ], m = ( 1, 0.001 ) | units.
    MSun)
```

For functions, variables, and methods, we use *snake_case*: lowercase separated by an underscore, and keep lines under 80 characters. Note that, in accordance with PEP 8, we prefer white spaces after commas (',') and double columns (':'), while all operators are surrounded by white space, such as in

```
mass_total = (1 | units.MSun) + (1 | units.MJupiter)
```

except for equal signs ('=') in function calls.

For classes, we use *PascalCase*, where all words begin with a capital letter

```
class MilkyWayGalaxy():
```

whereas constants are all in capitals with underscores separating the words

```
# This is the distance from the Sun to the Galactic center.
SUN_GALACTIC_DISTANCE = 8.5 | units.kpc
```

Comments in a code are complete sentences. But at the start of a function, we use docstrings

```
import numpy as np
from amuse.units import constants


def semi_to_orbital_period(semi_major_axis, mass_total):
    """
    Convert semi-major axis to orbital period
    """
    return 2 * np.pi * (semi_major_axis**3 / (constants.G * mass_total)).sqrt()
```

We use four spaces per indentation level. Note that portable document format (pdf) does not support white space, and they are not properly copy-pasted from the book to your `Python` environment.

For comparisons, we use "is" for singletons, and "==" for values

```
if x is None:
    pass
```

We regularly include Python snippets in the text. Sometimes they are written as standalone sentences or comments, and sometimes they are taken directly from one of the larger scripts in the example package. In the latter case, the code lines are associated with line numbers. These line numbers refer directly to the appropriate example script (although these could change if and when the examples are revised).

2.2 Internal Data Handling and Communication

Data is a collection of facts in the form of bits of information, often expressed in numbers, text, images, or sound. It represents information that can be quantified and used to understand patterns, trends, and relationships in the world around us. It serves as the foundation for decision making and leads to deeper insight. Data can be processed, manipulated, and analyzed to extract insights. As the amount and complexity of data, in science and beyond, continues to grow (probably exponentially), the need for improved data handling increases. We have no ready answer to the complexities of data analysis; we consider its handling a science, not an art.

2.2.1 Particles and Particle Sets

In our experience, most of the syntactic and mathematical features of Python (conditionals, loops, arithmetic operations, etc.—even the rigid use of indentation to define program structure) are broadly intelligible to most programmers, regardless of background. However, some more complex data structures may be unfamiliar to readers who are used to less structured programming languages, such as C or Fortran. Accordingly, we pause here to elaborate on this topic in AMUSE, using some important data structures as templates.

A Python class is a data structure together with a collection of associated functions, or methods, designed to manipulate the data. A particularly simple AMUSE class is a Particle, which contains only basic particle attributes, such as mass, position, velocity, etc., as well as a randomly generated key that uniquely identifies the particle in the AMUSE system. We can access its contents directly, and expand it to store all sorts of useful information.

```
from amuse.datamodel import Particle
from amuse.units import units

p = Particle()
p.mass = 1.0 | units.MSun
p.mass2 = 42 | units.kg
p.name = "Christine Chapel"
p.rank = "Nurse"
print(p)
```

The particle class is not very useful in and of itself, but it forms the basis for a vitally important class called a particle set, which is basically an array of particle objects, with conventions regarding content and a collection of methods providing information on and control over the data. Particle sets are used to transfer data from one code to another, and for eventual data processing and analysis. An element of a particle set is expected to have certain attributes, such as mass, position, velocity, radius, and possibly others. The built-in methods assume that these data exist.

You can generate a (rather useless, for now) particle set by using the following lines

```
from amuse.datamodel import Particles

particles = Particles(number_of_particles)
print(particles)
```

This prints the most fundamental ingredient in a particle set; the key, which identifies each particle. Note here the difference between the singular and the plural forms for declaring a particle and a particle set.

```
                 key
                   -
====================
 5512314219721684995
 8353633730176500107
 9653245888285946456
 9457276936209163492
14505666692805646583
14003077150553218257
====================
```

There is not much one can do with such a simple bunch of particles, except perhaps to use it as a very expensive random number generator. However, we can easily extend it in useful ways. We can create a particle set of 10 particles as follows:

```
from amuse.datamodel import Particles

particles = Particles(10)
```

We can then set up attributes in various ways:

```
particles.mass = 1 | units.MSun
```

sets all masses, while

```
particles[3].mass = 42 | units.kg
```

sets just the mass for the fourth particle (`Python` starts counting at 0). We could also have set the masses (or positions, etc.) in the `Particles` creation call, with

```
particles = Particles(10, mass=1 | units.MSun)
```

Some more useful ways of setting particle set attributes are presented below.

In some cases, one would like to expand the attribute list of a particle with an array. This can happen when one would like to add information about a chemical network to a particle, or some other attributes that normally is associated with particle data. Adding abundances, for example, for Hydrogen, Helium, Lithium, and Iron could be realized as follows

```
particles.add_vector_attribute("abundance", ("H", "He", "Li", "Fe"))
particles.abundance = [0.75, 0.24, 0.005, 0.005]
```

In the first line we add a vector attribute called "abundance" to the particle set, and specify the array of names for this attribute. In the second line, we initialize the values for all 10 particles with the same chemical composition.

Once the basic attributes are set, the methods associated with the `particle set` provide commonly used functionality. For example,

```
energy_kin = particles.kinetic_energy()
```

computes the total kinetic energy of the particle set, while

```
energy_pot = particles.potential_energy()
```

computes the total potential energy. Similarly,

```
com = particles.center_of_mass()
```

calculates the center of mass of the particle set, while

```
particles.scale_to_standard()
```

places the set in the center of mass frame and scales it to a standard configuration (see Section 4.1.2.1). Many AMUSE functions take a particle set as input and return analytical information on it.

You can ask `Python` to list the methods associated with a class by using the `dir` function:

```
dir(Particles)
```

which will provide a list of all the methods available in the Particles class. (Note the Python convention that names beginning and ending with double underscores are intended for private use, even though you can see them and in principle use them yourself —not recommended!) More extended information can be obtained with

```
help(Particles)
```

which provides a detailed description of the methods in the class.

Even here, though, Python asserts its inherent flexibility. Unlike classes in more strongly typed languages, Python classes are customizable. You can add your own variables at will for your own purposes—just do not expect them to be properly managed or preserved by the built-in methods! Such additional attributes are also ignored by the specific modules, and will not be copied back and forth between module and local script. Just keep in mind that they have to be handled explicitly by the script writer.

2.2.2 Grid-based Data

Technically, grids are conceptually quite different than particles, but in principle they can represent the same information, at least numerically. Grids are also used in gravity codes, such as in tree codes (Barnes & Hut 1986) and particle-mesh solvers (Hockney & Eastwood 1988), and they are quite common for hydrodynamics solvers. In principle the same information can be conveyed with particles, or with grids, but for a wide variety of reasons grids often turn out more complex because they are not necessarily Cartesian.

Voronoi grids, for example, represent a tessellated space, whereas a KDTree-type grid or a refined grid have different data structures. These differences in the fundamental data structure make it difficult to unify the grid-based data structures. We therefore project grids onto the already existing particle data structure.

Examples of grid-based codes include Athena and FLASH. Both use adaptive grids, although this is not how we initialize them. These grids are most easily initialized using a Cartesian grid, after which the code internally readjusts the grid structure to reflect the properties of the gas (often the density or temperature structure).

Grid-based hydrodynamical solvers or radiative transfer codes often handle the data locally in the particular code, rather than in the AMUSE script. An example is the Torch script (Wall et al. 2020a, b), which combines the grid-based magneto-hydrodynamics solver FLASH with AMUSE. The grid-based data in Flash are handled, saved, and stored in the code. The AMUSE script can, in principle, touch the data, but this leads to complex data consistency responsibility issues, performance loss, and unnecessary data replication.

A simple grid for the hydro-code Athena can be initialized as follows

```
from amuse.units import generic_unit_system
from amuse.community.athena import Athena

instance = Athena()
instance.parameters.x_boundary_conditions = ("periodic", "periodic")
instance.parameters.mesh_length = (8, 1, 1) | generic_unit_system.length
instance.parameters.mesh_size = (8, 1, 1)
grid = instance.get_extended_grid()
instance.initialize_grid()
print(grid.rho)
```

A more generic way of making a grid is

```
def make_grid(number_of_grid_cells,
              length,
              constant_hydrogen_density,
              inner_radius,
              outer_radius
              ):
    grid = Grid.create([number_of_grid_cells] * 3,
                       length.as_vector_with_length(3)
                       )
    grid.radius = grid.position.lengths()
    grid.hydrogen_density = constant_hydrogen_density
    grid.hydrogen_density[grid.radius <= inner_radius] = 0 | units.cm ** -3
    grid.hydrogen_density[grid.radius >= outer_radius] = 0 | units.cm ** -3
    return grid
```

We can plot the particles of a gaseous distribution simply using matplotlib, as demonstrated in Figure 2.1.

In Figure 2.2, we show three representations of a gaseous distribution (right), as a Voronoi tessellated structure (left), and as an adaptive Cartesian grid (middle).

We construct and plot the Voronoi tessellation from the particle gas `particle_data`, using the `scipy` package `Voronoi`.

```
from scipy.spatial import Voronoi, voronoi_plot_2d
voronoi_data = Voronoi(
    particle_data, furthest_site=False, qhull_options="Qbb Qc Qz Gc",
)
voronoi_plot_2d(voronoi_data, ax, show_vertices=False)
```

The refined grid is plotted conveniently using the YT package (see https://yt-project.org/doc/index.html).

```
AMR_data = yt.load("../data/lvl12-VorAMR_hdf5_chk_0001")
plt = yt.ProjectionPlot(AMR_data, "z", "bgpt")
plt.set_cmap(field=("flash", "bgpt"), cmap="arbre")
plt.annotate_grids(cmap="black_only")
```

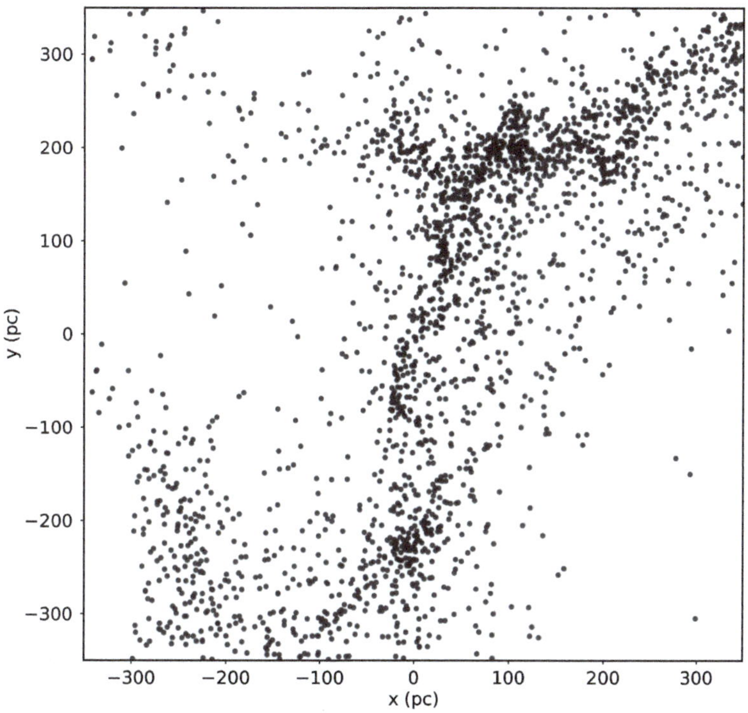

Figure 2.1. The calculation was performed using the Torch package in AMUSE. The projection is 700 × 700 pc. The tessellated, adaptive grid, and gaseous distributions are presented in Figure 2.2. The script to generate this figure can be found in `amuse.examples.voramr_logic_progression_plots`.

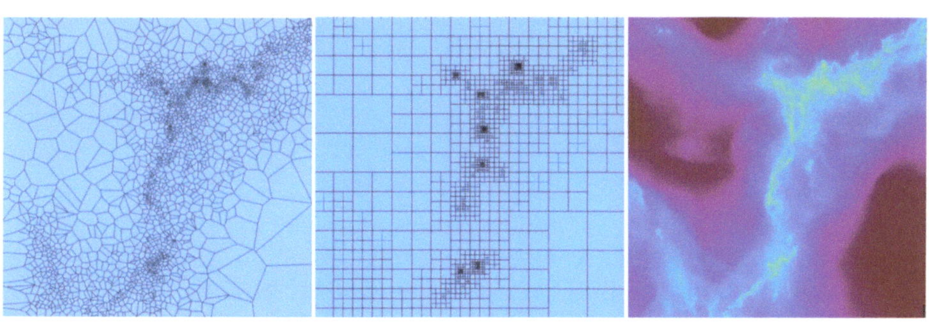

Figure 2.2. Voronoi tessellation (left) (Lewis et al. 2025), adaptive Cartesian grid (middle), and a gaseous representation (right) from a simulation performed using FLASH. The projection is 700 × 700 pc. in Figure 2.1 we show the particle gas distribution of the same filament. The script to generate this figure can be found in `amuse.examples.voramr_logic_progression_plots`.

Several codes make use of grids. Some of them solve for gravity, such as the various tree codes, and others solve for either hydrodynamics or radiative transfer. We will have more to say on these codes in their respective chapters: Chapters 4, 6 and 7.

2.2.3 Data Handling Between Code and Script

Let us imagine that particle information (mass, position, velocity, etc.) is stored in a particle set called `bodies`, which is part of local memory and lives in the user `Python` script. This data does not naturally live in the selected worker code, but the essential parts can be copied to the code as needed. For a stellar evolution code, this might be the mass and radius, and for an N-body code, this could also include position and velocity. This local information is not automatically updated after the code has been run using `{code}.evolve_model()`. Here, `{code}` may be the instantiation of any type of code for evolving stars or integrating equations of motion.

To access the new data, we could copy it directly from the remote code by referencing `{code}.particles`, which lives in the code interface and is updated by `{code}.evolve_model()`, but each access will cause AMUSE to open a one-time MPI or sockets connection to the running code where the remote data live, which can be slow in terms of compute time because it requires the construction of the connection. More importantly, it is less explicit. Note that some codes may have various types of particles. This is not uncommon in particle-based hydrodynamics solvers, which may make a distinction between `dm_particles` for dark matter, and `gas_particles` for the gas.

A more efficient way to communicate with a running code is realized by creating a *channel* to the remote data of the running worker code. A communication channel for (in this case) a gravity code is declared with

```
channel_from_gravity_to_framework = gravity.particles.new_channel_to(bodies)
```

Here, `gravity` is a pointer to the running code. To copy the particle data from the running code to the local `Python` framework, we say

```
channel_from_gravity_to_framework.copy()
```

Now, the masses, positions, and velocities of the remote data in the worker module are duplicated in the local `bodies` particle set in the framework (`Python`) script. The local data (in `bodies`) and the remote data (in `gravity.particles`) are consistent. The `copy()` command copies all data, but to tune performance or to prevent overwriting some local data, one can choose to copy only part of the data. For example, to update only the positions, but not the masses, velocities, or other attributes, we could write:

```
channel_from_gravity_to_framework.copy_attribute(["x", "y", "z"])
```

We could have arranged for the local data to be automatically synchronized with the worker data on each return from the worker code, but for the sake of user control and efficiency, we have opted instead for manual synchronization.

2.2.4 Storing and Recovering Data

We can store particle data in a file for further and future analysis. It is advisable to synchronize the local bodies with the code data (see Section 2.2.3) before writing them to a file. A snapshot can be saved to a file named `filename` as follows:

```
write_set_to_file(bodies, filename)
```

Optionally, a "`format={format}`" argument indicates the format in which the snapshot if stored—by default "`amuse`", which uses `hdf5`. This format is commonly used within the astronomical community; it provides a self-descriptive binary format with the potential of storing hierarchical data structures (The HDF Group 2010). Other recognized types include "`starlab`" (the internal StarLab `dyn` and `star` formats Portegies Zwart et al. 1999; Hut et al. 2010), "`csv`" (comma-separated-values), and plain text ("`txt`").

Time is not an attribute of the particle set, but rather of the code, in this case the N-body integrator. As a consequence, the age of a particle is not stored by default. This generally saves data and the historical context stems from the idea that all stars are born at the same time. Some codes do not even allow for particles to have different ages, but in AMUSE we are not bound by these limitations. To save the time of a snapshot, we can use:

```
write_set_to_file(bodies.savepoint(0 | model_time.unit), filename)
```

Here, the phrase `bodies.savepoint` ensures that the time is stored with the snapshot. A stored output file can be read back to memory:

```
bodies = read_set_from_file(filename)
```

If time was saved with the snapshot, it can be recovered:

```
model_time = bodies.get_timestamp()
```

For storing the ages of individual particles, one can simply add a new attribute to the particle set:

```
bodies.age = 0 | units.Myr
```

or to an individual particle:

```
Sun.age = 4.5678 | units.Gyr
```

Multiple snapshots can be stored in a single file. In that case, one can iterate over the individual snapshots using

```
many_snapshots = read_set_from_file(filename)
for snapshot in list(many_snapshots.history):
    print(f"The number of bodies in this snapshot is: {len(snapshot)}")
```

For runs with a large number of particles, or containing many snapshots, these files can become quite large. One solution is to split the output into smaller pieces and process them separately. Sometimes, however, this is not desirable. In such cases, one can keep the data on disk by keeping the file open while reading:

```
bodies = read_set_from_file(
    filename,
    copy_history=False,
    close_file=False,
)
```

Alternatively, one could write compound particle sets to a single file and recover them with:

```
gas, stars = read_set_from_file(
    filename,
    names=["gas", "stars"],
    copy_history=False,
    close_file=False,
)
```

Yet another way, which is often used in planetary systems, is to make a distinction in the particle set by introducing an extra attribute:

```
stars.name = "star"
planets.name = "planet"
moons.name = "moon"
bodies = stars + planets + moons
write_set_to_file(filename, bodies)
```

After reading the file, one can split the particle set:

```
bodies = read_set_from_file(filename)
stars = bodies[bodies.name == "star"]
```

and similarly for the other particle subsets.

2.3 Converting between Code Units and SI

2.3.1 Scaling Standard Units to Internal Code Units

Units are fully supported in AMUSE. Running one of the interfaced high-performance codes with units is not difficult. For intrinsic dimensionless solvers, including units results in the loss of the scale-free characteristics of the problem. This is only a problem in gravitational dynamics, which can be addressed elegantly as a scale-free problem using Hénon units (Hénon 1971). For stellar evolution, unit conversion is generally not an issue because both the Henyey and the parameterized stellar evolution codes already work in units that naturally convert to SI. Similar arguments hold for hydrodynamics solvers and radiative transfer codes, but they tend to be written for specific sets of units. We desire to achieve coherency and consistency in the use of units, and therefore introduce the converter class. If you are unfamiliar with `Python` classes, do not despair because we provide a quick overview in Appendix C.

The units in an N-body system are specified by providing any two of the mass, length, and time scales of the problem. As a practical matter, conversion to other systems requires the creation of a `converter`, which is a `Python` class specifically designed to keep track of units. We can construct a converter with mass unit 1 $M_\odot$ and spatial unit 1 au as follows:

```
converter = nbody_system.nbody_to_si(1 | units.MSun, 1 | units.au)
```

However, the following line

```
converter = nbody_system.nbody_to_si(72 | units.kg, 175 | units.cm)
```

which happens to be the weight and height would work equally well. In this last example, we really did use the International System of Units, whereas for most astronomical simulations, we are more likely to use solar masses, millions of years, and parsecs. Nevertheless, the conversion system is still generically called `nbody_to_si`.

A system in physical units can be converted back to *N*-body units using the same converter. For example, converting the positions back to *N*-body units can be accomplished with

```
positions_in_nbody_units = converter.to_nbody(bodies.position)
```

or back to SI units:

```
positions_in_si_units = converter.to_si(positions_in_nbody_units)
```

More information about alternative converters can be Section 2.3.2

2.3.2 Alternative Converters

Instead of using a mass and a time, or a mass and a length scale, to specify the set of units, it is also possible to use instead luminosity and density, or any other dimensionally independent quantities. The routine for specifying a different set of units can be imported by

```
from amuse.units.generic_unit_converter import ConvertBetweenGenericAndSiUnits
```

We can call this routine with (for example) luminosity `L` and density `rho`, which would look as follows:

```
converter = ConvertBetweenGenericAndSiUnits(L, rho, constants.G)
```

Alternatively, it can be called with any set of independent units:

```
converter = ConvertBetweenGenericAndSiUnits(
    1.0 | units.kpc,
    1.0e10 | units.MSun,
    constants.G,
)
```

or if `Rvir` is virial radius and `rho` is density,

```
converter = ConvertBetweenGenericAndSiUnits(Rvir, rho, constants.G)
```

2.3.3 Changing Parameters of the Simulation

Often the functionality of a community code needs to be modified; for example, to improve accuracy or set some other parameter. This is controlled by internal parameters within the code. Changing these internal parameters is always possible, but we want to avoid modifying and recompiling the code. For this reason, many internal code parameters can be queried and (re)set. The names of these parameters may be the same for some codes, but completely different for others. Fortunately, as described in Section 2.2.1, `Python` and the AMUSE interface have the ability to provide help on these matters. The command

```
dir(Ph4)
```

provides an extensive list of all the ingredients of the direct N-body code `Ph4` (see Chapter 4 for a more detailed description of N-body codes). This command can be given for each code, such as

```
dir(Mocassin)
```

The same command with the code `Bhtree` will provide a different list of functionalities for the tree code, and equally for the radiative transfer code `Mocassin`. More extended information can be obtained with

```
help(Ph4)
```

A list of parameters for `Ph4`, and their default values, can be obtained for the instantiation of a particular code, such as for the same gravity solver

```
gravity = Ph4()
```

we can get information about the parameter settings through

```
gravity.parameters?
```

A list of all available parameters (and their default settings) of this particular code is provided by

```
print(gravity.parameters)
```

but the same constructs such as help and dir can also be applied to the instantiation of the code, or to its member functions.

The parameters used in two seemingly similar codes may have different names. This sounds very confusing, but it reflects the diversity of the authors of the various interfaces. Our philosophy here is to try to reserve identical names for parameters that do exactly the same thing, but because many codes have slightly different functions or objectives, it is hard to be consistent and clairvoyant at the same time.

Some parameters control the actual working of a code. A smoothed-particles hydrodynamics code, such as Fi, allows the user to control the number of nearest neighboring particles to determine the local thermodynamical parameters. This can be tuned by setting the parameter from the default value of 32 to 64, one can write:

```
hydro.parameters.targetnn = 64
```

The number of free parameters varies a lot depending on the code. The stellar evolution code Seba has 11 parameters, the Ph4 N-body code has 20, and the smoothed-particles hydrodynamics code Fi has 60. Having more (or fewer) parameter does not necessarily say much about the reliability of the code, but it does say something about the effort we took in making these parameters accessible by the user from the Python interface.

The best way to learn about the various parameter settings, and the possibilities of a specific code, is by looking at good old-fashioned source code because all parameter settings are defined in the interface file of each community module in the AMUSE source code. For the Hermite package, for example, the parameters can be found in the method define_parameters() in the amuse.community.hermite.interface.Hermite class.

2.4 Tricks with Units

2.4.1 Setting Printing Strategies

Printing quantities with units can be awkward because the units often come out in undesired form: $1 M_\odot$ might be printed in grams, or 1 pc in centimeters. To prevent this unwanted behavior, you could explicitly convert to your desired units, e.g.

```
x = 10 | units.cm
print(f"x={x.in_(units.parsec)}")
```

but this is quite cumbersome and inconvenient for routine use. An alternative is to set a preferred printing strategy. For example, if your script performs all calculations in solar masses, solar radii, and megayears, you can indicate your preferred set of units with

```
set_printing_strategy(
    "custom",
    preferred_units=[units.MSun, units.RSun, units.Myr],
    precision=6,
    prefix="",
    separator="[",
    suffix="]",
    )
```

This will ensure that all numbers will be printed with six decimal places and with the units in brackets. But when you perform a galaxy collision simulation you may prefer

```
set_printing_strategy(
    "custom",
    preferred_units=[units.MSun, units.kpc, units.Gyr]
    )
```

To print in other unit systems, such as N-body units, a converter must be provided in the argument list; for example, `nbody_converter = converter` (see Section 2.3 for details).

2.4.2 The Command-line Argument Parser

Production scripts usually require tuning for application to a specific problem, and it is desirable to be able to change a script's input parameters, such as the number of stars, their masses or metallicity, the virial radius of a cluster, or the position of a black hole, without having to edit it every time. These parameters are traditionally set on the command line and subsequently parsed by the production script.

One solution in `Python` for managing command-line arguments is the argument parser

```
import argparse
```

Code example 2.1 presents a usage example of this command-line argument parser, where we allow the stellar mass `M`, metallicity `z`, and the end time of the simulation `model_time` to have the desired units.

```
def new_argument_parser():
    parser = argparse.ArgumentParser(
        formatter_class=argparse.ArgumentDefaultsHelpFormatter,
    )
    parser.add_argument(
        "-m",
        type=units.MSun,
        dest="mass",
        default=1.0 | units.MSun,
        help="stellar mass",
    )
    parser.add_argument(
        "-t",
        type=units.Myr,
        dest="time_end",
        default=4700.0 | units.Myr,
        help="end time of the simulation",
    )
    parser.add_argument(
        "-z",
        dest="metallicity",
        type=float,
        default=0.02,
        help="metallicity",
    )
    return parser


def main(**kwargs):
    stellar_minimal(**kwargs)


if __name__ == "__main__":
    arguments = new_argument_parser().parse_args()
    main(**arguments.__dict__)
```

Code example 2.1:. Command-line parser for a stellar evolution model and the main routine for calling the function listed in Code example 5.3 in Section 5.2.2. This snippet is taken from `amuse.examples.stellar_minimal`.

References

American National Standards Institute 1978, American National Standard Programming Language FORTRAN ANSI X3.9-1978,

Barnes, J., & Hut, P. 1986, Nature, 324, 446

Bezanson, J., Edelman, A., Karpinski, S., & Shah, V. B. 2017, SIAMR, 59, 65

Hénon, M. H. 1971, Ap&SS, 14, 151

Hockney, R. W., & Eastwood, J. W. 1988, Computer Simulation Using Particles (Bristol: Hilger)

Hut, P., McMillan, S., Makino, J., & Portegies Zwart, S. 2010, Starlab: A Software Environment for Collisional Stellar Dynamics, Astrophysics Source Code Library, ascl:1010.076

Kernighan, B. W., & Ritchie, D. M. 1988, The C Programming Language (2nd ed.; Englewood Cliffs, NJ: Prentice Hall)

Laozi, 2009, Tao Te Ching: An Illustrated Journey (London: Frances Lincoln Limited)

Lewis, S., et al. 2025, ApJ, 994, 69
Portegies Zwart, S. F., Makino, J., McMillan, S. L. W., & Hut, P. 1999, A&A, 348, 117
Klabnik, S., & Nichols, C. 2023, The Rust Programming Language (2nd ed.; San Francisco, CA: No Starch Press)
Stroustrup, B. 2013, The C++ Programming Language (4th ed.; Upper Saddle River, NJ: Addison-Wesley)
The HDF Group 2010, Hierarchical Data Format Version 5,
van Rossum, G. 1995a, Extending and Embedding the Python Interpreter CS-R9527, *CWI*
van Rossum, G. 1995b, Python Library Reference CS-R9524, *CWI*
van Rossum, G. 1995c, Python Reference Manual CS-R9525, *CWI*
van Rossum, G. 1995d, Python Tutorial CS-R9526, *CWI*
Van Rossum, G., & Drake, F. L. 2009, Python 3 Reference Manual (Scotts Valley, CA: CreateSpace)
van Rossum, G., Warsaw, B., & Coghlan, N. 2001, PEP 8: Style Guide for Python Code, Python Enhancement Proposal 8 (Wilmington, DE: Python Software Foundation)
Wall, J. E., Mac Low, M.-M., McMillan, S. L. W., et al. 2020a, ApJ, 904, 192
Wall, J. E., Tran, A., Mac Low, M.-M., et al. 2020b, Torch: Coupled Gas and N-body Dynamics Simulator, Astrophysics Source Code Library, ascl:2003.014
Warsaw, B. A. 2022, GNU Mailman Coding Style Guide (Boston, MA: Free Software Foundation)

Chapter 3

Initial Conditions

ın nız bogzarad
This too shall pass.

—Solomon – Israel Folklore Archive #126

Any simulation requires a starting point. We call these the initial conditions. Some initial conditions are simple, some are extraordinarily complex. The simplest ones are often the most interesting, and lead to the clearest understanding of the complex phenomena that follow through calculation. Many final conditions eventually turn into initial conditions for future simulations. In astrophysics, the ultimate initial conditions are probably the beginning of time and the Universe, whatever they were.

3.1 Starting Simulations

The generation of initial conditions for complex multiphysics simulations is a critical step in ensuring accurate and efficient computational modeling. The complexity grows with the demands and realism in the simulations. In multiphysics simulations, where multiple physical phenomena interact simultaneously, the choice of initial conditions has an unprecedented impact the simulation's stability, convergence, and the interpretability of the results.

The challenge lies in creating initial conditions that are consistent with the governing equations, boundary conditions, and physical constraints of the system, which, when evolved in time, teach you something about the underlying processes that govern the system's evolution.

3.1.1 Warning on Initial Conditions

Every simulation in AMUSE requires initial conditions to start a calculation. Typically, these entail choosing dynamical quantities such as particle mass, position,

doi:10.1088/2514-3433/adfc3cch3

and velocity, and stellar properties such as radii and luminosities, or gas properties such as density, temperature, and composition. These quantities are often the result of some prior calculation, or represent a restart from a previously stored state. In principle, any mathematically consistent set of initial choices is a valid way to start a simulation. In practice, however, we generally want to use some reasonable, physically motivated starting point for further study.

Some initial conditions are realistic, but most are not—and most of the time that is okay. Realistic initial conditions are the goal of (almost) all researchers, but they are hard to find and generally even harder to explain. Most studies start with somewhat unrealistic initial conditions, with the expectation (or at least the hope) that they capture the fundamental physics necessary to study the desired phenomenon, and that the details of the simulation will rapidly drive the system into some physically more meaningful configuration. In follow-up studies, we can then improve on the earlier choice of initial conditions, say by covering a larger part of parameter space or by zooming in on a region of interest. In some cases, for example the stellar density at the moment of core collapse in an idealized equal-mass star cluster (see Section 4.1.3.2), the evolution is quite insensitive to the details of the initial conditions. In others, however, such as star formation in an interstellar molecular cloud (Chapter 6), the outcome depends crucially on the initial state.

A simple example of unrealistic but widely accepted initial conditions are those used in gravitational N-body simulations of star clusters. Such simulations generally start by setting the number of stars N, and distributing them in space with specified radial (spherically symmetric) density and (truncated Maxwellian) velocity profiles. Most modern studies of star clusters incorporate some sort of (parameterized) stellar evolution, so the initial conditions must then also include stellar masses and radii. The entire system is then scaled (usually by adjusting the velocities) to virial equilibrium (see Section 4.1.2), before high-precision N-body integrators are used to study the dynamical evolution of the system. From an astronomical point of view, these initial conditions are not correct, but they can still be very useful in the study of gravitational dynamics.

More realistic initial conditions for this problem might start with an interstellar gas cloud, which collapses and fragments to form stars (see Section 6.7.4). Once massive stars begin to shine, they produce winds and radiation that cause the remaining gas to be blown away, terminating further star formation and locally leaving behind a star cluster. In this case, no ad hoc specifications of the positions and masses of cluster stars are required. But, of course, this only shifts the problem back in time because the initial conditions for the gas simulation must now specify the (clumpy) density distribution and the (turbulent) motion of the cloud. Those "initial conditions" in turn depend on the large-scale flow of gas in the galaxy, and so on. The fact is that there are no "true" initial conditions for this problem, just convenient milestones in a continuous process. Most generally adopted initial conditions have problems, but doing much better can turn garbage into science, and science into garbage.

AMUSE contains an extensive (and growing) collection of codes to generate initial conditions. Very often, one code provides the conditions from which another code continues. For example, when two stars collide, a stellar evolution code might provide the initial density and temperature profiles for the two colliding stars, while a gravitational dynamics code provides their relative positions and velocities. In this way, two codes are used to generate the conditions for starting a third code that simulates the hydrodynamics of the collision. In the following subsections, we discuss some common initial conditions used in astrophysical studies, but we focus on the simplest ones, those that are generated by a single rather simple routine.

3.1.2 Variations in Starting Points

One can imagine several approaches in the generation of initial conditions. A new starting point might be generated from a single free parameter, or it may be the result of a large-scale simulation. Both are found in astronomy; people still start by randomly selecting stellar masses from a power-law mass function (which has 3 degrees of freedom), while others take the end result of a large-scale cosmological simulation to extract a specific aspect that is subsequently modeled in higher resolution or by including different physical processes.

The few-parameter generated approach results in a clean experiment, with the advantage of easier interpretation, whereas the latter results in more complex relation between initial conditions and eventual simulation results. Which is preferable depends on your question.

We can separate the generation of initial conditions naively in the following categories (in order of increasing complexity):

- In steady-state initialization, we solve a stationary version of the problem to obtain a consistent initial state. Although it is not always clear exactly what consistent means, in our case it would reflect the masses, positions, and velocities of objects in space, such as a solar system, or as we demonstrate in Section 4.3.1 by generating a cluster of stars, or by realizing a molecular cloud that collapses under its own gravity (Section 6.5).
- By gradually introducing loads and boundary conditions over time, we assist a system to transition from simple initial values to more complex states. This ramping technique is often used when converting the end conditions from one physical domain to start a simulation in another. An example is the conversion of a thermal-equilibrium stellar evolution model to a dynamical equilibrium hydrodynamical model of the same star, as we demonstrate in Section 6.4.2 by simulating the collision between two stars.
- Experimental data or results from simplified models can provide interesting initial conditions. This data-driven approach has become more popular with the availability of large quantities of multimessenger data. We present a brief data-driven example in Section 9.1.4, where we generate initial conditions by mimicking observations of a giant molecular cloud.

- In multistep initialization, we use a series of simulations to progressively work up to the full complexity of the multiphysics problem. An example, provided in Section 8.7.3, explains how to simulate the evolution of planetary systems in a stellar cluster. In Section 10.3.3 (see also Figure 6.10), we go a step further by exploring the hydrodynamical collapse of a giant molecular cloud that forms stars with circumstellar disks. The disks, irradiated by the neighboring stars, subsequently collapse and form planetary systems that evolve together with the dynamical evolution of the cluster.

The last conditions discussed in this itemized list are the most complex, and often require a trial series of initial conditions and more elaborate simulations in order to execute the eventual calculation. One can hardly talk about "initial conditions" in these cases because they represent a morphable collection of data representing the system at an earlier epoch rather than a fixed starting point in time.

In this chapter, we elaborate a little on the simplest concepts of generating initial conditions. Along the sojourn through this book, you will find many other examples of the generation of starting conditions for simulating astronomical phenomena.

3.2 Creating Particle Distributions in Space

To generate a particle distribution, we will need to access the various functionalities of the particle set. In Section 2.2.1, we discussed the fundamental characteristics of this data structure. Here, we illustrate some of its important points, such as creating and storing point-mass information for a collection of particles in space.

3.2.1 Initial Condition for Gravitational Dynamics

For large systems, initial conditions are generally not specified exactly, and we must start with a random realization of some distribution function. In self-gravitating simulations of star clusters, the Plummer (1911) model is often adopted as an initial density profile. In AMUSE, a particle set representing a Plummer sphere of `number_of_stars` stars can be generated with

```
particles = new_plummer_model(number_of_stars)
```

The result is a `Particles` object, which contains the identities, masses, positions, and velocities of the `number_of_stars` particles generated. You could inspect the content of `particles` at this point by adding an additional line

```
print(particles)
```

This would result in something like (do absolutely try this yourself!):

```
      key        mass      radius          vx  ...          x           y           z
        -        mass      length length/time  ...     length      length      length
=========    ========    ========  ==========  ...  =========   =========   =========
120995...    1.00e-02    0.00e+00   7.666e-01  ...   1.95e-01   -2.01e-01    7.47e-02
 88511...    1.00e-02    0.00e+00   1.219e-01  ...  -1.65e+00   -1.55e+00   -2.92e+00
 19874...    1.00e-02    0.00e+00  -2.131e-01  ...   3.08e-02    1.20e+00   -1.18e+00
      ...         ...         ...         ...             ...         ...         ...
 18055...    1.00e-02    0.00e+00  -3.412e-01  ...  -1.90e+00    2.35e-01    2.04e+00
 46733...    1.00e-02    0.00e+00   6.214e-01  ...  -2.43e-01   -1.97e-02   -1.61e-01
 67731...    1.00e-02    0.00e+00  -1.089e-02  ...  -8.33e-01    7.62e-01    2.43e+00
=========    ========    ========  ==========  ...  =========   =========   =========
```

Note that the output format from your code will look somewhat different because we have abbreviated the output due to page limitations (and we did not specify the seed in the random number generator).

As noted in Section 2.2.1, the key is a unique identifier for each particle, followed by the mass, radius, velocity, and position. Except for the first (key) entry, these lines may appear in any order, and the columns may also appear in a different order. The first line identifies the parameter, while the second gives the unit. In this example, the units are generic (scaleless) because we have not specified any units explicitly.

The Plummer model represents a self-consistent equilibrium solution to Poisson's equation, and so is a valid description of an N-body system. But aside from the fact that it is simple and has no free dimensionless parameters in its description, it has no special claim to fame as a cluster model. Nevertheless, it is a very common choice.

Some initial condition generators are not considered standard, and require a community code to be built/installed before usage. For example, for "fractal" initial conditions (Goodwin & Whitworth 2004), we must have the amuse.community.fractalcluster package available. If this is not the case, using the new_fractal_cluster_model function will result in an error:

```
ModuleNotFoundError: No module named 'amuse.community.fractalcluster'
```

Figure 3.1 presents a small collection of various initial realizations common in N-body simulations: the Plummer sphere, a King model (with structure parameter $W_0 = 9$; King 1966), the top view of a galactic disk plus bulge model (for which we adopted Halogen with parameters $\alpha = 1$, $\beta = 5$, $\gamma = 0.5$; Zemp et al. 2015; Zemp 2014), and a fractal distribution (with fractal dimension 1.6; Goodwin & Whitworth 2004).

The various models are generated with calls similar to those discussed in Section 3.2. For the King model we write:

```
particles = new_king_model(N, W0=9.0)
```

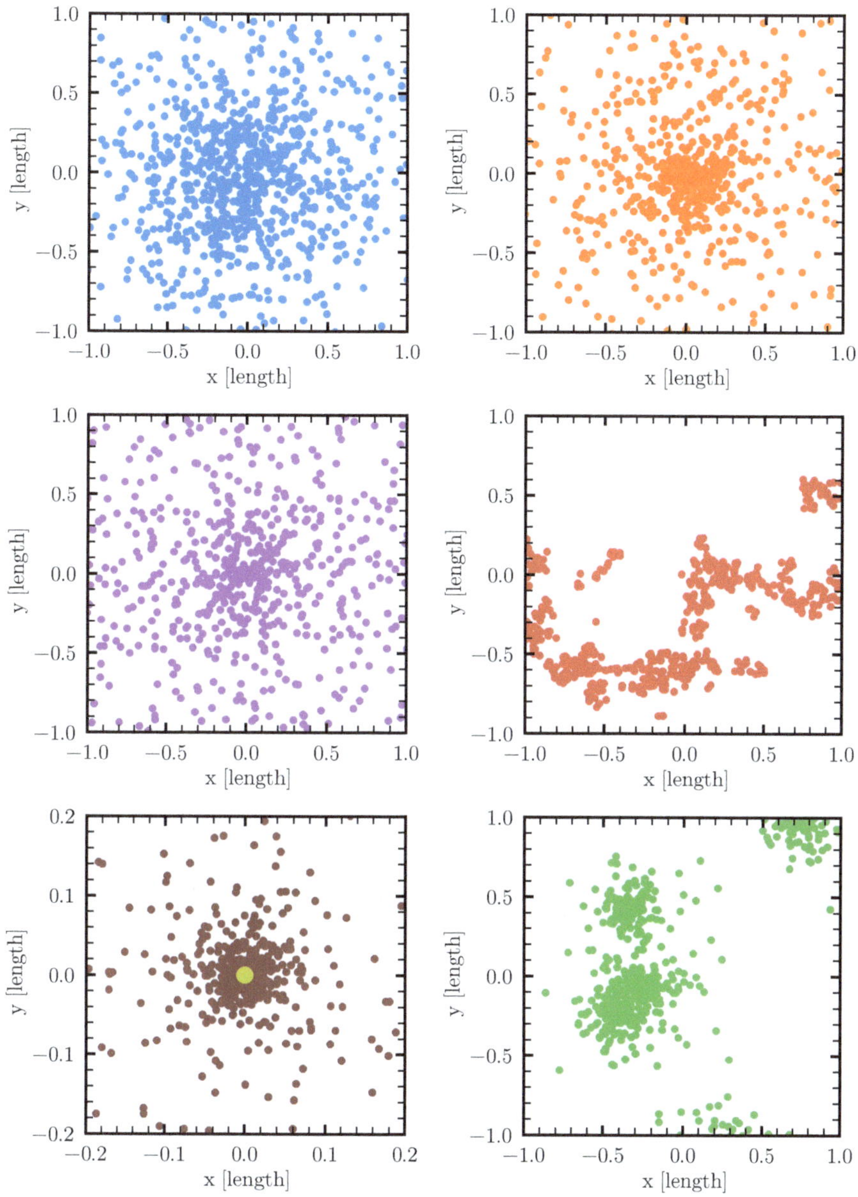

Figure 3.1. The inner parts, spanning 1 by 1 Hénon units (Hénon 1971; Heggie & Mathieu 1986), of projected distributions drawn from the Plummer model (top left), a King model with $W_0 = 9$ (top right), Halogen Galaxy model with $\alpha = 1$, $\beta = 5$, $\gamma = 0.5$ (middle left), a cluster with fractal dimension 1.6 (middle right), an Oort cloud with a central star (bottom left), and a hierarchical King model, each for $N = 1000$ particles. The source for making the Plummer distribution is presented in Code example 3.1. The script to generate this figure can be found in `amuse.examples.plot_density_distributions`.

The fractal distribution is generated with

```
particles = new_fractal_cluster_model(N=N, fractal_dimension=1.6)
```

For the fractal model (and for other special initial condition generators) one must include the appropriate file, which in this case can be loaded with

```
from amuse.ic.fractalcluster import new_fractal_cluster_model
```

These models have typically one free parameter. For the King model, this is the dimensionless depth of the potential (W_0), while for the fractal model it is the fractal dimension (F_d).

A Galaxy disk model can be constructed with

```
particles = new_halogen_model(N, alpha=1.0, beta=5.0, gamma=0.5)
```

which has three free parameters (see Equation 1 in Zemp et al. 2015, where these parameters are explained), A circumstellar disk can be generated using

```
from amuse.ext.protodisk import ProtoPlanetaryDisk
model = ProtoPlanetaryDisk(
    N,
    densitypower=1.5, Rmin=0.1, Rmax=1,
    q_out=1.0, discfraction=0.1,
).result
model.rotate(0., np.pi/6, np.pi/6)
```

This function has 10 arguments but the most important include the number of particles (N), the slope of the power-law projected density profile, the inner and outer disk radius (if defined in units these values are using the converter, see Section 2.3.1), the Toomre Q-parameter, and the mass fraction (here called `diskfraction`, indicating that the disk is 10% of the star's mass).

The best way to see what parameters a certain function in AMUSE has is by reading the actual routine. In this example, the routine lives in the amuse.ext.protodisk module. Every code you load has an associated source code, which you can read or copy. For this you will have to dive into the directory structure. It is probably best to take some time to make yourself familiar with AMUSE's directory structure.

The routine `ProtoPlanetaryDisk` returns a class whose parameter `result` is a particle set with additional information for performing hydrodynamical simulations, such as the internal energy, which can be used to calculate the temperature and pressure. A newly generated disk will be oriented in the x-y plane. We can rotate the disk, in azimuth (or yaw by rotation around the vertical z-axis), and elevation (or pitch, by rotation around the horizontal y axis). Here, we adopted an angle of $\pi/6$ radians for both rotation angles. The result is presented in the bottom left panel in Figure 3.1.

To illustrate how we can make somewhat more complex initial conditions, we generate a composite model by first making a King model for the center of masses of the stars, and subsequently making as many King models as center of masses were generated. Here is an example of five King models with one-fifth the total number of stars each, as presented in the bottom right panel in Figure 3.1.

```
model_com = new_king_model(5, W)
for i in range(5):
    model = new_king_model(int(number_of_stars / 5), W)
    # Shrink the cluster.
    model.position *= 0.1
    # Keep virialized.
    model.velocity *= 3.2
    # Move to the proper coordinates.
    model.position += model_com[i].position
    model.velocity += model_com[i].velocity
```

3.2.2 Initial Condition for Hydrodynamics

Like particle distributions for gravitational dynamical simulations, hydrodynamical simulations also require input density and other distributions. AMUSE provides most support for smoothed particle hydrodynamics (SPH) codes (see Chapter 6), and contains a variety of distribution-generating functions. Code example 3.1 illustrates the conversion of a one-dimensional stellar evolution model to a three-dimensional hydrodynamical particle distribution.

In Code example 3.2, in lines 25–27 the parameter `gas_particles` is returned from the function `convert_stellar_model_to_sph`. This is common practice when the called function returns a class, in this case a particle set. The return values are stored in a class containing the particle set `core_particle`, its radius (`core_r-adius`), and a `gas_particle` set. The resulting hydrodynamical models are plotted in Figure 3.2 (left panel); Code example 3.3 shows a portion of the plotting code.

```
import matplotlib.pyplot as plt
from amuse.units.nbody_system import length
from amuse.ic.plummer import new_plummer_model


def plot_plummer(number_of_particles=1000):
    fig = plt.figure(figsize=(5, 5))
    ax = fig.add_subplot(1, 1, 1)
    bodies = new_plummer_model(number_of_particles)
    ax.scatter(bodies.x.value_in(length), bodies.y.value_in(length))
    ax.set_xlim((-1, 1))
    ax.set_ylim((-1, 1))
    ax.set_xlabel("X")
    ax.set_ylabel("Y")
    plt.savefig("plummer.pdf")
    print("Saved figure in file plummer.pdf")
```

Code example 3.1: Routine to plot a Plummer sphere. Code fragment from `amuse.examples.plot_plummer`.

```
def stellar_model(
    number_of_particles=10000, mass_star=2.0 | units.MSun, time=0.0 | units.Myr
):
    star = Particle(mass=mass_star)
    stellar_evolution = Evtwin()
    se_star = stellar_evolution.particles.add_particle(star)
    print(f"Evolving {star.mass} to t= {time.in_(units.Myr)}")
    stellar_evolution.evolve_model(time)
    print(f"Stellar type: {se_star.stellar_type}")
    print("Creating SPH particles from the (1D) stellar evolution model")
    sph_particles = convert_stellar_model_to_sph(
        se_star, number_of_particles
    ).gas_particles
    stellar_evolution.stop()
    return sph_particles
```

Code example 3.2: Converting a one-dimensional stellar evolution model to a three-dimensional SPH particle representation. Code fragment from `amuse.examples.plot_hydro_density_distributions`.

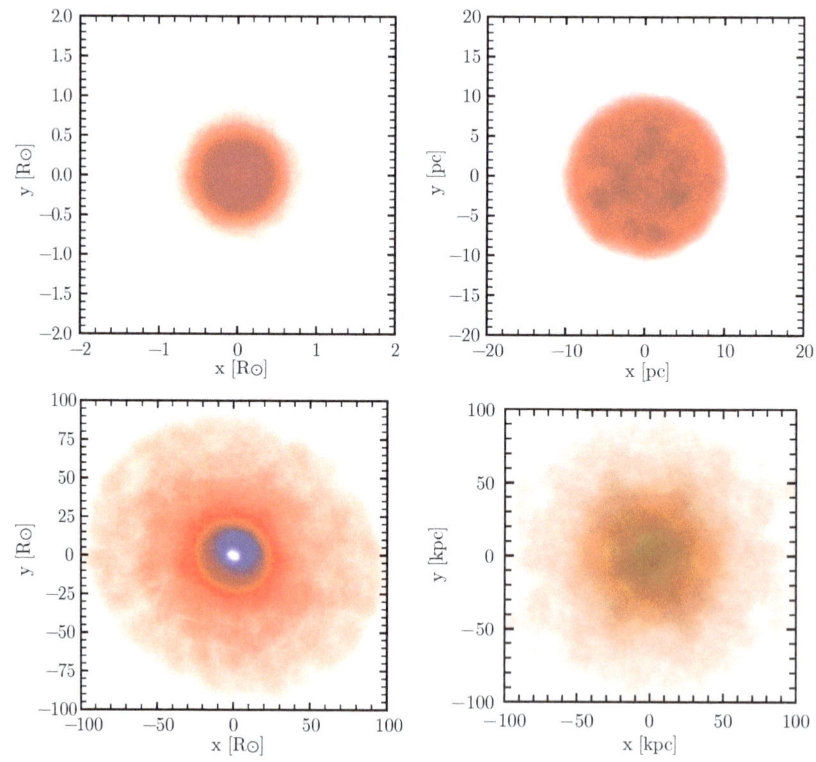

Figure 3.2. Some initial particle distributions for SPH codes. The top-row left panel shows a two solar mass star, the right a homogeneous spherical distribution with a turbulent velocity field, as is often used in simulations of molecular clouds. Particles are colored by their internal energy, from red (0.048 (km $s^{-1})^2$) to blue (1.82 (km $s^{-1})^2$). The bottom row shows a slightly tilted circumstellar disk with a total mass of 0.1% of the star's mass (left), and a 10^{10} $M_{\odot}$ galaxy model. The script to generate this figure can be found in `amuse.examples.plot_hydro_density_distributions`.

```
def plot_zams_stellar_model(number_of_particles=10000, mass_star=2.0 | units.MSun):
    length_unit = units.RSun
    sph_particles = stellar_model(number_of_particles, mass_star)
    figure = plt.figure(figsize=(6, 6))
    ax = figure.add_subplot(1, 1, 1)
    sph_particles_plot(
        sph_particles,
        min_size=500,
        max_size=500,
        alpha=0.01,
        view=(-2, 2, -2, 2) | length_unit,
    )
    ax.set_facecolor("white")
    ax.set_xlabel(f"x [{length_unit}]")
    ax.set_ylabel(f"y [{length_unit}]")

    save_file = "stellar_2MSunZAMS_projected.pdf"
    plt.savefig(save_file)
    print(f"Saved figure in file {save_file}\n")
```

Code example 3.3: Plotting the three-dimensional particle representation using the AMUSE `plot` module. Code fragment from `amuse.examples.plot_hydro_density_distributions`.

```
def gmc_model(
    number_of_particles=10000,
    mass_cloud=10000.0 | units.MSun,
    radius_cloud=10.0 | units.parsec,
):
    converter = nbody_system.nbody_to_si(mass_cloud, radius_cloud)
    sph_particles = molecular_cloud(
        target_number_of_particles=number_of_particles,
        convert_nbody=converter,
    ).result
    sph = Fi(converter)
    sph.gas_particles.add_particles(sph_particles)
    sph.evolve_model(1 | units.day)
    channel = sph.gas_particles.new_channel_to(sph_particles)
    channel.copy()
    sph.stop()
    return sph_particles
```

Code example 3.4: Converting a molecular cloud model to a three-dimensional SPH particle representation. Code fragment from `amuse.examples.plot_hydro_density_distributions`.

We do not have to start from a stellar model, of course. Code example 3.4 illustrates the creation of a three-dimensional hydrodynamical particle distribution describing a turbulent molecular cloud. As discussed in Section 3.2.1 in relation to the function `ProtoPlanetaryDisk()`, `molecular_cloud()` also returns a class whose parameter result contains the SPH particle set:

```
def plot_mass_function(masses, ximf, color, label, n_bins=51):
    "Plots a histogram of the masses provided, with a power law added if ximf > 0"
    mass_min = masses.min()
    mass_max = masses.max()
    log_min = np.log10(mass_min.value_in(units.MSun))
    log_max = np.log10(mass_max.value_in(units.MSun))
    bins = np.logspace(log_min, log_max, n_bins)
    print(bins)
    bin_number, bin_edges = np.histogram(masses.value_in(units.MSun), bins=bins)
    y = bin_number / (bin_edges[1:] - bin_edges[:-1])
    x = (bin_edges[1:] + bin_edges[:-1]) / 2.0
    for yi in y:
        yi = max(yi, 1.0e-10)

    plt.scatter(x, y, s=50, c=color, lw=0, label=label)

    if ximf < 0:
        c = (
            (mass_max.value_in(units.MSun) ** (ximf + 1))
            - (mass_min.value_in(units.MSun) ** (ximf + 1))
        ) / (ximf + 1)
        plt.plot(x, len(masses) / c * (x**ximf), c=color)
```

Code example 3.5: Routine to plot the generated Salpeter mass function. Code fragment from `amuse.examples.salpeter`.

```
from amuse.ic.molecular_cloud import new_molecular_cloud

sph_particles = new_molecular_cloud(
    target_number_of_particles=number_of_particles, convert_nbody=converter,)
```

Both routines can use the `convert_nbody` converter to scale between dimensionless N-body units and physical units. The result is a molecular cloud represented by N particles. The particles will have all have the same mass (`mass`) and internal energy (`u`), and positions and velocities consistent with a homogeneous gas sphere with a turbulent velocity field. (The system is integrated for a short time after creation to reduce random initial noise in the gas distribution.) The resulting density distribution is presented in Figure 3.2 (right panel). We will explore such models in more depth in Chapter 6.

We will not dwell here on more advanced code-coupling strategies, and defer discussion of radiative transport to Chapter 7. We postpone these advanced topics because they require multiple codes to be operational at the same time. Such coupling is discussed toward the end of this book, in Chapters 7 and 9.

3.3 Creating Mass Distributions

The default for `new_plummer_model` and `new_fractal_cluster_model` is to create stars of equal mass, but usually we want stars to have a range of masses, drawn from some distribution. This distribution is known as the initial mass function, and is one of the most important quantities in stellar astrophysics.

The initial mass function is often represented as a power law, meaning that the number of stars dN with masses between m and $m + dm$ is given by

$$\frac{dN}{dm} = A\, m^{\alpha} \tag{3.1}$$

for $M_{\rm min} < m < M_{\rm max}$. The choice $\alpha = -2.35$ (Salpeter 1955) is so common that it is has its own standard function call in AMUSE:

```
mass_zams = new_salpeter_mass_distribution(
    number_of_stars, mass_min=mass_min, mass_max=mass_max
)
```

which returns an array of `number_of_stars` masses randomly selected between `mass_min` and `mass_max`, distributed according to a Salpeter distribution, a power-law distribution with exponent $\alpha = -2.35$. The resulting mass function is presented in Figure 3.3. The naming of this mass `mass_zams` seems somewhat arcane, but refers to the mass of a star at the moment of hydrogen ignition (generally called the birth of the star) which is referred to as the zero-age main sequence (see also Section 5.1).

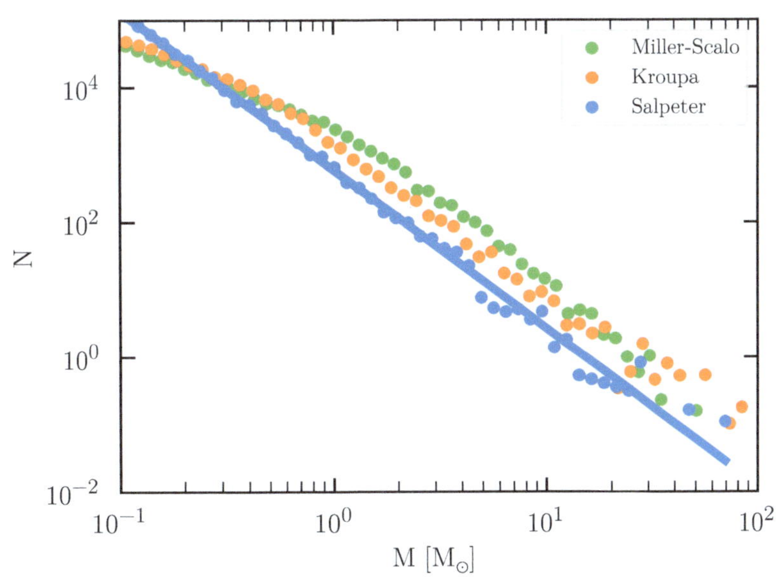

Figure 3.3. Initial mass function for 10^4 stars between $0.1\,M_\odot$ and $100\,M_\odot$. The Salpeter mass function is overplotted with an analytic power law with a slope of -2.35 (blue). Two other frequently used mass functions are shown in green and orange. The latter two are described by Miller & Scalo (1979) (in green) and Kroupa (2001) (orange). An example of how this is plotted is presented in Code example 3.5. The script to generate this figure can be found in `amuse.examples.salpeter`. The script to plot these curves is presented in Code example 3.5.

If you need a power-law mass function with a different slope, you can change the exponent by using a keyword argument, but we recommend that you use the more general version of the mass function generator:

```
mass_zams = new_powerlaw_mass_distribution(
    1000,
    mass_min=0.1 | units.MSun,
    mass_max=100 | units.MSun,
    alpha=-2.0,
)
```

which will return a list of 1000 stars with masses between 0.1 and 100 $M_{\odot}$, distributed according to a power law with $\alpha = -2$.

The masses can now be easily assigned to the previously generated Plummer distribution, by

```
particles.mass = mass_zams
```

or one can generate a new particle set:

```
particles = Particles(mass=mass_zams)
```

The particles can subsequently be assigned additional attributes, as already discussed. Note that changing the masses causes the system to have a different total mass and a different ratio of potential to kinetic energy, as anticipated in the function used to generate the positions and velocities, and the system generally needs to be rescaled (see Section 4.3.1).

There are several other popular mass functions for which we generated special routines, such as Miller & Scalo (1979)

```
particles.mass = new_miller_scalo_mass_distribution()
```

and Kroupa (2001):

```
particles.mass = new_kroupa_mass_distribution()
```

A generalization of these broken power-law mass distributions is available in the class `MultiplePartIMF`, which takes arrays of breakpoints and power-law slopes and allows the user to construct a mass function of personal choice.

3.4 Creating Systems of Multiple Stars and Planets

3.4.1 The Solar System

The initial conditions of the solar system are unknown. But if we are interested in the future dynamics of the solar system, we could start with the masses, positions, and velocities of the Sun and eight planets as they are observed at some moment in time, and then integrate the equations of motion forward in time. Table 3.1 presents such conditions from which such a simulation can be started.

The orbital parameters from this table are coded in the function `new_solar_system` (in the `amuse.ic.solarsystem` module). The parameters for moons, planetesimals, Trojans, Greeks, plutinos, the classic Kuiper belt, and the Oort cloud are not included in this table. For the moons, we have a separate function that generates initial conditions for all the solar system's moons, it is called `new_lunar_system_in_time` in the `amuse.ic.solar_system_moons` module. For the asteroids and other minor bodies, you will have to download the appropriate data and convert the ephemerides to Cartesian coordinates using your own routine (see the assignment in Section 3.6.2).

The source code to initialize the Sun, Venus, and Earth is presented in Code example 3.6. It starts by including two basic AMUSE functionalities—particles and units:

```
10 from amuse.datamodel import Particles
11 from amuse.units import units
```

A particle set is then declared in line 14:

```
14     particles = Particles(3)
```

Table 3.1. A Realization of the Solar System, up to the Last Known Significant Digit, Calculated for 2063 April 5 at 00:00 hours—Julian Date 2474649.5 Days (Corresponding to First Contact: http://www.startrek.com/article/origin-of-first-contact-day-explained)—Using the Ephemerides from http://ssd.jpl.nasa.gov/txt/p_elem_t2.txt, Estimating the Uncertainty in the Positions and Velocities Using the Dispersion Given by Folkner et al. (2011)

Name	Mass [M_{Jup}]	Position [au]			Velocity [km s^{-1}]		
Sun	1047.517	0.005717	−0.00538	-2.130×10^{-5}	0.007893	0.01189	0.0002064
Mercury	0.000174	−0.31419	0.14376	0.035135	−30.729	−41.93	−2.659
Venus	0.002564	−0.3767	0.60159	0.03930	−29.7725	−18.849	0.795
Earth	0.003185	−0.98561	0.0762	-7.847×10^{-5}	−2.927	−29.803	−0.000533
Mars	0.000338	−1.2895	−0.9199	−0.048494	14.900	−17.721	0.2979
Jupiter	1.000000	−4.9829	2.062	−0.10990	−5.158	−11.454	−0.13558
Saturn	0.299470	−2.075	8.7812	0.3273	−9.9109	−2.236	−0.2398
Uranus	0.045737	−12.0872	−14.1917	0.184214	5.1377	−4.7387	−0.06108
Neptune	0.053962	3.1652	29.54882	0.476391	−5.443317	0.61054	−0.144172

```
from amuse.datamodel import Particles
from amuse.units import units


def new_sun_venus_and_earth():
    particles = Particles(3)
    sun = particles[0]
    sun.name = "Sun"
    sun.mass = 1.0 | units.MSun
    sun.radius = 1.0 | units.RSun
    sun.position = (855251, -804836, -3186) | units.km
    sun.velocity = (7.893, 11.894, 0.20642) | (units.m / units.s)

    venus = particles[1]
    venus.name = "Venus"
    venus.mass = 0.0025642 | units.MJupiter
    venus.radius = 3026.0 | units.km
    venus.position = (-0.3767, 0.60159, 0.03930) | units.au
    venus.velocity = (-29.7725, -18.849, 0.795) | units.kms

    earth = particles[2]
    earth.name = "Earth"
    earth.mass = 1.0 | units.MEarth
    earth.radius = 1.0 | units.REarth
    earth.position = (-0.98561, 0.0762, -7.847e-5) | units.au
    earth.velocity = (-2.927, -29.803, -0.0005327) | units.kms

    sun.color = "#FFFF00"
    venus.color = "wheat"
    earth.color = "deepskyblue"

    particles.move_to_center()
    return particles
```

Code example 3.6: Code to generate initial conditions for Venus and Earth orbiting the Sun. Code fragment from `amuse.examples.sun_venus_earth`.

Attribute values are then set. The first particle in the array (`particle[0]`) is assigned to the Sun and the other two particles to Venus and Earth. (Note the details of setting three vector coordinates and units in a single line.) Then, writing

```
print(sun.position)
```

will print out the Cartesian coordinates of the Sun. After initialization, the particles are moved to the center of mass of the three-body system, using the (class-defined) function `move_to_center()` mentioned in the previous subsection. At the end of the function, the particle set is returned to the main script with

```
    return particles
```

A potentially easier way to achieve the initialization of the Sun, Venus, and Earth, would be to use the following few lines:

```
from amuse.ic.solarsystem import new_solar_system
solar_system = new_solar_system()
SunVenusEarth = solar_system[0] + solar_system[2] + solar_system[3]
SunVenusEarth.move_to_center()
```

The third particle in this array actually represents the center of mass of the Earth–Moon system. To separate these two objects, one could use this (somewhat verbose and slow, but more precise) version:

```
from amuse.ic.solar_system_moons import new_lunar_system
solar_system = new_lunar_system(Julian_date=2474649.5 | units.day)
Sun = solar_system[solar_system.name=="sun"]
Venus = solar_system[solar_system.name=="Venus"]
Earth = solar_system[solar_system.name=="Earth"]
SunVenusEarth = Sun + Venus + Earth
SunVenusEarth.move_to_center()
```

In this latter script one actually uses the mass, position, and velocity of Earth, rather than of the Earth–Moon system at Julian date 2474649.5.

The integration of the orbits can be carried out using the script in Code example 4.2 (see Section 4.4.1). Code example 3.7 presents a snippet of code to plot the planets' positions as they orbit the Sun. The results of the calculation are shown in Figure 3.4. We use `matplotlib.pyplot` (imported via convention as `plt`) to present the results, and later will show how to use an AMUSE-tailored version for plotting with units and hydrodynamical models. Note that here and throughout we provide references to the scripts that produced the figures and, for convenience, we save the plot in a file (`plt.savefile(...)`) in the current directory, and also display it on the screen (`plt.show()`).

Here, we probably should explain the two function calls on lines 97 and 99, which return arrays of the requested variables (here the Cartesian coordinates along the x and y axes) for the requested particle in astronomical units. The basic functioning of this routine, as a short stand-alone snippet, might look as follows:

```
time_series = read_set_from_file("stored_particle_time_series.amuse")
bodies = time_series.history[0]
Earth = bodies[bodies.name == "Earth"]
x = get_timeline_of_attribute_as_vector(time_series, Earth, "x")
```

We do explain this in more detail in Appendix A, though.

```
def plot_track(particles, output_filename):
    figure = plt.figure(figsize=(10, 10))
    plt.rcParams.update({"font.size": 30})
    ax = figure.add_subplot(1, 1, 1)
    ax.minorticks_on()
    ax.locator_params(nbins=3)

    x_label = "x [au]"
    y_label = "y [au]"
    plt.xlabel(x_label)
    plt.ylabel(y_label)

    ax.scatter([0.0], [0.0], color="#CCCC00", lw=8)
    for planet in particles:
        if planet.name.lower() in ["venus", "earth"]:
            ax.plot(
                planet.get_timeline_of_attribute_as_vector("x")[1].value_in(units.au
    ),
                planet.get_timeline_of_attribute_as_vector("y")[1].value_in(units.au
    ),
                color=planet.color,
            )
    ax.set_xlim(-1.3, 1.3)
    ax.set_ylim(-1.3, 1.3)

    plt.savefig(f"{output_filename}.pdf")
    print(f"\nSaved figure in file {output_filename}\n")
```

Code example 3.7: Plotting the orbits of Venus and Earth around the Sun using pyplot. This tells you more about `pyplot` than about AMUSE, but we present it here for completeness. Code fragment from `amuse.examples.sun_venus_earth`.

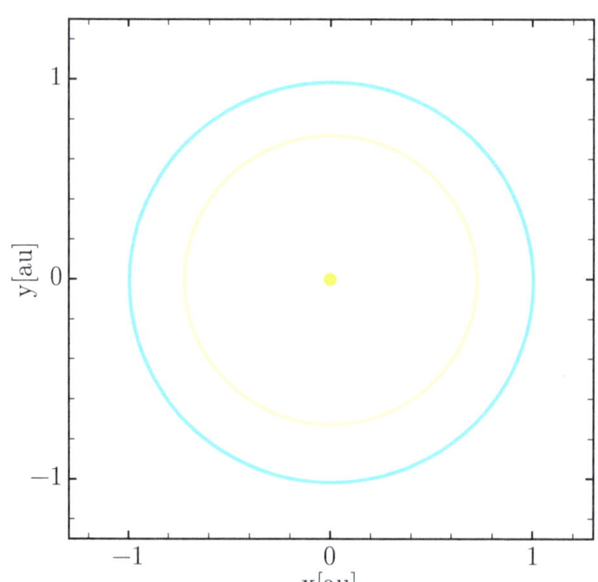

Figure 3.4. Orbital integration of the Sun, Venus, and Earth for 1 yr using the script in Code example 4.2. The script to generate this figure can be found in `amuse.examples.sun_venus_earth`.

3.4.2 Other Planetary Systems

In principle, any planetary system can be initialized in the above-described method, by providing the masses, positions, and velocities of the planets. Following planet formation through Oligarchic growth (Kokubo & Ida 2002)

```
from amuse.ic import make_planets_oligarch
from amuse.units import units
mass_star = 1 | units.MSun
radius_star = 1 | units.RSun
inner_radius_disk = 10 * radius_star
outer_radius_disk = 42 | units.au
mass_disk = 0.01 * mass_star
planetary_system = make_planets_oligarch.new_system(
    mass_star,
    radius_star,
    inner_radius_disk,
    outer_radius_disk,
    mass_disk,
)
print(planetary_system.planets)
```

and to print the planetary data

```
print(planetary_system.planets[0])
```

Integrating such a system can be achieved by

```
gravity = Ph4()
gravity.particles.add_particles(planetary_system)
gravity.particles.add_particles(planetary_system.planets[0])
```

The implementation of the oligarchic planet formation model was designed by Tremaine (2015) and we adopt the implementation by Hansen & Murray (2013).

```
from amuse.ic import make_planets_oligarch
mass_star = 1 | units.MSun
radius_star = 1 | units.RSun
inner_radius_disk = 10 * radius_star
outer_radius_disk = 42 | units.au
mass_disk = 0.01*mass_star
planetary_system = make_planets_oligarch.new_system(
    mass_star,
    radius_star,
    inner_radius_disk,
    outer_radius_disk,
    mass_disk,
)
```

3.4.3 Binaries and Hierarchical Multiples

Generating binaries and hierarchical multiple systems may be somewhat elaborate, but if done systematically it is not hard. It is best to start with the innermost (hardest) binary (or more than one if the system is highly hierarchical).

Let us start by making a simple binary, as in Code example 3.8. This might be a planet with moon, or a star with planet, or a galaxy with a dwarf companion. We assume that both objects are point masses, and that they have a Keplerian orbit.

The Kepler solver in this routine is presented in Code example 3.9.

After the binary is generated, one would like to see if it really turned out as expected. A simple way to achieve this is presented in Code example 3.10, where we calculate the semimajor axis and eccentricity from the binary's Cartesian coordinates.

```
def new_binary_orbit(stars, kepler):
    """
    Takes a Particles object of length 2 and returns a binary particle from the
    stars, using the values from `kepler`
    """
    if len(stars) != 2:
        raise ValueError("'stars' must consist of exactly two particles")
    rel_position = as_vector_quantity(kepler.get_separation_vector())
    rel_velocity = as_vector_quantity(kepler.get_velocity_vector())

    mu = stars[0].mass / stars.mass.sum()
    binary = Particle()
    binary.child1 = stars[0]
    binary.child2 = stars[1]
    binary.child1.position = mu * rel_position
    binary.child2.position = -(1 - mu) * rel_position
    binary.child1.velocity = -(1 - mu) * rel_velocity
    binary.child2.velocity = mu * rel_velocity
    print(binary.child1.position.in_(units.au))
    print(binary.child2.position.in_(units.au))
    return binary
```

Code example 3.8: Source for generating a binary of two objects with masses, positions, and velocities of the two stars star[0].mass and star[1].mass. Other orbital parameters can subsequently be calculated from a Kepler solver, as initialized in Code example 3.9. Source code is available in `amuse.examples.make_binary_star`.

```
def new_kepler(converter):
    """Starts and returns a new Kepler integrator with given converter"""
    kepler = Kepler(converter)
    kepler.initialize_code()
    kepler.set_longitudinal_unit_vector(1.0, 0.0, 0.0)
    kepler.set_transverse_unit_vector(0.0, 1.0, 0)
    return kepler
```

Code example 3.9: Initializing a Kepler object for system in the x–y-plane. Source code is available in `amuse.examples.make_binary_star`.

```
def calculate_orbital_elements(bi, kepler):
    """
    Takes a binary particle and a kepler code, returns the semi-major axis and
    eccentricity
    """
    comp1 = bi.child1
    comp2 = bi.child2
    mass = comp1.mass + comp2.mass
    pos = comp2.position - comp1.position
    vel = comp2.velocity - comp1.velocity
    kepler.initialize_from_dyn(
        mass,
        pos[0],
        pos[1],
        pos[2],
        vel[0],
        vel[1],
        vel[2],
    )
    a, e = kepler.get_elements()
    return a, e
```

Code example 3.10: Code snippet for verifying the binary's orbital parameters. Source code is available in `amuse.examples.make_binary_star`.

A higher order hierarchical multiple can be generated by iteratively using the generated binary as one of the stars in the binary generation routines.

3.5 Often Used Initial Conditions

For practical purposes, we have generated a series of often-used initial conditions. These are collected in a separate `Python` package with a number of generation functions. These functions take the form of `make_`, `add_`, `conv_` and `move_` for generating a new particle set, adding some attribute or feature, converting the particle set and moving it around. A complete list is presented in Section A.2.1.

These utilities can read an existing particle set from a file and write the extended or otherwise modified particle set to another file, after which it can be read in by one of the other utilities. In this way, it is possible to generate rather elaborate particle sets.

For example, a Plummer distribution of 100 stars in space of which 10 receive a system of eight planets following the Oligarchic growth model on the Sun's location in the Galaxy can be realized as follows.

```
python make_Plummer_model.py -N 100 -F plummer.amuse
python name_stars.py -f plummer.amuse --name PlanetarySystem \
  --nstars 10 -F plummerNamed.amuse
python add_planet_Oligarch.py -f PlummerNamed.amuse --name PlanetarySystem \
  --Nplanets 8 -F plummerWithPlanets.amuse
python move_all_particles.py -f plummerWithPlanets.amuse --pos -8300 0.0 27 \
  --vel 11.1 240 7.25 -F plummerWithPlanetsInGalaxy.amuse
```

This procedure results in a number of files `plummer.amuse`, `plummerNamed.amuse`, `plummerWithPlanets.amuse` and the final snapshot named `plummerWithPlanetsInGalaxy`. This latter snapshot can subsequently be run with a gravity code including a Galactic potential to study the evolution of the stellar system with planets.

By combining various routines, it is possible to create a broad range of particle sets that can be used as initial conditions. Of course, not all varieties of initial conditions are available, but we expect the list of initial condition routines to grow steadily with time.

3.6 Assignments

3.6.1 Create Your Own Initial Conditions Generator or Manipulator

The more routines we have for generating initial conditions, the more versatile AMUSE will become. This assignment requires some idea of how these generators and manipulators are constructed. Make a generator, constructor, manipulator for an application that is not yet available in AMUSE — for example, one to generate a gaseous circumstellar disk around a single star. By the time you read this, such a routine has probably already been added to the repository, so you should find another example for which you can make a generator. Test the code, and submit it to the git repository at https://github.com/amusecode/initial-conditions-generator.

3.6.2 Solar System Minor Bodies

Download the ephemerides of the solar system minor bodies (https://www.minorplanetcenter.net/iau/mpc.html), and generate a routine to convert the Kepler elements to Cartesian coordinates.

Take the initial conditions for the solar system (Table 3.1) and integrate the planets' orbits for 100 yr with the integrator of your choice (Table 4.1). Explain why you have selected this integrator, and discuss the effect of using another integrator:

1. Generate a snapshot for the solar system at the time of first contact.
2. Check the planets' positions and velocities from Table 3.1.

References

Folkner, W. M., et al. 2011, Proc. of the Journées 2010 "Systèmes de Référence Spatio-temporels" (JSR2010): New Challenges for Reference Systems and Numerical Standards in Astronomy ed. N. Capitaine (Saint-Mandé: Institut Géographique National) 43

Goodwin, S. P., & Whitworth, A. P. 2004, A&A, 413, 929

Hansen, B. M. S., & Murray, N. 2013, ApJ, 775, 53

Heggie, D. C., & Mathieu, R. D. 1986, Lecture Notes in Physics: The Use of Supercomputers in Stellar Dynamics, Proceedings of a Workshop Held at the Institute for Advanced Study, Princeton 267, ed. P. Hut, & S. L. W. McMillan; Berlin: Springer) 233

Hénon, M. H. 1971, Ap&SS, 14, 151

King, I. R. 1966, AJ, 71, 64

Kokubo, E., & Ida, S. 2002, ApJ, 581, 666

Kroupa, P. 2001, MNRAS, 322, 231
Miller, G. E., & Scalo, J. M. 1979, ApJS, 41, 513
Plummer, H. C. 1911, MNRAS, 71, 460
Salpeter, E. E. 1955, ApJ, 121, 161
Tremaine, S. 2015, ApJ, 807, 157
Zemp, M. 2014, Halogen: Multimass Spherical Structure Models for N-Body Simulations, Astrophysics Source Code Library, ascl:1407.020
Zemp, M., Moore, B., Stadel, J., Carollo, C. M., & Madau, P. 2008, MNRAS, 386, 1543

Part II

Fundamental Principles

AAS | IOP Astronomy

Astrophysical Recipes (Second Edition)

The art of AMUSE

Simon Portegies Zwart, Steve McMillan and Steven Rieder

Chapter 4

Gravitational Dynamics

Oh my God! — it's full of stars!

—Arthur C. Clarke

Here, we discuss how to solve relatively simple problems in Newtonian gravity using the monophysics solvers in AMUSE. Aside from the specific applications, this chapter gives an overview of how to perform simple experiments, what to be aware of, and how to address technical and algorithmic issues. This chapter should therefore be read as a general overview of the working of AMUSE with gravity as a focus. Even if you are not interested in gravity—for example, if your research is dominated by problems in radiative transfer or hydrodynamics—this chapter should still be your starting point.

4.1 In a Nutshell

4.1.1 Equations of Motion for a Self-Gravitating System

In the Philosophiæ Naturalis Principia Mathematica, Newton (1687) described his famous laws of motion in his chapter *Axiomata Sive Leges Motus* (the Axioms or Laws of Motion). He also laid the foundation for the general law of gravitation. In Newtonian gravity, the gravitational acceleration $\mathbf{a}_i$ of an object (subscript i) due to a group of objects (subscript j) is

$$\mathbf{a}_i \equiv \ddot{\mathbf{r}}_i = G\sum_{j\neq i}^{N} m_j \frac{\mathbf{r}_j - \mathbf{r}_i}{|\mathbf{r}_j - \mathbf{r}_i|^3}. \tag{4.1}$$

Here, G is the gravitational constant, and m_j and $\mathbf{r}_j$ are the mass and position of particle j. Equation (4.1) is a set of coupled, second-order, nonlinear, singular ordinary differential equations. Its solutions exhibit a bewildering variety of

doi:10.1088/2514-3433/adfc3cch4

complex, often chaotic motions, and many books have been written about how to numerically solve it (Aarseth 1985, 2003; Heggie & Hut 2003; Binney & Tremaine 2008). Furthermore, because gravity is a long-range force, this behavior can be seen on virtually all scales, from planets and moons to clusters of galaxies. Newton's equations of motion break down only for very large scales or very strong gravitational fields. They are, however, applicable on most scales considered in this book.

4.1.2 Gravitational Timescales

Timescales are critical for understanding fundamental processes in (astro)physics. Therefore, we start each chapter on a new physical domain in AMUSE with a brief overview of the relevant timescales. In the case of self-gravitating systems, these are the *dynamical timescale*, the *relaxation timescale*, the *dynamical friction timescale*, and (in extreme circumstances) the *gravitational radiation timescale*.

4.1.2.1 The Dynamical Timescale

The dynamical timescale is the scale associated with orbital motion in a gravitating system. Also known as the crossing time, it is generally considered as the time taken for a typical particle to cross the system,

$$t_{\rm dyn} = R/v, \tag{4.2}$$

where R is a characteristic dimension and v is a typical particle speed. In practice, R is often taken to be the mean radius of the system (defined more carefully below) and v is the root mean square velocity $\langle v^2\rangle^{\frac{1}{2}}$ of all particles relative to the center of mass.

For a two-body gravitating system, the dynamical timescale is simply the orbital timescale. More generally, the dynamical timescale is the timescale on which a self-gravitating system establishes virial (or dynamical) equilibrium, in which there is no net flux of particles across any radius (see the assignment on virial equilibrium in Section 4.7.5).

For a system in virial equilibrium, the virial theorem (Collins 1978) relates the (time-averaged) kinetic energy T and potential energy U of the system by

$$2T + U = 0. \tag{4.3}$$

For a system of mass M, we can write $T = \frac{1}{2}M\langle v^2\rangle$ and $U = -\frac{1}{2}GM^2/R_{\rm vir}$, where the latter expression defines the virial radius $R_{\rm vir}$. In that case, the virial theorem implies

$$\langle v^2\rangle = \frac{GM}{2R_{\rm vir}}. \tag{4.4}$$

Taking $R = R_{\rm vir}$ (and neglecting constants of order unity), we can then write the dynamical timescale as

$$t_{\rm dyn} = \sqrt{\frac{R_{\rm vir}^3}{GM}}, \tag{4.5}$$

which we recognize as a generalization of Kepler's third law.

In defining the dynamical timescale, the virial radius is often replaced with the easier-to-measure half-mass radius, $R_{\rm h}$, which is the radius of the sphere enclosing half of the total system mass. Although they are commonly used interchangeably, these two radii are distinct physical quantities. However, Spitzer (1987) notes that $R_{\rm vir} \simeq 0.8R_{\rm h}$ for a broad range of common dynamical models in virial equilibrium.

For an open star cluster with $R = 1$ pc and $M = 10^3 \mathrm{M}_\odot$, or a globular cluster with $R = 10$ pc and $M = 10^6 \mathrm{M}_\odot$, we find $t_{\rm dyn} \simeq 0.5$ Myr. For our Galaxy, with $R = 10$ kpc and $M = 10^{11} \mathrm{M}_\odot$, $t_{\rm dyn} \simeq 50$ Myr. Within AMUSE, we can easily perform such a calculation as follows:

```
import numpy as np
from amuse.units import constants

def dynamical_time_scale(mass, radius, G=constants.G):
    return np.sqrt(radius**3 / (G * mass))
```

Having imported the needed libraries and defined this function, you might try the following:

```
from amuse.units import units
mass = 10**11 | units.MSun
radius = 10 | units.kpc
time_dyn = dynamical_time_scale(mass, radius)
print(f'dynamical time scale={time_dyn}')
```

The first two lines initialize the parameters M and R. The function is then called to calculate the dynamical timescale, and the result is printed as

```
dynamical time scale=12.2404614048 1.21539420228e+14 * s
```

You will probably find that the time is printed in a rather strange set of units. This reflects the way in which AMUSE carries out its dimensional calculations, operating on quantities and their units separately until the result is needed. It is usually best to convert into a defined set of units with

```
print(f'dynamical time scale={time_dyn.in_(units.Myr)}')
```

This produces the more readable output

```
dynamical time scale=47.1433302078 Myr
```

You will see many more examples like the above code example in this book. We encourage you to copy and paste these code snippets into a `Python` terminal for execution and experimentation, but note that leading spaces as well as single quotes may not be copied correctly from the PDF version of this book into your `Python` environment. As an alternative to cutting and pasting from the book, you can find all the source code in the AMUSE repository in the module: `amuse.examples` (see also Section 1.4.3).

4.1.2.2 The Relaxation Timescale

The timescale on which the global characteristics (bulk system parameters and stellar orbital elements) of the system change is the relaxation timescale, t_r. In stellar systems, relaxation is driven by two-body encounters among stars, which change stellar orbital parameters and provide a means for energy and angular momentum to diffuse around the system, causing long-term structural evolution.

The two-body relaxation time is the timescale on which a "typical" particle has its orbit changed by (mostly distant) encounters with other particles. The relaxation time for a system of identical particles of mass m, number density n, and (three-dimensional) velocity dispersion $\langle v^2 \rangle$ is (Spitzer 1987)

$$t_r \simeq \frac{0.065\langle v^2 \rangle^{3/2}}{nm^2G^2 \ln\Lambda}, \tag{4.6}$$

where Λ is the ratio of the size of the system to the "strong encounter distance" $p_0 = Gm/\langle v^2 \rangle^{1/2}$, the impact parameter that would result in a 90° deflection for two typical cluster stars. For an idealized system of N identical objects in virial equilibrium, Spitzer (1987) finds $\Lambda = \gamma N$, with $\gamma = 0.4$. For more realistic systems, γ is probably considerably smaller: the value $\gamma = 0.11$ is in common use (Giersz & Heggie 1994; Heggie & Hut 2003).

A convenient global measure of relaxation in a system is the half-mass relaxation time, t_{rh}, which is obtained by replacing all terms in Equation (4.6) by their global averages and assuming virial equilibrium:

$$t_{rh} \simeq 0.138\frac{N}{\ln \gamma N}t_{dyn}. \tag{4.7}$$

In AMUSE, we could expand the earlier code fragment from Section 4.1.2.1 to

```python
import numpy as np
from amuse.units import constants, units

def dynamical_time_scale(mass, radius, G=constants.G):
    return np.sqrt(radius**3 / (G * mass))

def relaxation_time_scale(number_of_particles, mass, radius, G=constants.G):
    return (
        0.138 * number_of_particles / np.log(0.4 * number_of_particles)
        * dynamical_time_scale(mass, radius, G)
    )

number_of_particles = 500000
mass = 10**11 | units.MSun
radius = 10 | units.kpc
time_relax = relaxation_time_scale(number_of_particles, mass, radius)
print(f'relaxation time scale={time_relax.in_(units.Gyr)}')
```

For the Milky Way, with `number_of_particles` $\sim 10^{11}$ and $t_{\rm dyn}$ as derived above, this gives $t_{\rm rh} \simeq 2.7 \times 10^7$ Gyr, which indicates that relaxation is unimportant. In the jargon of the field, such systems are termed *collisionless*. However, for an open star cluster, with $N \sim 10^3$, $t_{\rm rh} \simeq 11$ Myr and we can expect considerable dynamical evolution in much less than the age of the Galaxy or the lifetime of the cluster. This is an example of a *collisional* dynamical system (note that the term has nothing to do with physical stellar collisions, which we will discuss in subsequent chapters).

4.1.2.3 The Dynamical Friction Timescale

Chandrasekhar (1943) showed that the gravity of a massive body moving through a uniform background distribution of low-mass objects creates a "wake"—an over-dense region—behind itself [see also Fellhauer & Lin (2007)]. The gravitational pull of this wake always acts oppositely to the massive body's motion, leading to dynamical friction that tends to slow it down. If the low-mass background has density ρ and is (reasonably) characterized by an isotropic velocity distribution $\varphi(v)$ (where $\int 4\pi v^2 \phi(v)\, dv = 1$), then the acceleration of the massive body due to dynamical friction against the background is

$$\mathbf{a}_{\rm df} = -\frac{16\pi^2 G^2 M \rho \ln\Lambda \mathbf{V}}{V^3} \int_0^V v^2 \phi(v)\, dv, \tag{4.8}$$

where M and $\mathbf{V}$ are the mass and velocity, respectively, of the massive body, and Λ is defined as before—except that we replace p_0 with the radius R of the massive body if $R > p_0$. Note that, unlike relaxation, this expression does not depend explicitly on either the number density or the mass of the low-mass background, just on the total density.

If (again reasonably) the velocity distribution of the background is thermal,[1] such that

$$\phi(v) = \frac{1}{(2\pi\sigma)^{3/2}} e^{-v^2/\sigma^2}, \tag{4.9}$$

where σ is the one-dimensional root mean square velocity of the low-mass particles, this expression becomes (assuming $M \gg m$)

$$\mathbf{a}_{\rm df} = -\frac{4\pi^2 G^2 M \rho \ln\Lambda \mathbf{V}}{V^3} \left[{\rm erf}(x) - \frac{2x}{\sqrt{\pi}} e^{-x^2} \right], \tag{4.10}$$

where $x = v/\sqrt{2}\sigma$ (Binney & Tremaine 2008). The term in square brackets is zero for $x = 0$, scales as x^3 for small x, and tends to 1 for large x, effectively equaling 1 for $x > 3$.

[1] More precisely, this generally means that the direction is isotropic and the speed is drawn from a Maxwell–Boltzmann velocity distribution (Maxwell 1860) associated with some temperature.

The dynamical friction timescale is

$$t_{\rm df} = \frac{V}{a_{\rm df}} = \frac{0.059\ V^3}{G^2 M \rho \ln\Lambda}, \tag{4.11}$$

where we have conventionally evaluated the bracketed expression in Equation (4.11) for $x = 1$. Combining Equations (4.6) and (4.11) (and neglecting differences in the definition of Λ), we find

$$t_{\rm df} = 0.91\left(\frac{V^2}{\langle v^2 \rangle}\right)^{3/2}\left(\frac{m}{M}\right) t_{\rm r}. \tag{4.12}$$

Thus, so long as V^2 is comparable to $\langle v^2 \rangle$, a massive body will sink toward the center of a galaxy or cluster on a timescale much shorter than the relaxation time of the background particles.

4.1.2.4 The Gravitational Radiation Timescale

Relativistic effects, caused (for example) by two nearby point masses on a close orbit, have a characteristic timescale that is usually much longer than any of the other dynamical timescales. We therefore tend to ignore them in N-body simulations, but it is still good to be aware of their magnitudes.[2] The timescale for orbital precession (often referred to as apsidal precession) is shortest, but it has limited effect on the global dynamical evolution of a stellar system, except in close resonant or von Zeipel–Lidov–Kozai resonance interactions (Naoz 2016). The relativistic apsidal precession (in radians per orbital period) for a two-body orbit of semimajor axis a, eccentricity e, and period P, is

$$\epsilon = 24\pi^3 \frac{a^2}{P^2 c^2 (1 - e^2)}. \tag{4.13}$$

Here, c is the speed of light. For the planet Mercury, this turns out to be about 5×10^{-7} radians per orbital period of $P \simeq 88$ days. It takes about a century for Mercury to make an extra loop around the Sun due to this effect.

The timescale for inspiral due to the emission of gravitational radiation is considerably longer, but more relevant for cluster dynamics. We can estimate this timescale using a post-Newtonian (PN) expansion of the orbital dynamics (Peters 1964). Ignoring several complications, we arrive at

$$t_{\rm gwr} \simeq 0.02\ \frac{c^5 a^4 (1 - e^2)^{7/2}}{G^3 M m (M + m)}, \tag{4.14}$$

where M and m are the masses of the orbiting bodies, and a and e are the Newtonian orbital semimajor axis and eccentricity, respectively. Adopting representative values ($M \simeq 1.44\ {\rm M}_\odot$ $m \simeq 1.39\ {\rm M}_\odot$, $a \simeq 2.8\ {\rm R}_\odot$, and $e \simeq 0.62$) for the famous Hulse–Taylor pulsar (Hulse & Taylor 1975), we find $t_{\rm gwr} \simeq 300$ Myr.

[2] Several integrators listed in Table 4.1 includes post-Newtonian corrections up to order 2.5.

The merger of such objects will result in a burst of gravitational waves. The first detection of such waves was made by the LIGO consortium on 2015 September 14 (Abbott et al. 2016).

4.1.2.5 Other Timescales

Many other timescales can be identified and defined. For example, timescales for specific processes, such as resonant relaxation (Rauch & Tremaine 1996) or the von Zeipel–Lidov–Kozai (von Zeipel 1910; Lidov 1962; Kozai 1962)[3] cycles in hierarchical triple systems, may be relevant in certain circumstances. In Sections 8.4.5 and 9.4.3, we present examples for which these additional timescales are important.

4.1.3 Star Cluster Dynamics

Stars form in clustered environments, and much of the story of stars in our Galaxy has to do with the dynamical evolution of star clusters. The N-body codes described in this chapter are central to this study. Because this topic plays such a key role in our narrative, we pause here to introduce some terminology, a few common diagnostics having to do with basic cluster dynamics, and an evolutionary timeline. We will return frequently to these concepts in our later discussion. The dynamics we describe are undoubtedly going on in, and may be modified by, the presence of a surrounding gas cloud, but the essential underlying processes are largely unchanged.

We, regrettably, can only scratch the surface of this interesting field of research, so we refer the reader to the extensive literature for more in-depth discussion (see, e.g., Spitzer 1987; Heggie & Hut 2003; Binney & Tremaine 2008).

4.1.3.1 Cluster Structure and Diagnostics

We begin with a newborn group of stars formed from an interstellar cloud (see Chapter 6). Both theory and observation (e.g., Gnedin et al. 2015) suggest that these stars are unlikely to have formed in precise virial (dynamical) equilibrium (Section 4.1.2.1). However, simulations indicate that the system will virialize and form a more or less spatially smooth cluster in virial equilibrium within a few dynamical times, erasing any substructure associated with the formation process. For typical cluster parameters, the timescale to form a single, well-defined cluster is a few million years. From this point on (at least, to the extent that we can neglect overall rotation), we are justified in treating the cluster as a roughly spherically symmetric object. Consequently, simulations of the evolution of clusters after the formation stage generally use idealized Plummer- (Plummer 1911) or King-like (King 1966) models (Section 3.2) as initial conditions.

By this stage, clusters are simple enough that they can be described by a few parameters. The virial (or, as we have seen, the half-mass) radius defines the overall cluster scale. Theoretical cluster models generically predict—and most observations show—the presence of a roughly constant-density central region known as the core.

[3] The Lidov paper was originally published in 1961 in Russian as Iskusstvennye Sputniki Zemli, #8, p. 5; the cited article is the English translation.

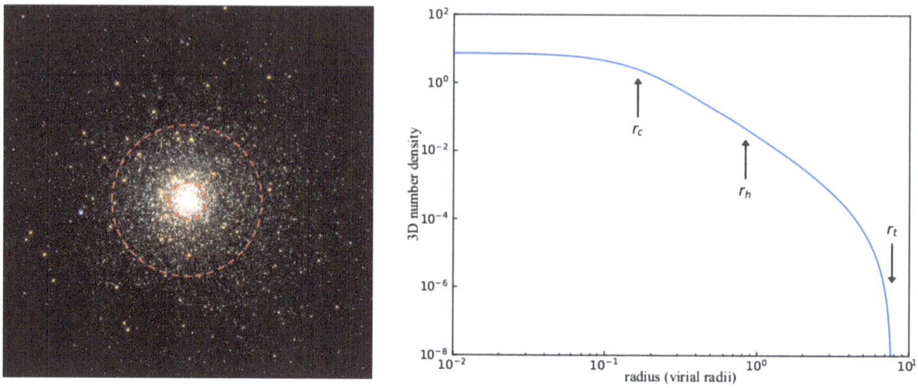

Figure 4.1. (a) The globular cluster M80 lies 10 kpc away and contains several hundred thousand stars. The cluster's core and half-mass radii are indicated by dashed red circles. The field of view is 3 arc minutes across. Original image credit: The Hubble Heritage Team (AURA/STScI/NASA/ESA). (b) The King model that best fits the cluster, with core, half-mass, and tidal radii marked.

Observers define the *core radius*, r_c, as the distance from the cluster center at which the surface brightness drops to half the central value. Because density is generally hard to measure in *N*-body models, theorists use a somewhat different definition, defining the core radius as the density-squared weighted root mean square stellar distance from the cluster center. (We also think that this sounds complicated!) Perhaps remarkably, the two definitions seem to broadly agree. The ratio of the core radius to the half-mass radius (or to the cluster's tidal radius in the ambient Galactic tidal field) is a key measure of cluster structure.

The cluster's *Lagrangian radii* are, by definition, the distances from the cluster center containing specified fractions of the total cluster mass. The 50% Lagrangian radius is the half-mass radius. Studying selected Lagrangian radii of key mass groups provides critical insight into the cluster's evolving density distribution. Figure 4.1(a) shows a Hubble Space Telescope (HST) image of the globular cluster M80. The cluster's core radius and half-mass radius are indicated by dashed red circles. Figure 4.1(b) shows the density profile of the best-fitting King model (with $W_0 = 7.5$; see Section 3.2). The cluster's tidal radius lies well outside the image shown in part (a); for many clusters, the tidal radius is determined from the best-fitting King model.

4.1.3.2 Essential Dynamics

As we just saw, dynamical friction causes the most massive stars in the cluster to sink toward the center (Spitzer 1969). The timescale can be very short, often less than the evolutionary lifetimes of the stars themselves (see Section 4.1.2.3). The most important consequences of this process are: (1) the surrounding lower-mass stars gain energy and expand outward as the massive stars lose energy and sink to the center of the cluster potential well; (2) the high density of the massive stars near the center means that the rate of both dynamical interactions (including binary formation) and physical collisions between stars (see Sections 8.5.2 and 8.5.4) is greatly increased, which influences the dynamical evolution of the system; (3) when stars lose mass—for example, due to winds or supernovae—the effect on the cluster is much stronger when the mass loss comes from the center rather than the periphery.

Mass segregation is often interpreted in terms of energy equipartition, the expected end-point of relaxation, which should lead to $\langle mv^2 \rangle$=constant for every stellar mass group. As a result, more massive stars slow down and sink toward the center of the cluster. However, perhaps surprisingly, the degree to which a system actually achieves equipartition is still not fully understood. Trenti & van der Marel (2013) reported that energy equipartition is never actually reached in HST observations and N-body simulations of clusters with realistic mass functions. On the other hand, simulations of more idealized two-component systems (containing just "light" and "heavy" stars; see below) indicate that equipartition is achieved, at least when the component mass ratio is not too large. The fine details of this fundamental process remain to be resolved.

On longer timescales, the escape of high-velocity stars (evaporation) and the internal effects of two-body relaxation, which transports energy from the inner to the outer regions of the cluster, lead to a phenomenon known as core collapse (Antonov 1962; Cohn 1980). During this phase, the central portions of the cluster accelerate toward infinite density while the outer regions expand. This process can be understood by recognizing that, according to the virial theorem [Equation (4.3)], a self-gravitating system has negative specific heat, so reducing its energy causes it to heat up. Hence, as energy moves from the (dynamically) warmer central core to the cooler outer regions, the core contracts and gets warmer, accelerating the contraction. The timescale for the process to reach completion (i.e., a core of formally zero size and infinite density) is $t_{cc} \sim 15 t_{rh}$ for a system of identical masses (Hénon 1961), and less in the case of a broad spectrum of masses (Inagaki & Saslaw 1985), for which the core-collapse timescale falls to $t_{cc} \sim 0.2 t_{rh}$ (Portegies Zwart & McMillan 2002).

Figure 4.2 illustrates some of these dynamical stages. It shows the evolution of a simple two-component stellar system, in which both components are initially distributed as the same $W_0 = 5$ King model. The mass ratio of the two components (1 and 2) is $\mu = m_2/m_1 = 3$, and the ratio of total masses is $M_2/M_1 = 1.6 \times 10^{-2}$. According to Equation (4.7), the half-mass relaxation time is $t_{rh} \simeq 300$ time units. Mass segregation, dynamical equipartition, core collapse, and dynamical binary formation and evolution in the more massive component can all be seen.

Several physical processes can counteract core collapse. As just mentioned, stellar mass loss (due to stellar winds or supernovae) tends to unbind the cluster, and mass segregation amplifies the effect on the core. In addition, binary systems, whether present initially in the cluster or formed dynamically (see Section 4.7.3), also play an important role. Briefly, "soft" binaries that have less binding energy than the mean stellar kinetic energy of the system tend to be destroyed by encounters with other stars. However, "hard" binaries (with binding energy greater than the mean) tend to become more tightly bound following an encounter, effectively heating the stellar system (Hills 1975; Heggie 1975). We will discuss in more detail how binary interactions and dynamics are handled in AMUSE in Section 8.7.1.

One important guideline in understanding global cluster dynamics is Hénon's principle (Hénon 1975): When two coupled dynamical processes operate on very

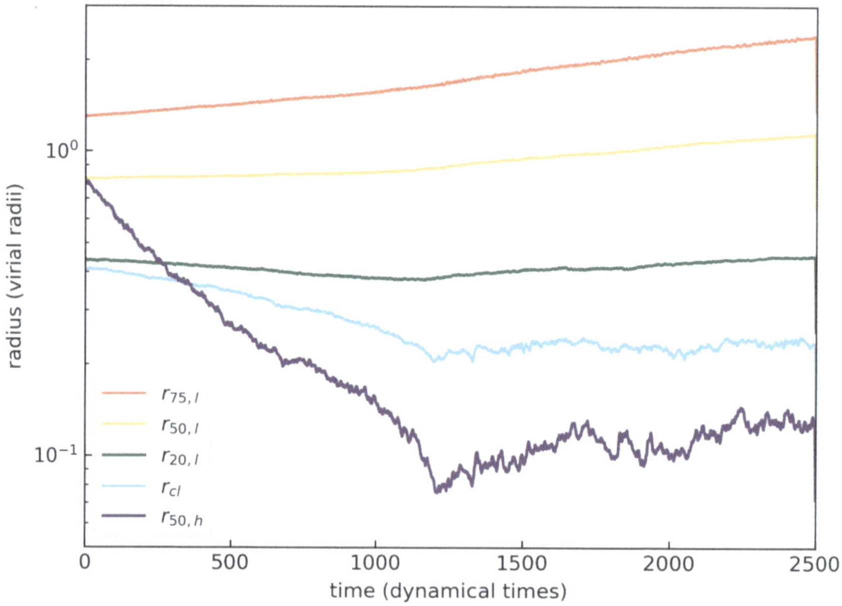

Figure 4.2. Evolution of a simple two-component system comprising 16k "light" (1) and 88 "heavy" (2) stars, with mass ratio $m_2/m_1 = 3$. The half-mass radius of the massive component, the core, and 20%, 50%, and 75% Lagrangian radii of the light component, are shown. The data have been smoothed using a square window of total width six time units. Mass segregation occurs during the first 600 time units; subsequently, both the massive system and the light core contract in quasiequilibrium. The system remains close to energy equipartition for $t > 600$. Eventually, after core collapse at $\sim$1200 time units, binary formation and dynamical heating cause both the massive system and the light core to expand. (Data courtesy of M. Brewer.)

different timescales—here, core energy production and the much slower outward conduction of energy through the half-mass radius—the faster process will always come into equilibrium with the demands of the slower one. As a result, the overall evolution of the cluster—long-term expansion and dissolution in the Galactic tidal field—is largely independent of the detailed physics operating in the core. Regardless of the precise core dynamics, a typical cluster is expected to dissolve in the Galactic tidal field in about 10 half-mass relaxation times (Lee & Ostriker 1987).

There is a considerable literature about the late stages of star cluster evolution, in what is often referred to as the postcollapse phase, during which so-called gravothermal oscillations may play a major role (Bettwieser & Sugimoto 1984; Makino 1996). Although this epoch is of considerable interest from a theoretical perspective, it is not clear what its observational relevance is in our relatively young universe.

4.1.4 Physics of the Integrator

Evaluating Equation (4.1) is not hard, nor is integrating the equations of motion, but it is important to appreciate the limitations of the adopted methods. It is sometimes said that gravity solvers, and N-body solvers in particular, have no underlying assumptions, and that all physical processes are properly taken into account. This, of course, is untrue. They solve the equations of motion in the nonrelativistic, point-mass approximation, quietly neglecting the spatial extent of the gravitating objects and the effects of their tidal interactions. For many studies, this does not pose a problem, but it is nevertheless important for the user to be aware of these shortcomings.

In solving the fundamental equations that govern the dynamical evolution of a self-gravitating system, we generally assume that the following conditions are met.

1. **Point-mass approximation**. An important and common underlying assumption is that all particles are point masses. Even if the particles should have nonzero sizes, their nonsphericity is generally ignored. The center of mass is assumed to be the geometric center of the particle. The consequences of this assumption are probably very small for most applications, but there certainly are cases where it is unwarranted. In Section 4.5.1, we will discuss a case in which the size of objects actually drives part of the physics, and it will turn out that dimensions of stars and planets are important drivers for interesting physical phenomena. In many hydrodynamical applications (discussed in Chapter 6), the finite size of objects motivates the performance of numerical simulations. Note that in Section A.5.2 we will explain the more complex aspects of interrupting one code to hand the control to another.
2. **Newton is correct**. The force in Equation (4.1) is pure Newtonian dynamics, ignoring relativistic corrections. The improvements made by Einstein (1916) and Verlinde (2017) are usually ignored. Some implementations of the N-body problem in AMUSE include additional terms that take relativistic effects into account, but most do not. Currently, there is no treatment of modified Newtonian dynamics (Milgrom 1983) in AMUSE, and no prescription for dark energy. These purported physical effects may be taken into account by modifying the spatial geometry or by introducing corrections to the force law, realized via the Bridge coupling strategy (see Chapter 9). Dark matter can be mimicked by introducing additional gravitating particles, although this is a coarse-grained approximation to the real physics.
3. **Gravity acts instantaneously**. In classical theories of gravitation, the "speed of gravity" is the rate at which a change in the distribution of energy or momentum of matter results in a change in the gravitational field it produces. In a more physically correct sense, the "speed of gravity" refers to the speed of a gravitational wave. In Newtonian N-body dynamics, this speed is infinite; in Einstein's theory of general relativity, it is the speed of light. This discrepancy can be safely neglected in many simulations because typical gas and stellar velocities are much smaller than the speed of light, except in the close proximity of massive black holes.

4.1.4.1 Numerical Accuracy

The digital computers on which we perform our calculations are often limited to only 16 decimal places.[4] This sounds like a lot for everyday calculations, but it is

[4] In 64 bit IEEE-754 compliant numerical representation, each floating-point number is presented by 1 bit for the sign, 11 for the exponent, and 52 for the mantissa.

quite limited if one considers that some simulations may require $\gtrsim 10^{16}$ operations (10 Petaflops), in which case round-off must play a potentially important role. This amount of computing corresponds roughly to the number of operations required to integrate the orbit of the Moon over the lifetime of the solar system. The effect of round-off is rarely studied in detail in N-body calculations, and the assumption is always made that it does not affect the long-term evolution of the system.

The solutions of an N-body problem are deterministic. Initial conditions define one and only one trajectory in phase space. Each initial point in phase space then maps to a unique final state of the system. In practice, we expect the calculated phase-space trajectory of an N-body simulation to differ from the true phase-space trajectory—N-body simulations give the wrong results (see Portegies Zwart & Boekholt 2018). Deviations from the trajectory, and the eventual location of the final state, can be caused by external perturbations, numerical round-off, and other sources of error. These sources of error may themselves be nondeterministic (e.g., in case partial forces are calculated in parallel). In Section 4.4.3, we present an example of how to measure the degree of chaos in an N-body system, by comparing a more accurate solution with a reprehensive solution.[5]

If computer errors are truly random and without systematic effects, the final state of an ensemble of inaccurate calculations should be centered around the true final state. Even though we do not know the shape or the dispersion of this distribution, the mean should still be consistent with the true solution. In a computer simulation, any initial point in phase space will map to a final state-point, but this is not the true final point. It is commonly believed that if we generate an ensemble of initial points in phase space and calculate them all to their final points in phase space, the distribution of final points will represent the true solution for the given initial distribution.

Every implementation of the N-body problem has its own characteristic numerical fingerprint that can be recognized by the growth of the numerical error. This is caused, in part, by specific choices of additional parameters, such as the time-stepping scheme or the introduction of a softening length that limits the depth of the potential. These choices are often hidden deep within the code, and it is widely assumed that they do not materially affect the outcome of the simulation. Despite these potential sources of error and uncertainty, our confidence in N-body simulations rests on the basic observed fact that all implementations lead to similar outcomes when applied to standard problems.

The detailed long-term evolution of a star cluster—virialization, mass segregation, core collapse, and long-term evolution (see Section 4.1.3)—appears to be substantially independent of both the initial model and the specific integration technique used. This has been repeatedly verified in a host of code comparisons over the past three decades (Chernoff & Weinberg 1990; Giersz & Heggie 1994; Fukushige & Heggie 1995; Giersz & Heggie 1996; Takahashi & Portegies Zwart 1998; Spinnato et al. 2003; Anders et al. 2012; Whitehead et al. 2013; Rodriguez et al. 2016). As a rule of thumb, we can trust predictions of the bulk or statistical properties of the system, but we should not take

[5] Solution to Newton's equations of motion for which the accumulation of numerical errors and the system's response exceeds the exponential growth of the initial offset δ (Portegies Zwart & Boekholt 2018).

individual stellar orbits too seriously, although there are clear counterexamples where the chaotic equations of motion are truly solved. In AMUSE, such arbitrarily precise calculations to a converged solution can be performed using the `Brutus` direct N-body code.

4.2 N-body Integration Strategies

In principle, the force calculation and the pushing of the particles are separate operations in a gravitational dynamics code, but they are coupled in most "pure" N-body implementations (see Section 4.2.2). Calculating the force in Equation (4.1) is straightforward, but the number of operations for the force on a single particle scales as the number of particles [$\mathcal{O}(N)$], and calculating the forces on all particles then scales as $\mathcal{O}(N^2)$, which becomes impractical for current computers without graphics processing unit (GPU) acceleration for $N \gtrsim 10^4$.

4.2.1 Global Structure of an N-body Code

Figure 4.3 presents a simple workflow for a basic N-body code. A handy programmer can probably rewrite this in about 100 lines of source code. Such codes can be (and have been) tuned for parallel operation, GPUs, and other high-performance/exotic hardware, and they form an excellent basis for numerical studies. Converting the flow chart in Figure 4.3 into a working program is not entirely trivial, but we refrain from going into more detail here because there are many excellent textbooks that do just that (e.g., Aarseth 2003; Heggie & Hut 2003).[6]

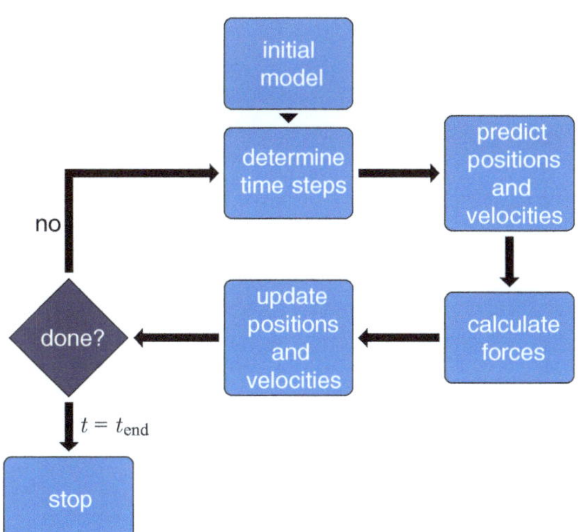

Figure 4.3. Workflow for a typical direct N-body code.

[6] For an illustration of how this problem can be addressed in a number of programming languages, see https://github.com/spzwart/NBabel.

The heart of an N-body code is the integration scheme, the mathematical time-discretization used to solve Equation (4.1). An important consideration in choosing an integration scheme is its order, which tells us how the algorithmic error (per step) δE scales with time step δt. For a scheme of order k, $\delta E \propto \delta t^k$. Typical schemes in common use are of second-, fourth-, or sixth-order. Higher-order schemes are preferred in direct-summation N-body calculations because a given accuracy can be achieved using longer time steps (remember here that the number of force operations per particle per step is $\mathcal{O}(N)$).

The program illustrated in Figure 4.3 uses a predictor–corrector scheme to advance the particles in time (see Section 4.2.3). There are many alternative approaches, but these schemes have become the de facto standard in astrophysical N-body simulations.

4.2.2 Types of *N*-body Code

Several families of algorithms have been developed to solve the N-body problem. Some scale as $\mathcal{O}(N)$ (Ambrosiano et al. 1988; Dehnen 2014a), some scale better, $\mathcal{O}(N \log N)$ (Barnes & Hut 1986), but some scale much worse, $\mathcal{O}(N^3)$ (Boekholt & Portegies Zwart 2015). The differences are rooted in optimization, where computational speed is, broadly speaking, traded for accuracy. We draw a distinction between *pure* N-body, *direct* N-body, and *approximate* N-body codes.

- **Pure *N*-body** codes solve Newton's equations of motion with no free physical parameters, although most have some capacity to flag special events, such as close encounters or binary dynamics. The only adjustable quantity in a pure N-body code is the time-stepping criterion used to integrate the equations of motion. Examples are `ph4` and `Hermite`, both based on the fourth-order Hermite predictor–corrector scheme (Makino & Aarseth 1992) (cf. Hut et al. 1995 for a lower-order approach). Tunable precision N-body codes are an interesting subset of pure N-body codes. They go beyond the standard double precision of digital computers, and allow energy conservation to be tuned to arbitrary precision. The code `Brutus` (Boekholt & Portegies Zwart 2015) has an extra free parameter, which controls the mantissa of the floating-point operations.
- **Direct *N*-body** codes solve Newton's equations of motion with additional parameters, often to speed up the code or to reduce round-off. Many such codes incorporate regularization techniques for resolving close encounters and close multiple subsystems, and/or neighbor schemes to reduce the force calculation time. The cost per step of these codes still scales as the square of the number of particles. `NBODY6` is a prime example (Aarseth 1985; Harfst et al. 2007).
- **Approximate methods** relax the strict N^2 procedure for computing the gravitational force. These include various flavors of tree codes, particle-mesh, and other multipole treatments. Such methods are very popular, mainly because they are very fast. The most common example is the Barnes–Hut tree scheme (Barnes & Hut 1986), in which forces from nearby particles are calculated directly, but more and more distant particles are grouped together into clumps

of increasing size, whose gravity is approximated by the first few terms of a multipole expansion. Very often, for computational efficiency, only the monopole term is retained—when calculating the orbit of the Sun in our Galaxy, it is reasonable to treat the Andromeda galaxy as a point mass. The result is a reduction of the $\mathcal{O}(N^2)$ complexity of a direct force calculation to $\mathcal{O}(N \log N)$. Examples of tree codes in AMUSE include `BHTree` (Barnes & Hut 1986), `MI6` (Iwasawa et al. 2015), and `Bonsai` (Bédorf et al. 2012; Bédorf et al. 2014). Another popular method is the fast multipole method (Greengard & Rokhlin 1987; Ambrosiano et al. 1988), or the equally fast but considerably more precise code `gyrfalcON` (Dehnen 2014a, 2014b), which scales even more favorably—$\mathcal{O}(N)$.

- **Hybrid methods** Hybrid N-body codes try to combine the advantages of high precision with the speed of tree codes. They often scale favorably compared to direct integration, but at a slightly reduced precision. One example, available in AMUSE is `Petar` (Wang et al. 2016). There are several scripts in AMUSE that exhibit similar advantages, such as the `Nemesis` (van Elteren et al. 2019) module and `LonelyPlanets` (Cai et al. 2019). In LonelyPlanets, we reduce these costs through a two-stage divide-and-conquer strategy (Cai et al. 2019; Stock et al. 2020). In stage one we simulate a dynamical evolution and stellar evolution of a cluster in the Galactic potential, in stage two we integrate the planetary systems with only the nearest perturbing cluster members. This strategy easily saves a factor of 10^4 in computer time when integrating a cluster of 1000 stars, of which 30 receive a system of 10 planets each. The cost of such a two-stage process is reduced by ignoring the effect individual planets have on the stars, but one loses the ability to follow the orbits of rogue planets in the cluster.

4.2.3 Discretization Strategies in N-body Simulations

A system of differential equations, such as Equation (4.1), can be discretized in time and solved using a variety of numerical techniques. Two basic strategies for obtaining a solution are *explicit* and *implicit* schemes. An explicit method provides a formula or procedure that allows the state of the system at time $t + \delta t$ to be determined entirely from the state of the system at time t. In an implicit method, the state of the system at time $t + \delta t$ also appears in the formula, leading to an implicit equation that generally must be solved iteratively.

Implicit methods tend to be more stable and (in principle) allow much longer time steps, although they are generally more expensive due to the iterative solution entailed. However, the longer time steps are often unattainable due to the chaotic and singular nature of the underlying equations—the characteristic timescales of the motion can differ by many orders of magnitude from particle to particle and from one time to another. As a result, explicit methods are generally favored over implicit methods for N-body applications.

Many explicit schemes exist, but by far the most commonly used approach in N-body simulations of stellar clusters is the predictor–corrector scheme (see Section 4.2.2). Historically, these methods appeared as the first iteration in implicit solutions to the discretized equations of motion, but in their modern incarnation they have become explicit solvers.

Broadly speaking, predictor–corrector schemes proceed in three steps:

- The predictor step uses whatever derivative information is available at time t (usually at least the acceleration, but often higher derivatives as well) to extrapolate all positions and velocities at time $t + \delta t$ as Taylor series (Taylor 1715).
- The accelerations (and higher derivatives) are then reevaluated using the predicted positions and velocities, and the new and old accelerations are combined to yield estimates of higher derivatives. For example, we can estimate the "jerk" (the derivative of the acceleration) as $j \simeq (a^{\text{new}} - a^{\text{old}})/\delta t$.
- The corrector step then refines the predicted positions and velocities using the additional derivative information, essentially adding extra terms to the Taylor series.

These schemes are compact, easy to program, and generally proceed with a single evaluation of the accelerations at each step (the new accelerations are reused in the next prediction), making them very efficient in terms of computer resources.

The second-order predictor–corrector scheme (often called the velocity-Verlet method, and sometimes implemented as the "Leapfrog" scheme; see Verlet 1967) is very widely used; we discussed in Chapter 1 how this integrator finds its origin in the Babylonian civilization between 350 and 50 BCE. Note that the terminology between Leapfrog, Verlet, and Babylonian schemes are a bit confusing. The velocity-Verlet does not correct the new positions. The predictor and corrector steps are usually reserved for methods that do at least one iteration, but possibly more. Although of low order, its simplicity and the fact that it is symplectic (see Section 4.2.3.2) make it an attractive choice in many applications. The fourth-order Hermite scheme (Makino 1991) is a natural generalization of the Verlet method (except that it is not symplectic). It uses accelerations and jerks, and has for some time been the method of choice in large N-body simulations. The last decade has seen the development and implementation of sixth- and eighth-order versions in large-scale integrators. Again, it goes beyond the scope of this book to dwell on this topic given that excellent texts already exist. We direct the interested reader to Nitadori & Makino (2008).

One disadvantage of explicit schemes is that errors on successive steps tend to be correlated, so the cumulative error grows faster than the random walk expected for uncorrelated errors (Brouwer 1937), for which the net error would grow as the square root of the number of steps. This can be solved with compensated Kahan–Babuška summation (Kahan 1965; Babuška 1969). In most applications, this more rapid error growth is an acceptable trade-off for the improved speed of the method.

The recently revived family of hybrid methods, which combine the good qualities of two or more integration algorithms (e.g., McMillan & Aarseth 1993), have very promising characteristics. Recent examples include `Pentacle` (Iwasawa et al. 2017) and `Petar` (Wang et al. 2016), which combine a Hermite direct code with a Barnes-Hut tree code. The implementation in `Petar` adds algorithmic regularization and PN effects to allow massive compact objects in a stellar cluster to be addressed.

4.2.3.1 Particle–Particle–Particle–Tree Schemes and Other Advanced Schemes
The development of a hybrid N-body integrator with the good energy conserving quality of the high-order explicit predictor–corrector scheme and the excellent performance of the hierarchical tree code approach is with the family of PP–PT (or P^3T, sometimes referred to as Pikachu) schemes (Iwasawa et al. 2015).

The PP-PT scheme combines direct N-body integration (particle–particle) for close interactions with the Barnes–Hut tree algorithm (particle–tree) for long-range interactions. The code `Petar` implements this scheme. `Petar` is designed for simulations of globular clusters, including stellar evolution, binary treatment, and PN effects. The algorithmic regularization (Hellström & Mikkola 2010) allows accurate treatment of close encounters and binary systems. `Petar` is particularly well suited for simulating dense stellar systems, such as globular clusters, galactic nuclei, and young star clusters. Its hybrid approach and use of modern computational techniques make it a powerful tool for studying complex astrophysical phenomena involving gravitational dynamics.

Finally, it is worth mentioning some recent Kepler-splitting methods, in which the problem is separated into N Kepler problems plus perturbations. The superposition principle is subsequently applied to knit the N^2 Kepler solutions together into a coherent N-body solution. The methods work, parallelize almost trivially, and the errors do not grow on long timescales (compared to the dynamical time), making such methods very suitable for studying the long-term dynamical evolution of the solar system (Wisdom & Holman 1991; Yoshida 1993; Duncan et al. 1998; Chambers 1999; Laskar & Robutel 2001; Saha & Tremaine 1992; Rein & Liu 2012; Liu et al. 2016).

4.2.3.2 Symplectic and Time-Reversible Schemes
In the context of long-time-scale calculations, the family of symplectic integrators are of particular interest. The defining property of a symplectic scheme is that it preserves phase-space area in a dynamical system. It follows that the phase-space volume element is preserved—the familiar Liouville theorem. Given that Hamiltonian systems also have these properties, symplectic integration of Hamiltonian systems yields numerical solutions that stay faithful to the true solutions, in the sense that physically conserved quantities are conserved numerically over long periods of time. Such schemes are also time-reversible, meaning that the initial conditions can be recovered (to machine accuracy) by running a simulation backward in time (see also Tremaine 2015). Note here that time reversibility should not be confused with a converged solution, which requires robust brute force calculations with `Brutus`.

In a symplectic scheme, it is common to split the underlying Hamiltonian into slowly and rapidly varying parts. For example, when discussing the orbits of minor bodies in the solar system, we could rewrite the Hamiltonian of the multibody system as

$$H = H_{\rm kep} + H_{\rm int}. \tag{4.15}$$

Here, we follow the example of Wisdom & Holman (1991). The leading term $H_{\rm kep}$ represents interactions between the minor bodies with the Sun, while $H_{\rm int}$ represents interactions among the minor bodies. (The latter term could, in turn, be further separated into changes on timescales that are short or long compared to the dynamical time.) Each component of the split Hamiltonian can be solved independently, using explicit schemes, and the solutions combined. We will return to symplectic schemes in Chapter 9, where we will adopt this approach as our standard for combining codes.

Explicit symplectic schemes can be very accurate, approaching machine precision with sufficiently small time steps, even when the equations are stiff. Their main drawback is that they require the time steps for all objects to be the same. Such shared time step schemes are often employed for planetary, galactic, and cosmological simulations, but are prohibitively expensive for star cluster simulations, which generally require individual and time-variable steps (Aarseth 1985). We further discuss the Babylonian or Verlet scheme in Section 9.1.1.

4.2.3.3 General Relativistic Effects

Most N-body codes in AMUSE adopt Newton's equations of motion, but there are a few exceptions that also solve for the general relativistic equations of motion. This is generally realized through a Taylor expansion of the Einstein–Infeld–Hoffman equations of motion (Einstein et al. 1938), called the PN expansions (Poisson & Will 2014). They elegantly balance accuracy with computational efficiency, allowing researchers to probe the relativistic dance of celestial bodies without the full complexity of Einstein's field equations.

The expansion unfolds as a series in powers of v/c, where v is the system's typical velocity and c is light speed. This approach modifies the spacetime metric and equations of motion, introducing relativistic corrections to Newtonian dynamics. As the expansion progresses through higher orders, it captures increasingly subtle effects, such as the perihelion precession (first-order Taylor expansion, or commonly called the 1-PN term) and gravitational waves. Many pseudorelativistic N-body codes incorporate the following terms:

1-PN: Terms up to $(v/c)^2$ leads to perihelion precession.

2-PN: Terms up to $(v/c)^4$ leads to gravitational time dilation.

2.5-PN: Includes radiation reaction effects.

3-PN and higher: More accurate but computationally intensive.

These expansions tend to break down when the orbital velocity approaches the speed of light, or in the close proximity to an event horizon. It is, however excellently suited for simulating orbits of black hole binaries in stellar clusters. The spacetime metric is expanded around the flat Minkowski metric, with corrections added at each order. The geodesic equation of motions are expanded, resulting in modified Newtonian equations of motion with relativistic corrections.

The underlying assumption for generating the PN expansions is that all black holes operate in pairs. Although a reasonable assumption so long as black holes are binary members, it breaks down for democratic triple (or higher-order multiple) black holes.

4.2.3.4 Parallelization and Hardware Acceleration

The N-body problem is relatively easy to parallelize. Several algorithms are known to perform well for each of the families of integration strategies mentioned in Section 4.2.2. However, we will not dwell on the details because many excellent papers and textbooks have already been written on these topics. The problem is also well suited to hardware acceleration GPUs. Although worthy competitors exist, the most common acceleration hardware currently used in the gravitational N-body community is the NVIDIA GPU, programmed using the CUDA language.[7] These tiny computers combine large amounts of computing power with relatively low prices, consistently leading the market in both teraflops per dollar and teraflops per watt. Many of the N-body codes listed in Table 4.1 are parallel, GPU-accelerated, or both.

4.3 The AMUSE Interface to Gravity Solvers

The preceding discussion of gravity solvers may mask a very basic fact: the codes all have very similar input and output. This is necessarily so, since they aim to solve the same physics. We take advantage of this in AMUSE to define a minimal Gravity Interface that can be used to access any gravitational solver in a homogeneous way.

A simple script to integrate Newton's equations of motion is presented in Code example 4.1. You can test it yourself with the following code:

```
import amuse.examples as ae
ae.run("gravity_minimal")
```

We will dissect the complete example in some detail here. The script starts by importing the required AMUSE packages:

```
from amuse.units import nbody_system
from amuse.ic.plummer import new_plummer_model
from amuse.community.hermite import Hermite
```

The first import, of `nbody_system`, makes a *unit system* available (see Section 2.3). The second import statement loads a function `new_plummer_model` that generates the initial conditions (see Chapter 3). Finally, the third import loads a class, `Hermite`, which is an example of a community code that implements the gravitational dynamics interface. All the codes that conform to this interface (we will encounter more of them later) implement in some way (as discussed in Section 4.2.2) a solver for Equation (4.1).

The code in Code example 4.1 is as simple as we could make it—and honestly, it does not do much. It can, however, be the basis for more rewarding investigations and it does contain in essential form all necessary ingredients of more complicated scripts. As we will see in later chapters, it can be expanded and used for far more

[7] See https://developer.nvidia.com/cuda-zone

```python
"""
Minimal example for gravity calculations in AMUSE
"""

from amuse.units import nbody_system
from amuse.ic.plummer import new_plummer_model
from amuse.community.hermite import Hermite


def gravity_minimal(number_of_stars=100, time_end=1 | nbody_system.time):
    """
    Initialises a Plummer sphere of length 'number_of_stars' and integrates until
    'time_end' with the Hermite integrator
    """
    stars = new_plummer_model(number_of_stars)

    gravity = Hermite()
    gravity.particles.add_particles(stars)

    energy_total_initial = gravity.kinetic_energy + gravity.potential_energy

    gravity.evolve_model(time_end)

    energy_total = gravity.kinetic_energy + gravity.potential_energy
    energy_error = (energy_total_initial - energy_total) / energy_total_initial

    print(
        f"model time = {gravity.model_time} "
        f"total mass = {stars.total_mass()} "
        f"total energy = {energy_total} "
        f"relative energy error = {energy_error}"
    )

    gravity.stop()


def main(**kwargs):
    gravity_minimal(**kwargs)


if __name__ == "__main__":
    main(number_of_stars=100, time_end=1 | nbody_system.time)
```

Code example 4.1: Minimal AMUSE code for solving a simple *N*-body problem. Source code is available in `amuse.examples.gravity_minimal`.

complex simulations. This minimal script, and many of the following scripts, is set up as follows:

- Generate a realization of the starting conditions (Section 4.3.1),
- Specify and initialize the (gravity) solver to use (Section 4.3.2),
- Evolve the model to a specific time (Section 4.3.4),
- Print diagnostics,
- Produce a figure of the results or analyze the data (Section 2.2.3),
- Shutdown the community code.

After loading in the appropriate packages, the Python interpreter processes the definition of the functions main and gravity_minimal and proceeds to the line

```
if __name__ == "__main__":
```

This piece of Python code tells the interpreter that the following lines are to be ignored when the script is not run standalone (see Appendix C). It is standard practice in Python to do this because it makes your code easily reusable: another script can import the function gravity_minimal without triggering execution of a run.

Here, the main function is called with its default parameters,

```
    main (number_of_stars=100, time_end=1 | nbody_system.time)
```

Gravitational N-body codes tend to employ dimensionless units or Hénon units (Hénon 1972; Heggie & Mathieu 1986). We briefly notice that in this system, the gravitational constant G, the total cluster mass M, and the cluster virial radius R_{vir} (see Section 4.1.2.1) each have a value of 1. For a cluster in virial equilibrium, the kinetic, potential, and total energies then are (Hénon 1971; Heggie & Mathieu 1986)

$$T = \frac{1}{4}, U = -\frac{1}{2}, E = -\frac{1}{4}. \tag{4.16}$$

N-body units are handy when we are interested in the outcome of a pure gravitational problem. Internally, N-body integrators have almost universally adopted them, as well as in N-body initial condition generating functions such as new_plummer_model, new_king_model, etc. (see Section 3.2). They greatly facilitate comparison between different codes. In addition, they afford the minor computational advantage of keeping the magnitudes of most physical variables close to unity, although they make little difference—at best, avoiding a single multiplication by G in each force calculation. On the downside, dimensionless units have caused confusion among generations of students and other noncognoscenti in their attempts to reconcile physical and N-body quantities. And as soon as other physical processes intrude, such as when combining gravitational dynamics with stellar evolution or hydrodynamics, N-body units rapidly lose their usefulness.

In reality, a typical N-body code does not care at all about units. It accepts a number with no questions asked, except perhaps whether it is a floating-point number or an integer. However, the AMUSE interface does care about dimensional and nondimensional units because it mediates the conversion between the units used in different modules. For this reason, any calculation involving N-body units that plans to use other modules with units must be calibrated by defining a converter specifying the N-body mass, length, and/or timescales in physical units (see Section 2.3). In Section 4.4 (Code example 4.3 in particular), we introduce the command-line argument parser argparse, and present an example of a script in which we use this with units (see Section 2.4.2).

```
import matplotlib.pyplot as plt
from amuse.io import read_set_from_file


def plot_gravity(
    filename="sun_venus_earth.amuse",
    length_unit=units.au,
):
    particles = read_set_from_file(filename)

    figure = plt.figure(figsize=(8, 8))
    ax = figure.add_subplot(1, 1, 1)

    colormap = ["yellow", "green", "blue"]  # specific to a 3-body plot
    size = [40, 20, 20]
    edgecolor = ["orange", "green", "blue"]

    for i, snap in enumerate(particles.history):
        ax.scatter(
            snap.x.value_in(length_unit),
            snap.y.value_in(length_unit),
            c=colormap,
            s=size,
            edgecolor=edgecolor,
        )
    ax.set_xlabel(f"x [{length_unit}]")
    ax.set_ylabel(f"y [{length_unit}]")

    save_file = "plot_gravity.png"
    plt.savefig(save_file)
    print(f"\nSaved figure in file {save_file}\n")
```

Code example 4.3: Minimal routine to plot *N*-body simulation data. Full source code is available in `amuse.examples.plot_gravity`.

4.3.1 Generating Initial Conditions

The main routine in Code example 4.1, where the real work is done, starts with

```
def gravity_minimal(number_of_stars=100, time_end=1 | nbody_system.time):
```

the declaration of the function that actually does the work. The next line (see also Section 3.2.1)

```
    stars = new_plummer_model(number_of_stars)
```

generates the initial density and velocity of the particles, in this case a Plummer model (Plummer 1911) with "`number_of_stars`" stars. In Section 3.2.1 we present an example output for this routine.

4.3.2 Specifying and Initializing the Gravity Solver

The next two lines in the code example initialize an instantiation of the *N*-body integrator, in this case the `Hermite()` integrator, and send the previously created generated particles to the integrator. Much of the magic in AMUSE is in this simple line:

```
gravity = Hermite()
```

The result is a detached process called `hermite_worker` that starts on your computer, running simultaneously with the `Python` program that spawned it. You can check that multiple processes are now running on your computer by typing

```
%> ps aux | grep hermite
```

on the command line (use `top -o cpu` on a Mac), which in our case shows something like (where we left out a few columns):

```
USER        PID \%CPU \%MEM TTY      STAT START   TIME COMMAND
spz       342887  0.0  0.0 pts/3    S+   10:49   0:00 hermite_worker 55101 127.0.0.1
spz       342888  0.0  0.0 pts/3    S+   10:49   0:00 hermite_worker 55101 127.0.0.1
spz       343018  0.0  0.0 pts/10   S+   10:50   0:00 grep --color=auto hermite
```

The two `hermite_worker` processes are the N-body integrator and its interface, both running at the host IP address (127.0.0.1). In an earlier version of AMUSE, the code would continue consuming 100% CPU time due to polling the AMUSE framework asking for things to do. Considering the modern desire to spend fewer cycles, and less electricity, we improved on this behavior, and have silenced the polling. The `Python` instance `gravity` is our portal to the N-body code.

Because the N-body code is a detached process, we must send it a termination signal from the AMUSE script at the end of the calculation. Otherwise, the N-body code will keep running in the background, until the `Python` interpreter cleans up ("garbage collects") the `gravity` variable. To stop `Hermite`, we terminate it by

```
gravity.stop()
```

This will send a kill signal to `Hermite`, and all data in the code will be lost.

In the example, `gravity` is an instance of a class implementing the AMUSE gravity interface. This object manages and transparently handles communication with the `hermite_worker` process or any other N-body code (see Section 2.2.3). Any of the other gravity solvers can be used in the same way by changing the call to the integrator. For example,

```
gravity = Bhtree()
```

instantiates and initializes the gravity solver `BHTree`, a Barnes–Hut tree code. Aside from this line, the interface is the same in all cases, as long as the other code is a gravity code.

4.3.3 Feeding Particles to the N-body Code

So far, we have not yet informed the N-body code which data it should use. These data come in the form of a particle set. In Section 4.3.1, we initialized a particle set that we called `stars`. We can send this particle set to the N-body code by adding the `stars` to the code already running

```
gravity.particles.add_particles(stars)
```

We now have two sets of particles—the `Python` particle-set `stars` and the set that now lives in the memory of the running code. The latter set can be addressed directly by querying `gravity.particles`. Once we tell the gravity solver to advance in time, these two particle sets will diverge. The particle set in `Python` will remain unchanged, but the copy resident in the integrator will be updated by the community code, here an instance of `Hermite`.

4.3.4 Evolving the Model to the Desired Time

The actual computational work is done in

```
gravity.evolve_model(time_end)
```

Here, the N-body code runs to time `time_end`, i.e. the solution to Equation (4.1) is advanced to the desired end time. After the code is done, we calculate the kinetic and potential energies.

To test the accuracy of the integrator, we print the final energies and the energy error since the system was initialized.

We are now ready to run the minimal script with the following arguments

```
%> python -m amuse.examples.gravity_minimal
```

which will produce the following (slightly edited to fit the text column) output:

```
T= 1.0 time M= 1.0 mass E= -0.250000001272 mass * length**2 * time**-2
       Q= -0.5068237292374607 dE= -5.089542761314269e-09

In this session you have used the AMUSE modules below.
Please cite any relevant articles:

   "AMUSE-Hermite (framework 2023.10.1.dev1+gb159086e4.d20240124)"
           ADS:1995ApJ...443L..93H (Hut, P., Makino, J. & McMillan, S., *
    Astrophysical Journal Letters* , **443**, L93-L96 (1995))


   "AMUSE (unknown version)"
           DOI:10.5281/zenodo.1435860
           ADS:2018araa.book.....P (Portegies Zwart, S. & McMillan, S.L.W., 2018)
           ADS:2013CoPhC.183..456P ** (Portegies Zwart, S. et al., 2013)
           ADS:2013A&A...557A..84P ** (Pelupessy, F. I. et al., 2013)
           ADS:2009NewA...14..369P (Portegies Zwart, S. et al., 2009)
```

This final message reports to the user the codes and components used and provides references that we hope they will cite if the code and script are run for production and/or publication purposes. If this message annoys you, it can be suppressed:

```
from amuse.support import literature
literature.TrackLiteratureReferences.suppress_output()
```

This concludes our first acquaintance with an AMUSE simulation script. Of course, some elements are still missing to be able to analyze and change our simulation. We want to be able to examine the data in more detail (Section 2.2.3), store the data for publication and analysis (Section 2.2.4), and change to the units used to communicate with the code (Section 2.3). Also, in order to change the numerics of the simulation, we need to be able to change solver parameters (Section 2.3.3) and to change to a different solver (Section 4.3.7).

4.3.5 Scaling the Simulation

So far, we have discussed only dimensionless N-body units. These are convenient for performing scale-free simulations, as are common in N-body simulations. In Section 2.3, we presented a more complete overview of how to declare and use converters in AMUSE. Here, we just reiterate the most critical parts needed to understand this chapter.

A `converter` is simply an object that contains knowledge of the units in a problem, which can be passed as needed as an argument to other functions requiring this information. For example, a unit converter can be provided when initiating an N-body solver:

```
gravity = Hermite(converter)
```

This establishes the connection between dimensionless and physical units. Now, we can feed particle sets to the N-body code in any units we choose, and retrieve dimensional information in any convenient set of units. The conversion is automatic as long as we stay within one system of units,

```
print(f"Earth's mass is: {(1 | units.MEarth).in_(units.kg)}")
```

or

```
print(
    f"The Sun's luminosity is about {(3.8e+33 | units.erg/units.s).in_(units.LSun)}"
)
```

4.3.6 Code Parameters

As we explained in Section 2.3.3, the working of various codes can be controlled and tuned through parameter settings. There we explained that we try to give these parameters the same name throughout a physical domain, but sometimes fail in this objective.

One case where we failed to be consistent is in the two direct N-body codes `ph4` and `Hermite`. The parameters to control the size of the time step in these codes are called `timestep_parameter` and `dt_param`, respectively. A comparable parameter in `BHTree` is called `time_step`. However, the latter is not inconsistent because `time_step` really sets the actual time step in `BHTree`. The differences in implementation are the main reason for failing to come up with a unique parameter name, which is a consequence of these codes being written by different authors with slightly different objectives. This forms a natural inconsistency in AMUSE.

A parameter where we succeeded in finding a common naming across various codes is the gravitational softening parameter ε (Aarseth 1963). This is a very important parameter, used in many N-body codes. Softening removes the singularity in the inverse-square force by replacing Equation (4.1) with

$$\mathbf{a}_i = G\sum_{j\neq i}^{N} m_j \frac{\mathbf{r}_j - \mathbf{r}_i}{[(\mathbf{r}_j - \mathbf{r}_i)^2 + \epsilon^2]^{3/2}}. \tag{4.17}$$

This is done to accommodate integration schemes that cannot handle point-mass potentials. Column 8 of Table 4.1 below lists whether or not a code is (or can be) softened. In Section 8.7.1, we describe a module that adds point-mass support to any otherwise softened code. The square of the softening parameter is generally referred to as `epsilon_squared` in all N-body codes. For example, ε^2 can be set to the squared mean interparticle spacing, in N-body units, with

```
gravity.parameters.epsilon_squared = 1./N**(2./3)
```

In case you prefer to look at good old-fashioned source code, these parameter settings are all defined in the interface file of each community module in the AMUSE source code. For the `Hermite` package, for example, the parameters can be found in the module `amuse.community.hermite.interface`, in the function `define_parameters()`.

4.3.7 Changing the Solver Used in the Simulation

Each community code in AMUSE can be imported with:

```
from amuse.community.{codename} import {Codename}
```

Note here that the first letter of the code name is capitalized, followed by lowercase. For example, to import the `BHTree` and `ph4` codes:

Table 4.1. Overview of the *N*-body Codes Incorporated in AMUSE

Name	Ref	Type	Order	Language	N	Soft.	Parallel	GPU
Brutus	1	Pure	2nd	C++	$\lesssim 10$	No	Yes	No
Hermite		Pure	4th	C++	$\lesssim 10^4$	n/y	Yes	No
ph4		Pure	4th	C++	$\lesssim 10^5$	n/y	Yes	Yes
HiGPUs	4	Pure	6th	C++	$\lesssim 10^4$	n/y	Yes	Yes
phiGRAPE	5	Pure	4th	F77	$\lesssim 10^5$	n/y	Yes	Yes
SmallN		Pure	4th	C++	$\lesssim 10$	n/y	Yes	No
Rebound	7	Direct	Var	C	$\lesssim 10^4$	No	Yes	No
Huayno	8	Direct	Sympl.	C	$\lesssim 10^3$	n/y	Yes	Yes
Mercury	9	Direct	Sympl.	F77	$\lesssim 10^4$	No	No	No
Octgrav	10	Approx.	TC 2nd	C++	10^4–10^7	Yes	Yes	Yes
Bonsai	11	Approx.	TC 2nd	CUDA	10^4–10^{11}	Yes	Yes	Yes
MI6	12	Approx.	PN 6th	C++	$\lesssim 10^6$	n/y	Yes	Yes
Mikkola	13	Direct	PN 4th	F77	$\lesssim 100$	No	No	No
Twobody		Direct	Kepler	Python	2	No	No	No
Sakura	15	Direct	Kepler	C++	$\lesssim 10^5$	n/y	Yes	No
Nbody6++	16	Direct	4th	F77	$\lesssim 10^5$	No	No	Yes
BHTree	17	Approx.	TC 2nd	C++	10^3–10^6	Yes	Yes	No
SecularMultiple	18	Approx.	orbit av.	C++	$\lesssim 10$	No	No	No
Hermite_GRX	19	Pure	4th	C++	$\lesssim 10^3$	n/y	Yes	No

Here, TC refers to tree code (see Section 4.2.2), and PN to the possibility of including post-Newtonian terms. References: 1 Boekholt & Portegies Zwart (2015), 4 Spera et al. (2012); Capuzzo-Dolcetta et al. (2013), 5 Harfst et al. (2007), 7 Rein & Liu (2012, 2011), 8 Jänes et al. (2014), 9 Chambers & Migliorini (1997), 10 Gaburov et al. (2010), 11 Bédorf et al. (2012), 12 Oshino et al. (2011); Iwasawa et al. (2015), 13 Mikkola (1983); Mikkola & Merritt (2008), 15 Jänes et al. (2014), 16 Aarseth (1985); Wang et al. (2015), 17 Barnes & Hut (1986), 18 Hamers & Portegies Zwart (2016), and 19 Portegies Zwart et al. (2022). The columns give the order of the integrator, the language used, the number of particles for which the code is most suitable and whether or not the code can handle softening, is parallelized, or is graphics processing unit enabled.

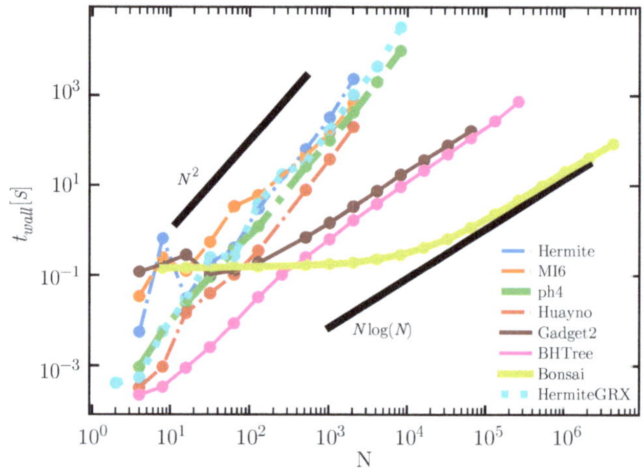

Figure 4.4. Wall-clock time as a function of N for a range of N-body solvers in AMUSE, integrating of a Plummer sphere for 1 N-body time unit. The runs were performed for 10 N-body time units. A script to generate this figure can be found at `amuse.examples.plot_nbody_performance`.

```
from amuse.community.bhtree import Bhtree
from amuse.community.ph4 import Ph4
```

Table 4.1 gives a brief overview of the current codes in AMUSE and their capabilities, and A.8 contains a complete (at the time of writing) list of the AMUSE modules. Note that while some codes, such as `ph4`, `SmallN`, `Huayno`, `BHTree`, and `Bonsai`, were added for strategic reasons, to fill a specific need, many others have been contributed by AMUSE users who used the framework to tackle some problem.

Figure 4.4 shows the performance of a variety of N-body solvers in AMUSE. Initial conditions are those of a Plummer sphere in virial equilibrium. No softening was used, but the runs were carried out for only 10 N-body time units; no close encounters occurred during this short time, and therefore no softening was necessary. We see two distinct performance behaviors in Figure 4.4. At the upper left, we have a group of direct N-body codes with $O(N^2)$ scaling. At right center, we see the ($N\log(N)$) behavior characteristic of tree codes. Code-to-code differences span about an order of magnitude and are the result of implementation details. The fastest code (bottom red curve) is the GPU-native `Bonsai`; the hardware characteristics of the GPU architecture explain its relatively poor performance for up to about 1000 particles but much better performance for larger numbers of particles.

4.4 Examples

4.4.1 Integrating the Orbits of Venus and Earth

In Section 3.4.1, we presented a plot of the integrated orbits of the Sun, Venus, and Earth (Figure 3.4). Now, we can improve our understanding of the scripts used to make this figure.

First, we generate the Cartesian coordinates for the Sun, Venus, and Earth, much as we did in Code example 3.7:

```
particles = new_solar_system()
SunVenusEarth = Particles()
SunVenusEarth.add_particle(particles[particles.name=='SUN'])
SunVenusEarth.add_particle(particles[particles.name=='VENUS'])
SunVenusEarth.add_particle(particles[particles.name=='EARTHMOO'])
SunVenusEarth.move_to_center()
print(SunVenusEarth)
```

Further down in Section 4.4.1.2, we show how you can make a realization of the solar system at a specific time.

Code example 3.6 presents the initialization routine, in which the masses, positions, and velocities of the Sun, Venus, and Earth are stored in a particle set. After initialization, the particle set is moved to the center of mass of the three-body system, to prevent the entire system from drifting, since we initialized it with the Sun at the origin. Subsequently, if we chose, we could give the solar system a position and systematic velocity in space by writing

```
particles.position += (-8.5, 0.0, 0.0) | units.kpc
particles.velocity += (11.1, -228.3, 7.25) | units.kms
```

These are the approximate position and velocity of the Sun relative to the Galactic center. These lines are best included after repositioning the solar system to its center of mass—otherwise the Sun, rather than the solar system's barycenter, would be placed at this location (although the difference is very small).

The integration routine presented in Code example 4.2 follows the orbital motion of Venus and Earth. In the first few lines, we load specific parts of AMUSE. Given the particle set in the argument list, we use the total mass and the distance to one of the particles (Earth—1 au) to initialize the converter. This converter is subsequently used to initialize the N-body code. We then add the particles to the N-body code and make special references to the two planets, Venus and Earth.

In the following lines, we initialize lists of x and y positions, which we update in the subsequent loop over time, with a time step of one day. In this case, we do not bother to initialize a channel to communicate between the framework and the running code. The function with a channel would be essentially the same, but would run slightly faster. After the time loop, we stop the N-body code and return the x- and y-position lists of the two planets.

Instead of storing the positions of Venus and Earth in arrays, we could write snapshots to a file at the end of each loop, using `write_set_to_file()`, and subsequently analyze the data with a separate script. To improve our diagnostics, we might also print or store the energy error produced by the N-body integrator at every diagnostic output interval.

```
from amuse.units import nbody_system
from amuse.community.huayno import Huayno


def integrate_solar_system(
    particles, time_step=1 | units.day, time_end=3 | units.yr,
):
    convert_nbody = nbody_system.nbody_to_si(
        particles.mass.sum(),
        particles[1].position.length(),
    )

    gravity = Huayno(convert_nbody)
    particles_in_code = gravity.particles.add_particles(particles)

    while gravity.model_time < time_end:
        gravity.evolve_model(gravity.model_time + (time_step))
        particles_in_code.new_channel_to(particles).copy()
        particles.savepoint(timestamp=gravity.model_time)

    gravity.stop()

    return particles
```

Code example 4.2: Program to integrate the orbits of Venus and Earth around the Sun. The full code example is in `amuse.examples.sun_venus_earth`.

4.4.1.1 Plotting the Results

We can subsequently plot the resulting data file from the simulation using the `pyplot` routines in the `matplotlib` package. An example code is presented in Code example 4.3. If we did not choose to store the data, we could instead simply have passed the arrays returned by `integrate_solar_system()` directly to the plotting routines. Alternatively, we could make a simple animation of the data, as presented in Code example 4.4.

Storing the animation as a gif file can be achieved by replacing line # 37 in Code example 4.4 by

```
anim.save("anim_gravity.gif", dpi=150, writer=animation.PillowWriter(fps=25))
```

4.4.1.2 Calculating Orbital Elements as Diagnostics

It is often useful to convert the six-dimensional Cartesian coordinates of a two-body system into Kepler orbital elements—semimajor axis, eccentricity, phase, etc. Because this is a common and important conversion, AMUSE offers a special routine to compute the orbital semimajor axis and eccentricity of any two-body system. A more detailed description of the use of Kepler elements, including discussion of some efficiency considerations, is presented in Section 8.3.2.

The two particles for which orbital elements are to be calculated must be stored in a separate particle set containing only those particles. For example, when we consider the solar system calculation in Section 4.4.1 (Code example 3.6), we can declare a subset of local particles `SunEarth` containing the Sun and Earth, respectively.

```
import matplotlib.pyplot as plt
from matplotlib import animation
from amuse.io import read_set_from_file
from amuse.units import units


def animate(x, y):

    def update(i):
        data = np.stack([x[i], y[i]]).T
        scat.set_offsets(data)

        return (scat,)

    number_of_snaps = len(x)
    fig = plt.figure(figsize=(8, 8))
    ax = fig.add_subplot(1, 1, 1)
    ax.set_xlim((-1.2, 1.2))
    ax.set_ylim((-1.2, 1.2))
    ax.set_xlabel("X [au]")
    ax.set_ylabel("Y [au]")

    colormap = ["#FFFF00", "wheat", "deepskyblue"]
    size = [40, 20, 20]
    edgecolor = ["orange", "wheat", "deepskyblue"]

    ax.set_facecolor("black")
    scat = ax.scatter(x[0], y[0], c=colormap, s=size, edgecolor=edgecolor)
    anim = animation.FuncAnimation(fig, update, frames=number_of_snaps, interval=30)
    anim.save("anim_gravity.mp4", dpi=150, writer=animation.FFMpegWriter(fps=25))
```

Code example 4.4: Minimal routine to animate *N*-body simulation data. Full source code is available in `amuse.examples.anim_gravity`.

```
SunEarth = Particles()
SunEarth.add_particle(particles[0])
SunEarth.add_particle(particles[2])
```

If desired, we could create a separate channel to this particle set:

```
channel_from_to_SunEarth = gravity.particles.new_channel_to(SunEarth)
```

The orbital elements can then be determined by calling

```
orbital_elements = orbital_elements_from_binary(SunEarth, G=constants.G)
```

This function is part of the extended AMUSE package, imported via

```
from amuse.ext.orbital_elements import orbital_elements_from_binary
```

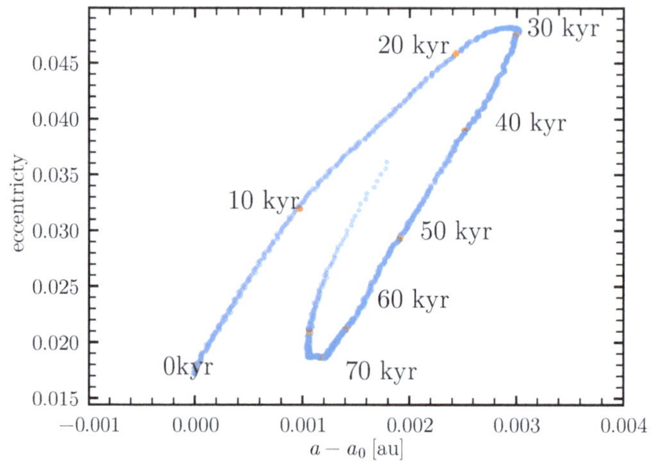

Figure 4.5. Earth's orbital semimajor axis (x-axis) and eccentricity (y-axis) as functions of time over a period of 80,000 yr. Initial conditions are taken from Table 3.1. The complete code example for the script that produced this figure is in `amuse.examples.earth_orbit_variation`. It should take a minute or so to run on a laptop.

The Kepler package underlying this function is written in standard N-body units and requires some notion of the value of Newton's constant G, which is provided as an argument. The routine returns a tuple containing the masses of the primary and secondary stars and the six Kepler elements: semimajor axis, eccentricity, inclination, argument of periapsis, line of the ascending node, and true anomaly.

We can integrate the inner solar system and study the variation in Earth's semimajor axis and eccentricity over an interval of 90,000 years. We present this evolution in Figure 4.5. Instead of initializing the solar system from the simplified initial conditions given earlier in Code example 3.6, we use AMUSE's built-in generator for any Julian date. To generate an initial realization of the solar system at the Julian date of 2,438,871.5 day, we write

```
particles = new_solar_system(Julian_date=2438871.5 | units.day)
```

Figure 4.6 presents the results of the numerical integration of the solar system over a much longer timescale. We will use this script in Section 4.4.1.3 to compare the variations in the eccentricity of Earth's orbit versus the temperature measurements at the Vostok station in Antarctica.

4.4.1.3 Temperature Record at Vostok Station

While the cycles in Earth's orbital elements are intrinsically interesting, it is even more interesting to compare them with observations. Instead of integrating just the innermost two planets, we include all eight planets in the solar system. In order to draw a comparison with observations, we integrate the solar system backward in

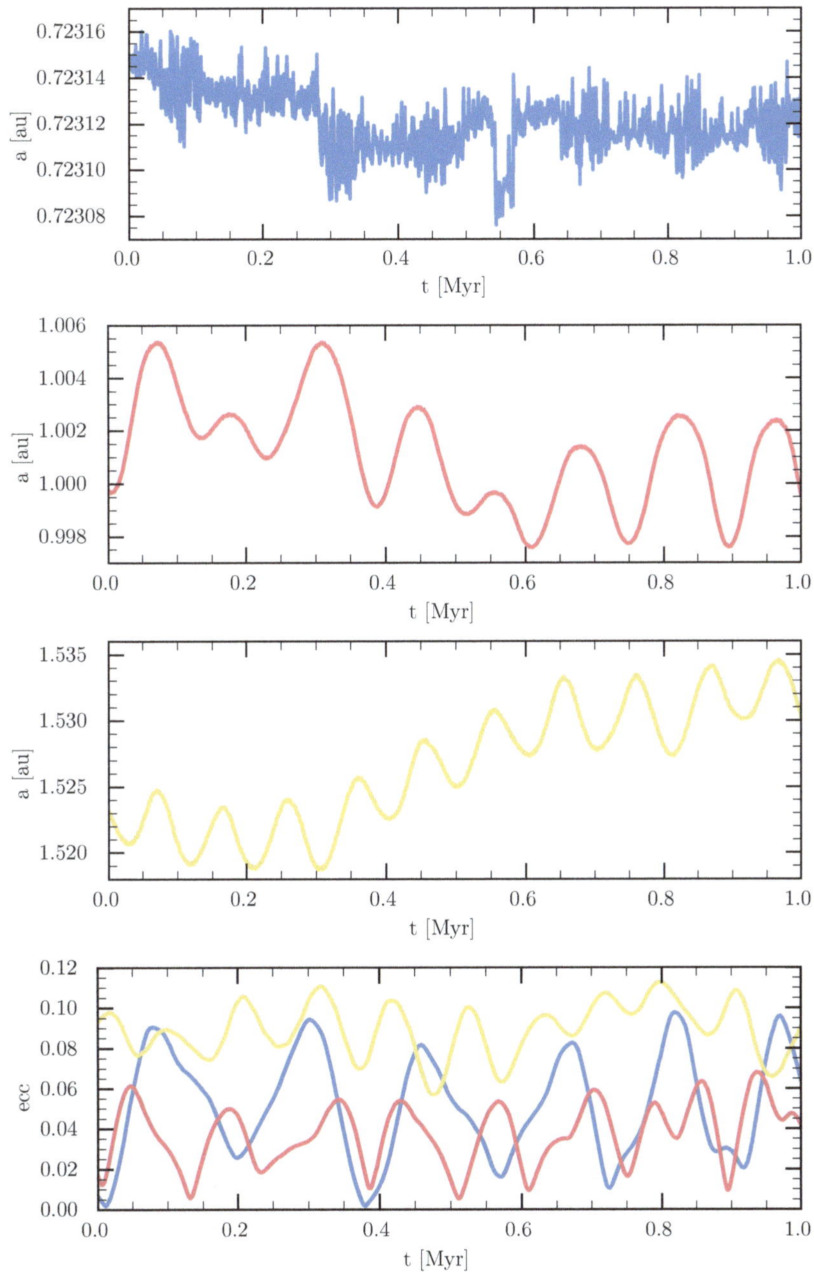

Figure 4.6. From top to bottom, the respective evolutions of the orbital semimajor axes of Venus (blue), Earth (red), and Mars (yellow) for the next million years. The bottom figure gives the orbital eccentricities of the three planets as functions of time, using the same color coding. The integration was performed using a script similar to that presented in Code example 3.7.

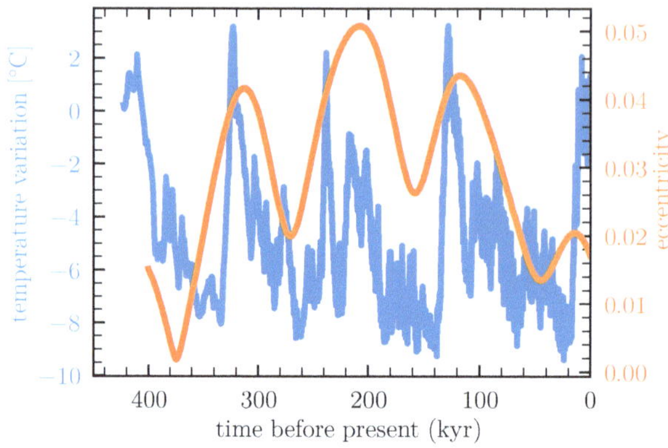

Figure 4.7. (Blue) Historical isotopic temperature record from the Vostok ice core (Jouzel et al. 1987; Petit et al. 1999). (Red) Orbital eccentricity evolution of the Earth over the last 400,000 yr. Integration was performed using `Huayno` with an accuracy parameter $\varepsilon = 10^{-3}$. The full code example of the script that produced this figure is in `amuse.examples.vostok_1999_temperature`. The Vostok ice-core temperature measurements are available via `amuse.examples.export("vostok_1999_temperature.data", "./")`.

time and check the results against the historical temperature record obtained from the ice at Vostok station in Antarctica. The backward calculation can be realized using one of the time-reversible N-body codes in AMUSE, or simply by reversing the velocities and integrating forward in time (the latter method was adopted here).

Figure 4.7 shows the measured changes in the temperature at Vostok station and the numerically computed changes in Earth's orbital eccentricity over the past 450,000 yr. The variations in eccentricity seem to correspond to variations in temperature (and CO_2) in Earth's atmosphere (Gildor & Tziperman 2000). It is interesting to note that the sharp rise in temperature seems to correlate with a rising eccentricity when it exceeds about 0.035. It has been suggested that these variations in the Earth's orbital eccentricity drive the 100,000 year ice-age cycle found in the historical temperature record from the Vostok ice core (Milankovitch 1941), but Earth's obliquity is important too.

4.4.2 Relativistic N-body Integration

Figure 4.8 shows an integration of the first stable three-body solution found by Montgomery (1998). This was the first stable nonhierarchical system to be discovered, and it caused quite a stir in N-body circles at the time. Subsequently, many stable "choreographies" of multiple bodies all following a single orbit were identified. Usually, this system is integrated in the Newtonian regime, but here we also show a solution where the system is relativistic. Even though we only integrate for a short time span, it is clear that the relativistic case the system is considerably less stable than in the Newtonian case.

This example was calculated using `hermite_grx` using the 2.5th order Taylor expansion to the Einstein–Infeld–Hoffman equations of motion with a scaling constant of $\gamma = 0.01c$. The stellar masses were $1\,M_\odot$ and the mutual distance 1 au.

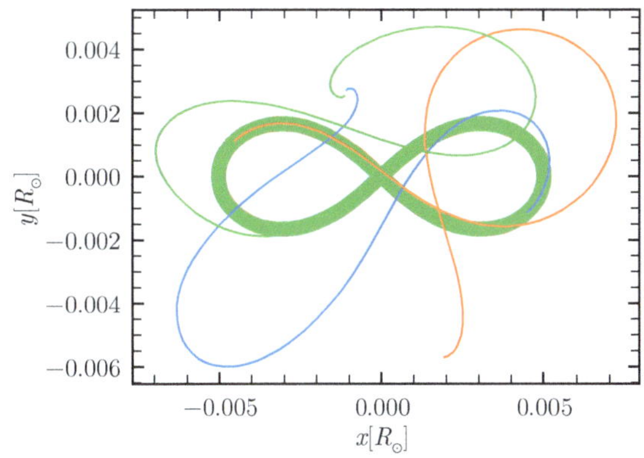

Figure 4.8. Montgomery's (1998) three-body problem, or the so-called stable figure 8 configuration in Newtonian gravity (thick green curve) for three black holes of $1\,M_\odot$. The three thin curves (green, orange, and blue) give the trajectories for the same initial configuration but when the velocities of the particles approach the speed of light. The calculations were conducted using `hermite_grx` with a scaling to the speed of light of $0.01c$. Whereas the Newtonian case shows the expected behavior predicted by Montgomery, the relativistic case ends in a merger between two, and eventually all three stars. The script to generate this figure can be found in `amuse.examples.montgomery`.

```
n_body_code = Hermitegrx
grav = n_body_code(converter)
pert = "2.5PN_EIH"
grav.parameters.perturbation = pert
grav.parameters.integrator = "RegularizedHermite"
grav.parameters.dt_param = 0.01
grav.parameters.light_speed = relative_light_speed * constants.c

```

Code example 4.5: Setting up the relativistic gravitational N-body code to integrate to 2.5th order PN terms with the regularized Hermite integrator. The full script can be found in `amuse.examples.montgomery`.

For a usual speed of light, this system is barely relativistic (represented by the thick green curve). But when we scale the speed of light to $\gamma = 0.01c$, the orbits deviate considerably from their Newtonian trajectories. Interestingly, one naively would expect the system to shrink, rather than the wider excursions as observable in Figure 4.8. In the relativistic case, two of the three stars collide (top middle). The setting up of the code is listed in Code example 4.5.

4.4.3 Fighting Exponential Divergence

All computers make errors. Larger computers just make more errors in a shorter amount of time than smaller computers. Normally, the 16 decimal places we work with in 64 bit IEEE-754 floating-point arithmetic hides the round-off nicely, but the problems may become apparent when dealing with large dynamic range or studying chaotic systems.

When we resolve planetary impact (on a scale of 10^3 km) in the local group of galaxies (100 kpc), the dynamic range exceeds a factor of 10^{14}, leaving only one or two digits for precision. This loss of precision when dealing with a wide dynamic range is mitigated by adopting different codes for different domains. For example, the solar system can be addressed in a dedicated code, whereas the local group is addressed in another code.

We do not have this luxury when studying chaos and can only solve this problem in two ways: either by adopting a statistical approach or using arbitrary precision. The first option is the most common, but the second has interesting possibilities.

`Brutus` (Boekholt & Portegies Zwart 2015) is a code that handles numerical round-off errors by simply hiding them. This is realized by combining an arbitrary precision library (Fousse et al. 2007) with the Bulirsch–Stoer integration strategy (Bulirsch & Stoer 1964).

The `Brutus` code is called similarly to any other N-body code in AMUSE, with the exception that it requires two additional parameters for tuning the precision and accuracy of the code. Precision can be tuned with the `word_length`, which is the number of bits with which the floating-point numbers in the code are represented. For regular double precision one would then use 64 bits.

The other parameter, called `bs_tolerance` (Bulirsch–Stoer tolerance), gives the accuracy at which the integration is performed. The smaller this parameter is set, the more accurate the calculation is performed. Although there is no strict linear relation between the `bs_tolerance` ($T_{\rm BS}$) and the `word_length` (L_w), we found an empirical expression that relates one to the other,

$$L_w = 32 + 4\left|\log_{10}(T_{\rm BS})\right|. \tag{4.18}$$

A word length of $L_w = 64$ bits then leads to a Bulirsch–Stoer of $T_{\rm BS} = 10^{-8}$.

The following snippet shows how to call `Brutus`.

```
gravity = Brutus()
gravity.parameters.bs_tolerance = bs_tolerance
gravity.parameters.word_length = word_length
gravity.parameters.dt_param = dt_param

gravity.particles.add_particles(bodies)
channel_to_framework = gravity.particles.new_channel_to(bodies)
```

The result of a calculation of a virialized three-body Plummer with two massive bodies and one minor object is presented in Figure 4.9. The orbits are integrated for 10 N-body time units, a tolerance of 10^{-16}, and a word length of 96 bits (24 decimal places, rather than the usual 16). A second calculation is performed with a tolerance of only 10^{-3} and a word length of 44 bits (11 decimal places).

The phase-space distance between the two solutions is shown as the solid line presented in Figure 4.10. The Lyapunov timescale (Lyapunov 1892) is given by the reciprocal of the slope of this dashed curve. We fit a slope of ~0.844, which corresponds to a Lyapunov timescale of about 1.2 N-body time units.

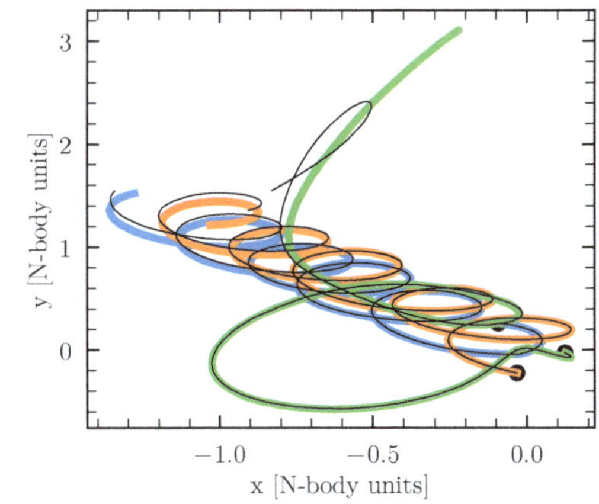

Figure 4.9. Orbital trajectories of a three-body problem (blue, orange, and green curves) calculated with 96-bit precision. The black curves give the result of identical initial conditions but calculated with 24-bit precision. The script, `amuse.examples.brutus`, takes about a minute to run.

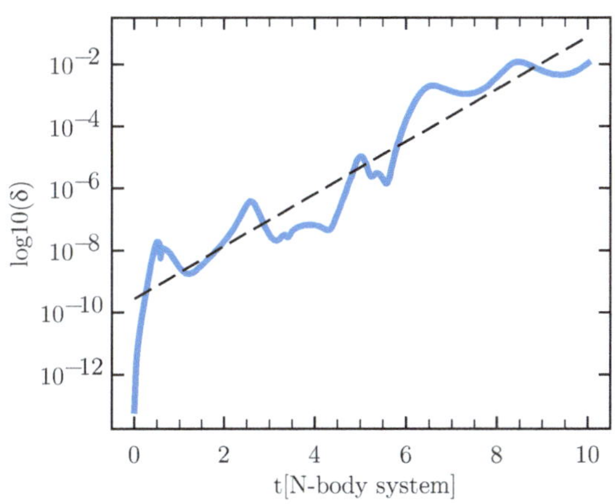

Figure 4.10. Phase-space distance between two calculations with 96-bit and 24-bit precision. The dashed line is a fit The script to generate this figure can be found in `amuse.examples.brutus`.

Running Brutus requires patience and determination given that it is arguably the slowest N-body code on the planet. However, for studying the chaotic behavior of self-gravitating systems, there is no practical alternative.

Instead of using arbitrary precise floating-point arithmetic, posits (or unum type III) seem to provide a promising alternative for higher precision at lower cost (Gustafson 2015). So far, however, there are no practical hardware implementations of posits, and the few available implementations go to a maximum of 64 bits.

4.4.4 Secular Multiples

When integrating planetary systems, sometimes resolving the individual orbital phases of all planets is simply too time consuming, particularly when the system is sufficiently stable that an approximate approach can suffice. In these cases, we can orbit-average the equations of motion. The basic idea behind such orbit averaging is the "wire" method, in which each planet is imagined to be spread out over its Keplerian orbit as though on a wire. Along that imaginary wire, the local density is proportional to the time spent at that orbital phase, with higher density near apocenter and lower density near pericenter. This approach can result in a substantial increase in computational speed. `SecularMultiple` is such a code (Hamers & Portegies Zwart 2016) in AMUSE.

In `SecularMultiple`, the secular (i.e. orbit-averaged) evolutions of hierarchical multiple systems are composed of binary orbits, in much the same way as is done in `Sakura` (see Table 4.1 or Section A.8), with an arbitrary number of bodies. For application domains, one can think of a hierarchical triple system with two stars in an "inner" binary orbited by a third body, a planetary system with a central star and several planets, or a hierarchical quintuple system with multiple planets. The code is based on an expansion of the Hamiltonian of the system in terms of ratios of orbital separations, which are assumed to be small. Subsequently, the Hamiltonian is averaged analytically over all orbits and the equations of motion are solved numerically. Speed is the main advantage of this approach: the code runs faster than any direct N-body integrator while still capturing the secular evolution. Individual terms in the expansion can easily be switched on or off to provide a better understanding of their effects on the evolution of the system. A downside is that the system must be hierarchical, since nonsecular effects are not captured. In particular, systems in which mean-motion resonances are important (such as tightly packed planetary systems) should be avoided.

We illustrate `SecularMultiple` for a "2 + 2" quadruple system: two binaries orbiting one another. As in other N-body codes in AMUSE, `SecularMultiple` works with particle sets. However, in addition to specifying parameters for the individual particles, the hierarchy of the system must also be specified. The declaration of particle attributes in a hierarchical quadruple system is illustrated in Code example 4.6. These attributes connect particles to one another. In this case, a body is a particle with a mass, and with attribute `is_binary` set `False`. Binaries are also part of the particles set, but with a True value for `is_binary`. Each binary has two children, `child1` and `child2`, indicating other members (bodies or other binaries) of the particle set, as well as attributes describing its orbital elements.

The example script has four bodies, `particles[0]` to `particles[3]`, and three binaries. The first binary $\mathcal{A}$ is `particles[4]`, the second binary $\mathcal{B}$ is `particles[5]`, and the outer orbit $\mathcal{C}$ of the $\mathcal{A}$-$\mathcal{B}$ binary is `particles[6]`. We adopt masses 1.0 $M_\odot$ and 0.8 $M_\odot$ for binary $\mathcal{A}$ with $a_\mathcal{A} = 1.0$ au, $e_\mathcal{A} = 0.1$, and $i_\mathcal{A} = 75^\circ$. Binary $\mathcal{B}$ has primary and secondary masses of 1.1 $M_\odot$ and 0.9 $M_\odot$, $a_\mathcal{B} = 1.2$ au, $e_\mathcal{B} = 0.1$, and $i_\mathcal{B} = 80^\circ$. The outer binary, $\mathcal{C}$, has $a_\mathcal{C} = 100$ au, $e_\mathcal{C} = 0.3$.

Running `SecularMultiple` is no different from running any other N-body code in AMUSE, using the function `evolve_model()` in the event loop. Figure 4.11 presents the evolution of the orbital separation and eccentricity of the

```python
def initialize_multiple_system(
    number_of_bodies,
    masses,
    semimajor_axis,
    eccentricity,
    inclination,
    argument_of_pericenter,
    longitude_of_ascending_node,
):
    """
    Initializes a system of multiples
    """

    number_of_binaries = number_of_bodies - 1
    particles = Particles(number_of_bodies + number_of_binaries)
    for index in range(number_of_bodies):
        particle = particles[index]
        particle.mass = masses[index]
        particle.is_binary = False
        particle.radius = 1.0 | units.RSun
        particle.child1 = None
        particle.child2 = None

    for index in range(number_of_binaries):
        particle = particles[index + number_of_bodies]
        particle.is_binary = True
        particle.semimajor_axis = semimajor_axis[index]
        particle.eccentricity = eccentricity[index]
        particle.inclination = inclination[index]
        particle.argument_of_pericenter = argument_of_pericenter[index]
        particle.longitude_of_ascending_node = longitude_of_ascending_node[index]

        # Specify the `2+2' hierarchy:

        if index == 0:
            particle.child1 = particles[0]
            particle.child2 = particles[1]
        elif index == 1:
            particle.child1 = particles[2]
            particle.child2 = particles[3]
        elif index == 2:
            particle.child1 = particles[4]
            particle.child2 = particles[5]
    binaries = particles[particles.is_binary]

    return particles, binaries
```

Code example 4.6: Snippet to set up a "2+2" hierarchical quadruple system composed of two binaries that orbit each other. The full script can be found in `amuse.examples.hierarchical_quadruple`

various components of this quadruple system. The top panel shows the semimajor axes (solid curves) and periapsis distances (dotted lines) of the three orbits $\mathcal{A}$, $\mathcal{B}$, and $\mathcal{C}$; the bottom panel shows the inclinations of $\mathcal{A}$ and $\mathcal{B}$ relative to $\mathcal{C}$. High Lidov–Kozai-like eccentricity oscillations occur in the $\mathcal{A}$ and $\mathcal{B}$ orbits, although they are more complex than the equivalent isolated hierarchical triple systems and the

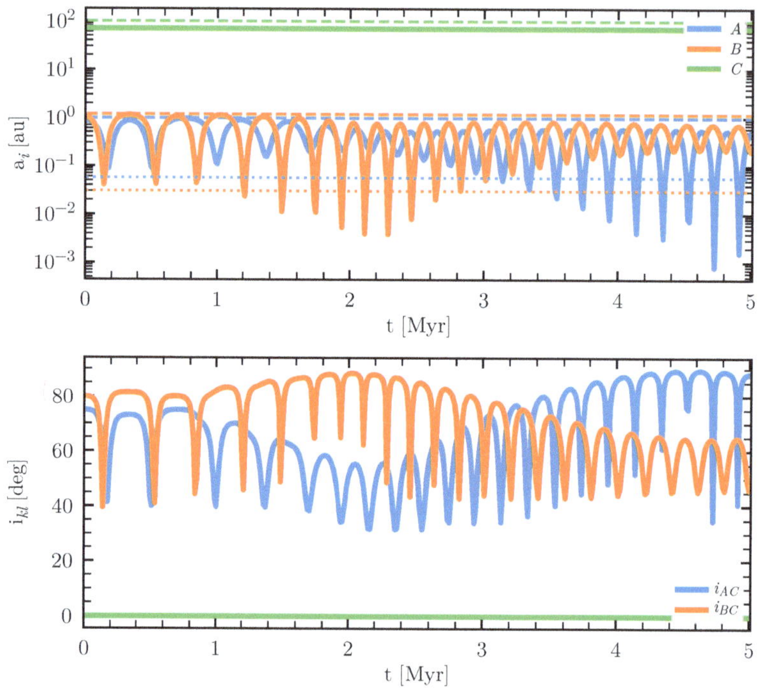

Figure 4.11. Example evolution of a 2 + 2 quadruple system. Top panel: The semimajor axes (dashed lines) and periapsis distances (solid curves) of orbits $\mathcal{A}$, $\mathcal{B}$, and $\mathcal{C}$ as a function of time. Bottom panel: The inclinations of $\mathcal{A}$ and $\mathcal{B}$ relative to $\mathcal{C}$. Canonical values expected for Lidov–Kozai oscillations in the three-body test-particle quadrupole-order approximation are shown in the top panel as dotted lines. The dynamics are more complex in this quadruple system, compared to the situation of two uncoupled triple systems. The script that produced this figure is `amuse.examples.hierarchical_quadruple`. It takes about half a minute to run.

maximum eccentricities are considerably higher. (To study the difference due to the fourth body in the system, we can easily reduce the semimajor axis of one of the binaries $\mathcal{A}$ or $\mathcal{B}$ by a factor of 100, effectively reducing it to a point mass. This quenches the effect of one of them being a binary and recovers the classic Lidov–Kozai effect in the remaining triple system.)

In addition to point-mass secular Newtonian dynamics, other physical processes can be included in `SecularMultiples`. Such as PN effects up to and including order 2.5, or tidal evolution (including misaligned spins). The Mardling & Aarseth (2001) stability criterion for hierarchical triples is also implemented and checked.

4.4.5 Merging Galaxies

There are at least three methods whereby gravity is calculated in SPH codes. These include direct summation, the tree algorithm, and self-consistent mean fields using fast Fourier transforms (Hockney & Eastwood 1988). Direct summation has the advantage of allowing accurate energy conservation, which can be important for problems near a stability limit (e.g., for a binary star system near its innermost stable circular orbit). These direct-summation methods are typically employed on GPUs using the same algorithms as for N-body codes (Portegies Zwart et al. 2007).

The most common engine for computing self-gravity in an SPH code is a tree code. Such codes are well suited to simulating galaxy mergers. In Section 3.2, we discussed how to generate a galaxy model using `Halogen` (Zemp et al. 2008; Zemp 2014). Here, we will use `GalactICs` (Kuijken & Dubinski 1995), which is designed to set up a self-consistent galaxy model with a disk, bulge, and dark halo. In principle, this code can be used to construct realistic models with both gas and particles for the Milky Way, Andromeda, or any other galaxy, but it often requires some delicate tuning to get the proper structure.

We use `GalactICs` to generate two identical disk galaxies with `n_halo` particles in each halo, `n_bulge` particles in each bulge, and `n_disk` particles in each disk. We eventually scale the models to virial equilibrium. The first galaxy is created as follows:

```
converter = nbody_system.nbody_to_si(mass_galaxy, radius_galaxy)
galaxy1 = new_galactics_model(
    n_halo,
    converter,
    do_scale=True,
    bulge_number_of_particles=n_bulge,
    disk_number_of_particles=n_disk,
)
```

The second galaxy is constructed by simply making a copy of the first. Both galaxies are placed in an interacting orbit; the hydrodynamics solver is then provided with the particles and integrates the equations of motion without taking hydrodynamics into account.

```
converter = nbody_system.nbody_to_si(1.0e12 | units.MSun, 100 | units.kpc)
dynamics = Gadget2(converter, number_of_workers=4)
dynamics.parameters.epsilon_squared = (100 | units.parsec)**2
set1 = dynamics.particles.add_particles(galaxy1)
set2 = dynamics.particles.add_particles(galaxy2)
dynamics.particles.move_to_center()
```

For plotting purposes, we want to keep track of only the disk and bulge particles. This is accomplished by selecting only those particles added after the halo was generated (the order of creation is halo, bulge/disk). For this reason, we save pointers to the particles in the running code in the local parameters `set1` and `set2` for the first and second galaxies, respectively. Here, we take advantage of the fact that `add_particles` always returns a pointer to the last object added (in this case the bulge/disk particles).

Figure 4.12 presents two snapshots of the interacting galaxies, one at the start of the simulation and the other 200 Myr later when the two galaxies have just passed

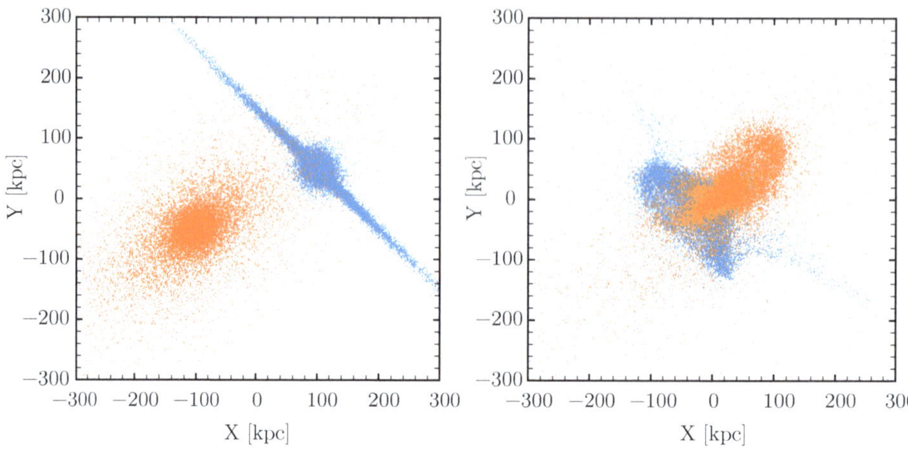

Figure 4.12. Collision between two galaxies, just before they interact (left) and 200 Myr later (right), when both galaxies have been distorted by the interaction. The calculation was performed using `Gadget2` (see Chapter 6 and Springel 2000, 2005) as a gravity solver, without hydrodynamics. Both galaxies are made of 20,000 halo particles, with 10,000 for each the bulge and the disk. In this figure, we only show the bulge and disk particles. The simulation was performed using `amuse.examples.merge_two_galaxies.` `It` should take a couple of minutes to run on a laptop.

each other. The two galaxies are identical copies, but they are rotated using the particle-set attribute `rotate`. For the galaxy on the left in Figure 4.12 (red), this is realized with

```
galaxy1.rotate(0.0, numpy.pi/2, numpy.pi/4)
```

The galaxies are subsequently translated to a new position about 200 kpc apart and given a velocity toward each other of about 10 km s^{-1}. We refrain here from a detailed analysis of the resulting merger, but it may be rewarding to simulate the cosmic collision between the Milky Way and the Andromeda galaxy.

4.5 Choosing the Right Code

It is not always trivial to choose the right code for your project. There is generally an optimal choice, but one can wonder if there is a wrong choice. Deciding on what code to select is not trivial, and the optimization matrix may be complex. The arguments that tip the balance in favor of one code can be based on the physics, precision, accuracy, long-term stability, energy error, angular momentum conservation, performance, and so on.

The first question to ask is to what extent the gravity in your problem should be coupled to another physical domain. We cannot really make a firm recommendation because each code can be coupled to other solvers in AMUSE (see Section 8 and Chapter 9), although there are a few details to consider.

Each code in AMUSE has an application domain. Once the calculations leave this domain, the code stops working in some way. It could crash, but worse it could become inefficient or continue without producing sensible results. To mitigate this problem, we have introduced what we call *stopping* conditions.

4.5.1 Stopping Conditions

Stopping conditions are criteria which, once reached, allow the particular code triggering the condition to stop in an appropriate state. Control is then returned to the AMUSE script, in which action may be taken. These actions include stopping the particular code, handing the data to another already existing code, storing data and performing diagnostics, or starting a new code to perform further calculations. Stopping conditions are a fundamental and powerful way to mitigate all sorts of problems that may arise when combining different solvers operating on a wide variety of domains. They are fundamental to AMUSE, and each domain has its own set of stopping conditions.

The most common stopping conditions for gravity are listed in Table 4.2. These include detection of collisions, escapers, if an interaction is still taking place, and binary (pair) formation. There are also possible limits that can be imposed using stopping conditions, such as the number of steps in the integrator or a time-out for the calculation.

A stopping condition, for example to detect a collisions, can be used as follows:

```
gravity = Ph4()
collision_detection = gravity.stopping_conditions.collision_detection
collision_detection.enable()
```

Here, the N-body code will stop as soon as two stars approach each other within the sum of their radii (however defined). The interrupt is called in the event loop where the N-body code is evolved,

Table 4.2. Overview of the Stopping Conditions Available in Most N-body Codes

code.stopping_conditions.	Default Value	Default Unit
...number_of_steps	1	None
...timeout	4.0	s
...collision_detection	bool	False
...escaper_detection	bool	False
...interaction_over_detection	bool	False
...pair_detection	bool	False

```
while gravity.model_time < time_end:
    gravity.evolve_model(gravity.model_time + time_step)
    if collision_detection.is_set():
        resolve_collision(collision_detection, gravity, converter)
```

The appropriate information about the event can be queried from `collision_detection` through accessor functions. If particles are associated with the event, these are stored as a particle set in

```
collision_detection.particles
```

and can be used as such.

Note that although the state of the N-body code is consistent, it will not have completed its evolution to the appropriate end time `gravity.model_time + time_step)`, but has stopped at same time between `gravity.model_time)` and `gravity.model_time + time_step)`. In this example, the routine `resolve_collision()` resolves the collision by merging the masses of the two colliding stars, and communicates their center of mass properties to the new particle that replaces the two colliding particles. In principle, this can be realized in a few lines of code, or one may invoke a hydrodynamics solver to effectively resolve the stellar collisions.

4.6 Validation

Quickly prototyping a small script that runs a large scientific simulation is easy with AMUSE, but how do you guarantee that everything works correctly? You cannot, and we cannot either. However, we can at least guarantee that the coupling of the code is done correctly and that the units are properly converted and transferred.

Some codes are easier to validate than others. Direct gravitational solvers have to conserve energy, at least if they are run as isolated, standalone modules. Checking for energy conservation is therefore always a good thing to do. For stellar evolution, there is no conserved energy, but you would surely like the total mass of the star plus the total mass lost in a wind and a supernova to be conserved. Each methodology has its own conserved quantities. As a researcher and user of any scientific code, you should always explicitly check them.

Time-stepping in any given code is generally managed internally, but as a user you can in principle change the time steps as you like. However, in addition to the fact that we do not recommend fiddling with the community source code, we also do not recommend overriding the safety settings for things like time step control. Some of the community modules are very sensitive to such changes and may simply fail, but others will always produce some output. These latter codes are the most dangerous: "garbage in, garbage out." If a code receives completely wrong information, in the form of nonsensical initial conditions or parameter settings, then the best thing that can happen is that it crashes. The worst outcome is that the code will continue to run, producing reasonable-looking nonsense.

4.6.1 Error Propagation and Validation

The solar system is about 4.5 Gyr old, which means that Earth has made more than four billion orbits around the Sun. During each integration step in a simulation, the program makes tiny mistakes (e.g., round-off errors at about the 15th digit, or larger errors due to algorithmic limitations), which slowly accumulate with time. The accumulated energy error may eventually grow to exceed the total energy of the system, at which point we would be forced to conclude that the integration has failed to represent the physical system we are trying to model.

For lack of a satisfactory alternative, the energy error is often used as the sole diagnostic for quantifying the quality of an N-body integration—but beware: a gravitational system (probably) responds exponentially to small errors, drifting away from whatever the true solution is and probably rendering the numerically determined position and velocity of any particle meaningless. Even worse are the small phase-space errors introduced by the integrator which, due to the chaotic nature of the system, grow exponentially. This causes your system to quickly diverge from what you intended to calculate, and you will not notice it because the energy error may remain small.

The error in the energy may grow randomly (according to Brouwer's law; see Section 4.2.3) or, more likely, systematically and slowly. Therefore, the energy error is sometimes a rather poor indicator of the quality of an N-body integration (partly for reasons discussed in Section 4.1.4.1), but so long as it does not grow beyond a certain fraction of the total energy, the result may be sufficiently accurate for scientific interpretation.

The boundary above which we consider the energy error acceptable is quite arbitrary and context-dependent. If no number is quoted, you should, as a scientist, doubt the quality of the calculation (even though the animations and figures may look great). As a rule of thumb, for direct N-body simulations, the cumulative energy error over the duration of the calculation should remain below 1% of the total energy of the entire system, and below 10^{-4} of the total energy of the system over an interval of one dynamical time. Many researchers would regard these limits as far too high.

4.6.1.1 Error Behavior and Analysis

The errors in N-body integrations can be measured objectively by checking conserved quantities, such as energy and angular momentum. In Code example 4.1, we demonstrated how energy conservation can be checked at runtime. This provides a useful first-order diagnostic, although for the solar system (for example), it informs you mainly about the integration errors for the most massive bodies. It is far less informative about errors in the integration of minor bodies.

The error behavior of N-body codes depends both on the algorithm used to calculate the forces and how the equations of motion are integrated. Figure 4.13 shows the relative energy error for three N-body solvers—a (fourth-order) direct-summation Hermite solver, a similar high-order solver but including the Einstein–Infeld–Hoffman equations of motion for general relativity, and a (second-order)

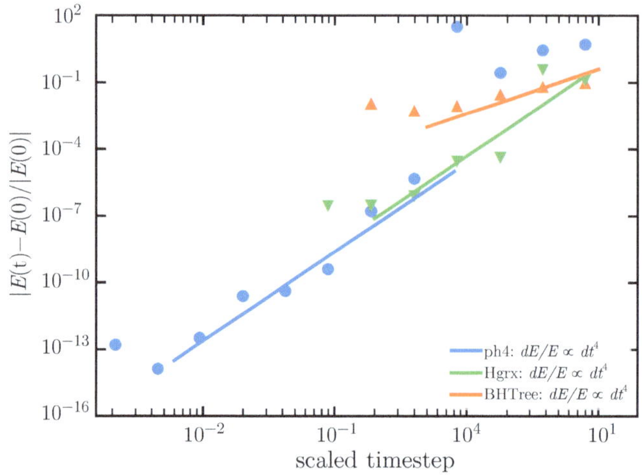

Figure 4.13. Energy error of the integration for one dynamical time of $N = 1000$ equal-mass particles initially distributed in a Plummer sphere. The integration was carried out using the direct N-body integrator `ph4` (blue bullets), `Hermite_GRX` (green triangles), and with the `BHTree` tree code (orange triangles). The three solid lines show the global trends in energy conservation of the N-body codes (these are not fits). The `Hermite_GRX` run was performed for only 0.1 N-body time units, to reduce the computational cost. The script to generate this figure can be found in `amuse.examples.run_nbody_precision`. The `ph4` and `BHTree` runs take a few seconds each, but the runs with `Hermite_GRX` are considerably more expensive, taking about an hour to run on a laptop.

Barnes–Hut tree code with a Leapfrog solver. The general relativistic run was performed with a speed of light one hundred times higher than the maximum velocity of any of the particles, which is rather arbitrary, but demonstrates the point that the scaling is identical to the Newtonian fourth-order integrator.

Each solver behaves somewhat differently, but this is not unexpected. The direct N-body integration can reach very small energy errors, below 10^{-13}. For smaller values of the time-step parameter $\eta \lesssim 0.005$, the error starts increasing again. From this point, numerical round-off starts to drive the energy error in the calculation. We refer to Press et al. (1992), which excellently quantifies this point. For a time step parameter of unity (and larger), the error suddenly increases to a value of $\mathcal{O}(1)$. Once the time step parameter becomes too large (apparently around order unity), the integrator is unable to follow individual trajectories accurately. The integrator does not break completely, but fails to acquire convergence in the solution. This is reflected in the large energy error, and is one of the reasons why it is important to always check the energy conservation in runs.

The relativistic integrator `Hermite_GRX` shows a similar behavior as the Newtonian integrator, but it is much more expensive to run (the force in general relativity also depends on the velocity and therefore scales $\propto N^3$). The numerical round-off in the relativistic equations of motion starts to become noticeable at a much larger value of $\eta \lesssim 0.1$ compared to the Newtonian calculation, because the number of operations per performance calculation is orders of magnitude larger. It is therefore not very useful to run relativistic calculations with extremely small values of η (except, of course, when one runs with an arbitrarily precise integrator, such as `Brutus`).

With the tree code, the error saturates because the error at short time steps is dominated by the tree force evaluation algorithm, not by the integration scheme. However, this comparison is unfair because in `BHTree` this parameter is defined differently. In `BHTree`, the time step is either an absolute fraction of the local crossing time (when running in dimensionless N-body units) or it is an absolute value in units of (millions of) years. In this case, the calculations are performed using dimensionless N-body units, and η for `BHTree` should be interpreted as a fraction of the dynamical crossing time. The other parameter that can be used to tune the precision performance of a tree code such as `BHTree` is the opening angle θ. In this example, we used the default value of $\theta = 0.75$.

4.7 Assignments

4.7.1 Orbital Trajectories

Take the initial conditions for the solar system (Table 3.1) and integrate the planets' orbits for 100 yr with the integrator of your choice (Table 4.1). Explain why you have selected this integrator, and discuss the effect of using another integrator.

1. Draw graphs of the energy error as a function of time, and of the change in the planet's semimajor axis, da/a, as a function of time.
2. Draw a graph of the evolution of the semimajor axes and eccentricities of Venus, Earth, and Mars, and compare them with the results in Figure 4.6.
3. Run the same initial conditions with two other integrators, and explain the differences in the results.

   Compare the evolution of the semimajor axes and eccentricities for Venus, Earth, and Mars with those presented in Figure 4.6. In what quantity are the differences most pronounced?

4.7.2 Vostok

In Section 4.4.1.3, we discussed the variation in the eccentricity of Earth's orbit over the last 450,000 yr. Although the variations in the average temperature measured at the Vostok station are not entirely attributable to the apparent cycle in Earth's eccentricity, it is evident that some periodicity appears in Earth's orbit.

These variations are probably due to the influence of only a few (maybe only one) other planets. It is unlikely that Mercury or Neptune drives this cycle, but the N-body code we used to create Figure 4.7 can help us find out which planet is responsible for the variations in Earth's orbit.

Perform this analysis and find the planet that dominates the evolution of Earth's orbit around the Sun.

4.7.3 Dynamical Binary Formation

(a) In cluster dynamics, it is common to define an energy scale, called kT (as in thermodynamics), by $K = 3/2NkT$, where N is the number of stars and K

is the total kinetic energy of the system. This turns out also to be an interesting energy threshold for binary systems (see Section 4.1.3.2). Run an N-body simulation with $N = 100$ stars using a code of your choice (see Table 4.1), until the appearance of a binary with a binding energy exceeding 100 kT. You can take Code example 4.1 as a basis for the N-body simulation. The simplest way to do this is to run in steps of (say) one dynamical time, and compute the relative binding energies of all pairs of stars at the end of each step. This is fairly inefficient, but infrequent and fast enough with 100 stars that it should be doable in `Python`.

Be careful here because several of the N-body codes in AMUSE use softened potentials and are unable to reach such hard binaries, while others will have great difficulty integrating hard binaries.

Use a Plummer distribution as the initial density profile and take all the masses to be the same.

(b) Perform the run with a different N-body code, using exactly the same initial realization of the initial conditions.

(c) Is the moment at which the first 100 kT binary appears the same, or does it depend on the code?

(d) What changes when a mass function is introduced? Adopt a Salpeter mass function (see Equation (3.1)) with $\alpha = -2.35$ (Salpeter 1955). Remember to rescale the system to virial equilibrium (see Section 4.3.1) after setting the stellar masses.

Explain why the system should be rescaled to virial equilibrium.

In an N-body integration performed in dimensionless N-body units, the only parameter to specify for the mass function is the range over which the masses should be varied (the total mass is normalized), assuming that the power-law slope remains the same. How does the moment of the formation of the first hard binary depend on the range of masses?

4.7.4 L_1 Lagrangian Point

Lagrangian points in a self-gravitating system of two bodies are locations at which there is either a peak or a saddle point in the effective potential surface (de Lagrange 1772) in the frame rotating with the bodies. An N-body code can easily be adapted to draw such an equipotential surface. Figure 4.14 presents an example for the case of the Sun and a 0.2 $M_{\odot}$ mass companion at a distance of 1 au. The script to generate these equipotential surfaces can easily be adapted to any binary system, including those with nonzero eccentricity.

Interplanetary travel is expensive in terms of time and fuel. Moving through Lagrange points provides an efficient way to travel from one celestial body to another (although they can also lead to objects becoming trapped in a local orbit, as is the case with Jupiter's Trojan satellites). In Figure 4.14, the most efficient way to travel from the star to the left to the one to the right is through L_1. This trajectory can be calculated most easily by starting at L_1 and allowing a test mass to "fall" into the potential well to the left or the right.

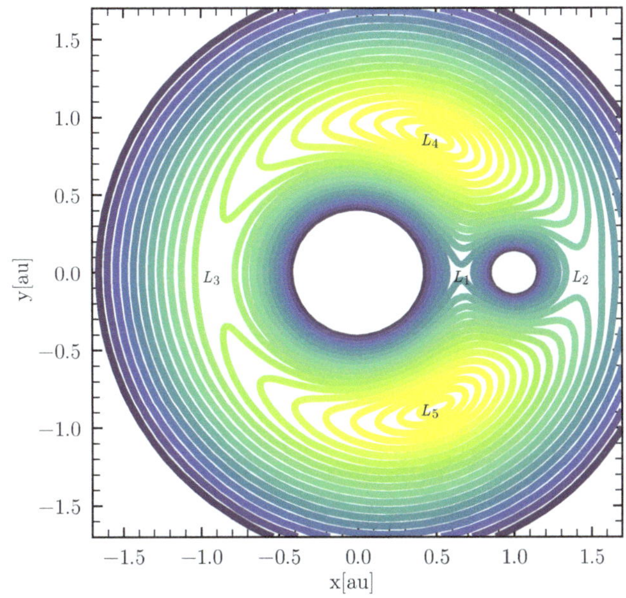

Figure 4.14. Equipotential surfaces for a two-body system. In this example, we adopted a primary mass of 1 $M_\odot$ with a 0.2 $M_\odot$ companion in a circular orbit at a distance of 1 au (to the right). The classical five Lagrange points are indicated by $L_1 - L_5$. The script to generate this figure can be found in `amuse.examples.lagrange_points`.

(a) Calculate the most energy-efficient orbit from Earth to the Moon in an isolated system, but with Earth and Moon on the proper relative orbit. Plot this trajectory in Cartesian coordinates.

(b) Repeat the calculation for Earth and the Moon in orbit around the Sun, with the planet Jupiter added to further perturb the system. Are the two trajectories the same?

Hint: Start by initializing a particle on the first Lagrangian point of the Earth–Moon system and give it a small velocity toward Earth. Repeat with a small velocity in the direction of the Moon.

4.7.5 Virial Equilibrium

(a) Calculate the evolution of the virial ratio for an N-body system with a mass function between 0.1 and 100 $M_\odot$ using the codes `ph4`, `Huayno`, and `BHTree`, starting with an initial virial ratio of $Q = 0.2$ (Section 4.3.1), and continue until the system virializes. Perform the calculation with $N = 100$, increasing by factors of 2 up to $N = 1600$, and make a plot of the first moment in time for which $Q = 0.5$ as a function of N for each of the three codes.

The virial ratio will oscillate due to random close encounters, but the virialization timescale can be estimated by studying the way in which the oscillations damp out and the viral ratio approaches its equilibrium value of 0.5.

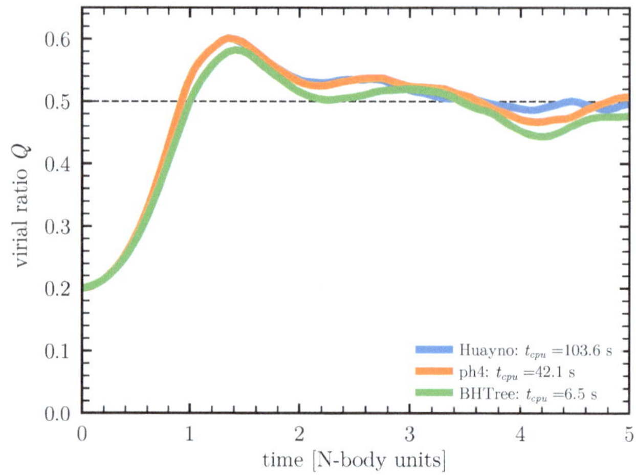

Figure 4.15. Virial ratio Q as a function of time for 1000 equal-mass particles initially distributed in a Plummer sphere with $Q = 0.2$. The integrations were performed using `ph4` (blue line), `Huayno` (red), and `BHTree` (yellow). The script that produced this figure is in `amuse.examples.gravity_to_virial`. It should run in a couple of minutes on a laptop.

(b) Figure 4.15 presents an example $q(t)$ for three N-body codes, using 1000 stars of equal mass.

Answer the following questions:

(i) Why does the system with equal-mass stars virialize on a timescale of about an N-body time unit?
(ii) What is the effect of introducing a mass function on the virialization timescale?
(iii) Why does the way in which the system virializes depend on the code?
(iv) What is the effect of increasing the number of stars on the virialization timescale?
(v) Why do the two direct N-body systems remain slightly above virial after about 1 N-body time unit?

References

Aarseth, S. A. 2003, Gravitational N-body Simulations (Cambridge: Cambridge: Cambridge Univ. Press)

Aarseth, S. J. 1963, MNRAS, 126, 223

Aarseth, S. J. 1985, Multiple Time Scales, ed. J. U. Brackbill, & B. I. Cohen (New York: Academic Press) 377

Abbott, B. P., Abbott, R., Abbott, T. D., et al. 2016, PhRvL, 116, 241102

Ambrosiano, J., Greengard, L., & Rokhlin, V. 1988, CoPhC, 48, 117

Anders, P., Baumgardt, H., Gaburov, E., & Portegies Zwart, S. 2012, MNRAS, 421, 3557

Antonov, V. A. 1962, Solution of the Problem of Stability of Stellar System Emdenas Density Law and the Spherical Distribution of Velocities (Leningrad: Vestnik Leningradskogo Universiteta)

Babuška, I. 1969, Proc. IFIP Congress Information Processing 68, ed. A. J. H. Morrell (Amsterdam: North-Holland) 11

Barnes, J., & Hut, P. 1986, Natur, 324, 446

Bédorf, J., Gaburov, E., & Portegies Zwart, S. 2012, JCoPh, 231, 2825

Bédorf, J., Gaburov, E., Fujii, M. S., et al. 2014, Proc. of the Int. Conf. for High Performance Computing, Networking, Storage and Analysis, SC '14 (Piscataway, NJ: IEEE) 54

Bettwieser, E., & Sugimoto, D. 1984, MNRAS, 208, 493

Binney, J., & Tremaine, S. 2008, Galactic Dynamics (2nd ed.; Princeton: Princeton Univ. Press)

Boekholt, T., & Portegies Zwart, S. 2015, ComAC, 2, 2

Brouwer, D. 1937, AJ, 46, 149

Bulirsch, R., & Stoer, J. 1964, NuMat, 6, 413

Cai, M. X., Portegies Zwart, S., Kouwenhoven, M. B. N., & Spurzem, R. 2019, MNRAS, 489, 4311

Capuzzo-Dolcetta, R., Spera, M., & Punzo, D. 2013, JCoPh, 236, 580

Chambers, J. E. 1999, MNRAS, 304, 793

Chambers, J. E., & Migliorini, F. 1997, AAS/DPS Meeting, 29, 27.06

Chandrasekhar, S. 1943, ApJ, 97, 255

Chernoff, D. F., & Weinberg, M. D. 1990, ApJ, 351, 121

Cohn, H. 1980, ApJ, 242, 765

Collins, G. W. 1978, The Virial Theorem in Stellar Astrophysics (Tucson, AZ: Pachart Publication)

de Lagrange, J.-L. 1772, Essai sur le Problème des Trois Corps (Paris: Royal Academy of Sciences)

Dehnen, W. 2014a, ComAC, 1, 1

Dehnen, W. 2014b, gyrfalcON: N-body Code, Astrophysics Source Code Library, ascl:1402.031

Duncan, M. J., Levison, H. F., & Lee, M. H. 1998, AJ, 116, 2067

Einstein, A. 1916, AnP, 354, 769

Einstein, A., Infeld, L., & Hoffmann, B. 1938, AnMat, 39, 65

Fellhauer, M., & Lin, D. N. C. 2007, MNRAS, 375, 604

Fousse, L., Hanrot, G., Lefèvre, V., Pélissier, P., & Zimmermann, P. 2007, ACM Trans. Math. Softw, 33, 13–es

Fukushige, T., & Heggie, D. C. 1995, MNRAS, 276, 206

Gaburov, E., Bédorf, J., & Portegies Zwart, S. 2010, OCTGRAV: Sparse Octree Gravitational N-body Code on Graphics Processing Units, Astrophysics Source Code Library, ascl:1010.048

Giersz, M., & Heggie, D. C. 1994, MNRAS, 268, 257

Giersz, M., & Heggie, D. C. 1996, MNRAS, 279, 1037

Gildor, H., & Tziperman, E. 2000, PalOc, 15, 605

Gnedin, N. Y., Glover, S. G. O., Klessen, R. S., & Springel, V. 2015, Star Formation in Galaxy Evolution: Connecting Numerical Models to Reality, ed. Y. Revaz, P. Jablonka, R. Teyssier,, L. Mayer, et al. (Berlin: Springer)

Greengard, L., & Rokhlin, V. 1987, JCoPh, 73, 325

Gustafson, J. L. 2015, The End of Error: Unum Computing (Boca Raton, FL: CRC Press)

Hamers, A. S., & Portegies Zwart, S. F. 2016, MNRAS, 459, 2827

Harfst, S., Gualandris, A., Merritt, D., et al. 2007, NewA, 12, 357

Heggie, D. C. 1975, MNRAS, 173, 729

Heggie, D. C., & Mathieu, R. D. 1986, The Use of Supercomputers in Stellar Dynamics 267, ed. P. Hut, & S. L. W. McMillan; Berlin: Springer) 233

Heggie, D., & Hut, P. 2003, The Gravitational Million-Body Problem: A Multidisciplinary Approach to Star Cluster Dynamics (Cambridge: Cambridge Univ. Press)
Hellström, C., & Mikkola, S. 2010, CeMDA, 106, 143
Hénon, M. 1961, AnAp, 24, 369
Hénon, M. 1972, IAU Colloq. 10: Gravitational N-Body Problem ed. M. Lecar (Dordrecht: Reidel) 406
Hénon, M. 1975, IAU Symp. 69, Dynamics of the Solar Systems ed. A. Hayli (Dordrecht: Reidel) 133
Hénon, M. H. 1971, Ap&SS, 14, 151
Hills, J. G. 1975, AJ, 80, 809
Hockney, R. W., & Eastwood, J. W. 1988, Computer Simulation Using Particles (Bristol: Hilger)
Hulse, R. A., & Taylor, J. H. 1975, ApJL, 195, L51
Hut, P., Makino, J., & McMillan, S. 1995, ApJL, 443, L93
Inagaki, S., & Saslaw, W. C. 1985, ApJ, 292, 339
Iwasawa, M., Portegies Zwart, S., & Makino, J. 2015, ComAC, 2, 6
Iwasawa, M., Oshino, S., Fujii, M. S., & Hori, Y. 2017, PASJ, 69, 81
Jänes, J., Pelupessy, I., & Portegies Zwart, S. 2014, A&A, 570, A20
Jouzel, J., Genthon, C., Lorius, C., Petit, J. R., & Barkov, N. I. 1987, Natur, 329, 403
Kahan, W. 1965, Commun. ACM, 8, 40
King, I. R. 1966, AJ, 71, 64
Kozai, Y. 1962, AJ, 67, 591
Kuijken, K., & Dubinski, J. 1995, MNRAS, 277, 1341
Laskar, J., & Robutel, P. 2001, CeMDA, 80, 39
Lee, H. M., & Ostriker, J. P. 1987, ApJ, 322, 123
Lidov, M. L. 1962, Planet. Space Sci., 9, 719
Liu, L., Wu, X., Huang, G., & Liu, F. 2016, MNRAS, 459, 1968
Lyapunov, A. M. 1892, Commun. Math. Soc. Krakow, 2, 1
Makino, J. 1991, ApJ, 369, 200
Makino, J. 1996, ApJ, 471, 796
Makino, J., & Aarseth, S. J. 1992, PASJ, 44, 141
Mardling, R. A., & Aarseth, S. J. 2001, MNRAS, 321, 398
Maxwell, J. C. 1860, PMag, 19, 19
McMillan, S. L. W., & Aarseth, S. J. 1993, ApJ, 414, 200
Mikkola, S. 1983, MNRAS, 203, 1107
Mikkola, S., & Merritt, D. 2008, AJ, 135, 2398
Milankovitch, M. 1941, Zavod za Udzbenike i Nastavna Sredstva: Kanon der Erdbestrahlung und seine Andwendung auf das Eiszeiten-problem (Belgrade: Königlich Serbische Akademie) (English Translation, 1998, Canon of Insolation and the Ice Age Problem)
Milgrom, M. 1983, ApJ, 270, 365
Montgomery, R. 1998, Nonli, 11, 363
Naoz, S. 2016, ARA&A, 54, 441
Newton, I. 1687, Philosophiae Naturalis Principia Mathematica **(Vol 1,; London: Royal Society)**
Nitadori, K., & Makino, J. 2008, NewA, 13, 498
Oshino, S., Funato, Y., & Makino, J. 2011, PASJ, 63, 881
Peters, P. C. 1964, PhRv, 136, 1224
Petit, J. R., Jouzel, J., Raynaud, D., et al. 1999, Natur, 399, 429
Plummer, H. C. 1911, MNRAS, 71, 460

Poisson, E., & Will, C. M. 2014, Gravity (Cambridge: Cambridge Univ. Press)
Portegies Zwart, S. F., Boekholt, T. C. N., Por, E. H., Hamers, A. S., & McMillan, S. L. W. 2022, A&A, 659, A86
Portegies Zwart, S. F., & Boekholt, T. C. N. 2018, CNSNS, 61, 160
Portegies Zwart, S. F., & McMillan, S. L. W. 2002, ApJ, 576, 899
Portegies Zwart, S. F., Belleman, R. G., & Geldof, P. M. 2007, NewA, 12, 641
Press, W. H., Teukolsky, S. A., Vetterling, W. T., & Flannery, B. P. 1992, Numerical Recipes in C: The Art of Scientific Computing (Cambridge: Cambridge Univ. Press)
Rauch, K. P., & Tremaine, S. 1996, NewA, 1, 149
Rein, H., & Liu, S. F. 2012, A&A, 537, A128
Rein, H., & Liu, S.-F. 2011, REBOUND: Multi-purpose N-body Code for Collisional Dynamics, Astrophysics Source Code Library, ascl:1110.016
Rodriguez, C. L., Morscher, M., Wang, L., et al. 2016, MNRAS, 463, 2109
Saha, P., & Tremaine, S. 1992, AJ, 104, 1633
Salpeter, E. E. 1955, ApJ, 121, 161
Spera, M., Capuzzo Dolcetta, R., & Punzo, D. 2012, HiGPUs: Hermite's N-body Integrator Running on Graphic Processing Units, Astrophysics Source Code Library, ascl:1207.002
Spinnato, P. F., Fellhauer, M., & Portegies Zwart, S. F. 2003, MNRAS, 344, 22
Spitzer, L. 1969, ApJL, 158, L139
Spitzer, L. 1987, Dynamical Evolution of Globular Clusters (Princeton, NJ: Princeton Univ. Press)
Springel, V. 2000, GADGET-2: A Code for Cosmological Simulations of Structure Formation, Astrophysics Source Code Library, ascl:0003.001
Springel, V. 2005, MNRAS, 364, 1105
Stock, K., Cai, M. X., Spurzem, R., Kouwenhoven, M. B. N., & Portegies Zwart, S. 2020, MNRAS, 497, 1807
Takahashi, K., & Portegies Zwart, S. F. 1998, ApJL, 503, L49
Taylor, B. 1715, Methodus Incrementorum Directa et Inversa (London: Typis Pearsonianis)
Tremaine, S. 2015, ApJ, 807, 157
Trenti, M., & van der Marel, R. 2013, MNRAS, 435, 3272
van Elteren, A., Portegies Zwart, S., Pelupessy, I., Cai, M. X., & McMillan, S. L. W. 2019, A&A, 624, A120
Verlet, L. 1967, PhRv, 159, 98
Verlinde, E. 2017, ScPP, 2, 016
von Zeipel, H. 1910, AN, 183, 345
Wang, L., Spurzem, R., Aarseth, S., et al. 2015, MNRAS, 450, 4070
Wang, L., Spurzem, R., Aarseth, S., et al. 2016, MNRAS, 458, 1450
Whitehead, A. J., McMillan, S. L. W., Vesperini, E., & Portegies Zwart, S. 2013, ApJ, 778, 118
Wisdom, J., & Holman, M. 1991, AJ, 102, 1528
Yoshida, H. 1993, CeMDA, 56, 27
Zemp, M. 2014, Halogen: Multimass Spherical Structure Models for N-Body Simulations, Astrophysics Source Code Library, ascl:1407.020
Zemp, M., Moore, B., Stadel, J., Carollo, C. M., & Madau, P. 2008, MNRAS, 386, 1543

AAS | IOP Astronomy

Astrophysical Recipes (Second Edition)

The art of AMUSE

Simon Portegies Zwart, Steve McMillan and Steven Rieder

Chapter 5

Stellar Structure and Evolution

> How often have I said to you that when you have eliminated the impossible, whatever remains, however improbable, must be the truth?
>
> Sir Arthur Conan Doyle

Stars evolve. From the moment of birth, a star's internal structure and appearance change with time. Some of these changes occur very slowly and hardly affect the local environment, but others are swift and have far-reaching repercussions. In this chapter, we review the basic evolution of stars and show how relatively simple stellar-evolution calculations can be conducted.

5.1 In a Nutshell

Stellar evolution is a very different game from Newtonian gravity. Whereas the physical equations underlying gravitational dynamics can be described and solved using basically high-school mathematics, solving the equations for stellar structure and evolution is challenging even for experts. In this chapter, we present an overview of stellar structure and evolution, although it will barely be enough to give an appreciation of the complexities of stellar-evolution codes. We start by reviewing some fundamental timescales in stellar evolution, followed by a rudimentary review of the structure equations and some prescriptions for stellar evolution. More extensive discussion can be found in Kippenhahn & Weigert (1967) and Eggleton (2006).

A star is a gravity-confined nuclear fusion reactor (Bethe 1939). Stars are born when parts of a molecular cloud are compressed by turbulence or other external forces (such as supernovae), collapse under their own weight, and fragment into a collection of prestellar clumps. These clumps continue to contract and accrete, and their central temperatures increase. Some of the excess heat is radiated away, but some of it goes to increase the pressure, ultimately halting the gravitational contraction. These hydrodynamical

doi:10.1088/2514-3433/adfc3cch5

processes are discussed in more detail in Chapter 6. A clump becomes a star when its central temperature and density reaches the point where nuclear fusion can start. This moment is generally associated with the birth of the star. At this point, the star is said to lie on the *zero-age main sequence* (ZAMS). From then on, the star evolves, as nuclear fusion in its core[1] drives irreversible structural change.

Figure 5.1 shows images of two red supergiant stars, Antares (Ohnaka et al. 2017) and Betelgeuse (Haubois et al. 2009), showing their complex surface temperature structure. Such observations have only recently become possible. By now, apart from the Sun (see Figure 5.5), a total of 26 stars have been resolved. A more theoretically oriented view of the internal structure of three stars of different masses is presented in Figure 5.2.

During its lifetime, a star experiences several evolutionary phases, each associated with distinct burning regimes in the central nuclear furnace. This evolution is often presented as a trajectory in temperature–luminosity space, in what is called a Hertzsprung–Russell (H–R) diagram. Figure 5.3 is an H–R diagram illustrating the evolution of three stars of different masses. Colors indicate various stellar-evolutionary stages. Very roughly speaking, we can divide the evolution of a star into three distinct phases: the main sequence, the post-main sequence (or giant phase), and the remnant phase, when the star has turned into a compact object. Along the main sequence, stars are generally comparable in size to the Sun. Giants can be much larger—as large as several astronomical units (see Figure 5.1). Remnants are much smaller than the Sun—comparable to or much smaller than Earth.

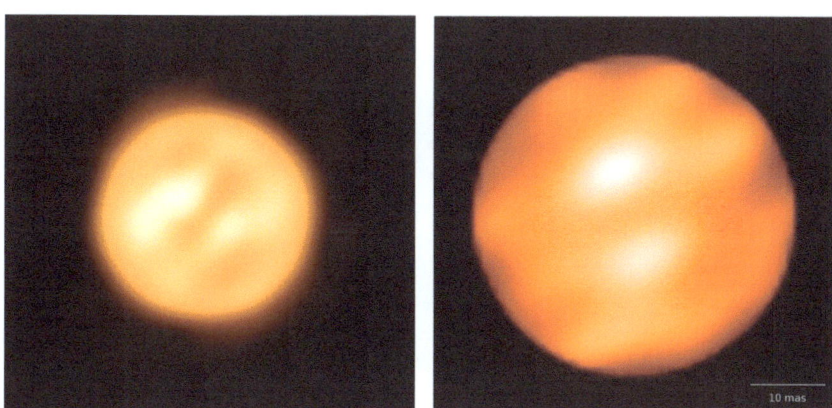

Figure 5.1. Stars are finite objects with structure on their surface, as demonstrated here with the observations of Antares (left, image with the ESA VLTI; image credit: ESA/Ohnaka et al. 2017) and Betelgeuse (right, image by Observatoire de Paris; image credit: NASA/Haubois et al. 2009). Both stars are similar in size—about 4.1–5.5 au (Dolan et al. 2016).

[1] The stellar core is the central portion of the star where nuclear fusion occurs. The transition from fusion to nonfusion is typically quite sharp; see Figure 5.4.

Figure 5.2. Schematic structure of a low-mass star (left), an intermediate-mass star (middle), and a high-mass star (right). They are not to scale, but the various colored regions in the stellar interior indicate the nuclear-burning region (blue), the convective zone (red), and the radiation-dominated region (yellow).

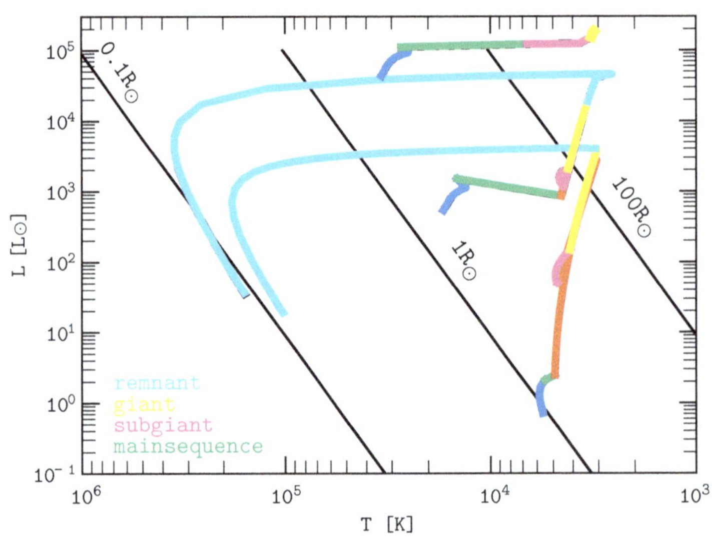

Figure 5.3. H–R diagrams (luminosity versus temperature) for three stars of masses 1 $M_\odot$ (bottom set curves), 5 $M_\odot$, and 20 $M_\odot$ (top curves), from the ZAMS to the moments they turn into compact remnants. The colored text in the bottom left identifies the evolutionary stage of the star along its track (with the caveat that a star at the tip of the asymptotic giant branch should probably not be identified as a remnant). The black diagonal lines indicate the run of luminosity and temperature for blackbodies with radii 0.01 $R_\odot$ (left), 1 $R_\odot$, and 100 $R_\odot$ (rightmost line). The figure was made using the script `amuse.examples.plot_stellar_evolution_track`.

To guide the eye in Figure 5.3, we plot lines of constant stellar radius r. The radius can be estimated from the star's effective temperature ($T_{\rm eff} \approx T$) and luminosity by approximating the stellar spectrum as a blackbody, resulting in

$$L = 4\pi r^2 \sigma T_{\rm eff}^4, \tag{5.1}$$

where σ is the Stefan–Boltzmann constant. This is a considerable simplification because the stellar spectrum is generally not described particularly well as a blackbody, but it is a reasonable first approximation to the true state of the star.

5.1.1 Stellar Timescales

Three fundamental timescales are relevant for stellar evolution. These are the *dynamical timescale* $t_{\rm dyn}$, the *Kelvin–Helmholtz timescale* $t_{\rm KH}$, and the *nuclear timescale* $t_{\rm nuc}$ (Bradt 2014). They are timescales on which distinctly different physical processes operate.

5.1.1.1 The Dynamical Timescale

The stellar analog to the dynamical timescale is defined analogously to the gravitational dynamical timescale discussed in Section 4.1.2. For a star of mass m and radius r, the dynamical timescale is

$$t_{\rm dyn} = \sqrt{\frac{r^3}{Gm}}. \tag{5.2}$$

This is the dynamical (orbital) timescale for a particle orbiting at the stellar surface, but it is also (approximately) the time taken for a sound wave to cross the star. The dynamical timescale for the Sun (`m = 1 | units.MSun` and `r = 1 | units.RSun`) can be calculated as follows:

```
import numpy as np
from amuse.units import units
from amuse.units import constants

def dynamical_time_scale(mass, radius, G=constants.G):
    return np.sqrt(radius**3/(G * mass))

print(
    f"solar dynamical time scale = "
    f"{dynamical_time_scale(1 | units.MSun, 1 | units.RSun).in_(units.minute)}"
)
```

which turns out to be about $t_{\rm dyn} \simeq 27$ minutes. If a comet hits the Sun, the other side of the star learns about the impact in about half an hour.

The related *convective timescale* $t_{\rm conv}$ is the timescale on which material convectively bubbles up from the stellar interior toward the surface. In convection, the stellar luminosity is carried outward by physical motion of the stellar gas rather than by radiation. It can be characterized by the mean distance Δr traveled by a typical rising gas parcel (usually assumed to be related to the local pressure scale height $P/|\nabla P|$) and the mean speed $\bar{v}$ with which the parcel rises:

$$t_{\rm conv} = \frac{\Delta r}{\bar{v}}. \tag{5.3}$$

For the Sun, $\Delta r \simeq 0.1 R_\odot$ and $\bar{v} \simeq 11$ m s^{-1}. We then obtain $t_{\rm conv} \simeq 0.2$ years.

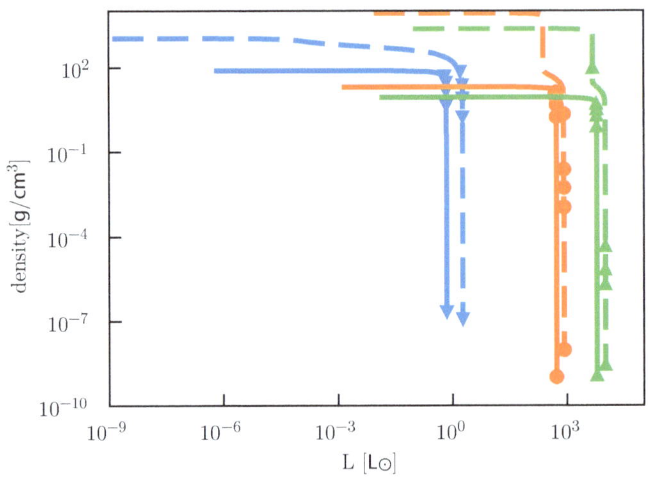

Figure 5.4. Density as a function of total luminosity in the stellar interior, showing the difference between the stellar core and envelope. Results are shown for three stellar masses: 1 $M_{\odot}$ (blue), 5 $M_{\odot}$ (red), and 10 $M_{\odot}$ (yellow). The solid curves are for stellar ages of 1 Myr; the dashed curves are for 11 Gyr, 90 Myr, and 20 Myr, for the 1 $M_{\odot}$, the 5 $M_{\odot}$, and 10 $M_{\odot}$ stars, respectively. Bullets and triangles are drawn at 20% intervals of the enclosed mass of the star. The script that produced this figure is `amuse.examples.plot_stellar_-structure`. It should take a couple of minutes to run on a laptop.

5.1.1.2 The Thermal Timescale

The Kelvin–Helmholtz (or thermal) timescale is the timescale on which thermal energy leaves a star. It can be written as

$$t_{\rm KH} = \frac{E_{\rm th}}{L}, \tag{5.4}$$

where $E_{\rm th}$ is the star's total thermal energy and L is its luminosity, the total energy leaving the stellar surface per unit time.

Like star clusters, stars in equilibrium (i.e., having no bulk motion) also are governed by the virial theorem [see Equation (5.5)], with the total thermal energy E_{th} replacing the total kinetic energy T (W. J. M. Rankine 1853)[2]

$$2E_{\rm th} + U = 0. \tag{5.5}$$

Thus, neglecting (as usual) extraneous factors of order unity, we can write the Kelvin–Helmholtz timescale as

$$t_{\rm KH} = \frac{Gm^2}{rL}. \tag{5.6}$$

[2] W. J. Macquorn Rankine already stated in 1850 "*that the sum of all the energies of the universe, actual and potential, is unchangeable.*" https://archive.org/details/miscellaneoussci00rank/page/202/mode/1up?view=theater.

For the Sun (`m = 1 | units.MSun`, `r = 1 | units.RSun` and `L = 1 | units.LSun`), we have

```
def Kelvin_Helmholtz_timescale(mass, radius, luminosity, G=constants.G):
    return G * mass**2 / (radius * luminosity)
```

in which case $t_{\mathrm{KH}} \simeq 31$ Myr.

5.1.1.3 Nuclear Fusion and the Nuclear Timescale

The collapse of a protostellar clump toward the main sequence proceeds on a Kelvin–Helmholtz timescale. During its slow contraction, a star follows the pre-main sequence or *Hayashi track* (Hayashi 1961). During this period, the star decreases in luminosity but stays at roughly the same surface temperature. The central temperature and pressure increase until nuclear fusion begins to contribute considerably to the total energy budget. The central density and temperature then exceed $\rho_c \gtrsim 7$ g cm^{-3} and $T_c \gtrsim 10^7$ K.

On the main sequence, where the star spends most of its lifetime, hydrogen (H) fuses into helium (He) through a variety of channels that may be summarized schematically by

$$4(^1\mathrm{H}) \rightarrow {}^4\mathrm{He} + 2\nu_e + 26.7\ \mathrm{MeV}, \tag{5.7}$$

where ν_e represents an electron neutrino that streams out of the core, through the star, and into interplanetary space. The 26.7 MeV represents the net energy liberated as a result of this reaction. Overall, a fraction $\epsilon \simeq 0.7\%$ of the total nuclear mass is converted into energy in the form of gamma-rays. That energy ultimately contributes to the star's luminosity. Assuming that the innermost 10% of the star's mass is fusing, and assuming the above efficiency, we can write a timescale for nuclear fusion in a main-sequence star:

$$t_{\mathrm{nuc}} = 7 \times 10^{-3}\,\frac{0.1mc^2}{L} = 7 \times 10^{-4}\,\frac{mc^2}{L}. \tag{5.8}$$

In AMUSE, a simple estimate of the nuclear timescale for the Sun (`m = 1 | units.MSun` and `L = 1 | units.LSun`) gives

```
def nuclear_time_scale(mass,  luminosity, c=constants.c):
    return 7e-4 * mass * c**2 / luminosity
```

From now on, we will leave out the obvious imports of `Python` packages. This shows that the Sun has $t_{\mathrm{nuc}} \simeq 10$ Gyr. This timescale is the Sun's approximate nuclear-burning lifetime. More careful calculations yield 12.462 ± 0.002 Gyr (based on calculations you can do yourself in Section 5.5).

Once the hydrogen in the stellar core has fused into helium, a star starts to fuse helium into carbon (C):

$$3^4\mathrm{He} \rightarrow^{12} \mathrm{C} + 7.3\ \mathrm{MeV} \tag{5.9}$$

(the reaction is somewhat more complicated than indicated here, but this is the outcome). Subsequent fusion stages form oxygen (O) and heavier elements (such as Neon, Ne) by alpha capture:

$$^{4}\text{He} +^{12}\text{C} \rightarrow {}^{16}\text{O} + 7.2\ \text{MeV}, \tag{5.10}$$

$$^{4}\text{He} +^{16}\text{O} \rightarrow {}^{20}\text{Ne} + 4.7\ \text{MeV}, \tag{5.11}$$

etc. This progression is sometimes called the *alpha ladder*. How far reactions proceed depends on the mass of the star. The Sun will never make it past oxygen, but the most massive stars will keep producing heavier elements up to iron and nickel. Fusion eventually stops at nickel (^{56}Ni), after which the reaction becomes endothermic: ^{56}Ni does not produce energy when fused.

Figure 5.4 presents the luminosity generated in the stellar interior as a function of time for three stellar masses and at two different moments in time. In each of these cases, the core is easily seen in the sudden drops in density and luminosity.

5.1.2 Physics of the Interior

Unlike the case with self-gravitating systems, we cannot easily solve the fundamental physical equations governing stellar structure and evolution. That would require us to resolve detailed nuclear reaction networks, the ways in which matter and energy flow around the star, and the details of how radiation interacts with matter.

The cross sections for interactions between radiation and matter are in fact quite well known, and nuclear reaction rates can in principle be measured in the laboratory. The reaction rates used in stellar-evolution calculations are generally extrapolated from laboratory measurements made at higher temperatures and much lower densities. The fusion reactor ITER[3] will have an operational temperature of 1.5×10^8 K and an electron number density of $n_e \sim 10^{18} - 10^{21}$ m^{-3}, corresponding to an equivalent hydrogen density of $\sim 10^{-12} - 10^{-9}$ g cm^{-3}. For comparison, the temperature and density of the solar core are roughly 1.5×10^7 K and 150 g cm^{-3}, respectively.

5.1.2.1 Underlying Assumptions in Stellar-evolution Models

Like many gravity solvers, stellar-evolution codes make a number of simplifying assumptions. Some are hidden in the mathematical or numerical models, while others are more explicit. Instead of solving for the microphysics, we generally solve for the macroscopic state of the star, which can only be realized by neglecting several aspects of the physics. The neglect of these effects does not (seem to) affect the zeroth-order calculation for most stars, and to some extent can be included as a first-order correction. Here, we list some of the most important assumptions.

1. **Spherical symmetry**. The assumption of spherical symmetry is a sweeping approximation that reduces the stellar structure problem from a complex three-dimensional set of partial differential equations to a one-dimensional set of ordinary differential equations. In the process, we lose the possibility of

[3] See https://www.iter.org

including stellar rotation, internal magnetic activity, and any other nonradial internal motion. In this approximation, the star is viewed as a series of one-dimensional arrays of macroscopic parameters, such as density, temperature, mean molecular weight, etc., stored in radial bins running from the center to the surface.

This greatly simplifies the calculation, but most actual stars are observed to have magnetic fields of varying strengths, do rotate, and exhibit a wealth of nonspherical features. One only has to look at an image of the Sun, with its many sunspots and prominences, to realize that spherical symmetry is only an approximation (see Figure 5.5). (On the other hand, solar activity—even at its maximum—is only a tiny fraction of the Sun's energy budget).

2. **Local thermodynamic equilibrium**. Radiation plays an important role in transporting energy through a star. Let us temporarily ignore the possibility of convection (see below), and assume that all energy generated in the stellar core moves through the stellar interior in the form of radiation. If the total luminosity within a sphere of radius r is $L(r)$, then the energy flux across radius r is

$$F(r) = \frac{L(r)}{4\pi r^2}. \tag{5.12}$$

In equilibrium, all the luminosity produced inside the star eventually reaches the surface and escapes into space.

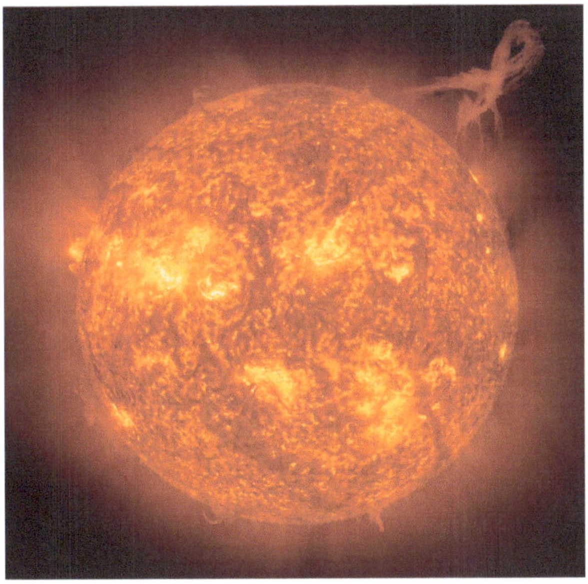

Figure 5.5. Surface image of the Sun (image from the SOHO mission at EIT 304 Å on 2000 January 18). The light and dark orange patches (as opposed to the yellow areas, which are regions of solar activity) show granulation, the result of temperature variations due to convection just below the surface. Courtesy of SOHO/[EIT] consortium. SOHO is a project of international cooperation between ESA and NASA.

The outward energy flux F is driven by a temperature gradient within the star. Basically, if the temperature decreases with radius, then at any point in the star, there will be slightly more photons moving outward from within than moving inward from outside. However, a system in thermodynamic equilibrium would have constant temperature throughout, and therefore cannot have temperature gradients. Therefore, a star cannot be in perfect thermodynamic equilibrium. Instead, we must assume that the star is very close to thermodynamic equilibrium at any radius, but there is a slight radial temperature gradient driving the outward flux. This assumption is called *local thermodynamic equilibrium.*

The mean free path of a photon is much smaller than the stellar radius, so each photon interacts frequently with local stellar material on its way out. This supports the assumption that the gas and the radiation are in local thermodynamic equilibrium: at any given point within the star, the radiation and the gas have the same temperature, even though the temperature decreases with radius. Note that these assumptions do not necessarily apply in other circumstances, such as in young neutron stars or black holes.

3. **Convection and mixing**. Radiation is not the only energy transport mechanism operating in most stars. Convection is another important mechanism that plays a vital role in determining stellar structure and evolution. Stars of very low mass tend to be fully convective, the Sun has a convective envelope, and massive stars tend to be convective in their central regions. The imprint of convection on the surface of the Sun is illustrated in Figure 5.2, and is nicely visible in Figure 5.5.

When the energy flux through part of a star cannot be transported efficiently by radiation (e.g., when the flux becomes very large, or in a region where the opacity is very high), the energy is instead transported by convection—turbulent motion in the stellar interior. Stellar material is heated, rises, cools, and sinks back, forming convective cells that clearly cannot be properly described in a one-dimensional spherical stellar model: How can two parcels of gas pass one another on a line?

Turbulence gives rise to many numerical complications and is not realistically resolved in current state-of-the-art one-dimensional stellar structure and evolution models. Rather, it is handled in an ad hoc fashion, assuming that convective velocities are sufficient to carry the required energy flux and that the radial extent of a convective cell is a fixed fraction of the pressure scale height, $\Delta r = \alpha P/|dP/dr|$, where α is a parameter that can be adjusted to calibrate models against the Sun. It is generally assumed that any region of a star that becomes convectively unstable is mixed by the convection process.

Convection is particularly important in the earliest stellar-evolution phases (e.g., along the Hayashi track before the star reaches the ZAMS) and in much later phases (e.g., when the star reaches its maximum radius along the asymptotic giant branch; AGB), but some regions of most stars will become convectively unstable at some time during their evolution.

Magnetic fields are generally considered to have a relatively small effect on the overall structure and evolution of most stars (even in white dwarfs or neutron stars), and are therefore generally neglected. If some form of magnetism is included in the stellar-evolution code, it often follows from perturbations or through simplified models rather than full magnetohydrodynamic (MHD) calculations (see, e.g., Section 10.2.1, where we adopt a MHD model for a collapsing giant molecular cloud). Some stellar-evolution codes include such equations for the evolution of toroidal and poloidal magnetic field components derived from mean-field MHD (Takahashi & Langer 2021). These implementations often use Alfvén's theorem to formulate a conservative form of angular momentum transfer due to the Lorentz force.

Stellar material can be efficiently redistributed by convection. For example, nuclear-processed elements can be transported to the stellar surface, or fresh hydrogen from the stellar envelope can be brought into the core. The latter process is generally identified with convective overshoot (where the momentum of the convected material carries it beyond the unstable region, into the overlying stable gas) and has the effect of prolonging the lifetime of a star.

4. **Mass loss by stellar winds**. In addition to radiating energy into space, stars also lose mass. The amount of mass lost is very hard to calculate from first principles. Consequently, stellar winds are generally implemented as ad hoc prescriptions calibrated against observations (which are also uncertain). One problem here is that the stars that lose the most mass generally have the shortest lifetimes. As a result, the number of examples against which approximate methods can be calibrated is quite limited. Our imperfect theoretical understanding of the physics and the expense of a detailed numerical prescription means that mass loss in stellar-evolution codes is often parameterized, rather than solved explicitly.
5. **Equation of state**. Although it is not, strictly speaking, an assumption, any calculation of stellar structure requires an equation of state to relate the gas pressure (P) to the temperature (T) and density (ρ).

In many cases, an ideal gas equation of state may be used. This relation is simple: at sufficiently high temperatures, the gas is fully ionized and is described well as an ideal gas, in which case (Clapeyron 1834; Clausius 1857)

$$P = nkT = \frac{\rho kT}{\mu}. \tag{5.13}$$

Here, n is the number of particles per unit volume, k is the Boltzmann constant (not to be confused with the Stefan–Boltzmann indicated with σ in Equation (5.1)), and μ is the mean molecular weight. For a fully ionized gas with hydrogen fraction (by mass) X, helium fraction Y, and heavy element fraction $Z = 1 - X - Y$, because each hydrogen atom contributes one electron, each helium atom contributes two electrons, and (we assume) each heavy atom of atomic weight A contributes on average $A/2$ electrons, the total number density is

$$n = \frac{\rho}{m_H}\left(2X + \frac{3}{4}Y + \frac{1}{2}Z\right), \tag{5.14}$$

where $m_{\rm H}$ is the mass of a hydrogen atom. Setting $Z = 1 - X - Y$, we find

$$\mu = 2m_{\rm H}\left(1 + 3X + \frac{1}{2}Y\right)^{-1}. \quad (5.15)$$

For solar ZAMS composition, $X = 0.7$, $Z = 0.02$, and $\mu \simeq 0.62m_{\rm H}$.

Other pressure terms may be important in some circumstances. In hot stars, radiation pressure must be taken into account, so the expression for the pressure becomes

$$P = \frac{\rho kT}{\mu} + \frac{1}{3}aT^4, \quad (5.16)$$

where $a \equiv 4\sigma/c$ is the radiation constant. This equation of state is a good description of most of the interiors of most stars for most of their lives. At very high densities, quantum mechanical electron degeneracy pressure may dominate:

$$P_e = \left(\frac{3}{8\pi}\right)^{2/3} \frac{h^2}{5m_e} n_e^{5/3}. \quad (5.17)$$

Here, n_e is the electron number density and m_e is the electron mass. Degeneracy pressure is particularly important in the cores of evolved stars and remnants. Finally, near the stellar surface, the gas may be only partly ionized and the ionization fraction of each species must be computed from the Saha–Langmuir equation (Saha 1921; Kingdon & Langmuir 1923). In addition, molecules may form, further complicating the calculation of μ.

5.1.3 Final Stages of Stellar Evolution

A star continually fights its own gravitational pull, but gravity inevitably wins, sooner or later causing the central parts of the star to turn into a compact remnant—a white dwarf, neutron star, or black hole.

In low- and intermediate-mass stars—those with masses at the end of the giant branch of less than about 8 $M_\odot$—the core grows until it reaches a state where electron degeneracy pressure prevents further contraction, and fusion ends somewhere along the alpha ladder (see Section 5.1.1.3). How far the star ascends the alpha ladder depends on the temperature and density in the stellar interior. Figure 5.6 shows the central temperature as a function of the core density. Stars are born in the lower left corner, at low temperature and density. In this regime, the star fuses hydrogen (see Section 5.7). As the star ages, its central density and temperature increase, allowing it to ascend the ladder. Only the most massive stars reach the top rung, fusing ^{52}Fe to ^{56}Ni. The portion of the script that carries out the evolutionary calculation is shown in Code example 5.1.

The core density and temperature of the most massive star in Figure 5.6 grow steadily, whereas the stars of lower mass make some interesting excursions during their evolution. The differences between these tracks are caused by the various energy transport mechanisms and changes in the equation of state in the stellar

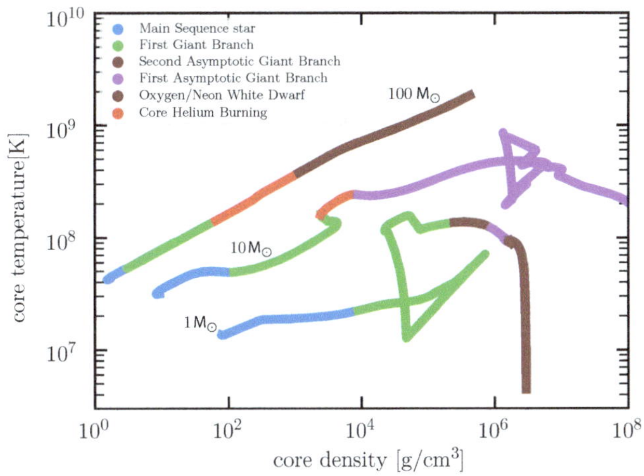

Figure 5.6. Central temperature as a function of core density for a 1 $M_\odot$ (bottom), a 10 $M_\odot$ (middle), and a 100 $M_\odot$ star (top) of solar metallicity. The colors identify various evolutionary phases (see legend), starting with the main sequence at the lower left to giants and white dwarfs to the right. The code to produce this plot can be found in `amuse.examples.stellar.calculate_core_temperature_density` (see Code example 5.1). To plot the results, it may be more practical to use the precalculated data in the simple plotting script `amuse.examples.plot_core_temperature_density`, because the Henyey stellar-evolution code `MESA` (see Table 5.1) used to generate this figure takes a few hours for each single star. In particular, the 1 $MM_\odot$ star tends to be slow along the AGB just before the green part of the curve; magnifying this part of the evolution will reveal interesting sinusoidal variations in the core density and core temperature of the star.

```
    while not stellar_remnant(stellar):

        star.evolve_one_step()

        nzones = star.get_number_of_zones()
        rhoc = star.get_density_profile(nzones)[0]
        Tc = star.get_temperature_profile(nzones)[0]

        time.append(stellar.model_time)
        rho_core.append(rhoc)
        T_core.append(Tc)
        stellar_type.append(star.stellar_type)
        si = star.stellar_type.value_in(units.stellar_type)

        with open(filename, "wb") as file:
            pickle.dump([time, rho_core, T_core, stellar_type], file)

```

Code example 5.1: Calculating the core temperature and density of a stellar model (used to make Figure 5.6). The full script can be found in `amuse.examples.calculate_core_temperature_density`.

interior. For the Sun and other low-mass stars, the alpha ladder stops when carbon and helium fuse to form oxygen; for more massive stars, it ends at oxygen and/or neon (see Equation (5.10)). Eventually, nuclear fusion stops because (due to electron degeneracy) the contraction of the stellar core cannot produce the density and temperature required to reach the next rung of the alpha ladder. As the star runs out of fuel, instabilities arise and propagate to the outer layers, eventually ejecting them into space in the form of a dense stellar wind. This final phase is called the post-AGB. The leftover cinder of processed material becomes a white dwarf, the composition of which depends on how far the star ascended the alpha ladder.

The evolution of a higher-mass star ends much more violently. Once all available fuel is spent, and the stellar core is mainly composed of iron-group (Scandium ^{45}Sc to Nickel ^{56}Ni) elements, further fusion is not possible. As the core mass grows due to nuclear fusion in the overlying layers, the only remaining route is collapse. The resulting phenomenon is called a supernova, although this term is also often applied to the core-collapse process. A wide variety of supernovae have been identified observationally. Broadly speaking, we can recognize two basic types, II and Ib/c, but many subcategories have been defined (Smartt 2009). To make a rough distinction, type II supernovae are the result of core collapse in a massive hydrogen-rich star, as just described, leading to the formation of a neutron star or black hole (Chiu 1964). Type Ib/c supernovae result from the collapse of helium-rich (Wolf–Rayet) or stripped stars and may lead to black hole formation, although this remains uncertain (Georgy et al. 2009; Langer 2012; Laplace et al. 2020). There is a hypothetical class of homogeneously evolving stars (Laplace et al. 2020); though never observed, they are an interesting potential aberration of stellar-evolution calculations. Type Ia supernovae may mark the collapse of a massive white dwarf, possibly formed via a white-dwarf merger, or through accretion from a companion star.

Supernovae are not necessarily symmetrical, and the compact object may receive a velocity kick as a result of the explosion. This is often referred to as a *natal kick* of the neutron star or black hole in a type Ib/c or type II supernova. These velocity kicks are uncertain, but may be as high as 1000 km s^{-1} (Helfand & Tademaru 1977; Lyne & Lorimer 1994; Kapil et al. 2023).

5.2 Simulating Stellar Evolution

The most detailed stellar-evolution codes solve the underlying (one-dimensional) differential equations for stellar structure and evolution (for details, see Eggleton 2006). They advance the nuclear burning and stellar structure equations at every time step, and by taking energy transport and nuclear energy generation into account, they converge to a time-dependent solution. These methods are accurate, within the simplifying assumptions listed earlier, but relatively slow. As a result, a second class of much more rapid methods has appeared. They use fits to the more detailed results, generally expressed in the form of look-up tables (Meynet et al. 1994; Meynet & Maeder 2005; Brott et al. 2011; Köhler et al. 2015), fitting formulae (Eggleton et al. 1989; Hurley et al. 2000), or other parameterizations (Varshavskii & Tutukov 1975; Tutukov & Fedorova 2001; Spera et al. 2015). We refer to the first class of stellar-evolution codes as direct solvers. The second class, we call parametric stellar-evolution prescriptions; both are available in AMUSE.

Direct stellar-evolution solvers use so-called Henyey stellar-evolution models (Henyey et al. 1959, 1964) providing a one-dimensional representation of a star using radial distributions of stellar mass, density, composition, etc. Hybrid Henyey codes exist, in which the basic differential equations are expanded to include atmosphere models, rotation, and some form of convective overshoot and mixing, but those models are, strictly speaking, still one-dimensional (Schaerer et al. 1996; Stasińska & Schaerer 1997). Research on true three-dimensional stellar-evolution code is still very limited (Bazán et al. 2003; Rizzuti et al. 2023).

Parametric stellar-evolution methods come in two varieties: those adopting functional forms of fits to earlier calculated stellar-evolution tracks, and look-up tables. Fitting formulae adopt polynomials with mass and sometimes composition—typically expressed as metallicity—as free parameters (see Section 5.2.3), and yield the stellar radius, temperature, and luminosity as functions of time since the ZAMS. Look-up tables tend to interpolate between tabulated values for stellar mass, radius, luminosity, etc., as a function of time and metallicity. Parametric stellar-evolution models generally have very limited internal structure information. Some include some notion of a core–envelope structure, and possibly additional parameters such as the radius of gyration, but their structural content is minimal.

Look-up parametric stellar-evolution methods (indicated in Table 5.1 as *parametric*) generally interpolate from precomputed tables of stellar-evolution tracks (mass, radius, temperature, luminosity, etc.). In some sense, the look-up and parametric stellar-evolution codes are equivalent, but they differ in their computational requirements in terms of both memory management and CPU performance. In general, however, both are much faster than Henyey codes. The evolution of a 1 $M_\odot$ star from ZAMS until it leaves the AGB takes about a second for a parametric code and up to a day for a Henyey code.

5.2.1 Initial Conditions for Stars

A star is born (hydrogen fusion starts) with a certain mass and metallicity, at which point we call it a ZAMS star. For a parameterized stellar-evolution code (see Section 5.2), the specification stops there; the future evolution of the star is completely determined via fitting formulae and/or tables.

For a Henyey code (see Section 5.2) these parameters generally also describe the star, but a number of additional parameters must also be specified. In this type of code, the nuclear reaction network, the hydrodynamics, and the radiative transport are generally solved in a number of shells—a star in a Henyey code is layered like an onion (Steig 2008). One of the parameters in such a code is the number of layers (onion shells) to use in the code's description of stellar structure. In general, we must specify the run of temperature, density, etc. throughout the (one-dimensional, spherically symmetric) star. The `MESA`, `EVTwin`, and `GENEC` stellar-evolution codes can generate initial stellar models based on mass and metallicity, but we can also have these codes start from completely arbitrary structures, under our control. It is possible, for example, to start with a pre-main-sequence star, in which case the evolution starts somewhere on the Hayashi track (see Chapter 5), or with a giant planet. Another interesting initial stellar model is the hydrogen-exhausted core of a giant without an envelope, which could be interpreted as a zero-age helium

Table 5.1. Stellar-evolution Codes Incorporated in AMUSE

Name	Ref	Type of Code	Language	Rotation Evolution	Binary Structure	Internal	Parallel
MESA	1	Henyey	F95	Yes	Yes	Yes	Yes
EVTwin	2	Henyey	F77	No	Yes	Yes	No
GENEC	3	Henyey	F90	Yes	No	No	No
SSE/Bse	4	Parametric	F77	No	Yes	No	No
SeBa	5	Parametric	C++	No	Yes	No	Yes
Toy	6	Scaling relations	C++	No	No	No	No

1: Paxton et al. (2010, 2011); 2: Eggleton et al. (2011); Eggleton (1971, 2006); 3: Maeder & Meynet (1988); Ekström et al. (2012); Groh et al. (2019), https://martseba.github.io/; 4: Hurley et al. (2000); 5: Portegies Zwart & Verbunt (1996); Portegies Zwart & Yungelson (1998); Toonen et al. (2012); Portegies Zwart & Verbunt (2012); 6: (Hut et al. (2003)).

main-sequence star (these days often called a stripped star because they naturally follow from an episode of type B mass transfer; Kippenhahn & Weigert 1967). Yet another interesting study are chemically homogeneous stars, which are sometimes considered as progenitors for black hole mergers (Brott et al. 2011).

Code example 5.2 presents a simple script to initialize a star of mass M and metallicity z, and plots the associated density profile. Figure 5.7 shows the density structure of an $M = 2\ \mathrm{M}_\odot$ star simulated using two of the Henyey codes, `EVTwin` and `MESA`. In addition, we plot the structure of two 1 $\mathrm{M}_\odot$ stars that merge into a single star at a very early stage in their evolution (see Chapter 5 and Code example 5.5 for details). Such a collision would provide a good reason to start with a nonstandard stellar initial model. In this case, one could use `MakeMeAMassiveStar` (MMAMS; Lombardi et al. 2002; Gaburov et al. 2008) to generate a stellar initial model.

```
def get_density_profile(code=Evtwin, mass=2.0 | units.MSun, metallicity=0.02):
    stellar = code()
    stellar.parameters.metallicity = metallicity
    stellar.particles.add_particle(Particle(mass=mass))
    radius = stellar.particles[0].get_radius_profile()
    rho = stellar.particles[0].get_density_profile()
    stellar.stop()
    return radius, rho
```

Code example 5.2: Script to initialize a single ZAMS star using `EVTwin`. Code fragment from `amuse.examples.initialize_single_star`.

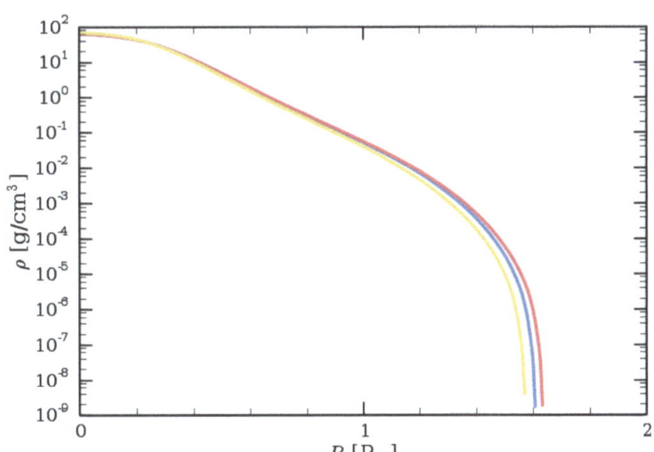

Figure 5.7. Density profile of a star on the ZAMS. All three curves are for a single 2 $\mathrm{M}_\odot$ star. Two curves are calculated using the stellar structure and evolution codes `EVTwin` (blue curve) and `MESA` (red). The yellow curve is generated from a merger of two 1 $\mathrm{M}_\odot$ stars, which are combined conservatively using `MMAMS` for constructing a 2 $\mathrm{M}_\odot$ in dynamical equilibrium, whose structure is then calculated using `MESA`. The merger calculation is performed using the script in Code example 5.5. The script to generate this figure can be found in `amuse.examples.stellar.merger_stellar_density_profile`.

```
def merge_two_stars(
    mass_primary=5 | units.MSun,
    mass_secondary=3 | units.MSun,
    time_collision=1.0 | units.Myr,
):
    """
    Merges two stars using the Make Me a Massive Star code and returns the
    resulting density profile.
    """
    primary = Particle(mass=mass_primary)
    secondary = Particle(mass=mass_secondary)

    stellar = Mesa()
    primary = stellar.particles.add_particle(primary)
    secondary = stellar.particles.add_particle(secondary)

    stellar.evolve_model(time_collision)

    stellar.merge_colliding(
        primary.copy(), secondary.copy(), Mmams, return_merge_products=["se"]
    )
    radius = stellar.particles[0].get_radius_profile()
    rho = stellar.particles[0].get_density_profile()
    stellar.stop()
    plot_density_profile(radius, rho)
```

Code example 5.5: Simple routine to merge two stars (see `amuse.examples.merge_two_stars`). The script should run in about a minute on a laptop.

It is interesting to note that the profiles calculated by the three codes are not the same. The largest differences are in the outer parts of the stars, and probably these do not make a huge difference to global stellar evolution. But they may affect quite dramatically how the star appears in an H–R diagram. A larger star translates to a dimmer and cooler stellar surface.

5.2.2 Stellar-evolution Modules in AMUSE

Currently, in AMUSE we can choose between three Henyey codes, two parametric prescriptions for stellar evolution, and three binary evolution codes. Table 5.1 presents some fundamental parameters for these codes.

Just as we can easily run a simple N-body simulation in AMUSE, we can also run a simple stellar-evolution calculation. The snippet in Code example 5.3 looks quite similar to the code presented in Code example 4.1 of Section 4.1.1, and indeed, the interface function names are deliberately chosen to be similar from one module to another. Code example 5.3 omits the details of parsing the command line to obtain the desired parameters and passing them to the `main` function. To incorporate that, we could include the `Python` argument parser, as discussed in Appendix A (see Section 2.4.2).

Instead of declaring a gravitational N-body code, as we did in Section 4.1.1, here we initialize the stellar-evolution code, and add a single particle of mass `m`, as follows:

```
from amuse.units import units
from amuse.datamodel import Particle
from amuse.community.seba import Seba


def stellar_minimal(
    mass=1.0 | units.MSun, metallicity=0.02, time_end=4700 | units.Myr
):
    stellar = Seba()
    stellar.parameters.metallicity = metallicity
    star = stellar.particles.add_particle(Particle(mass=mass))

    initial_luminosity = star.luminosity
    time_step = 0.1 | units.Myr

    while stellar.model_time < time_end:
        stellar.evolve_model(stellar.model_time + time_step)

        print(
            f"at T={stellar.model_time.in_(units.Myr)} "
            f"L(t=0)={initial_luminosity}, "
            f"L (t={star.age.in_(units.Myr)})={star.luminosity.in_(units.LSun)}, "
            f"m={star.mass.in_(units.MSun)}, "
            f"R={star.radius.in_(units.RSun)}"
        )

    luminosity = star.luminosity

    print(
        f"Time={stellar.model_time.in_(units.Myr)} "
        f"L={np.log10(luminosity.value_in(units.erg / units.s))}"
    )

    stellar.stop()
```

Code example 5.3: Minimal routine for evolving a single star of mass M and metallicity z to some end time `model_time`. This snippet was taken from the source code in `amuse.examples.stellar_minimal`.

```
stellar = Seba()
stellar.particles.add_particle(Particle(mass=mass))
```

The default metallicity for most codes is the solar value: $z = 0.02$, but as can be seen, this parameter is easily changed, which we explain in Section 5.2.3.

The stellar-evolution code chosen here is `SeBa`, which is a parametric stellar-evolution code (see Table 5.1). The single star is added to the repository of stars in the running `SeBa` instantiation. In the line

```
initial_luminosity = stellar.particles.luminosity
```

we fetch the stellar luminosity directly from the stellar-evolution code. We do something similar in the eventual print statement

```
print(
    f"at T={stellar.model_time.in_(units.Myr)} "
    f"L(t=0)={initial_luminosity}, "
    f"L (t={star.age.in_(units.Myr)})={star.luminosity.in_(units.LSun)}, "
    f"m={star.mass.in_(units.MSun)}, "
    f"R={star.radius.in_(units.RSun)}"
)
```

where we print some of the stellar properties after the star has been evolved to `model_time`. This is mainly a matter of convenience. For efficiency, we might alternatively have used a `copy()` function on a predefined communication channel, as discussed in Section 2.2.3.

Stars in AMUSE are categorized into types, from pre-main sequence to black hole. This parameter is called the `stellar_type`. It is convenient for referring to the evolutionary state, but it should only be used as a rough measure of the stellar structure. We provide a complete list of AMUSE stellar types in Appendix A.

5.2.3 Changing the Metallicity of a Star

Stellar-evolution codes are tuned to the Sun. Many stars, probably most, have a different metallicity. For historical reasons, metallicity is a code parameter, rather than a characteristic of the star. We can change the metallicity of the stellar-evolution code by setting the appropriate parameter. As a consequence, all stars in that particular code will be evolved according to that specific metallicity. In order to run multiple stars with different metallicities, we would have to instantiate multiple stellar-evolution codes, as we will demonstrate later in Section 5.2.6.

Changing the stellar metallicity in any of the stellar-evolution codes is easy. For example, we can start `SeBa` with one tenth solar metallicity ($0.1z_\odot$) by

```
stellar = Seba()
stellar.parameters.metallicity = 0.002
```

or equivalently, `EVTwin`

```
stellar = Evtwin()
stellar.parameters.metallicity = 0.002
```

Both codes can be provided with a star and evolved

```
stellar.particles.add_particles(Particles(mass=1 | units.MSun))
stellar.evolve_model(1 | units.Myr)
```

which for `EVTwin` leads to a radius of $2.93\,\mathrm{R}_\odot$.

For a different metallicity, it is often better to start the stellar-evolution code from a pre-main-sequence star because prebuilt models are only stored in the AMUSE repository for solar metallicity.

5.2.4 Starting a Star on the Pre-main-sequence

Henyey stellar-evolution codes can as easily start on the pre-main-sequence, or at any stellar stage, so long as there is a consistent input stellar-evolution model. The Henyey codes will construct such a model if it is not readily available, although this may consider a considerable amount of computing time.

We can, for example for example initiate the MESA stellar-evolution code at solar metallicity.

```
stellar = Mesa(version="2208")
```

generates an instantiation of MESA. Here, we adopted MESA version 2208, as we explicitly indicated as an argument. Now, give it a 1 $M_\odot$ star to work with and evolve it for 1 Myr:

```
stellar.pre_ms__stars.add_particles(Particles(mass=1 | units.MSun))
stellar.evolve_model(1 | units.Myr)
```

Or give it a 1 $M_\odot$ pre-main-sequence star to work with, and evolve it for 1 Myr:

```
stellar.native_stars.add_particles(Particles(mass=1 | units.MSun))
stellar.evolve_model(1 | units.Myr)
```

At an age of 1 Myr, after the ZAMS, we can obtain the radius of the 1 $M_\odot$ star with

```
print(f"Stellar radius={stellar.particles.radius}")
```

returns

```
Stellar radius=[0.893773591284] RSun
```

for the star that was evolved for 1 Myr after the ZAMS, and

```
Stellar radius=[9.03388865301] RSun
```

for the pre-main-sequence star at an age of 1 Myr after the onset of Hydrogen burning. The pre-main-sequence star is, as expected, considerably larger than the star that started on the ZAMS. Even after 50 Myr, the pre-main-sequence star is still somewhat larger than the ZAMS stellar radius. But the pre-main-sequence star, by that time, has lost a little mass (about four Earth masses).

We can do the same for EVTwin, which gives

```
Stellar radius=[0.895857558887] RSun
```

for the `1 | units.MSun` star of solar metallicity. Starting on the pre-main-sequence in EVTwin regrettably does not work in the same way as in MESA, but requires a separate function call:

```
stellar = Evtwin()
stellar.new_particle_method(pms=True)
stellar.particles.add_particles(Particles(mass=1 | units.MSun))
stellar.evolve_model(1 | units.Myr)
```

5.2.5 Improving the Stellar-evolution Solver

We can improve the stellar-evolution solver by adding two new features to the code. The first is to store the results for later use. This is accomplished in much the same way as when we evolved a cluster of stars in the gravity solver and graphed the results (similar to the snippet presented in Code example 4.3). After storing the stellar data at the end of each step, we can plot it immediately in the same script, or save it and use a second small `Python`/AMUSE script to visualize the results.

Evolving a single star is somewhat limited. The second improvement is to evolve an entire population to study the isochrones of the stellar system—shown as curves on the H–R diagram showing stars of the same age. We can generate a list of masses from a stellar mass function (see Section 3.3, Equation (3.1)). Any distribution will do, but from an observational perspective, the Salpeter (1955) mass function, a power law with index `alpha = -2.35`, is popular (see Section 3.3). AMUSE contains a special function, `new_salpeter_mass_distribution`, to initialize masses according to this distribution, but here we use a more general power-law mass function to accomplish the task. The code

```
m = new_powerlaw_mass_distribution(N, Mmin, Mmax, alpha=-2.35)
stellar.particles.add_particles(Particles(mass=m))
```

returns an array of `N` masses chosen randomly between `Mmin` and `Mmax` with a Salpeter power-law distribution, and it then adds them to the stellar-evolution solver. For a more general discussion on stellar mass functions, we refer the reader to Section 3.3.

Figure 5.8 is an H–R diagram of 1000 stars drawn from a Salpeter mass function with initial masses between 0.1 $M_{\odot}$ and 100 $M_{\odot}$, evolved to an age of 4.5 Gyr. Due to the random sampling, the minimum mass turned out to be 0.100008472297 $M_{\odot}$ and the maximum is 7.85091223641 $M_{\odot}$. We assumed solar metallicity and evolved the stars using `SeBa` and `MESA`. The Henyey code crashed on several of the stars, particularly those of mass $\gtrsim$1.2 $M_{\odot}$; we have omitted these stars from Figure 5.8. In Section 8.3.3, we discuss how AMUSE deals with a crash of one of the underlying codes. Table 5.2 presents a small performance comparison between the four AMUSE stellar-evolution modules.

5.2.6 Evolving an Inhomogeneous Stellar Population

In the stellar-evolution prescriptions implemented in AMUSE, time and metallicity are properties of the code, rather than of the star. This is somewhat confusing, but it is how the Henyey codes are structured. Changing the metallicity, in particular, requires the code to load a new set of opacity tables describing the stellar interior, and this is simply too time-consuming to do efficiently every time we switch our

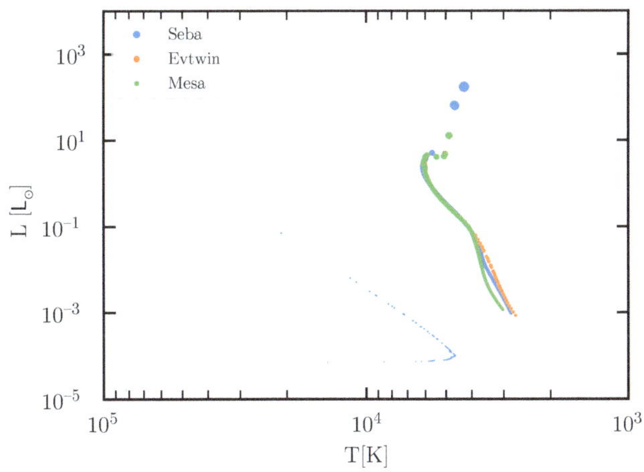

Figure 5.8. H–R diagram of 3000 stars with solar metallicity and initial masses between 0.1 $M_\odot$ and 100 $M_\odot$, evolved to an age of 4.5 Gyr. The stars were evolved using `SeBa` (blue), `MESA` (green), and `EVTwin` (red); the symbol sizes are proportional to the square root of the stellar radius. Only `SeBa` managed to evolve the stars through the white-dwarf phase. This figure was produced with script `amuse.examples.stellar_isochrone`. Using `MESA` and `EVTwin`, the evolution of stars to the remnant stage is time-consuming; this script will take many days to run, even on a multicore workstation.

Table 5.2. Comparison of the Four Available Stellar-evolution Codes in AMUSE

Name	t_{CPU}	$M/M_\odot$	$R/R_\odot$	$L/L_\odot$	T/K
Mesa	0:35.29	0.996	4.11	8.01	4792
EVTWIN	3:20.07	0.993	18.6	98.0	4212
SSE/BSE	0:0.63	0.999	2.58	3.40	4884
Seba	0:0.62	0.999	2.58	6.40	4878

Notes: Here, we calculated the evolution of a 1 $M_\odot$ solar-metallicity star to an age of 11.75 Gyr. For `EVTwin`, the 1 $M_\odot$ star turned into a white dwarf slightly after this, and therefore takes considerably more CPU time than the other Henyey codes. The runs were performed on a 2.4 GHz Core-i7 processor. CPU times are measured in hours:minutes.

attention from one star to another. An analogy might be the time in an N-body code, which also is an attribute of the code rather than of the individual particles. Thus, evolving a stellar population of a single age or metallicity is easy, but a multiage or multimetallicity population requires a little extra work.

The easiest way to evolve a population of stars of different ages is by instantiating as many stellar-evolution codes as there are stellar age groups, or maybe even one code for each individual star. Code example 5.4 presents a small script for evolving several stars, each to a different age and with its own stellar-evolution code. When simulating a limited number of stars, the method in Code example 5.4 is probably adequate because it is simple and transparent. Performance is not really an issue, because each stellar-evolution code can run independently, which makes for an embarrassingly parallel problem. For large numbers of stars, however, the memory requirements of this approach may become prohibitive.

```
def evolve_with_different_stellar_model():
    times = [10, 100, 1000] | units.Myr
    stars = Particles(mass=[1, 2, 4] | units.MSun)
    stellars = [Seba(), Mesa(), Sse()]
    channels = []
    for star, stellar in zip(stars, stellars):
        stellar.particles.add_particle(star)
        channels.append(stellar.particles.new_channel_to(stars))

    for time, channel, stellar in zip(times, channels, stellars):
        stellar.evolve_model(time)
        channel.copy()

    for time, star in zip(times, stars):
        print(f"Time= {time} M= {star.mass}")

    for stellar in stellars:
        stellar.stop()
```

Code example 5.4: Example code to initiate three stellar-evolution codes to evolve three stars, each to a different age. This code fragment is taken from the source code in `amuse.examples.multiple_stellar_code`.

5.2.7 Enforcing Stellar Mass Loss/Gain

The stellar-evolution codes in AMUSE have predetermined mass-loss prescriptions. They are generally not very elaborate because stellar mass loss is not understood particularly well. For this reason, and in order to allow the user to enforce mass loss or gain from other sources (imposed by hand in your script), we have enabled user-specified mass changes. If you invoke this feature, it is important to realize that you will have to switch off the standard mass-loss procedures. Built-in winds on the RGB and on the AGB must be turned off via:

```
stellar.parameters.RGB_wind_scheme=0
stellar.parameters.AGB_wind_scheme=0
```

This will disable the mass-loss prescription in the instantiation of the stellar-evolution code, and all stars evolved with this code are consequently affected.

Stellar mass loss can subsequently be imposed on a particular particle `i` in the code by

```
stellar.particles[i].mass_change = dmdt
```

where `dmdt` is the mass-loss rate of star `i`. The other stars are unaffected and will not lose any mass. For example, for an Eta-Carinae-like luminous blue variable star, one might say

```
dmdt = -2.e-2 | units.MSun/units.yr
```

to model the remarkably high mass-loss rate of that star. A positive value of `dmdt` indicates that the star will gain mass; a negative value means mass loss.

Henyey stellar-evolution codes are generally not prepared for large changes in the stellar mass, and may respond unexpectedly when the user makes an adjustment. One of the major issues here is that the internal time step is calculated on the assumption that mass loss is handled internally. Incorporating a much larger mass-loss rate—for example, when simulating Roche-lobe overflow and calculating the mass loss from the radius of the Roche lobe relative to the stellar radius—may cause the internal time step to become too large to accommodate the externally imposed rate. It may even cause the stellar-evolution code to crash—which might be the best possible outcome given that the code's response to an excessive mass change may shift the evolution onto a new and entirely different track. This evolutionary track is not necessarily wrong, it could represent a viable solution to the underlying differential equations, but one may wonder if it really represents an observable star (stars are not onions; after Steig 2008).

Changing the mass-loss rate of a particular star therefore also requires changing the stellar-evolution time step of that star. This might be realized for star `i` by allowing it to lose less than `dm = -0.02 | units.MSun` within one time step, as follows:

```
new_time_step = min(dm/dmdt, stellar.particles[i].time_step)
stellar.particles[i].time_step = new_time_step
stellar.evolve_model()
```

Here, we start by calculating the new time step, after which we overwrite the star's time step with a smaller one and evolve the stars. The stellar-evolution code will try to evolve the star to the specified next time step. However, this may fail because it also has the requirement that the star remain in thermal and dynamical equilibrium, in which case the actual time step taken by the code may be different.

In our example of evolving Eta Carinae as a 90 $M_\odot$ ZAMS star, we obtain an initial evolution time step of 1.3 yr with `MESA`. When changing the mass-loss rate as just described and reducing the time step to 0.5 yr, checking the progress of the code indicates that the star was actually evolved for only 0.495 yr. After the first step is taken with the enhanced mass-loss rate and the reduced time step, subsequent time steps will be determined internally again. To guarantee that the star evolves according to the desired mass-loss rate and with subsequent small time steps, these parameters will have to be set again before the next step is taken.

In general, changing the time step will affect the calculation results, but often not by much. For example, we might imagine evolving a solar-metallicity 90 $M_\odot$ star for 1.3 yr in a single step or in multiple steps. The radius of the star when evolved in a single step is 13.423 $R_\odot$, versus 13.434 $R_\odot$ when evolved for the same amount of time, but in 10 equal-sized steps. Such a difference is seemingly unimportant, although sometimes a difference in size of just 1 $R_\oplus$ might make a difference.

5.2.8 Accessing Stellar Interiors

The direct (Henyey) stellar-evolution codes in AMUSE solve for the entire structure of the stellar interior. The internal structure of a star is stored in zones, each of which represents a grid point at which the stellar structure equations are solved. Modern stellar-evolution codes typically use between 200 and 2000 zones. The number of zones for star `i` in the stellar-evolution code `stellar` can be queried by

```
n_zones = stellar.particles[i].get_number_of_zones()
```

Stellar interiors are accessed using the particles attribute function `get_xxx_profile()`, which returns an array of data corresponding to the stellar internal structure parameter `xxx`, which may be `mass`, `radius`, `temperature`, `luminosity`, `density`, `entropy`, `thermal_energy`, etc.

For example, the internal temperature structure can be obtained by

```
temperature_profile = stellar.particles[i].get_temperature_profile(n_zones)
```

The returned array (here called `temperature_profile`) contains the temperature profile of the star, starting at the center (the first element of the array) and running all the way to the photosphere (the last element). To see the available options, it is probably best to use the `inspect` class functionality in `Python`. For example, for `EVTwin`:

```
import inspect
from amuse.community.evtwin import Evtwin
print(inspect.getsource(Evtwin))
```

Note that the zones in the AMUSE stellar-evolution codes are not distributed linearly in mass, radius, or any other parameter, but are spaced to ensure optimal functioning of the code. It is therefore important to always obtain the mass or radius profile along with the desired parameter. The code snippet in Code example 5.1 shows how to access the internal density and temperature profile of a star. Other internal stellar parameters can be accessed in similar ways. The number of access functions for the stellar interior is quite large, mainly because stellar evolution is a complex process, involving a wide variety of physics.

We encourage the reader to experiment with querying the internal stellar structure, and even changing it. To reduce the temperature in the outermost shell of a 10 $M_\odot$ star, one could say

```
T_profile[-1] = 0.9 * T_profile[-1]
stellar.particles[i].set_temperature_profile(T_profile)
stellar.particles[i].time_step = 0.5 * stellar.particles[i].time_step
stellar.evolve_model()
```

Here, we also reduce the time step (arbitrarily cutting it in half), in an attempt to allow the stellar-evolution code to adjust to the change. We emphasize that it is very easy to crash a stellar-evolution code by fiddling with the stellar structure because the changes can easily cause the code to fail to converge. A typical example of output you should expect, after 117 lines of debugging messages, will eventually look like:

```
CodeException: Exception when calling function "evol-
ve_for", of code
"MesaInterface", exception was "lost connection to code"
```

Seeing how a code crashes and understanding why (here because it was forced on the code by small changes to the stellar interior) is an excellent way to appreciate the complexity of stellar-evolution calculations.

5.2.9 Modeling Stellar Mergers

Stellar structure is critical when we merge two stars. A fast and convenient approximate approach to determining the interior structure of a merger product is provided by the `MakeMeAStar` code (MMAS; Lombardi et al. 2003) and its extension MMAMS (Gaburov et al. 2008). These codes use data on the interiors of the two colliding stars to construct a new self-consistent stellar model, sorting mass shells by buoyancy to create the merger product, which can then be recommitted into a Henyey stellar-evolution code to continue the evolution. Mass loss in the merger event is calibrated against a set of independent hydrodynamical simulations (mainly using smoothed particle hydrodynamics; see Section 6.2.1). The result is a fast method for obtaining a first-order estimate of the structure of the product of a low-velocity, roughly head-on merger of two stars.

Code example 5.5 presents a simple script in which two main-sequence stars of masses `Mprim` and `Msec` are merged using `MMAMS`. Here, the primary and secondary masses are sent to the merger routine, which creates and evolves the two stars to time `t_coll` and then uses `MakeMeAMassiveStar` to merge them. The two lines

```
primary = stellar.particles.add_particle(bodies[0].as_set())
secondary = stellar.particles.add_particle(bodies[1].as_set())
```

initialize the primary and secondary stars. The particles in `bodies` contain only mass; they have no position or velocity information because we never initialize those quantities.

Note that functions such as `new_plummer_model()`, which generate initial conditions, generally return a particle set, but `Particles(...)` only returns a list. Thus, before adding the particles to the `stellar` instantiation, they first have to be converted into a particle set. In addition, the return value of `add_particles` is a pointer to the particle last added in the community code. (This was also true in the numerous previous examples using `add_particles`, but we ignored this fact in those cases.) In this case, it is convenient to maintain explicit access to the primary and secondary stars in the community code, rather than referring to them as `stellar.particles[0]` and `stellar.particles[1]`.

After the initialization of the two stars, we evolve them to time `t_coll`. Because both the `primary` and `secondary` point directly to community code memory (which happens unseen here, via the MPI channel or a socket), they contain the same internal properties of the star as in the module memory. After the evolution step, we merge the two stars using the routine merge_colliding in MakeMeAMassiveStar. We work here with copies of the stars because both will be removed from the stellar particle set when the merger product is created. On return, the merger product is in stellar_particles[0]. The phrase return_merge_products=["se"] ensures that it has a Henyey structure describing its internal properties. The function merge_two_stars then returns the radius and density attributes of this structure. This script was used to generate Figure 5.7. In Section 5.4.2, we expand on this method by performing simulations to try to reconstruct an observed blue straggler.

This may look like a bit of dark AMUSE magic, but keep in mind that all it does is hand two stars to the merger routine and return two arrays describing the internal structure of the merger product.

5.2.10 Interrupting Stellar Evolution

As we discussed in relation to gravitational dynamics in Section 4.5.1, certain events in stellar-evolution calculations may surpass a particular code's domain of reliability. An example is the supernova explosion of a massive star, which can happen quite suddenly. The Henyey codes are not designed to continue after the supernova. We can flag and handle these events using stopping conditions, much as we discussed for gravitational dynamics in Section 4.5.1 (a more general discussion is presented in Section A.5.2). The evolution of the remnant can then be continued using another code that is better suited to the task.

We enable the stopping condition in the stellar code by

```
detect_supernova = stellar.stopping_conditions.supernova_detection
detect_supernova.enable()
```

In the event loop, we can use the following snippet:

```
if detect_supernova.is_set():
for ci in range(len(detect_supernova.particles(0))):
    print(detect_supernova.particles(0))
    code_body = Particles(particles=detect_supernova.particles(0))
    local_body = particles_in_supernova.get_intersecting_subset_in(bodies)
```

Here, code_body is a pointer to the exploding star in the stellar-evolution code stellar, while local_body is the same star in the local particle set. The former has the most up-to-date data from the stellar-evolution code, but the latter may contain additional information used by other modules or additional local calculations.

5.3 Binary Evolution

Stars often appear in pairs or higher multiplicities. Many of the codes in AMUSE suitable for handling single stars can also handle binaries. In all cases, however, binary evolution is something of a fix to what a single star does because there are quite a number of assumptions underlying the evolution of binary stars. That said, binary evolution is relatively easy compared to triple evolution. In Appendix B.2.1, we briefly discuss triple evolution, as implemented in TRES, but here we limit ourselves to the fundamental principles of binary evolution.

Instead of following binary evolution through one of the available codes, one could write an AMUSE script to evolve a binary star using any of the available stellar-evolution codes. In that way, one can easily test to what degree differences in the various stellar-evolution codes lead to different binary evolution outcomes. Here is a brief recipe for such a binary evolution prescription.

Two stars evolve basically independently of each other so long as their mutual distance is considerably larger than the sum of their radii. Once the distance at pericenter [$q = a(1 - e)$] becomes smaller than about 5 times the radius of the larger star, tidal effects start to become important. Note that binaries with orbital periods $P_{\rm orb} \gtrsim 20$ year probably never reach this stage, except if they have very high eccentricities. For two 1 $M_\odot$ stars, a 20 year period relates to an orbital distance of about 10 au. Those wide binaries are only subject to stellar wind mass loss (cf. Section 5.3.1) and possibly the effect of a supernova (cf. Section 5.3.6).

5.3.1 Stellar Winds

On the main sequence, most stars only lose a little mass, but in massive stars or stars in their late evolutionary stages, mass loss can be quite copious. Such mass loss drives the expansion of the orbit by conservation of angular momentum and energy.

If the total amount of mass lost by the most massive star is $\eta = (M - M_0)/M_0$ (Livio & Warner 1984), we write the change in orbital separation from a_0 to a as (Portegies Zwart & Verbunt 1996)

$$\frac{a}{a_0} = \left(\left[\frac{M}{M_0}\right]^{\eta} \frac{m}{m_0}\right)^{-2} \left(\frac{M_0 + m_0}{M + m}\right)^{2n_j - 1}. \tag{5.18}$$

Note that we allow the secondary star to accrete a small fraction ($m - m_0 < M_0 - M$) of the mass lost by the primary. Here, we have introduced the parameter n_j, which represents the amount of angular momentum that is carried with the material that is lost from the binary system. For an isotropic wind $n_j = 0$, but, for example, in the case of nonconservative mass transfer from a donor star to an accretor $n_j > 0$, due to the amount of angular momentum that is carried with the lost mass (see Section 5.3.4).

5.3.2 Tidal Circularization and Synchronization

Before mass transfer starts (if it starts), tidal effects tend to circularize the orbit, and synchronize the eventual donor star. Both tidal circularization and synchronization are effective, and full synchronization is generally achieved before one of the stars initiates mass transfer. Tidal circularization proceeds on a timescale $t_{circ} \sim 18,000$ yr (Duquennoy et al. 1992), during which the eccentricity $e$damps to zero (Zahn 1977)

$$\frac{de}{dt} = \frac{e}{t_{circ}}\left(\frac{P_{\mathrm{orp}}}{[1\mathrm{day}]}\right)^{-16/3}. \tag{5.19}$$

As the orbit circularizes, the binary's angular momentum is conserved and the orbit widens, with

$$(1 - e^2)a = \mathrm{constant}. \tag{5.20}$$

The circularization timescale is short compared to most other processes that drive the evolution of a binary.

We often work with orbital separation (or semimajor axis a), but sometimes it is simply more convenient to use the orbital period (P_{orb}). We can convert one to the other using Kepler's third law (Kepler 1609)

$$P_{\mathrm{orb}} = 2\pi\left[\frac{a^3}{G(M + m)}\right]^{1/2}. \tag{5.21}$$

Here, G is Newton's constant and the masses are M for the primary and m for the secondary.

The orbital period is probably the shortest timescale in the problem. Binary evolution calculations are often treated in a secular regime. In fact, it is often the thermal timescale of the most massive most evolved star (typically the donor) that sets the timescale on which binary evolution proceeds.

5.3.3 Roche Lobe

The Roche lobe is named after Roche's description of the equipotential surface of two stars in a circular orbit. The shape is symmetric for rotation along the axis connecting the two stars (Paczyński 1971). In Section 4.7.4 and Figure 4.14, we presented an example of using an N-body code to calculate the equipotential surfaces of a binary system. This method is not limited to circular orbits, as in the original formulation of the problem. Luckily, Eggleton (1983) wrote an elegant one page paper on a fitting formula calculating the radius of a sphere with approximately the same volume within the zero-equipotential surface of a star with mass M orbiting a companion of mass m in a circular orbit with semimajor axis a:

$$r_L \simeq \frac{0.49a}{0.5 + q^{-2/3}\ln(1 + q^{1/3})}, \tag{5.22}$$

where $q = m/M$.

When a star of radius R exceeds its Roche radius ($R > r_L$), it starts to transfer mass to its companion, the consequences of which can be quite profound and lead to a wealth of curious phenomena.

5.3.4 Stable Mass Transfer

When one of the stars fills its Roche lobe, mass may start to flow from its surface to the companion (accreting) star. If mass transfer proceeds stably, the variation in orbital parameters is then calculated using Equation (5.18), otherwise we use the common-envelope prescription Equation (5.23).

The choice of what equation to use is generally a Boolean in binary population synthesis studies, although hybrid approaches have been seen. The main problem with mass transfer in binaries is the number of free parameters associated with the process. In our simple description of the process, we already encountered four such hard to constrain parameters: α, γ, η, and n_j. Each of these parameters is dimensionless, has a value probably of order unity, and is barely constrained within an order of magnitude. These parameters may depend on other parameters in the system, such as the stellar masses, orbital separation, and even on internal stellar parameters. One of the main scientific problems in constraining and determining these parameters hides in the complexities of their coupled effect, and the lack of our ability to perform laboratory experiments. As a professional guess, it may be best to take unity for each of these parameters, see what happens, and then vary them individually, or to stay away from binary evolution population synthesis altogether. The last option is probably scientifically the most honest one.

5.3.5 Common-envelope Evolution

One of the most dramatic events that can follow Roche-lobe overflow is common-envelope evolution. We could spend an entire chapter (maybe even a book) on this topic and all the associated folklore, its relative importance, and the many solutions a plethora of researchers have come up with. The common envelope may be the most controversial topic in modern astrophysics.

We limit our discussion here to a single important question, the degree to which the post-common-envelope binary shrinks while expelling the envelope of the Roche-lobe filling (donor) star. Here, we assume the donor to be the most massive of the two, and be composed of a core (with mass M_c) and an envelope (M_e). The envelope is gravitationally bound to the core, and can be lifted if sufficient energy is deposited in it. Where and how this energy is deposited in the envelope is parameterized with a free parameter α. The internal structure and binding energy of the envelope is parameterized with the free parameter λ. By equating the binding energy of the Roche-lobe filling star's envelope with that of the binary before and after the common envelope, we arrive at the following equation (Webbink 1984):

$$\frac{a}{a_0} = \frac{M_c}{M}\left(\frac{1 + 2a_0}{\alpha\lambda R}\frac{M_e}{m}\right)^{-1}, \tag{5.23}$$

which can be used to calculate the post-common-envelope orbital separation a from the initial orbital separation a_0. The attractive aspect of this equation is that all our ignorance is hidden in the two products $\alpha\lambda$. Decoupling α and λ is practically difficult, but probably quite essential. For a review of this topic, we refer to Ivanova et al. (2013).

5.3.6 Supernova Explosions

A supernova explosion of one of the stars in a binary system has two effects: 1) the exploding star loses mass suddenly, and 2) the resulting compact object may receive a velocity kick. For clarity and simplicity, we assume that the mass loss is instantaneous compared to the orbital period, after which calculating the orbital elements of the postsupernova binary is a better of solving the Kepler equation.

Here, we do not present the derivation, but just the result (Hills 1983)

$$\frac{a}{a_0} = \left(1 - \frac{\Delta m}{M_0}\right)\left[1 - \frac{2a_0}{r}\frac{\Delta m}{M_0} - 2\frac{v_0 v_k}{v_{c,\,0}^2}\cos(\theta) - \left(\frac{v_k}{v_{c,0}}\right)^2\right]^{-1}. \tag{5.24}$$

The eccentricity after the supernova then becomes

$$e^2 = 1 - \frac{a_0(1 - e^2)^2}{1(a + e_0\cos(\theta_0))^2}\frac{\left(v_{0,\,x}^2 + v_{k,\,x}^2 + v_{k,\,y}^2 + 2v_{0,\,x}^2 v_{k,\,x}^2\right)}{G(M_0 - \Delta m)}. \tag{5.25}$$

Here, the angles ϕ and θ represent the kick velocity vector from the companion star in the orbital plane, and perpendicular to the orbital plane. A derivation for arbitrary hierarchy was derived in Pijloo et al. (2012).

5.3.7 Magnetic Braking

Low-mass stars may have magnetically coupled winds, which cause a relatively large loss of angular momentum, even though only a little mass is lost (Rappaport et al. 1983). The amount of angular momentum lost by magnetic braking $\dot{J}_{\rm mb}$ in terms of the binary's angular momentum J_b can be written in terms of the radii and masses of the two stars.

$$\frac{\dot{J}_{\rm mb}}{J_b} = \frac{3.8 \times 10^{-30} R_s^{4-\gamma}(M + m) r^\gamma \omega^2}{M a^2}. \tag{5.26}$$

Here, R_s is the radius of the magnetic (primary) star in units of $R_\odot$, and $\gamma \simeq 2.5$. Usually, magnetic braking is applied to white dwarfs with a main-sequence companion between 0.6 and 1.5 $M_\odot$.

5.3.8 Gravitational-wave Radiation

General relativistic effects become important for binary evolution once two massive compact object find themselves in a tight orbit. The orbit shrinks at a rate of (Peters 1964)

$$\frac{da_{\rm gr}}{dt} = -\frac{64}{5}\frac{G^3 M m(M + m)}{c^5 a^3 (1 - e^2)^{7/2}} f_a, \tag{5.27}$$

with the Taylor-series expansion of the Einstein-Infeld-Hoffman equation (Einstein et al. 1938) to second order in the eccentricity

$$f_a = 1 + \frac{73}{24}e^2 + \frac{37}{96}e^4 + \mathcal{O}(e^6). \tag{5.28}$$

In addition, the circularization of the orbit may be presented in terms of a Taylor-series expansion in e

$$\frac{de_{\rm gr}}{dt} = -\frac{304}{15}e\frac{G^3Mm(M+m)}{c^5a^4(1-e^2)^{(5/2)}}f_e, \tag{5.29}$$

with

$$f_e = 1 + \frac{121}{304}e^2. \tag{5.30}$$

5.3.9 Initializing and Evolving a Binary

In AMUSE, these equations for binary evolution are implemented in some form in each of the stellar-evolution codes; these codes are suitable for evolving binaries.

To initialize a binary, one must first initialize the individual stars, and then make a new particle set containing the binary properties, semimajor axis, and eccentricity. The binary particle is then given a pointer to each of the component stars. The current stellar-evolution codes are all binary-aware, but the implementation in the Henyey stellar-evolution codes differ slightly from those in the parameterized codes. Code examples 5.6 and 5.7 present a simple example of how to create and evolve a binary in AMUSE, and how to create a time series of orbital parameters.

To evolve a binary, we often ignore the inclination, argument of pericenter, line of the ascending node, and the mean anomaly. This effectively means that we are studying the evolution of an orbit-averaged system. As a consequence, one cannot trivially couple these binary evolution codes to a dynamical N-body code. To achieve such a coupling, as discussed in Section 8.7.2, one has to assume that binary evolution proceeds on discrete steps comprising multiple orbits during which the orbital phase remains constant. This sounds rather complex, and it is clear what one ideally should do, but it often works in practice. We are also unsure if there is a right way to evolve a binary (or multiple) including stellar physics and gravitational dynamics, except if one would solve all the differential equations systematically and

```
def create_double_star(mass_primary, mass_secondary, semi_major_axis, eccentricity):
    primary_stars = Particles(mass=mass_primary)
    secondary_stars = Particles(mass=mass_secondary)

    stars = Particles()
    primary_stars = stars.add_particles(primary_stars)
    secondary_stars = stars.add_particles(secondary_stars)

    double_star = Particles(semi_major_axis=semi_major_axis, eccentricity=
    eccentricity)
    double_star.child1 = list(primary_stars)
    double_star.child2 = list(secondary_stars)
    return double_star, stars
```

Code example 5.6: Routine to initialize a single binary star, as called in Code example 5.7.

```
def evolve_double_star(
    mass_primary,
    mass_secondary,
    semi_major_axis,
    eccentricity,
    time_end,
    number_of_steps,
):
    double_star, stars = create_double_star(
        mass_primary, mass_secondary, semi_major_axis, eccentricity
    )
    time = 0 | units.Myr
    time_step = time_end / number_of_steps

    code = Seba()
    code.particles.add_particles(stars)
    code.binaries.add_particles(double_star)

    channel_from_code_to_model_for_binaries = code.binaries.new_channel_to(
    double_star)

    t = [] | units.Myr
    a = [] | units.RSun
    e = []
    while time < time_end:
        time += time_step
        code.evolve_model(time)
        channel_from_code_to_model_for_binaries.copy()
        t.append(time)
        a.append(double_star[0].semi_major_axis)
        e.append(double_star[0].eccentricity)
    code.stop()
    return t, a, e
```

Code example 5.7: Routine to evolve a single binary star with `mass_primary`, `mass_secondary`, `semi_major_axis`, and `eccentricity`. The full script can be found in `amuse.examples.binary_evolution`. It should run in a few seconds on a laptop.

synchronously. In principle, it is possible to do this in AMUSE, but the code would be very slow, and without any guarantee that a valid answer would be obtained.

5.3.10 Initial Binary Fraction

One interesting parameter is the fraction of binaries one expects in a system. If you study binaries only, the most common assumption is that all stars are member of a binary system. At a later stage, you can then simply sprinkle in a population of evolved single stars to match the observed statistics. For cluster simulations, however, the binary fraction is important from the beginning.

The fraction of binaries among a certain primary mass seems to increase with mass: more massive stars are more likely to have a bound companion compared to lower-mass stars. We calculate the binary fraction among ZAMS stars from the following equation

$$f_{\rm bin}(M) = \sqrt{R_0}M/m_{\rm max}, \qquad (5.31)$$

where R_0 is a random number uniformly distributed between 0 and 1, M is the mass of the primary stars (see Section 5.3.11), and $m_{\rm max}$ is the maximum mass in the selected mass function. One could interpret $f_{\rm bin}(M)$ as a probability for binarity for a population with a primary mass close to M, or treat it as an individual probability of the star having a binary companion.

5.3.11 Initial Distribution Functions for Binaries

Binaries in parameterized evolution codes are initialized at birth with the masses of the two stars (M and m, or $q \equiv m/M$), the orbital separation (a), and eccentricity (e). The parameters for an individual binary are randomly taken from distribution functions for each of these parameters. We do not know the shape of these distribution functions from first principles, and observed binary populations are riddled with selection effects.

To randomly select initial conditions for binary population synthesis using the distribution functions, we recommend following these steps:

1. Select the primary mass M from an initial mass function (see Section 3.3).
2. Select the orbital period based on the primary mass-dependent distribution. For example, we could use the function provided in Moe & Di Stefano (2017). This distribution peaks at different periods for different primary masses For solar-type stars, it peaks at $\log(P/{\rm days}) \simeq 5$ at about $a \simeq 50$ au.
3. Select the mass ratio ($q \equiv m/M$) based on the primary mass and orbital period using the prescription given by Moe & Di Stefano (2017). This distribution has three components: low-mass primary stars in short orbits tend to have small companion masses, whereas high-mass stars tend to have massive companions, with $q \to 1$ for short-period high-mass stars.
4. The eccentricity (e) is generally distributed thermally. The observed excess of circular orbits for short-period binaries may be a natural consequence of tidal circularization.
5. Finally, the binary fraction is not uniform across all masses, but low for low-mass primary stars and approaching unity for high-mass primaries. For M-dwarfs ($M \sim 0.08$ to $0.60\,{\rm M}_\odot$), one could adopt values from Winters et al. (2019). For clarity, we fit a simple power law to the observed binary fractions, of the form

$$f_{\rm binary} = AM^{\alpha}. \qquad (5.32)$$

   Here, $A = 0.46 \pm 0.03$ and $\alpha = 0.22 \pm 0.03$ with a fit reliability $R^2 \simeq 0.96$. Of course, a binary fraction is capped between 0 and 1.

We recommend implementing these relations using rejection sampling or inverse transform sampling to generate random values from these distributions (Devroye 1986). The minimum orbital period is set to correspond to Roche-lobe filling on the ZAMS.

By following these steps and using the provided distribution functions, you can generate a reasonable set of initial conditions for binary population synthesis that is statistically consistent with the observed characteristics of binary star systems.

5.3.12 Rejuvenation due to Collision or Accretion

When a star accretes material, by mass transfer from a companion or following a collision, a number of important changes occur. It starts to burn its nuclear fuel at a higher rate, and becomes hotter and brighter. Possibly the most important change, however, is the possibility that some of the accreted material finds its way into the stellar core, where nuclear fusion occurs. As a result, the accreting star may appear younger than the surrounding stars in a stellar cluster. Such "rejuvenated" stars are called *blue stragglers* (see Section 5.4.2). We will discuss the physics of stellar collisions in more detail in Section 6.4.2.

Rejuvenation is an important aspect of stellar and binary evolution, but different codes adopt very different approaches to the process. A Henyey code manages accretion onto a star by adding fresh gas to the outermost shell in the one-dimensional stellar structure. The internal solvers for the structure equations of the star will then resolve this addition and arrive at a new structure with the appropriate burning rates, etc., and ultimately mix the new gas into the interior. In a parameterized code, stars evolve along precalculated tracks and new tracks cannot easily be generated. Therefore, these codes incorporate an approximate prescription for on what track, and where on that track, an accreting star should reappear. This, of course, is a complete fudge, but it works reasonably well in practice. We warn the reader that these prescriptions are chosen because they seem reasonable and often work, but they are generally not tested against detailed calculations.

The two parameterized codes in AMUSE employ similar approaches toward rejuvenation, although they differ in detail. Following accretion, the star in `SSE` is restarted on the ZAMS appropriate to the new mass. In `SeBa`, the evolution of a star is separated into partial tracks that represent the different stellar-evolution phases (see Table A.1). After accretion, a star is placed at the same relative position between the beginning and end of the same track of a star with the increased mass. The relative age of the star along this track is reduced according to the fraction of accreted mass relative to the original mass of the star. In this way, a star can accrete small amounts of mass in consecutive phases of accretion.

For example, suppose that a 5 $M_\odot$ star halfway through the main sequence stage —at an age of 0.5 $t_{\rm tams}(5\ M_\odot)$—accretes 0.1 $M_\odot$ following a phase of rapid mass transfer. Here, $t_{\rm tams}$ refers to the *terminal-age main sequence*, the point where the star moves from the main sequence to the subgiant branch on the H–R diagram. According to the prescription in `SSE`, this new 5.1 $M_\odot$ star will start its evolution as a zero-age star at this mass. In `SeBa`, this same star will be restarted at a fraction of $(0.5\ M_\odot – 0.1\ M_\odot\ /5\ M_\odot) \times t_{\rm tams}(5.1\ M_\odot) = 0.48\ t_{\rm tams}(5.1 M_\odot)$.

A nice aspect of parameterized stellar-evolution codes is their predictable behavior. The Henyey codes behave less predictably, but they provide a better, if more expensive, answer to this complicated problem.

5.3.13 Reading and Writing Binary Evolution Files

The data structure for stellar binaries (and higher-order hierarchies) is somewhat more elaborate than for a single particle set. Storing stellar multiples is therefore also somewhat more elaborate. Writing stellar binaries can be realized by separately storing the individual stars (here called `stars`) and the binaries (here called `binaries`). We write them in two separate files.

```
write_set_to_file(stars, "individual_tars.hdf5", "amuse")
write_set_to_file(binaries, "stars_binary_pairs.hdf5", "amuse")
```

We can read the data from the duplicate files with the reverse operation,

```
stars = read_set_from_file('individual_stars.hdf5','amuse')
binaries = read_set_from_file('stars_binary_pairs.hdf5','amuse')
```

5.4 Examples

5.4.1 Response of a Star to Mass Loss

An important issue in stellar evolution is the response of a star to changes in its mass. This response is generally expressed as the dimensionless adiabatic or the thermal (Kelvin–Helmholtz) response, $\zeta_{\rm ad}$ or $\zeta_{\rm th}$, where we define, for mass loss dm

$$\zeta = \frac{d \ln r}{d \ln m} \tag{5.33}$$

(Hjellming & Webbink 1987), where $r(m)$ is the radius in the star containing mass m and the expression is evaluated at the surface of the star (where $m = M$). The response ζ can be compared with the change in the size of the Roche lobe in order to determine the rate at which the star loses mass due to Roche-lobe overflow (Passy et al. 2012).

In binary population synthesis codes, one often adopts a simple parameterization of ζ—it is generally held constant, independent of dm/dt, mass, or the evolutionary stage (Ge et al. 2010). In `SeBa`, for example, $\zeta_{\rm th} = 0$ for main-sequence stars less massive than $0.4\,{\rm M}_\odot$, $\zeta_{\rm th} = 0.55$ for stars more massive than $1.5\,{\rm M}_\odot$, and $\zeta_{\rm th} = 0$ for all other stars, except for those in the Hertzsprung gap, for which $\zeta_{\rm th} = -2$, independent of mass. The negative sign here indicates that the star shrinks as a result of mass loss. In these parameterizations, $\zeta_{\rm th}$ is assumed to be independent of dm/dt, which is not realistic because the response of the star to mass loss must depend on the mass-loss rate.

In AMUSE, we can run a direct stellar-evolution code, `EVTwin` or `MESA`, to calculate ζ and also study the dependence of ζ on the rate of mass loss. Code example 5.8 presents a simple routine to calculate the ζ coefficients for a particular star. This function is called by the main routine in Code example 5.9, in which a star of initial mass M is evolved. Figure 5.9 presents the evolution of ζ as a function of

```
def calculate_zeta(star, metallicity, dmdt):
    stellar = Mesa()
    stellar.parameters.metallicity = metallicity
    stellar.particles.add_particles(star)
    radius_old = star.radius
    star.mass_change = dmdt
    mass_step = 0.01 * star.mass
    star.time_step = mass_step / dmdt
    stellar.particles.evolve_one_step()
    radius_new = stellar.particles[0].radius
    dlnr = (radius_new - radius_old) / radius_old
    dlnm = (stellar.particles[0].mass - star.mass) / star.mass
    zeta = dlnr / dlnm
    stellar.stop()
    return zeta
```

Code example 5.8: Routine to calculate $\zeta_{\rm th}$ for a star. This function is called by the main function Code example 5.9.

```
def massloss_response(
    mass=1.0 | units.MSun, metallicity=0.02, dmdt=-0.01 | units.MSun / units.yr
):
    stellar = Mesa()
    stellar.parameters.metallicity = metallicity
    bodies = Particles(mass=mass)
    stellar.particles.add_particles(bodies)
    stellar = turnon_massloss(stellar)
    channel_to_framework = stellar.particles.new_channel_to(bodies)
    copy_argument = ["age", "mass", "radius", "stellar_type"]
    while stellar.particles[0].stellar_type < Second_Asymptotic_Giant_Branch:
        stellar.particles.evolve_one_step()
        channel_to_framework.copy_attributes(copy_argument)
        star = stellar.particles.copy()
        zeta = calculate_zeta(star, metallicity, dmdt)
        print(
            "Zeta=",
            zeta[0],
            bodies[0].age,
            bodies[0].mass,
            bodies[0].radius,
            dmdt,
            bodies[0].stellar_type,
        )
    stellar.stop()
```

Code example 5.9: Main function to evolve a single star of mass M in time. Here, we use the internal stellar-evolution code parameter `Second_Asymptotic_Giant_Branch = 6 | units.stellar_type`, which is implemented as a part of the `units` package. This function calls Code example 5.8 to calculate ζ. The full script can be found in `amuse.examples.massloss_response`.

time for a few stars and mass-loss rates. We used MESA in these cases and continued the calculations until the code became impractically slow, which generally happened somewhere on the first AGB (see Figure 5.3).

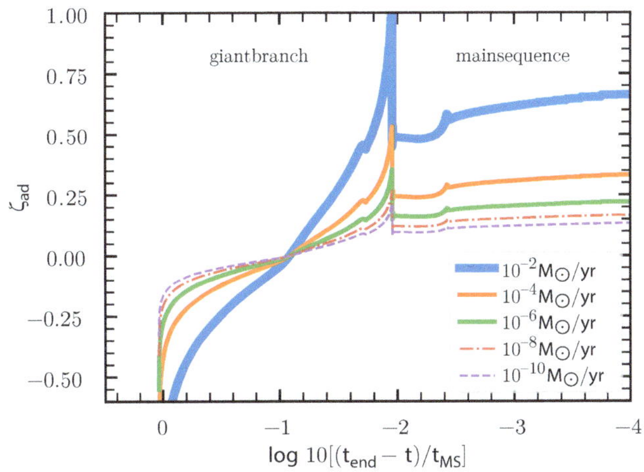

Figure 5.9. Response of an $M = 1\ M_\odot$ star to mass loss, expressed as a function of time (relative to the main sequence lifetime). We performed the calculation for mass-loss rates $dm/dt = 10^{-2}\ M_\odot\ yr^{-1}$ (top curve), $dm/dt = 10^{-4}\ M_\odot\ yr^{-1}$, $10^{-6}\ M_\odot\ yr^{-1}$, $10^{-8}\ M_\odot\ yr^{-1}$, and $10^{-10}\ M_\odot\ yr^{-1}$ (bottom curve). This figure was created using a script based on `amuse.examples.stellar.massloss_response`. The calculation took several hours on a laptop, so we include the data file (`Zeta_M1MSun_for_various_dmdt.data` in the `data` directory) for a 1 $M_\odot$ star and plot this using the `amuse.examples.plot_stellar_massloss_response` module.

5.4.2 Blue Stragglers in M67

M67 was discovered between 1772 and 1779 by Johann Gottfried Koehler, who described it as a “rather conspicuous nebula in elongated figure, near Alpha of Cancer.” Charles Messier described the cluster on 1780 April 6 as a “cluster of small stars with nebulosity below the southern claw of the Crab.” He gave the cluster its current identification M67. In 1783, William Herschel counted a total of 20 stars. Today, M67 is known to host nearly 1900 stars (Uribe et al. 2007) and goes under many names, among them HR3515, OC1549.0, C0847 + 120, NGC2682, and [KPR2004b]212.

The age of a cluster is often measured by the distribution of stars in the H–R diagram. Key to this age determination is the main sequence turn-off—the temperature and luminosity at which stars leave the relatively hot main sequence and start to cross the Hertzsprung gap to cooler regions. This turn-off point is associated with a certain stellar mass and hence age. Stars more massive than the turn-off mass have already started to cross the Hertzsprung gap, whereas lower-mass stars are still on the main sequence.

For the cluster M67, the turn-off mass is $\sim 1.4\ M_\odot$ (Boffin et al. 2015, p.51). There is also a population of stars that lie along the main sequence, but are brighter and hotter than the main sequence turn-off. These stars are called blue stragglers, and were first discovered in the globular cluster M3 (Sandage 1953). Blue stragglers are commonly visible in any star cluster for which a turn-off can be clearly determined, but their origin remains somewhat elusive. Table 5.3 lists some fundamental and derived parameters for five blue stragglers in M67. Their temperatures and

Table 5.3. Observed Parameters for Some M67 Blue Stragglers

name	$T_{\rm eff}$ [K]	V mag	L [$L_\odot$]	$\log g$
S984(F134)	6170	12.259	10.55	3.9
S975	6820	11.078	3.558	4.4
S997	6675	12.127	9.350	4.4
S1082	7050	11.226	4.078	4.5
S2204	6650	12.892	18.91	4.6

Notes. M67 is about 4 Gyr old (Shetrone & Sandquist 2000). The luminosity is calculated from $10^{(V-\beta)/2.5}$. Here, $\beta = V - M_v = 9.7$ is the distance modulus.

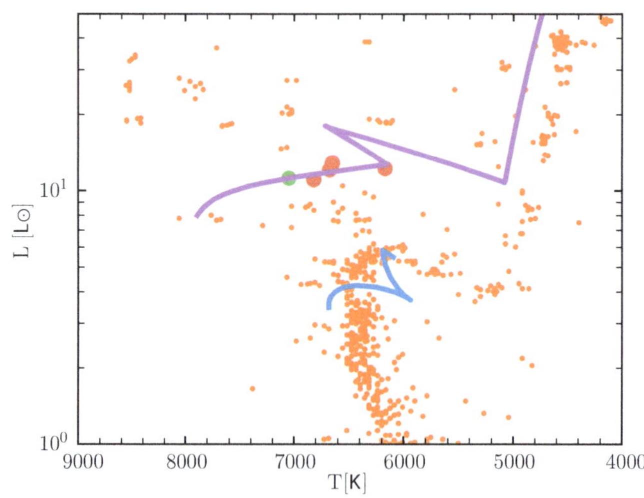

Figure 5.10. H–R diagram for the star cluster M67. The data are from http://webda.physics.muni.cz/ (Eggen & Sandage 1964). The red (the stars S984, S975, S997, and S2204) and green (S1082) bullets are the blue stragglers listed in Table 5.3. The other bullet points (orange) are the M67 stars. Two lines indicate the evolutionary track for a star near the turn-off of 1.4 $M_\odot$ (blue) and 1.7 $M_\odot$ (purple), both at twice the solar metallicity ($Z = 0.02$). This figure was generated using the script `amuse.examples.plot_M67withBSS_tracks`. It should take 5–10 s to run on a laptop.

luminosities can be compared directly with the output of any stellar-evolution module in AMUSE.

Because they are brighter and hotter than the turn-off, it has been suggested that blue stragglers result from stellar mergers during binary evolution or following stellar collisions in the dense central region of star clusters. Their higher masses and younger ages are then direct consequences of their collisional history. Here, we discuss their possible association with stellar evolution; we will return to the possibility of a collisional origin in Section 8.5.2.

Figure 5.10 presents a historical H–R diagram for M67, including the five blue stragglers listed in Table 5.3. We overplot two evolutionary tracks to demonstrate how the turn-off stars evolve to an age of 4 Gyr. The upper 1.7 $M_\odot$ evolutionary track runs nicely through the blue stragglers, but the tip of the curve is off the plot. This indicates that these stars would already have evolved to the white-dwarf stage, so they would be absent in this observation if they really were 4 Gyr old. The other track (blue curve) gives the evolutionary track of a 1.4 $M_{\odot}$ star to the same age. These stars are still visible in the cluster near the end of the turn-off or early Hertzsprung gap.

5.5 Validation

Stellar-evolution codes are hard to validate. One major problem is that we cannot observe the full evolution of any star. Generally, there are two observable parameters that can be tested directly—the effective temperature (or something equivalent) and the luminosity. Recent advances in observational techniques and the discovery of planets orbiting many stars have given us new insights into stellar atmospheres, pulsations, etc., but these are often not solved for in stellar-evolution codes (however, see also Christensen-Dalsgaard et al. 1996; Aerts et al. 2010).

The equations of stellar structure and evolution are generally quite well-behaved. There are few significant nonlinear effects, and there is no fundamental reason to think that solving the various differential equations should not provide the right answer (mathematically speaking), so long as the star is in thermal and dynamical equilibrium. Subsequent validation is carried out by comparing stellar-evolution tracks with star cluster isochrones, or even with the positions of individual stars in the H–R diagram. This is the basic reason why stellar-evolution theory has been so successful.

The only star for which a more thorough validation can be carried out is the Sun, because its surface parameters are quite well determined. In Figure 5.11, we compare the solar temperature and luminosity with the results of the four stellar-evolution codes currently available in AMUSE. Although the differences are not large, they are still significant, particularly considering the small error bars on the observed values, which are smaller than the plotted symbols (see Williams 2013).

It remains challenging to compare individual stellar-evolution codes, particularly if they give slightly different results for the same input parameters (mass, metallicity, and age). Often, the differences are fairly small and are not a major concern for many coupled problems that involve hydrodynamics, radiative transport, or gravitational dynamics. The global mass loss of a population of stars in the central portion of a star cluster is the result of a complicated combination of dynamical and stellar evolution, and may depend only to first order on the specific stellar-evolution code used. It remains important, however, to be aware of these small differences and recognize that a certain subspace of initial parameters may give different results, depending on the adopted stellar-evolution code.

5.6 Assignments

5.6.1 Stellar Comparison

Rewrite the simple script `plot_solar_comparison` used in Figure 5.11 to compare the computed temperatures and luminosities of some other stars. The best stars to try this on are the closest and brightest—Rigel, Betelgeuse (see

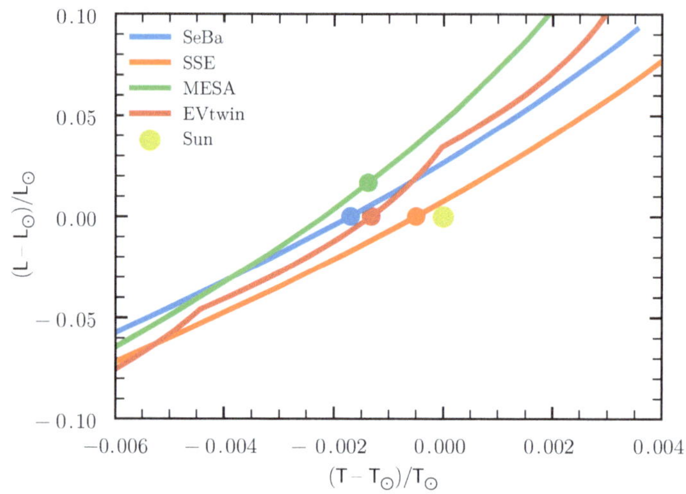

Figure 5.11. Comparison of the effective temperature and luminosity for the Sun, calculated using the four stellar-evolution codes in AMUSE. The calculations were performed for a star with solar metallicity and an initial mass of 1 $M_\odot$, evolved from 4 Gyr to 5.2 Gyr (curves from left to right). The points indicate the moments (using each code's internal time resolution) when the Sun is best represented by the models. These occurred at 5.18 Gyr, 5.20 Gyr, 4.33 Gyr, and 4.30 Gyr for `SeBa`, `SSE`, `MESA`, and `EVTwin`, respectively. The solar temperature and luminosity are taken from Allen's Astronomical Quantities (Cox 2000). This figure was generated using the script `amuse.examples.plot_solar_comparison`. It should take about 20 s to run on a laptop.

Figure 5.1), the red supergiant R Sculptoris, and both stars in the Spica binary system (see Table 5.4). Note that, even for these stars, the direct stellar-evolution codes may be unable to converge to a solution, in which case they will simply break.

5.6.2 Ages of the M67 Blue Stragglers

The five blue stragglers in M67 listed in Table 5.3 show a remarkable coherence in terms of their mass. In Section 5.4.2 we demonstrated that each of them lies close to the main sequence location of a 1.7 $M_\odot$ star of twice solar metallicity. Their ages, however, are clearly not the same because each of them is positioned differently along the 1.7 $M_\odot$ evolutionary track.

Using any of the stellar-evolution codes in AMUSE, determine the ages of each of the blue stragglers in Table 5.3, and perhaps even improve on the listed mass estimate, assuming that they were formed as normal ZAMS stars.

5.6.3 Constructing Isochrones

Write an AMUSE script to construct a series of 10 isochrones equally spaced in time from 1 Gyr to 10 Gyr for stars of 0.2 $M_\odot$ to 20 $M_\odot$, with solar metallicity. Plot the isochrones in the temperature (K)–luminosity ($L_\odot$) plane.

Table 5.4. Observed and Measured Parameters for Some Nearby Bright Stars

Name	Ref	Mass [$M_\odot$]	Age [Myr]	T [K]	L [$L_\odot$]	R [$R_\odot$]	[Fe/H]
Rigel	1	21 ± 3	8 ± 1	12, 100	$1.0–1.40 \times 10^5$	78.9 ± 7.4	−0.06
Spica A	2	11.4 ± 1.2	12.5	22,400	12,100	7.40 ± 0.57	–
Spica B	3	7.21 ± 0.75	12.5	18,500	1,500	3.64 ± 0.28	–
Betelgeuse	4	7.7–20	7.3	3140–3641	$9–15 \times 10^5$	950–1200	0.05
R Sculptoris	5	1.3 ± 0.6	–	2640 ± 80	5500	355 ± 55	–

References 1: Przybilla et al. (2006, 2010); 2: Lyubimkov et al. (1995); Harrington et al. (2009); 3: Palate et al. (2013); Tkachenko et al. (2016); 4: Ramírez et al. (2000); Mohamed et al. (2012); 5: Wittkowski et al. (2017).

Convert the H–R diagram into a color–magnitude diagram. Luminosity can be converted to magnitudes in the usual way, by taking distance modulus and interstellar reddening [$E(B - V) = 0.04$] (Taylor 2007) into account,

$$m_{\rm star} = m_{\odot} - 2.5 \log \left[\frac{L_{\rm star}}{\rm L_{\odot}} \left(\frac{\rm pc}{d_{\rm star}} \right)^2 \right]. \tag{5.34}$$

Here, the apparent magnitude of the Sun is $m_{\odot} = -26.73$, and its distance is $d_{\odot} \simeq 2.25 \times 10^{-8}$ pc (Allen 1955; Cox 2000). Color ($B - V$) can be converted to temperature using the following polynomial fit:

$$\log T(K) \simeq [14.551 - (B - V)]/3.684. \tag{5.35}$$

Address the following questions:

- What is the best stellar-evolution module to use for this task?
- What age best fits the observed star cluster M67? The data for M67 can be found at http://www.univie.ac.at/webda/cgi-bin/ocl_page.cgi?cluster=M67.
- What distance do you have to assume for the best fit?
- What age or metallicity spread do you need in order to obtain a better comparison between the simulated cluster and the observations?
- What additional parameters could you introduce in order to improve the fit?

References

Aerts, C., Christensen-Dalsgaard, J., & Kurtz, D. W. 2010, Asteroseismology (Berlin: Springer)

Allen, C. W. 1955, Astrophysical Quantities (London: Athlone Press)

Bazán, G., Dearborn, D. S. P., Dossa, D. D., et al. 2003, Conf. Ser. 293, 3D Stellar EvolutionASP ed. S. Turcotte, S. C. Keller, & R. M. Cavallo (San Francisco, CA: ASP) 1

Bethe, H. A. 1939, PhRv, 55, 434

Boffin, H. M. J., Carraro, G., & Beccari, G. 2015, Ecology of Blue Straggler Stars (Berlin: Springer)

Bradt, H. 2014, Astrophysics Processes (Cambridge: Cambridge Univ. Press)

Brott, I., de Mink, S. E., Cantiello, M., et al. 2011, A&A, 530, A115

Chiu, H. Y. 1964, AnPhy, 26, 364

Christensen-Dalsgaard, J., Dappen, W., Ajukov, S. V., et al. 1996, Sci., 272, 1286

Clapeyron, É 1834, Mémoire sur la puissance motrice de la chaleur (Paris: Jacques Gabay)

Clausius, R. 1857, AnP, 176, 353

Cox, A. N. 2000, Allen's Astrophysical Quantities (New York: AIP Press)

Devroye, L. 1986, Non-uniform Random Variate Generation (New York: Springer)

Dolan, M. M., Mathews, G. J., Lam, D. D., et al. 2016, ApJ, 819, 7

Duquennoy, A., Mayor, M., & Mermilliod, J. C. 1992, Binaries as Tracers of Star Formation, ed. A. Duquennoy, & M. Mayor (Cambridge: Cambridge Univ. Press) 52

Eggen, O. J., & Sandage, A. R. 1964, ApJ, 140, 130

Eggleton, P. P. 1983, ApJ, 268, 368

Eggleton, P. P., Tout, C., Pols, O., et al. 2011, STARS: A Stellar Evolution Code, Astrophysics Source Code Library, ascl:1107008

Eggleton, P. 2006, Evolutionary Processes in Binary and Multiple Stars (Cambridge: Cambridge Univ. Press)
Eggleton, P. P. 1971, MNRAS, 151, 351
Eggleton, P. P., Fitchett, M. J., & Tout, C. A. 1989, ApJ, 347, 998
Einstein, A., Infeld, L., & Hoffmann, B. 1938, AnMat, 39, 65
Ekström, S., Georgy, C., Eggenberger, P., et al. 2012, A&A, 537, A146
Gaburov, E., Lombardi, J. C., & Portegies Zwart, S. 2008, MNRAS, 383, L5
Ge, H., Hjellming, M. S., Webbink, R. F., Chen, X., & Han, Z. 2010, ApJ, 717, 724
Georgy, C., Meynet, G., Walder, R., Folini, D., & Maeder, A. 2009, A&A, 502, 611
Groh, J. H., Ekström, S., Georgy, C., et al. 2019, A&A, 627, A24
Harrington, D., Koenigsberger, G., Moreno, E., & Kuhn, J. 2009, ApJ, 704, 813
Haubois, X., Perrin, G., Lacour, S., et al. 2009, A&A, 508, 923
Hayashi, C. 1961, PASJ, 13, 450
Helfand, D. J., & Tademaru, E. 1977, ApJ, 216, 842
Henyey, L. G., Wilets, L., Böhm, K. H., Lelevier, R., & Levee, R. D. 1959, ApJ, 129, 628
Henyey, L. G., Forbes, J. E., & Gould, N. L. 1964, ApJ, 139, 306
Hills, J. G. 1983, ApJ, 267, 322
Hjellming, M. S., & Webbink, R. F. 1987, ApJ, 318, 794
Hurley, J. R., Pols, O. R., & Tout, C. A. 2000, MNRAS, 315, 543
Hut, P., Shara, M. M., Aarseth, S. J., et al. 2003, NewA, 8, 337
Ivanova, N., Justham, S., Chen, X., et al. 2013, A&ARv, 21, 59
Kapil, V., Mandel, I., Berti, E., & Müller, B. 2023, MNRAS, 519, 5893
Kepler, J. 1609, Astronomia Nova Vol. 1 (Heidelberg: G. Voegelinus)
Kingdon, K. H., & Langmuir, I. 1923, PhRv, 22, 148
Kippenhahn, R., & Weigert, A. 1967, ZA, 65, 251
Köhler, K., Langer, N., de Koter, A., et al. 2015, A&A, 573, A71
Langer, N. 2012, ARA&A, 50, 107
Laplace, E., Götberg, Y., de Mink, S. E., Justham, S., & Farmer, R. 2020, A&A, 637, A6
Livio, M., & Warner, B. 1984, Obs, 104, 152
Lombardi, J. C., Thrall, A. P., Deneva, J. S., Fleming, S. W., & Grabowski, P. E. 2003, MNRAS, 345, 762
Lombardi, J. C., Warren, J. S., Rasio, F. A., Sills, A., & Warren, A. R. 2002, ApJ, 568, 939
Lyne, A. G., & Lorimer, D. R. 1994, Natur, 369, 127
Lyubimkov, L. S., Rachkovskaya, T. M., Rostopchin, S. I., & Tarasov, A. E. 1995, ARep, 39, 186
Maeder, A., & Meynet, G. 1988, A&AS, 76, 411
Meynet, G., & Maeder, A. 2005, A&A, 429, 581
Meynet, G., Maeder, A., Schaller, G., Schaerer, D., & Charbonnel, C. 1994, A&AS, 103, 97
Moe, M., & Di Stefano, R. 2017, ApJS, 230, 15
Mohamed, S., Mackey, J., & Langer, N. 2012, A&A, 541, A1
Ohnaka, K., Weigelt, G., & Hofmann, K. H. 2017, Natur, 548, 310
Paczyński, B. 1971, ARA&A, 9, 183
Palate, M., Koenigsberger, G., Rauw, G., Harrington, D., & Moreno, E. 2013, A&A, 556, A49
Passy, J.-C., Herwig, F., & Paxton, B. 2012, ApJ, 760, 90
Paxton, B., Bildsten, L., Dotter, A., et al. 2010, Modules for Experiments in Stellar Astrophysics, Astrophysics Source Code Library, ascl:1010.083
Paxton, B., Bildsten, L., Dotter, A., et al. 2011, ApJS, 192, 3

Peters, P. C. 1964, PhRv, 136, 1224
Pijloo, J. T., Caputo, D. P., & Portegies Zwart, S. F. 2012, MNRAS, 424, 2914
Portegies Zwart, S. F., & Verbunt, F. 1996, A&A, 309, 179
Portegies Zwart, S. F., & Verbunt, F. 2012, SeBa: Stellar and Binary Evolution, Astrophysics Source Code Library, ascl:1201.003
Portegies Zwart, S. F., & Yungelson, L. R. 1998, A&A, 332, 173
Przybilla, N., Butler, K., Becker, S. R., & Kudritzki, R. P. 2006, A&A, 445, 1099
Przybilla, N., Firnstein, M., Nieva, M. F., Meynet, G., & Maeder, A. 2010, A&A, 517, A38
Ramírez, S. V., Sellgren, K., Carr, J. S., et al. 2000, ApJ, 537, 205
Rankine, W. J. M. 1853, Proc. of the Philosophical Society of Glasgow, Vol. 3, ed. W. J. Millar (London: Richard Griffin & Co.) 276
Rappaport, S., Verbunt, F., & Joss, P. C. 1983, ApJ, 275, 713
Rizzuti, F., Hirschi, R., Arnett, W. D., et al. 2023, MNRAS, 523, 2317
Saha, M. N. 1921, RSPSA, 99, 135
Salpeter, E. E. 1955, ApJ, 121, 161
Sandage, A. R. 1953, AJ, 58, 61
Schaerer, D., de Koter, A., Schmutz, W., & Maeder, A. 1996, A&A, 310, 837
Shetrone, M. D., & Sandquist, E. L. 2000, AJ, 120, 1913
Smartt, S. J. 2009, ARA&A, 47, 63
Spera, M., Mapelli, M., & Bressan, A. 2015, MNRAS, 451, 4086
Stasińska, G., & Schaerer, D. 1997, A&A, 322, 615
Steig, W. 2008, Shrek! (New York: Square Fish)
Takahashi, K., & Langer, N. 2021, A&A, 646, A19
Taylor, B. J. 2007, AJ, 133, 370
Tkachenko, A., Matthews, J. M., Aerts, C., et al. 2016, MNRAS, 458, 1964
Toonen, S., Nelemans, G., & Portegies Zwart, S. 2012, A&A, 546, A70
Tutukov, A. V., & Fedorova, A. V. 2001, ARep, 45, 882
Uribe, A., Barrera-Rojas, R. S., & Brieva, E. 2007, Conf. Ser. 371, Statistical Challenges in Modern Astronomy IVASP ed. G. J. Babu, & E. D. Feigelson (San Francisco, CA: ASP) 403
Varshavskii, V. I., & Tutukov, A. V. 1975, AZh, 52, 227
Webbink, R. F. 1984, ApJ, 277, 355
Williams, R. D. 2013, Sun Fact Sheet (Washington, DC: NASA)
Winters, J. G., Henry, T. J., Jao, W.-C., et al. 2019, AJ, 157, 216
Wittkowski, M., Hofmann, K. H., Höfner, S., et al. 2017, A&A, 601, A3
Zahn, J. P. 1977, A&A, 57, 383

AAS | IOP Astronomy

Astrophysical Recipes (Second Edition)
The art of AMUSE
Simon Portegies Zwart, Steve McMillan and Steven Rieder

Chapter 6

Hydrodynamics

Reality continues to ruin my life.

—Bill Watterson

Gas and fluid dynamics play an important role in many astrophysical processes. When addressing these physical phenomena, we refer to the collection of solvers by the generic term *hydrodynamics*. Gas dynamics forms the basis for our understanding of star and planet formation; it drives X-ray sources and accretion onto supermassive black holes, and it dictates the appearance of galaxies. In AMUSE, hydrodynamics is addressed numerically using two quite distinct methods: the Lagrangian formalism with smoothed particles, and Eulerian grid-based finite-volume formalisms. These approaches are complementary. Hybrid approaches that combine the advantages of both methods are also available in AMUSE. Simulating hydrodynamics is an art, but one with a solid mathematical background.

6.1 In a Nutshell

Hydrodynamics deals with the motion of fluids driven by internal and external forces; for example, due to pressure gradients and gravity. Although fluid dynamics is important for everyday life, we discuss it here in the context of astronomical phenomena. This means we will largely ignore many important fluid properties, such as viscosity and surface tension, that are critical to Earth-bound applications.

In AMUSE, we confine ourselves to a continuum description of the gas—we do not attempt to describe the microscopic behavior of its constituent particles. In doing so, we are assuming that we can sensibly define thermodynamic quantities, such as density and temperature, on the spatial scales of interest. In numerical simulations, the spatial variation of fluid properties may be represented in a variety of different ways—on a regular grid, or an irregular tessellation, or in terms of a

doi:10.1088/2514-3433/adfc3cch6

collection of macroscopic particles that sample the fluid at different locations. In all cases, if the sampling scale is λ, the gas has number density n, and we are interested in a physical flow with characteristic length scale L, then we require $\lambda \ll L$ and $n\lambda^3 \gg 1$ (i.e., a resolution element should contain many particles but still be much smaller than the length scales of interest). Fortunately, these requirements are often met and we can make reliable quantitative statements about the fluid flow.

Some astrophysical processes—for example, jets, stellar winds, and supernova explosions—have symmetries that may allow a good approximation by reducing the dimensionality of the problem, but often no such simplification exists. In addition, there is a wide range of instabilities for which the typical length scale depends on the dimensionality of the problem, as in the Rayleigh–Taylor (dense fluid on top of a low-density fluid) and Kelvin–Helmholtz (shear between two fluids) instabilities. For these reasons, we address here only the hardest problem—fluid dynamics in three dimensions.

Solving the fundamental equations of hydrodynamics is one of the hardest numerical problems in computational science. Three-dimensional fluid flows are very computationally demanding. The underlying equations are nonlinear and the dynamic range may exceed algorithmic capabilities by orders of magnitude. At the same time, there is only a very limited set of analytic solutions against which numerical implementations can be tested (Coggeshall 1991). Laboratory experiments are possible, but it is often hard to compare them quantitatively with simulations, especially in astrophysics. There are a number of excellent resources on hydrodynamics, ranging from basic but comprehensive introductions (Landau & Lifshitz 1959) to more astronomically oriented texts (Shore 2007; Regev et al. 2016).

6.1.1 Underlying Equations

Astrophysical fluids can be described by a relatively small number of variables. A fluid element has position $\mathbf{x}$ and velocity $\mathbf{v}$, as well as internal thermodynamic properties, such as density ρ, composition, pressure P, temperature T, entropy s, and so on. For astrophysical systems, the composition is usually no more complicated than specifying just the hydrogen mass fraction X, the helium fraction Y, and/or a "metal" fraction $Z = 1 - X - Y$ that accounts for all the heavier elements (note that in stellar evolution in Chapter 5, we indicate this parameter with a lower case z). The connection among the thermodynamic variables is provided by the equation of state (see Section 6.1.1.3). The dynamics of a fluid can be described by three independent partial differential equations, which express the conservation of mass, momentum, and energy.

6.1.1.1 Conservation of Mass

The rate of change of the mass m in some volume V is equal to the rate at which mass flows into or out of V across its surface S:

$$\frac{dm}{dt} = -\int_S \rho\mathbf{v} \cdot d^2\mathbf{S}, \tag{6.1}$$

where $d^2\mathbf{S}$ is the (outward-pointing) vector element of area on S. Writing

$$m = \int_V \rho \, d^3V, \tag{6.2}$$

and using the divergence theorem on the right side of Equation (6.1), we find

$$\frac{dm}{dt} = \int_V \frac{\partial \rho}{\partial t} \, d^3V = -\int_V \nabla \cdot (\rho \mathbf{v}) \, d^3V. \tag{6.3}$$

Because this must be true for any volume V, we immediately obtain the *continuity equation*

$$\frac{\partial \rho}{\partial t} + \nabla \cdot (\rho v) = 0. \tag{6.4}$$

In the frame of a moving fluid element, we can recast this in terms of the so-called Lagrangian (or comoving) derivative

$$\frac{d}{dt} \equiv \frac{\partial}{\partial t} + \mathbf{v} \cdot \nabla, \tag{6.5}$$

to write

$$\frac{d\rho}{dt} + \rho \nabla \cdot v = 0. \tag{6.6}$$

6.1.1.2 Conservation of Momentum

Newton's second law (see also Equation (4.1)), applied to a fluid element, is simply

$$\rho \frac{d\mathbf{v}}{dt} = \mathbf{F}, \tag{6.7}$$

where $\mathbf{F}$ is the net force per unit volume and "d" represents a Lagrangian derivative. In an astrophysical context, the most important forces contributing to the right side of Equation (6.7) are pressure ($\mathbf{F} = -\nabla P$) and gravity ($\mathbf{F} = \rho \mathbf{g} = -\rho \nabla \phi$), so

$$\frac{d\mathbf{v}}{dt} = -\frac{\nabla P}{\rho} + \mathbf{g}. \tag{6.8}$$

Equation (6.8) is known as the *Euler equation.* Other forces (e.g., radiation pressure) can be included in an analogous way. Expanding out the Lagrangian derivative, we find

$$\frac{\partial \mathbf{v}}{\partial t} + \mathbf{v} \cdot \nabla \mathbf{v} = -\frac{\nabla P}{\rho} + \mathbf{g}. \tag{6.9}$$

Note the nonlinear term on the left side, which is the origin of many of the most intractable problems in computational fluid dynamics.

Thus far, we have neglected internal viscous stresses ($\mathbf{F} = \rho \, \mathbf{f}_{\text{visc}}$) in the Euler equation. By adopting a specific model for including viscous stresses in Equation (6.8) and Equation (6.9), we obtain the *Navier–Stokes equation*. For an incompressible flow, the viscous term can be written

$$\mathbf{f}_{\text{visc}} = \nu \, \nabla^2 \, \mathbf{v}, \tag{6.10}$$

where ν is the kinematic viscosity. Although physical viscosities are often too small to be important for large-scale astrophysical flows, similar *artificial viscosity* terms are used in many simulations to suppress short-wavelength features that would otherwise fall below the resolution of the simulation (Von Neumann & Richtmyer 1950; see Section 6.1.2.2).

6.1.1.3 The Equation of State

We now have four equations (continuity and Euler) involving five fluid quantities: $\mathbf{v}$, ρ, and P. In order to solve this system of equations, we need to close it. In principle, this can be accomplished by adopting an equation of state, which connects ρ and P.

We discussed some possible equations of state in Section 5.1.2.1. Many of the hydrodynamics modules in AMUSE adopt a simple expression of the form

$$P = A(s)\rho^{\gamma}, \tag{6.11}$$

where γ is the adiabatic index and s is specific entropy. For a monatomic adiabatic ideal gas, $\gamma = \frac{5}{3}$ and s is constant, so P is a function of ρ only. In these cases, the entropy s can only change in shocks (see Section 6.1.2.2), so the flow is adiabatic except when shocks are encountered (Longair 2011). Sometimes, the index γ is allowed to vary in a density-dependent way to model heating and cooling processes. An isothermal equation of state is modeled simply by forcing the temperature to the desired value at the end of every step.

In reality, however, the true equation of state usually involves other thermodynamic variables, such as temperature $T = T(P, \rho)$ or entropy $s = s(P, \rho)$, and we need additional equations to complete the system. For example, to determine the temperature, we may need to take into account heating and cooling processes in the gas, adding an energy equation to close the set of equations:

$$\Delta e = Q - P\Delta V. \tag{6.12}$$

Here, e is the internal energy per unit mass of the gas, $V = 1/\rho$ is specific volume, and $Q = T\Delta s$ is the net heating, which may have contributions from a variety of sources, such as dissipation, nuclear reactions, cosmic rays, radiative heating and cooling, etc. If the gas cools at a rate $\mathcal{C}$ and is heated at a rate $\mathcal{H}$ (both per unit mass), then the energy equation is

$$\frac{de}{dt} = \mathcal{H} - \mathcal{C} - \frac{P}{\rho} \; \nabla \cdot \mathbf{v}. \tag{6.13}$$

The simplified equations of state described previously are approximations to the solutions of this equation.

The full equation of state is determined by a number of processes, including the degree of ionization of the medium, its chemistry and composition, the presence of nuclear reactions, heating by cosmic rays, and various radiative processes. The most sophisticated (and costly) treatments model these processes explicitly; the temperature and ionization structure of the gas is determined by equating the heating and cooling rates, or it may be provided by an additional module that calculates the radiative structure of the system (see Chapter 7). Once the temperature T and ionization fraction of the gas have been determined, the pressure P in a star-forming region of density ρ is accurately given by

$$P = \frac{\rho k T}{\mu} + \frac{1}{3} a T^4 \tag{6.14}$$

(cf. Equation (5.16)), where k is the Boltzmann constant, μ is the mean molecular weight with the dimension of mass (rather than another common definition in which it is normalized to the mass of hydrogen), and a is the radiation constant (this parameter a has nothing to do with the semimajor axis used in Chapter 4).

6.1.2 Turbulence and Shocks

Fluids are commonly characterized by several dimensionless parameters that describe the overall state of the flow. Two of the most important are the *Reynolds number*, which quantifies the effect of viscosity, and the *Mach number*, which specifies the speed of the flow.

6.1.2.1 Reynolds Number

The Reynolds number Re is defined as the ratio of inertial to viscous forces in a fluid flow. If the flow has length scale L, velocity scale v, and kinematic viscosity ν, then dimensional analysis leads to the expression

$$Re = \frac{v\,L}{\nu}. \tag{6.15}$$

For low Reynolds number, $Re \ll 1$, viscous forces dominate and the flow is smooth. Fluids with $Re \gg 1$ tend to be turbulent.

For $Re \to \infty$, the flow behaves as a perfect inviscid fluid ($\nu \to 0$), but this does not mean that viscosity is irrelevant. Turbulent flow is characterized by motion on a wide range of scales, from the largest to the smallest turbulent eddies. The nonlinear term in the Euler equation sees to it that energy in the large-scale motion with $Re \gg 1$ is systematically transferred to smaller and smaller scales, eventually reaching a scale where $Re \sim 1$ and viscosity can efficiently dissipate the energy. These small scales are often out of reach for numerical simulations, with the result that turbulence acts as an effective viscosity, and high Reynolds number flows are very difficult to model numerically.

6.1.2.2 Mach Number

The Mach number $\mathcal{M}$ is the ratio of the fluid speed to the local sound speed

$$\mathcal{M} = \frac{v}{c_s}. \tag{6.16}$$

For an adiabatic flow, each fluid element has constant entropy and the sound speed is

$$c_s = \sqrt{\frac{\gamma P}{\rho}} = \sqrt{\frac{\gamma k_B T}{\mu}}, \tag{6.17}$$

where γ is the adiabatic index. Thus, the Mach number depends on temperature through c_s. The last equality of Equation (6.17) assumes an ideal gas.

Flows with $\mathcal{M} \lesssim 0.3$ are well described by the simplified incompressible fluid equations. Supersonic flows ($\mathcal{M} > 1$) lead to shock waves. Simply put, given that information in a fluid travels at the speed of sound, material in front of a supersonic disturbance cannot receive any warning before the disturbance arrives, so it cannot "get out of the way." As a result, material piles up in front of the disturbance as it propagates through the fluid, forming a shock. Astrophysics abounds with situations in which shock waves can form. Sometimes, as with stellar outflows and supernovae, the disturbance may move at thousands of kilometers per second through the surrounding medium. In other circumstances, such as the collision of two interstellar clouds, the speed of the disturbance may be quite modest—only a few kilometers per second—but still much greater than the local sound speed in a cool gas.

As illustrated in Figure 6.1, in the frame moving with the shock, high-speed material (index 1) flows in from the left while low-speed material (index 2) flows out to the right. In the vicinity of a shock, the Mach number is always $\mathcal{M} > 1$ before the shock and $\mathcal{M} < 1$ behind it. Within this simple one-dimensional geometry, we can write the equations describing the flow in flux-conservative form:

$$\begin{aligned}
\frac{\partial \rho}{\partial t} &= -\frac{\partial}{\partial x}(\rho u), \\
\frac{\partial}{\partial t}(\rho u) &= -\frac{\partial}{\partial x}(\rho u^2 + P), \\
\frac{\partial E}{\partial t} &= -\frac{\partial}{\partial x}[u(E + P)],
\end{aligned}$$

where u is the fluid speed and $E = \rho e + \frac{1}{2}\rho u^2$ is the total energy density of the fluid, with e the specific internal energy.

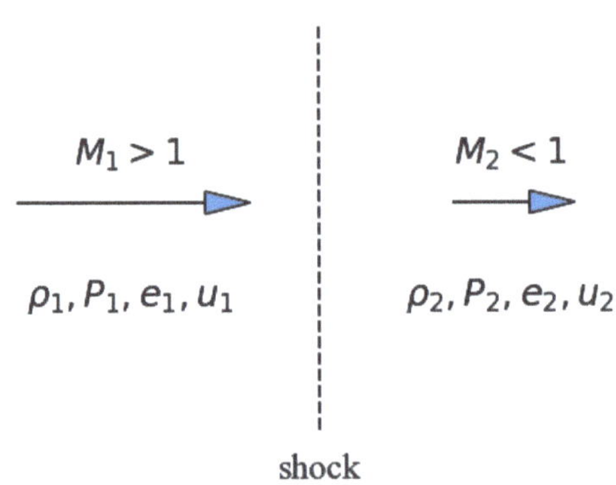

Figure 6.1. Geometry of a plane shock wave, in the frame of the shock.

Integrating these equations across the shock discontinuity and assuming a steady state, for a stationary shock in its frame of reference, we obtain the *Rankine–Hugoniot conditions* (Rankine 1870; Hugoniot 1885), expressing conservation of mass, momentum, and energy across the shock:

$$\begin{aligned}\rho_1 u_1 &= \rho_2 u_2,\\ \rho_1 u_1^2 + P_1 &= \rho_2 u_2^2 + P_2,\\ \rho_1 u_1\left(e_1 + \frac{1}{2}u_1^2 + P_1/\rho_1\right) &= \rho_2 u_2\left(e_2 + \frac{1}{2}u_2^2 + P_2/\rho_2\right).\end{aligned}$$

Note that we can use the first condition to simplify the third and give Bernoulli's principle (Bernoulli 1738): $e_1 + 0.5u_1^2 + P/\rho_1 = e_2 + 0.5u_2^2 + P/\rho_2$.

These equations can be solved for the jump in various quantities of interest across the shock, expressed here in terms of the preshock Mach number $\mathcal{M}_1 = u_1/c_{s1}$ (Landau & Lifshitz 1959):

$$\begin{aligned}\frac{\rho_2}{\rho_1} &= \frac{u_1}{u_2} = \frac{(\gamma+1)\mathcal{M}_1^2}{(\gamma-1)\mathcal{M}_1^2+2},\\ \frac{P_2}{P_1} &= \frac{2\gamma\mathcal{M}_1^2-(\gamma-1)}{\gamma+1},\\ \frac{T_2}{T_1} &= \frac{\left[2\gamma\mathcal{M}_1^2-(\gamma-1)\right]\left[(\gamma-1)\mathcal{M}_1^2+2\right]}{(\gamma+1)^2\mathcal{M}_1},\\ \mathcal{M}_2^2 &= \frac{2+(\gamma-1)\mathcal{M}_1^2}{2\gamma\mathcal{M}_1^2-(\gamma-1)}.\end{aligned}$$

The thickness of the shock is ultimately determined by the viscosity of the fluid (see Equation (6.10)), which turns preshock kinetic energy into postshock thermal energy and entropy. For systems without large-scale viscosity or explicit heating and cooling mechanisms, shocks are the only regions where the gas can increase its entropy. In many astrophysical contexts, the fluid viscosity is very low, so the shock is well-approximated as a discontinuous jump in fluid properties from left to right, as just described. However, in simulations, numerical and artificial viscosity can exceed the true physical viscosity by many orders of magnitude, spreading out the shock and limiting the resolution of the flow.

6.1.3 Hydrodynamical Timescales

To further develop our intuition, we define some timescales relevant to astrophysical phenomena. In Table 6.1, we provide an overview of various hydrodynamical regimes and their characteristic parameters, such as temperature and density. The range of parameters spans many orders of magnitude, and the physics may vary considerably across this range of astronomical objects, but practically all of these can be (and are) addressed with the hydrodynamical models described in this chapter.

Table 6.1. Physical Characteristics of Some Typical Regimes in Which Hydrodynamics Plays a Role

Typical Environment	T [K]	n [cm^{-3}]	h [pc]	State of Hydrogen
Cold neutral medium	50–100	20–50	100–300	Atomic
Warm neutral medium	10^4	0.2–0.5	300–400	Atomic
Warm ionized medium	8000	0.2–0.5	10^3	Ionized
H II regions	8000	0.2–0.5	70	Ionized
Stellar coronæ	10^6–10^7	10^{-4}–10^{-2}	10^{-7}	Ionized
Stellar interiors	10^4–10^7	10^{29}	10^{-9}	Ionized
Molecular clouds	10–20	10^2–10^6	80	Molecular
Circumstellar disks	3000–7000	10^{11}	10^{-5}–10^{-3}	Molecular

Estimates of the relevant fundamental parameters are given, including the typical temperature (T), number density (n), scale height (h), and the state of the gas. For the circumstellar disks, we adopt the midplane densities and radial scales.

6.1.3.1 Free-fall Timescale

As with a self-gravitating stellar system, we can define a collapse (or free-fall timescale for a gaseous cloud). This timescale can be written

$$t_{\rm ff} = \frac{1}{\sqrt{G\rho}}. \tag{6.18}$$

Note that, because $\rho \sim M/R^3$, this is essentially the dynamical timescale defined in Equation (4.5). For a 1 $M_\odot$ molecular cloud in a cubic parsec volume, $\rho \simeq 7 \times 10^{-23}$ g cm^{-3} and $t_{\rm ff} \simeq 15$ Myr.

The self-gravity of a gas is often ignored in hydrodynamical simulations, particularly when the gravity is dominated by another object, as in a relatively low-mass circumstellar disk, or when the gas is extremely hot and expanding, as in an expanding supernova shell. In other circumstances, of course, such as star formation, self-gravity is critical.

6.1.3.2 Sound-crossing Timescale

The sound-crossing timescale of a region of gas with length scale R is

$$t_{\rm sound} = \frac{R}{c_s}, \tag{6.19}$$

where the sound speed c_s is given by Equation (6.17). For a star in virial equilibrium (see Section 5.1.1.3), $t_{\rm sound} \approx t_{\rm ff}$.

6.1.3.3 Viscous Timescale

We can estimate the shortest unstable wavelength of a flow if we know the kinematic viscosity ν. The timescale on which a disturbance of wavenumber k is damped is

$$t_{\rm visc} \sim \frac{1}{k^2\nu}. \tag{6.20}$$

This can be seen by estimating $|\mathbf{v}|/|\mathbf{f}_{\rm visc}|$ in Equations (6.8) and (6.10), or by writing $L = 1/k$, $v = L/t_{\rm visc}$ and requiring that the Reynolds number $Re \sim 1$. Stabilization occurs if this timescale is shorter than the growth timescale of an instability. Comparing with the instability growth times given by Equations (6.25) and (6.21), we see that viscosity results in an upper limit on the unstable wavenumber k.

6.1.4 Hydrodynamical Instabilities

Hydrodynamical flows are subject to a number of instabilities, the best known being the Kelvin–Helmholtz, Rayleigh–Taylor, and Jeans instabilities. The first two operate in (for example) boundary layers between fluids; the second and third occur in external or self-gravitating potentials, respectively. The growth of these instabilities is eventually arrested by viscosity, the effect of which increases as spatial scales become small and velocities become high.

6.1.4.1 Rayleigh–Taylor Instability

The Rayleigh–Taylor instability arises when a dense fluid (density ρ_1) lies above a less dense fluid (density ρ_2) in a gravitational field (which defines "above"). The growth timescale for a mode with wavenumber k in a uniform field g is

$$t_{\rm RT} = \left(\frac{\rho_1 - \rho_2}{\rho_1 + \rho_2}\; gk\right)^{-1/2}. \tag{6.21}$$

The instability is driven by the fractional difference between the two densities. The timescale $(gk)^{-1/2}$ in Equation (6.21) is the time needed for a point on the boundary to accelerate (with acceleration g) through a distance comparable to the wavelength $(1/k)$. Shorter wavelengths grow more quickly, and they become nonlinear sooner, but the long wavelengths are ultimately crucial for mixing the fluids.

6.1.4.2 Shear and Kelvin–Helmholtz Instabilities

The Kelvin–Helmholtz instability arises in the interface layer between fluids with different velocities. Consider a fluid of constant mean density ρ, which is subject to shear due to the presence of another fluid with which it shares a boundary at $z = 0$. Near the boundary, the relative flow velocity is

$$\mathbf{v} = v(z)\ \mathbf{e}_x = Sz\ \mathbf{e}_x, \tag{6.22}$$

where $S = dv/dz\,|_0$ is the local shearing rate of the fluid. Equation (6.22) can be viewed as a first-order Taylor expansion of the actual shear flow near $z = 0$.

Now consider two parcels of fluid, one at $z = 0$ and the other at $z = \varepsilon$, where ε is small. The kinetic energy (per unit volume) in the two parcels is

$$E = \frac{1}{2}\rho v(0)^2 + \frac{1}{2}\rho v(\epsilon)^2 = \frac{1}{2}\rho S^2\epsilon^2. \tag{6.23}$$

Moving in the frame of reference of the fluid at $z = 0$, the momentum of the lower parcel is zero, while a parcel in the upper layer has a momentum (per unit volume) of $\rho v = \rho S\varepsilon$. Now imagine switching the two parcels, so the top one becomes the bottom, and vice versa, mixing and equalizing their momenta. If total momentum is conserved in the process, then the momentum of each of the two parcels becomes $\frac{1}{2}\rho S\epsilon$ and they both move with the average speed $\frac{1}{2}S\epsilon$. The kinetic energy of the switched parcels is then

$$E_s = \frac{1}{4}\rho S^2\epsilon^2. \tag{6.24}$$

Thus, the new flow consisting of the two switched parcels has lower energy than the original: the difference is $\Delta E = E - E_s = 1/4\rho S^2\varepsilon^2 > 0$. This energy is available for amplifying initial perturbations and driving instabilities. This is the fundamental reason why shear instabilities are so common: the shear itself is a reservoir of kinetic energy that can be used to drive the instability. Figure 6.2 presents an actual example of observed Kelvin–Helmholtz instability.

If gravity can be neglected, then the shear flow is demonstrably unstable on all length scales. For a shear flow involving fluids of densities ρ_1 and ρ_2 and velocities v_1 and v_2, the growth timescale for wavenumber k is (Bradt 2014)

$$t_{\rm KH} = \frac{\rho_1 + \rho_2}{\sqrt{\rho_1\rho_2}}\frac{1}{k\Delta v}. \tag{6.25}$$

Here, $\Delta v \equiv |v_1 - v_2|$ and $\lambda = 2\pi/k$. When the densities are comparable, $t_{\rm KH} \sim 1/k\Delta v$, so the growth timescale is comparable to the time taken for the fluids to move one wavelength relative to one another. Again, the shortest wavelengths grow most rapidly. In Section 6.6.2, we present an example to test the AMUSE hydrodynamics solvers on a Kelvin–Helmholtz instability.

Figure 6.2. Kelvin–Helmholtz instability observed in the atmosphere of Jupiter, as photographed by the *Juno* mission on 2017 December 16 at 12:43 EST at a distance of 13,345 km. Image Credit: NASA/JPL-Caltech/SwRI/MSSS/Gerald Eichstadt/Sean Doran © CC NC SA.

6.1.4.3 Jeans Instability

The Rayleigh–Taylor and Kelvin–Helmholtz instabilities appear along the boundaries between fluids. The Jeans instability can arise in a single fluid, so long as the fluid is sufficiently dense.

If the free-fall timescale t_{ff} (Equation (6.18)) in the fluid exceeds the sound-crossing timescale t_{sound} (Equation (6.19)), pressure forces can overcome gravity and the fluid is stable against small perturbations. However, if $t_{sound} > t_{ff}$, the fluid is unstable against gravitational collapse. The resulting length scale is called the Jeans length and can be estimated by conflating Equations (6.18) and (6.19), which results in

$$\lambda_J = \frac{2\pi}{k_J} \sim \frac{c_s}{\sqrt{G\rho}}. \tag{6.26}$$

A more careful analysis of the collapse, based on a study of the stability of one-dimensional sound waves (see Binney & Tremaine 1987), gives

$$\lambda_J = c_s\sqrt{\frac{\pi}{G\rho}}. \tag{6.27}$$

All scales larger than λ_J are unstable against gravitational collapse, whereas smaller scales are stable. An associated mass, called the Jeans mass, is conventionally taken to be the mass contained within a sphere with diameter equal to the Jeans length:

$$M_J = \frac{4}{3}\pi\rho\left(\frac{1}{2}\lambda_J\right)^3 = \frac{\pi^{5/2}}{6}\frac{c_s^3}{\sqrt{G^3\rho}}. \tag{6.28}$$

For a typical dense molecular cloud with temperature 10 K and density 10^3 hydrogen molecules cm^{-3}, the sound speed is $c_s = 260$ m s^{-1}, so $\lambda_J = 1$ pc, and $M_J = 26\ M_\odot$.

6.1.5 Physics of the Integrator

6.1.5.1 Underlying Assumptions

Several assumptions must be made about the state of the fluid, depending on accuracy requirements and our knowledge of the underlying physics. We mention here just some of the most important assumptions made in astronomical fluid dynamical simulations.

1. **Fluid approximation**. As mentioned previously, to derive the fluid equations (Equations (6.9) and (6.8)) solved by all of the hydrodynamics modules in AMUSE, it is necessary to assume that we can find a length scale that is small compared to any scale of interest, but large enough to contain sufficient particles for density, temperature, etc. to be well defined. For example, imagine a giant molecular cloud of mass $10^4\ M_\odot$ and radius 3 pc. The mean density is $\rho \approx 6.0 \times 10^{-21}$ g cm^{-3}. If the composition is pure molecular hydrogen, $m_{H2} = 3.35\times 10^{-24}$ g, then a cubic meter contains 1.8×10^{12} molecules, more than enough to accurately define local thermodynamic

quantities. At the same time, 1 m is far smaller than any scale of interest for studies of interstellar gas flow or star formation.

Note that, even if the gas can be represented well as a continuous fluid, a similar problem recurs in simulation methods that treat the fluid as an ensemble of particles (see Section 6.2.1). It is important to ensure that the number of particles is large enough that the Poisson noise introduced by their finite number is small, so that the particles accurately represent the underlying distribution function. In principle, we could use a huge number of small particles to represent a cloud, but this would be numerically infeasible. However, if we make the particles too massive, we run the risk of averaging too much, thereby artificially smoothing the solution. One of the most important ingredients of hydrodynamical simulations is choosing the proper resolution to address the problem of interest.

2. **Classical fluid**. All of the AMUSE hydrodynamics modules assume that the fluid at any point can be described by its velocity, temperature, and composition, and that quantum mechanical and relativistic effects can be ignored.
3. **Local thermodynamic equilibrium**. We assume that every fluid element is in local thermodynamic equilibrium, i.e., all particles in the element, can interact locally and reach equilibrium with each other (see Section 5.1.2). As a consequence, there is only one degree of freedom: the temperature. This assumption is violated in shocks, in which the temperature is formally discontinuous (as are the related thermodynamic quantities). All AMUSE modules contain prescriptions to handle such sharp features, either by incorporating explicit shock-capturing schemes or by spreading shocks out over a few resolution elements.

All of these assumptions are built into the hydrodynamics modules in AMUSE. It is probably worth your time to check that they actually apply before launching into a large-scale AMUSE simulation.

6.2 Hydrodynamics in AMUSE

The hydrodynamics modules in AMUSE can be divided into two broad categories, each having its own fan base and sets of advantages and disadvantages. Experts will argue (interminably) the merits of each. Here, we try to take an agnostic view.

6.2.1 Types of Hydrodynamics Code

1. **Lagrangian methods: SPH.** In the Lagrangian description of a fluid flow, we follow the motion of a fluid element as described by Equation (6.8). One strength of Lagrangian methods is that they preserve Galilean invariance, i.e., the laws of motion are the same in all inertial frames.

   In most numerical implementations of this approach, the fluid is represented by particles. Each particle represents a finite mass of the fluid; density and other thermodynamic quantities are computed by kernel estimation from nearby particles. Because each particle may be thought of as being smoothed out over some region, these methods are generically referred to as smoothed particle hydrodynamics (SPH). The amount of smoothing determines the accuracy and resolution of the method, so it must be chosen carefully. Larger smoothing means that many particles contribute to the density at any given point, reducing statistical fluctuations but also decreasing resolution. Interactions among particles include gravity, implemented much as in an N-body code (see Section 4.2), as well as pressure forces due to a particle's neighbors.

SPH codes are relatively straightforward to implement. The use of particles naturally makes them well-suited to astrophysical phenomena involving complex morphologies and/or poorly defined boundaries. They tend to be very stable, although the results can be adversely affected by noise. One significant problem is the lack of a natural prescription for dissipation and diffusion, which are generally controlled by free parameters. Because they put the particles where the mass is, SPH codes generally provide a poor description of low-density regions in a flow. The smoothing formulation leads to difficulties in resolving environments with high-density or temperature contrasts. In particular, shocks may be difficult to resolve, although there are many variations on the basic SPH method in which such complex phenomena can be handled effectively (e.g., Monaghan & Gingold 1983). Variations and extensions to the fundamental SPH description allow for the study of complex processes, including, for example, magnetohydrodynamics, turbulence, and various relativistic effects (Iwasaki & Inutsuka 2011; Price et al. 2018).

2. **Eulerian methods: Grid-based and finite-volume schemes.** Instead of following the motion of individual fluid elements, the equations of motion (as expressed in Equation (6.9)) can instead be solved on a grid. Most high-performance Eulerian grid-based codes rely on so-called Godunov (1959) schemes, which solve the underlying partial differential equations using a conservative finite-volume method at each intercell boundary. The original Godunov method was accurate only to first order in space and time, but higher-order methods have since been developed (van Leer 1979; Colella & Woodward 1984), and are now the norm.

   In the simplest implementation of a Eulerian scheme, a fixed three-dimensional grid is imposed on the system and the fluid dynamical equations are solved on the grid vertices. The geometry can vary from run to run, but it remains fixed at runtime. Common grid choices include Cartesian, spherical, cylindrical, or toroidal coordinates. Ideally, the problem under study has some symmetry, allowing it to map naturally onto some grid topology, and this symmetry should not change greatly during the simulation.

   In contrast to Lagrangian methods, grid-based methods are not Galilean-invariant, because the algorithm to calculate the fluxes has preferred directions. This may pose a problem when studying Kelvin–Helmholtz or other shear instabilities. For such studies, grid codes tend to give different solutions for the same relative velocity between the two layers when the entire system is moving (see Section 6.1.4.2). This can be solved (at considerable expense in computer resources) by reducing the truncation error in the flux calculation algorithm and increasing resolution until the hydrodynamical instabilities are resolved (Godunov & Kulikov 2014). This is rarely done because of its computational cost. As a consequence, Eulerian fixed-grid codes are not really suitable for studying discontinuities and multiphase media. These problems seem to be solved with moving-mesh methods, but further study is required.

The fixed-grid topology allows grid-based hydrodynamics codes to perform excellently on large parallel architectures. In addition, it is relatively straightforward to extend the fluid dynamical equations to include radiative transfer, magnetic fields, or relativistic effects. Despite the resolution problems introduced by adopting a fixed grid, these methods are still quite popular.

Fixed-grid methods can be improved by allowing the grid to be refined, depending on local resolution requirements. The resulting adaptive mesh refinement (AMR) codes are suited particularly well to addressing the development of discontinuities, such as shocks and gravitational collapse, whose appearance in a simulation is hard to predict in advance (MacNeice et al. 2000; Adams 2014). The main disadvantages of grid-based methods are their relatively large memory requirements and the complexity of coding and optimization. In addition, the grid topology may be reflected in the solution, and boundary conditions may produce unwanted artifacts in the flow.

3. **Unstructured mesh schemes.** Particle- and grid-based methods are slowly being superseded by moving-mesh methods that solve the fluid dynamical equations on unstructured grids that evolve with the fluid (Robertson et al. 2010; Springel 2010). In many ways, these methods combine the best properties of Lagrangian and Eulerian methods. They are efficiently parallelizable by domain decomposition, and are suitable for solving systems with steep temperature and/or density gradients. A major advantage of these methods is that the solution is independent of grid structure, removing one of the major disadvantages of AMR methods.

Table 6.2 lists the most important hydrodynamics codes currently available in AMUSE. Here, we draw a distinction between particle-based (TreeSPH), fixed grid (grid), AMR codes, and moving mesh (MM). Each of the codes is parallelized, because hydrodynamics is expensive in terms of computing time, and high resolution requires considerable amounts of computer memory. AMUSE further contains an interface to `FLASH` (Fryxell et al. 2000), which requires registration or permission to use. These codes are not routinely distributed with AMUSE, but the interface is available within the repository. In our estimation, the best codes to use are `Fi` or `Phantom`. Ideally, we would like to interface with `BonsaiSPH` (Bedorf & Portegies Zwart 2020), but we still have not resolved how to address multiple GPUs across a distributed memory machine in the AMUSE interface.

We can run a hydrodynamics code in AMUSE in much the same way as an N-body or stellar evolution simulation would be run. Because the two main

Table 6.2. Hydrodynamics Codes Distributed within AMUSE

Name	Ref.	Type of Code	Language	Parallel	Boundaries
Gadget2	1	TreeSPH	C	Yes	Ciiopr
Fi	2	TreeSPH	F95	Yes	Ciiopr
Phantom	3	TreeSPH	F95	Yes	Ciiopr
Capreole	4	grid	F95	Yes	–iopr
Athena3D	5	AMR	C	Yes	–iopr
mpiAMRVAC	6	AMR	F95	Yes	–iopr
FLASH	8	AMR	F90	Yes	–i-pr
Arepo	9	MM	C	Yes	Ciiopr

Each of these codes compiles by default with an adiabatic equation of state. An isothermal equation of state is implemented, but requires the code to be recompiled, except for `Fi` where it can be set as a code parameter. The global code characteristics are indicated in the third columns, where TreeSPH indicates the SPH codes for which the interacting forces are calculated using a tree-code, grid and AMR refer to Cartesian or refined grid methods, and MM refers to moving mesh. The last column gives the available boundary conditions accessible directly from AMUSE: "cont," "inflow," "interface," "outflow," "periodic," and "reflective" (see Section 6.2.3). For Athena3D see also https://trac.princeton.edu/Athena/wiki/AthenaEqns. References: 1: Springel (2000, 2005); 2: Hernquist & Katz (1989); Pelupessy et al. (2004); Gerritsen & Icke (1997); 3: Price et al. (2018); 4: Mellema et al. (1991); 5: Gardiner & Stone (2005, 2008); Stone et al. (2008, 2010); 6: Meliani et al. (2007); Keppens et al. (2012); van der Holst et al. (2012); Porth et al. (2014); 8: Fryxell et al. (2000); 9: Pakmor et al. (2016).

algorithms (grid and SPH) used to solve the fluid equations differ so greatly, the user must decide in advance which one to use. Ideally, we would like to be able to switch back and forth between the two approaches as needed. Regrettably, that is not currently possible in AMUSE because we do not know how to uniquely convert a particle distribution to an adaptive grid, and generating a low-noise particle distribution from an adaptive mesh is challenging. The SPH interface is developed better and used most widely, so most of the discussion that follows will assume that choice. In Section 6.2.3, we will present some simple examples using grid codes.

6.2.2 Smoothed Particle Hydrodynamics

Code example 6.1 presents a simple function for setting up and solving a hydrodynamics problem in AMUSE. This script does not do much, but in the next few paragraphs we will demonstrate that, as with gravity and stellar evolution, it is possible to do some interesting experiments with just hydrodynamics.

The function first sets up a virialized gas Plummer (1911) sphere of specified mass and radius, realized as `N` particles (see Section 3.2):

```
converter = nbody_system.nbody_to_si(Mtot , Rvir)
gas = new_plummer_gas_model(N, convert_nbody=converter)
```

Note that, because the function new_plummer_gas_model is written in N-body units, we need a `converter` (see Section 2.3) to define the physical units used. Most hydrodynamics solvers work in CGS units, unlike the N-body and stellar units used in previous chapters. However, as discussed in Section 2.3, AMUSE keeps track of such details, so we do not have to do it explicitly.

```
def hydro_minimal(
    number_of_particles=100,
    total_mass=1 | units.MSun,
    virial_radius=1 | units.RSun,
    time_end=6 | units.hour,
):
    converter = nbody_system.nbody_to_si(total_mass, virial_radius)
    gas = new_plummer_gas_model(number_of_particles, convert_nbody=converter)

    hydro = Gadget2(converter)
    hydro.gas_particles.add_particles(gas)
    energy_total_init = (
        hydro.kinetic_energy + hydro.potential_energy + hydro.thermal_energy
    )
    hydro.evolve_model(time_end)
    write_set_to_file(hydro.particles, "hydro.amuse")

    energy_kinetic = hydro.kinetic_energy
    energy_potential = hydro.potential_energy
    energy_thermal = hydro.thermal_energy
    energy_total = energy_kinetic + energy_potential + energy_thermal
    virial_ratio = (energy_kinetic + energy_thermal) / energy_potential
    energy_difference = (energy_total_init - energy_total) / energy_total
    com = hydro.gas_particles.center_of_mass()
    print(
        f"T= {hydro.get_time()} M= {hydro.gas_particles.mass.sum()} "
        f"E= {energy_total} Q= {virial_ratio} "
        f"dE= {energy_difference} CoM= {com.in_(units.RSun)}"
    )

    hydro.stop()
```

Code example 6.1: Minimal routine for evolving a Plummer gas cloud of mass `total_mass` and virial radius `virial_radius` to some end time `time_end`. Note that we left out the various `import` statements. The full example can be found in `amuse.examples.hydro_minimal`.

We initialize the SPH code `Gadget2` as our hydrodynamics code and commit the gas particles, much as we did in Chapter 4:

```
hydro = Gadget2(converter)
hydro.gas_particles.add_particles(gas)
```

We are now ready to evolve the hydrodynamics solver to some end time `time_end`:

```
hydro.evolve_model(time_end)
```

Eventually, we could dump the data to a file with the name `filename`

```
write_set_to_file(hydro.particles, filename)
```

We might improve this hydrodynamical calculation by running it in small time increments and printing diagnostics after each, just as we did for the gravity solver (Code example 4.1). Hydrodynamics simulations are often used in this way to make animations. Many of these can be quite instructive, but beware of convincing-looking animations that actually present erroneous results. Validating results is an important but difficult aspect of these simulations.

6.2.3 Grid-based Methods

Setting up a grid code is not much different from setting up an SPH code, except that defining the boundary conditions and generating the initial conditions are somewhat more elaborate. The fundamental calling sequence is similar to the SPH version. First, we declare the converter and start the code:

```
converter = nbody_system.nbody_to_si(1|units.MSun 1|units.RSun)
hydro_code = Athena(converter, number_of_workers=16)
```

Grid-based hydro codes parallelize quite well, and the calculations are often computationally intensive. We therefore recommend always running them in parallel. In this example, we run the code using 16 cores.

We can tune the working of the hydrodynamics solver by specifying some of the fundamental parameters, such as the adiabatic index, for which we adopt $\gamma = 5/3$ for the ideal monatomic gas. Another important parameter is the Courant–Friedrichs–Lewy condition, which is used to control the convergence while solving the hyperbolic partial differential equations in the finite-difference scheme (see Courant et al. 1928), for which we choose a value of 0.3,

```
hydro_code.parameters.gamma = 5/3
hydro_code.parameters.courant_number=0.3
```

More detail on the important parameters used by this AMUSE module (and others) are presented in Appendix A.

We further specify the grid resolution in each of the three Cartesian coordinates, and the length of the computational domain

```
hydro_code.parameters.nx = resolution
hydro_code.parameters.ny = resolution
hydro_code.parameters.nz = resolution

hydro_code.parameters.length_x = grid_size
hydro_code.parameters.length_y = grid_size
hydro_code.parameters.length_z = grid_size
```

Here, `resolution` is an integer value, and `grid_size` represents the width of a grid cell (with units). We subsequently specify the boundary conditions, which are attributes of the running hydro code and can be set as parameters. For example, to apply open outflow boundary conditions along the Cartesian coordinates,

```
hydro_code.x_boundary_conditions = ("outflow", "outflow")
hydro_code.y_boundary_conditions = ("outflow", "outflow")
hydro_code.z_boundary_conditions = ("outflow", "outflow")
```

Referring to the left and right boundaries in the appropriate axes. Alternatives to the `"outflow"` boundary condition are `"cont"`, `"inflow"`, `"interface"`, `"periodic"`, and `"reflective"`, although not all of these may be defined for each of the available hydrodynamics solvers in AMUSE. The keywords refer to the different ways in which grid boundaries are handled, and they can be set separately for each of the three Cartesian coordinates. In this case, we opt for "outflow" boundary conditions. Table 6.2 presents a brief overview of the boundary conditions supported by each code; Appendix A gives more detail on the boundary conditions.

After the hydrodynamics solver has been initialized, we construct the grid, which is created with a specified geometry and filled with distributions of density, momentum, and energy. Initializing a three-dimensional grid can be realized with the built-in function from the `datamodel` package

```
from amuse.datamodel import new_regular_grid
grid = new_regular_grid((100, 100, 100), [10, 10, 10] | units.RSun)
```

which generates a three-dimensional grid of 10^6 ($=100 \times 100 \times 100$) cells, with total lengths of 10 $R_\odot$ in each of the three Cartesian coordinates (the computational domain runs from zero to 10 $R_\odot$ in each). Densities (`grid.rho`), energies (`grid.energy`), and momenta (`grid.momentum`) can then be set in the individual grid cells. Global grid properties can be added to each of the grid cells in much the same way as we add properties to a particle set, e.g.,

```
grid.rho = 1.e-10 * units.MSun/units.RSun**3
```

Other parameters can be set similarly.

Filling the grid with useful initial conditions is complicated, and only a limited number of built-in functions exist to assist you in the task. One of these functions is `fill_grid_with_spherical_cloud`, which adds a homogeneous spherical density distribution to `grid`. Thus, to add a homogeneous spherical cloud with density equal to the mean solar density,

```
from amuse.ext import cloud
cloud.fill_grid_with_spherical_cloud(
    grid,
    center = [5.0, 5.0, 5.0] | units.RSun,
    radius = 1 | units.RSun,
    rho = 1.0 | units.MSun / units.RSun**3,
    rhovx = 0.0 | units.kg / (units.s * units.m**2),
    rhovy = 0.0 | units.kg / (units.s * units.m**2),
    rhovz = 0.0 | units.kg / (units.s * units.m**2),
    energy = 1.e+40 | units.erg / units.RSun**3
)
```

Sometimes, one needs to add additional attributes, such as chemical composition or magnetic fields, to the grid. This can be accomplished by adding a vector attribute to each of the grid cells. The function `add_global_vector_attribute` adds a vector quantity to a local particle set or grid. For example, to add a three-dimensional vector called `composition` with components `X`, `Y`, and `Z` describing the composition in hydrogen, helium, and metals (as in stellar evolution, Section 5.2.1, where according to the common terminology one uses lower cases for the same variables.), we could write:

```
grid.add_global_vector_attribute("composition", ["X","Y","Z",])
grid.composition = (0.7, 0.29, 0.01)
```

The grid itself is eventually copied to the running hydro code using a channel.

```
channel = grid.new_channel_to(hydro_code.grid)
channel.copy()
```

Running the hydrodynamics solver to some end time `t_end` is no more elaborate than running any other code:

```
channel = hydro_code.grid.new_channel_to(grid)
dt = t_end/10.
while hydro_code.model_time<t_end:
     hydro_code.evolve_model(hydro_code.model_time + dt)
     channel.copy()
     write_set_to_file(grid, filename, ''amuse'')
hydro_code.stop()
```

A fully working example of this procedure can be found in `amuse.examples.supernova_grid`.

6.2.4 Managing Shocks and Discontinuities

Efficient and accurate treatment of shocks is a key component of all modern hydrodynamics simulation codes (see Section 6.1.2.2). Simply put, finite-difference schemes do not handle discontinuities well, and when the discontinuity is too steep (spread over too many resolution elements), a shock wave produces unphysical density oscillations behind the advancing front.

Simple finite-difference codes (like the SPH codes in AMUSE) rely on artificial viscosity to prevent the steep gradients and unphysical oscillations associated with shocks. There are many variations, but the basic idea goes back to Von Neumann & Richtmyer (1950), who replaced the pressure P in the Euler (Equation (6.8)) and energy (Equation (6.13)) equations by $P + q$, where

$$q = \begin{cases} \alpha \rho h^2 (\nabla \cdot \mathbf{v})^2, & \nabla \cdot \mathbf{v} < 0 \\ 0, & \nabla \cdot \mathbf{v} > 0. \end{cases} \tag{6.29}$$

Here, h is the particle smoothing length in an SPH calculation or the local grid spacing in a grid code. The method artificially increases the local pressure in regions of strongly convergent flows to limit large density gradients. The parameter $\alpha \sim 1$ is tuned to spread shocks out over a few (4–5) resolution elements, maintaining stability but losing precision.

More sophisticated *shock-capturing schemes* (incorporated into the grid-based AMUSE modules) leverage our knowledge of basic one-dimensional shock physics to model the flux of mass, momentum, and energy from one grid cell to the next (see Section 6.6.1). These schemes provide much better resolution of fine structure, albeit at the expense of a considerable increase in code complexity. Comparable schemes have been introduced into the SPH formalism (Inutsuka 2002; Murante et al. 2011), with only limited success.

6.2.5 Initializing a Grid from a Particle Distribution

Sometimes it is easier to generate a particle distribution with certain properties and then convert it to a grid. For example, we might start by generating an initial Plummer sphere (see Chapter 4) particle distribution, as described in Section 6.6.2. We can then convert this distribution to a grid by using the function `convert_SPH_to_grid`:

```
from amuse.ext.sph_to_grid import convert_SPH_to_grid
sph_code = setup_sph_code(Fi, N, Rvir, rho, u)
grid = convert_SPH_to_grid(sph_code, (100, 100, 100))
```

Here, we adopt the SPH code `Fi` with `N` particles distributed in a sphere with radius `Rvir`, density `rho`, and internal energy `u`. The SPH code is then used to fill the grid cells from the particle distribution. Bear in mind that some AMUSE data structures naturally map onto grids. For example, the one-dimensional internal structure of a star in a Henyey stellar evolution code (see Section 6.2) is easily translated into a three-dimensional grid for use in a hydro code.

Figure 6.3 presents the result of a grid-based hydrodynamical simulation of an exploding core-halo "star". Initially, the star had a central 1.4 $M_{\odot}$ core, of radius 0.1 $R_{\odot}$, surrounded by a homogeneous spherical halo of mass 3 $M_{\odot}$ and radius 1 $R_{\odot}$. We then uniformly injected 10^{51} erg of energy into the core, causing it to explode. The adopted grid resolution was fixed at 256^3. The symmetric blobs visible at the corners are artifacts of this Cartesian grid.

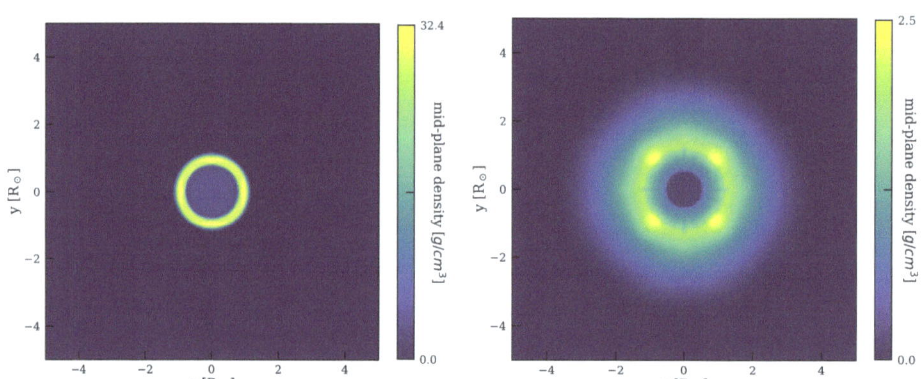

Figure 6.3. Midplane density distribution of a 4 $M_{\odot}$ core-halo "star" whose 1.4 $M_{\odot}$ core explodes at time 0 with an energy of 10^{51} erg. Resolution is 256 × 256 × 256 grid cells. The two snapshots are taken at 100 s (left) and 300 s (right) after the explosion. The calculation was performed using `Athena`. The figure was made using the script `amuse.examples.supernova_grid`. It takes a couple of hours to run on a laptop.

Figure 6.4 presents the result of an SPH calculation using exactly the same initial conditions as for the grid calculation just described Figure 6.3. The presentation in Figure 6.4 is slightly different from Figure 6.3, because generating an image from a particle distribution entails interpolating the particle densities onto the image plane, rather than simply taking slices from a grid, but the two images show interesting similarities and discrepancies. In this case, we first performed the calculation using one script, then wrote the data to a file, and later made the plot using another script.

To make the figure, we first construct a regular grid of x- and y-coordinates:

```
def setup_grid(N):
    x,y = numpy.indices((N+1,N+1))
    x = L*(x.flatten()-N/2.)/N
    y = L*(y.flatten()-N/2.)/N
    z = 0.*x
    vx = 0.*x
    vy = 0.*x
    vz = 0.*x
    x = units.parsec(x)
    y = units.parsec(y)
    z = units.parsec(z)
    vx = units.kms(vx)
    vy = units.kms(vy)
    vz = units.kms(vz)
    return x, y, z, vx, vy, vz
```

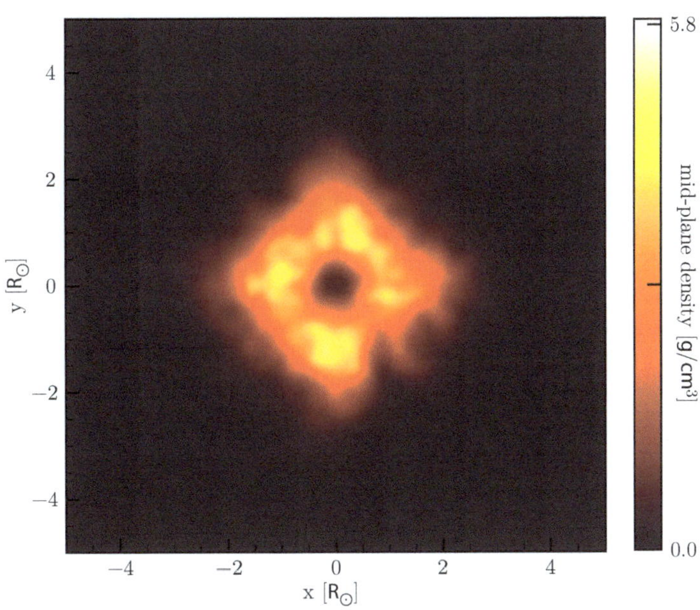

Figure 6.4. Projected density distribution of the same core-halo exploding object as in Figure 6.3. This run was performed using the SPH code `Gadget2` with 10^5 equal-mass particles. The image is taken 300 s after the injection of 10^{51} erg of energy into the core. The data to make the figure were generated using the script `amuse.examples.supernova_sph`, requiring a few minutes to run. The figure itself was made using `amuse.examples.plot_supernova_sph`.

and fill this using interpolated values from a running SPH code:

```
def make_hydromap_and_show_picture(sph_particles, N=100, L=10):
    hydro = Gadget2(converter)
    hydro.gas_particles.add_particles(sph_particles)
    x, y, z, vx, vy, vz = setup_grid(N)
    rho,rhovx,rhovy,rhovz,rhoe = hydro.get_hydro_state_at_point(x,y,z,vx,vy,vz)
    rho = rhoe.reshape((N+1,N+1))
    pyplot.imshow(numpy.log10(rho.value_in(units.erg/units.RSun**3)))
    pyplot.show()
```

When generating initial conditions for a grid-based hydrodynamics code, we would ideally like to be able to run the same initial conditions using an SPH code. The translation from a grid to a particle representation is not unique, but the routine `convert_grid_to_SPH` performs such a conversion. The resulting particle distribution still reflects the structure of the grid—even in an SPH code, the structure of the grid is not erased until the system has evolved for a time comparable to the free-fall timescale (Section 6.1.3.1).

6.3 Extending the Hydrodynamics Solver

Many hydrodynamical studies focus on the evolution of a fluid flow with specified initial conditions, as discussed in Section 6.5 and Section 6.4.2 below. In some cases, however, we need to understand what happens when gas enters or leaves the system. In this section, we explore how such processes are implemented in AMUSE.

6.3.1 Generating New SPH Particles

New gas may be introduced as part of the boundary conditions in the computational domain (e.g., the inflow of a gas stream from one side), or from a source within the simulated volume (e.g., a stellar wind or a supernova). In the SPH formalism, new gas is introduced simply by initializing new particles on some surface or in some volume around a source. The location of the source and the state of the gas are typically provided by another code, such as a stellar evolution calculation that generates a wind from an evolving star.

The code fragment below demonstrates how to create `n_new` new SPH particles in an existing system, add them to the existing gas particles (`gas_particles`), and subsequently synchronize the gas particles in the main `Python` script with the particles in the hydro code,

```
new_gas = Particles(n_new)
give_particles_some_properties(new_gas)
gas_particles.add_particles(new_gas)
gas_particles.synchronize_to(hydro.gas_particles)
```

The astronomical magic is hidden in the routine `give_particles_some_properties()`, which provides the newly created particles with position, velocity, temperature, etc. We do not spell out this function in detail, but an example is provided in Code example 6.2. The function `synchronize_to` ensures that the

```
def new_sph_particles_from_stellar_wind(stars, mass_per_gas_particle):
    new_sph = Particles(0)
    for i, star in enumerate(stars):
        number_of_gas = int(-star.Mwind / mass_per_gas_particle)
        if number_of_gas == 0:
            continue
        total_mass_of_gas = mass_per_gas_particle * number_of_gas
        star.Mwind += total_mass_of_gas
        new_gas = Particles(number_of_gas)
        new_gas.mass = mass_per_gas_particle
        new_gas.h_smooth = 0.0 | units.parsec

        dx, dy, dz = uniform_unit_sphere(number_of_gas).make_xyz()
        new_gas.x = star.x + (dx * star.radius)
        new_gas.y = star.y + (dy * star.radius)
        new_gas.z = star.z + (dz * star.radius)
        for j, gas_particle in enumerate(new_gas):
            r = gas_particle.position - star.position
            r = r / r.length()
            v_wind = (
                constants.G
                * star.mass
                / (gas_particle.position - star.position).length()
            ).sqrt()
            gas_particle.u = 0.5 * (v_wind) ** 2
            gas_particle.vx = star.vx + r[0] * star.terminal_wind_velocity
            gas_particle.vy = star.vy + r[1] * star.terminal_wind_velocity
            gas_particle.vz = star.vz + r[2] * star.terminal_wind_velocity
        new_sph.add_particles(new_gas)
    return new_sph
```

Code example 6.2: Routine to generate new SPH particles modeling the stellar wind of a set of stars. This code example is a fragment from the script `amuse.examples.hydro_outflow_particles`. The full script models a colliding-wind binary system and will take several weeks to run on a laptop. In Chapter 9, we present an example of how this code works when coupled to gravity.

particle set in the main script is consistent with the particles in the `hydro` module. Note that we could have accomplished the same result by adding `new_gas` directly to the `hydro_particle` set:

```
hydro.gas_particles.add_particles(new_gas)
```

Code example 6.2 presents an example of how to set the characteristics of the SPH particles. In this case, we are modeling the colliding winds of a binary star system. The positions and velocities of these stars typically come from a gravity code, while the stellar parameters are determined by a stellar evolution code. The amount of mass lost by each star is included in the star particles as a free parameter (`star.MWind`), which can be set in the main script from the stellar evolution code. The attribute `MWind` is not standard, but is temporarily added to the stellar particle set just for this purpose:

The new particles created in Code example 6.2 can be incorporated into the SPH code in two basic ways. At one extreme, they could be given high radial velocities

```
def v_terminal_teff(star):
    t4 = (np.log10(star.temperature.value_in(units.K)) - 4.0).clip(0.0, 1.0)
    return (30 | units.km / units.s) + ((4000 | units.km / units.s) * t4)
```

Code example 6.3: Determining the terminal velocity of an SPH particle for a star with a given temperature. This expression was used by Pelupessy & Portegies Zwart (2012) to determine the terminal velocity of the winds of massive stars.

but low temperatures, in which case the SPH code will establish a wind-like outflow that will shock and thermalize as it impacts the surrounding medium. Alternatively, at the other extreme, they could be given low velocities and high temperatures, consistent with the postshock thermalized gas, as calculated by the routine in Code example 6.3. The wind speed and/or shocked temperatures can be taken from a stellar evolution code. In reality, the proper prescription lies somewhere in between.

We must make these choices because SPH is incapable of resolving the small-scale details of the true outflow, as the wind streams out, shocks, and thermalizes, so we must inject the energy as some combination of kinetic and thermal energy. The details can be resolved by some AMR grid models, but at the cost of very short time steps. The best choice between pure kinetic and pure thermal energy input is generally resolution-dependent. It is interesting to try both extreme methods to see the effect of these different strategies on the results of the simulation. In Section 6.4.3, we discuss an example in which a star accretes from a companion wind.

6.3.2 Removing SPH Particles

It is sometimes also necessary to remove gas from a system; for example, when it leaves the computational domain, falls into a black hole, or is accreted by another object. Removing gas is not much harder than adding gas, as discussed in Section 6.3.1. We simply delete the unwanted gas particles from the local particle set, via `particles_to_remove`, and synchronize the particle set in the hydro code:

```
removed_particles = particles_to_remove.copy()
gas_particles.remove_particles(particles_to_remove)
gas_particles.synchronize_to(hydro.gas_particles)
```

We store a copy of the removed particles in a local particle set called `removed_particles`. Later, we might use this particle set to check that we properly removed all particles, and that conserved quantities, such as mass, are actually conserved.

In many applications, particles are removed when they are accreted by a sink particle, which swallows gas particles when they approach within some specified distance. A sink particle has mass, position, radius, velocity, and possibly more parameters, such as some notion of the number of gas particles accreted, the accreted angular momentum, or the chemical composition of the accreted material. A sink particle can be declared just like any other particle (refer to Section 4.3.1 for an example). Here is a simple example of a nonmoving sink particle with zero mass at the origin of the coordinate system:

```python
from amuse.units import units, nbody_system
from amuse.ic.gasplummer import new_plummer_gas_model


def accrete_particles(sinks, gas):
    removed_particles = Particles()
    for s in sinks:
        xs, ys, zs = s.x, s.y, s.z
        radius_squared = s.radius**2
        insink = gas.select_array(
            lambda x, y, z: (x - xs) ** 2 + (y - ys) ** 2 + (z - zs) ** 2
            < radius_squared,
            ["x", "y", "z"],
        )
        if len(insink) == 0:
            return insink
```

Code example 6.4: A sink particle consumes gas particles. This is a fragment from the script `amuse.examples.hydro_sink_particles`.

```python
sink = Particle()
sink.mass = 0 | units.MSun
sink.radius = 1 | units.au
sink.position = (0, 0, 0) | units.au
sink.velocity = (0, 0, 0) | units.kms
sink.accreted_mass = 0 | units.MSun
```

The radius of the sink particle is arbitrarily set here to `1 | units.au`, but in practice this radius reflects the distance from which it accretes.

Code example 6.4 presents a simple example of how we might decide which particles should be accreted by a collection of sink particles and removed from the calculation. Although simple, it shows the fundamental way in which gas accretion can be managed. This method is used, for example, in molecular-cloud-collapse simulations, when sinks are formed in a Jeans unstable clump. The equations of motion of the sink can be integrated using an N-body code, while at the same time it can continue to accrete from the surrounding gas. The techniques needed to couple the gravity solver with a hydro code are discussed in Chapters 8 and 9. In Section 6.4.3, we present a simple example of a small experiment with a sink particle; a more elaborate example is discussed in Chapter 10.

Stars (or other objects) may emit and accrete gas particles at the same time. For example, a protostar might accrete matter from the inner edge of a surrounding disk while simultaneously emitting jets perpendicular to the disk. A simple way to realize this is to place the sink over the center of mass of the star, with an accretion radius smaller than the radius at which wind particles are released.

6.3.3 Sinks and Dark Matter Particles

In the previous section, we discussed the possibility of allowing sinks to form and to accrete material. Once a sink forms, it behaves like a point particle that interacts gravitationally but not hydrodynamically with the rest of the system, unlike a parcel of gas. Sinks are widely used in hydrodynamics codes as a way to avoid the need to follow star formation to small and very costly scales once a pocket of gas becomes Jeans unstable and starts to collapse. Combining the collapsing mass into a sink that continues to accrete but no longer behaves as a fluid particle realizes substantial savings in computer time. In many ways, sink particles in the AMUSE `hydro` modules are the fluid analogs of the binary and multiple systems managed by the `multiples` module (see Section 8.7.2)—except that interactions among sinks are far easier to handle, as we will see (Section 6.3.4).

Most hydrodynamics solvers can accommodate sinks because they are designed to include nongas particles in the calculation. Some codes use built-in "sink particles" to follow the formation and accretion of protostars. Others include "dark matter particles" intended to model dark matter in galactic or cosmological contexts, but they too can be used to model sinks. When they exist in an SPH code, AMUSE refers to them as `dm_particles` to distinguish them from the `gas_particles` we have already encountered. Dark matter particles can be quite handy and are used in a wide variety of applications, including compact objects or as the degenerate core of a star on the giant branch. In the latter case, the giant's envelope would be comprised of gas particles. The more generic particle set `particles` is also available, and contains both the `gas` and the `dm` particles.

On adding a new particle to an SPH code, it is important to indicate whether it is a `gas` or a `dm` particle. The former is evolved according to the fluid equations of motion, possibly including self-gravity; the latter is affected only by gravity. Used as sinks, dark matter particles can continue to accrete from the surrounding gas, under the control of the top-level AMUSE script.

Dark matter particles in a hydro code are integrated according to Newton's equations of motion. In most codes, the integration itself is realized using a second-order Leapfrog scheme, and the gravitational potential is softened (see Section 4.1.4). This is probably sufficient for many hydrodynamical problems in which one is not interested in the subtleties of the gravitational dynamics. However, in Chapter 4, we discussed far more accurate gravity solvers, and it may be desirable to integrate the dark matter particles using one of those methods. This requires code coupling, which we discuss in Chapter 9.

6.3.4 Cooling the Gas Using an External Routine

Heating and cooling are computed and applied by the `evolve_model()` member function in the cooling class, which adjusts the internal energy and temperature of the gas according to a simple physical model, forcing the gas toward an isothermal equation of state on a timescale consistent with actual cooling processes (see Section 6.3.4). It was used by Lützgendorf & Helm (2016) (see also Section 10.4) to cool the hot winds of the S-stars orbiting the supermassive black hole in the Galactic center. The results are applied directly to particles or grid cells (Gerritsen & Icke 1997); the hydro code then takes care of the consequences.

The calling sequence between `hydro_model and cooling_model` is

```
cooling = cooling_model()
hydro = hydro_model()
dt = 0.05 | units.Myr
while time < t_end:
    time += dt/2
    channel_to_cooling.copy()
    cooling.evolve_model(time)
    channel_from_cooling_to_hydro.copy()
    time += dt/2
    hydro.evolve_model(time)
    channel_from_hydro_to_cooling.copy()
    cooling.evolve_model(time)
    channel_to_framework.copy()
```

Note that we advance the cooling for half a step, then use that temperature structure to advance the hydro for a full step, and then complete the cooling step. This interlaced stepping strategy results in second-order coupling between the cooling and hydro codes. It is inspired by the second-order code coupling strategy discussed in Chapter 9. The same time-stepping strategy was adopted when combining stellar evolution and gravitational dynamics in Chapter 8. The choice of time step `dt` is a matter of experimentation and some art. Overly large time steps will lead to errors, but small time steps can make the code very slow. The proper time step is determined using convergence arguments (see Section 6.6 for a discussion).

6.3.5 Stopping Conditions

Much in the same way as the gravity codes (Section 4.5.1 and the stellar evolution modules (Section 5.2.10), hydrodynamics solvers also have their own stopping conditions.

The most common stopping conditions for hydro are listed in Table 6.3. These include detections for the upper or lower limits on local density, binary hardness,

Table 6.3. Overview of the Stopping Conditions Available in Most N-body Codes

`Stopping_conditions_`	Default Value	Default Unit
...`maximum_density`	Max	Mass / (length**3)
...`minimum_density`	−1.0	Mass / (length**3)
...`maximum_internal_energy`	Max	Length**2 / (time**2)
...`minimum_internal_energy`	−1.0	Length**2 / (time**2)
...`number_of_steps`	1	None
...`out_of_box_size`	0.0	Length
...`out_of_box_use_center_of_mass`	1	None
...`timeout`	4.0	s
...`pair_detection`	4.0	s

and for particles leaving the computational domain, but also limits on the number of steps, and computational time-out.

6.4 Some Simple Experiments

6.4.1 Circumstellar Disk with a Bump

Planets form in disks, but these disks tend to be lumpy. In this example, we show that a lump disperses in a few orbits. We start our simulations with an axisymmetric 1 $M_{\odot}$ disk, with an isothermal equation of state, around a 10 $M_{\odot}$ star. The density falls off toward the outer edge of the disk as a power law with exponent -3/2. The inner and outer edges are at 10 au and 100 au. These choices are largely motivated by the numerical cost of extending the disk closer to the central star. After generating the disk, we introduce a small perturbation in the form of a bump in a circular orbit at a distance of 50 au from the star. The bump was generated using a Plummer gas sphere with a characteristic radius of 5 au and a mass of 0.5 $M_{\odot}$. Code example 6.5 presents a snapshot of the routine used to generate the disk and bump:

The disk is constructed from a cylindrical slice in which all particles are in circular Keplerian orbits around the central star, with a radial density profile described by a power law and a scale height consistent with the adopted Safronov–Toomre Q-parameter (see below; Safronov 1960; Toomre 1964). This is realized using the routine

```
disk = ProtoPlanetaryDisk(
    n_disk,
    convert_nbody=converter,
    densitypower=1.5,
    Rmin=1,
    Rmax=Rmax/Rmin,
    q_out=1.0,
    discfraction=1.0,
).result
```

which takes as arguments the number of disk particles, a converter, a density power law (here 1.5, lower density at the outer rim), the minimum and maximum disk radii, the Safronov–Toomre Q-parameter, and the mass fraction of the disk. The scaling parameters in this argument list (`Rmin`, `Rmax`, and `discfraction`) are relative to the converter parameters (in this example, 1 $M_{\odot}$ and 10 au).[1]

The Safronov–Toomre parameter provides a stability criterion for a differentially rotating disk. It can be derived from the Jeans stability criterion by comparing the self-gravity of the disk with its thermal pressure, and can be expressed in terms of the radial velocity dispersion σ_r (or the sound speed for gaseous disks), the epicyclic frequency κ, and the surface density Σ:

$$Q = \frac{\sigma_r \kappa}{\pi G \Sigma}. \tag{6.30}$$

[1] Instead of generating a standard protoplanetary disk, we could have achieved a similar initial condition using the more general routine to construct a uniform cylinder using `uniform_unit_cylinder`.

```
def new_disk_with_bump(
    mass_star=10 | units.MSun,
    n_disk=100,
    mass_disk=1.0 | units.MSun,
    radius_min=1.0 | units.au,
    radius_max=100.0 | units.au,
    mass_bump=0.1 | units.MSun,
    radius_bump=5.0 | units.au,
    dist_bump=10 | units.au,
):

    converter = nbody_system.nbody_to_si(mass_star, radius_min)
    disk = ProtoPlanetaryDisk(
        n_disk,
        convert_nbody=converter,
        densitypower=1.5,
        Rmin=1,
        Rmax=radius_max / radius_min,
        q_out=1.0,
        discfraction=mass_disk / mass_star,
    ).result
    com = disk.center_of_mass()

    # determine bump's local velocity

    inner_particles = disk.select(
        lambda r: (com - r).length() < dist_bump, ["position"]
    )
    M_inner = mass_star + inner_particles.mass.sum()
    v_circ = (
        (constants.G * M_inner * (2.0 / dist_bump - 1.0 / dist_bump)).sqrt().
    value_in(units.kms)
    )

    # initialize bump

    Nbump = int(n_disk * mass_bump / mass_disk)
    bump = new_plummer_gas_model(
        Nbump, convert_nbody=nbody_system.nbody_to_si(mass_bump, radius_bump)
    )
    bump.x += dist_bump
    bump.velocity += [0, v_circ, 0] | units.kms

    disk.add_particles(bump)
    disk.move_to_center()
    return disk
```

Code example 6.5: Routine to generate a disk with a Plummer model bump. This code example is a fragment from the script `amuse.examples.hydro_disk_with_bump`.

Disks tend to be stable for $Q > 1$. In Code example 6.5, we generate the disk with $Q = 1$, which results in a barely stable, rather flat disk. (Later, in Section 8.6.1, we will present an example of a much more stable thick disk with $Q = 25$.)

The disk is generated using 50,000 SPH particles, with 25,000 more for the bump, and a softening length of $\varepsilon = 10\,\text{au}$. We then select an SPH code for the

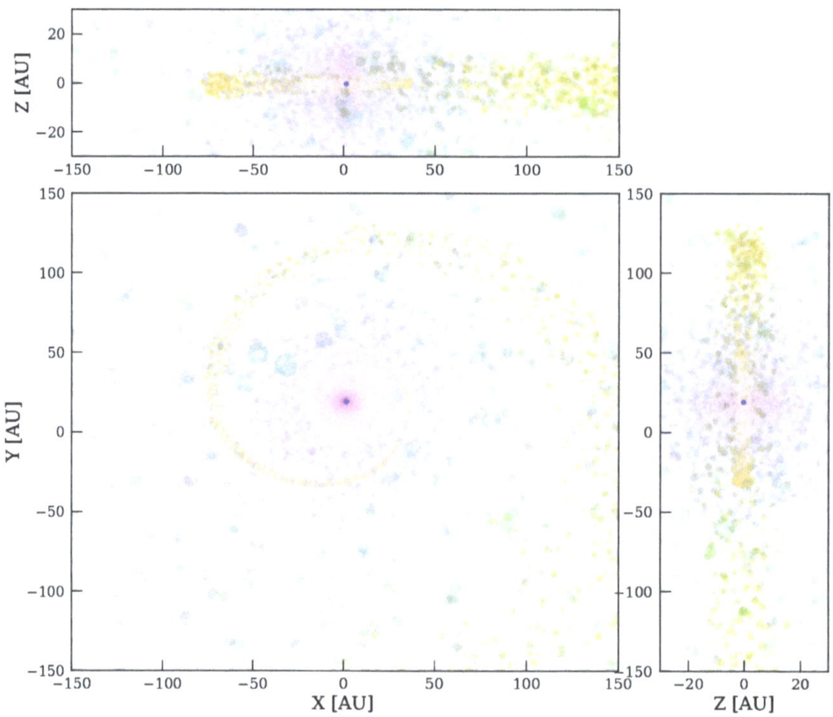

Figure 6.5. Face-on and edge-on views of the disk with a bump after 2000 yr. Individual SPH particles (75000 in total) are represented as semitransparent bullet points. The sizes represent their height above the disk, relative to the z scale height, and the colors are generated from the pressure, velocity, and temperature represented by red, green, and blue, respectively. The lack of blue in the spiral arm, which is a result of the disrupted bump, is a result of the relatively low temperature of the bump with respect to the rest of the disk. The calculation was performed using the script `amuse.examples.hydro_disk_with_bump`, the figure was made with `amuse.examples.plot_hydro_disk_with_bump`.

hydrodynamics, in this case `Gadget2`. To keep things simple, we include the central star as a dark matter particle in the hydro code. We subsequently integrate the equations of motion of the star and the hydrodynamics of the gas for 2000 yr, which is long compared to the $\sim$ 100 yr Keplerian orbital period of the bump around the star. Figure 6.5 presents snapshots of the disk plus bump 2000 yr after the start of the simulation. By this time, the bump has been spread out over the disk, forming a nice spiral arm. We will study the survival time of the bump in Assignment 6.7.3

6.4.2 Colliding Stars

Collisions play an important role in the evolution of the stars in dense star clusters (Hills & Day 1976), as well as for the formation of the Moon and other solar system objects (Benz et al. 1986). In a pioneering study, Benz & Hills (1987) performed SPH simulations of the collisions of stars in a star cluster environment. Their simulations followed collisions of identical low-mass main-sequence stars, modeled as simple

$n = 3/2$ polytropes (Chandrasekhar 1933), on both head-on and grazing trajectories. Their main conclusion was that grazing collisions result in higher degrees of mixing of the inner and outer stellar layers than do head-on collisions.

Instead of two polytropes, we adopt here a full Henyey stellar evolution model of two 0.6 $M_\odot$ stars evolved to 12 Gyr using MESA (see Section 6.2). The two evolved stellar models were converted to SPH particle representations using the routine presented in Code example 8.11 and described in Section 8.5.4. Our SPH realization has a mass resolution of $20M_{\rm Earth}$ per SPH particle (significantly finer resolution than the original calculations). Our head-on collision has zero velocity at infinity, so the relative speed of the two stars is $v = v_{\rm esc} = \sqrt{2Gm/r} \simeq 440$ km s^{-1}. Here, $m = m_1 + m_2$ is the total mass and r is the initial separation. Following Benz & Hills (1987), we start the two stars touching, so $r = r_1 + r_2$. The grazing collision has initial relative speed $v = 1.67v_{\rm esc} \simeq 740$ km s^{-1}. Because Benz and Hills neglected to mention the adopted impact parameter, we simply placed the stars in contact with a transverse offset of $0.5(r_1 + r_2)$ (see their Table 1).

The results are summarized in Figure 10 of Benz & Hills (1987). Figure 6.6 presents a similar figure showing the results of our simulations. The leftmost four bars indicate the initial structure of the stars. The second set of bars indicates the degree of mixing in the merger product following the head-on collision. The rightmost set shows the result of a grazing collision. Comparing our Figure 6.6 with Figure 10 of Benz & Hills (1987), we observe differences in the degree of mixing. The head-on case compares reasonably well with the original calculation. The innermost quartile of the merger product is refreshed with about 15% of material from the next quartile out, whereas the outermost region (fourth quartile) is enriched with roughly 20% of material from below (third quartile). Overall, the degree of mixing in this case is less than reported by Benz and Hills. In contrast, the grazing collision seems to mix the stars more effectively in our simulation than in the original calculations. Virtually all of the mass in the first quartile of the incoming stars is lifted to higher zones, more than 50% of it finding its way into the outermost two quartiles of the merger product.

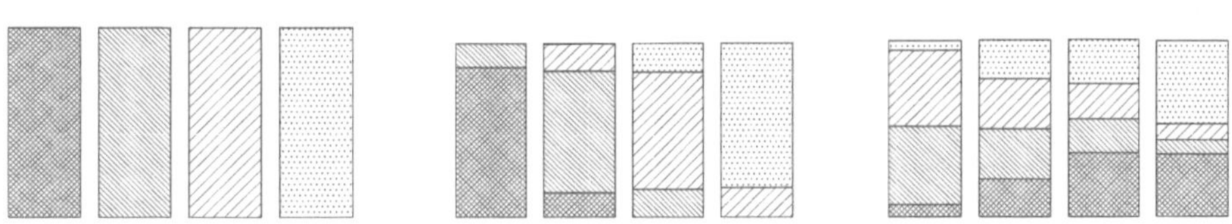

Figure 6.6. Recomputation of Figure 10 in Benz & Hills (1987), in which two 12 Gyr old 0.6 $M_\odot$ main-sequence stars collide with zero velocity at infinity. The left set of four vertical bars depicts the initial (precollision) structure of the stars; the various shading schemes represent the inner 25%, 50%, 75%, and the outer 25% of the mass. The middle set shows how the original stellar material is rearranged following a head-on collision, while the right-hand set shows the result of a grazing collision. The figure was made using the script `amuse.examples.plot_1987ApJ...323..614B` which depends on two other modules `star_to_sph` and `1987ApJ...323..614B` in the `amuse.examples` package. Running the hydrodynamics of the collision takes a few hours on a laptop.

6.4.3 Accreting from the Wind of a Companion

In this example, we measure the amount of mass accreted by a 1 $M_\odot$ star from the wind of a 2 $M_\odot$ companion. Both stars have reached an age of 1.494 Gyr, at which time the 2 $M_\odot$ primary has lost some 0.0752 $M_\odot$ of its initial mass and has grown into a supergiant of radius 0.99 au and surface temperature 3300 K. Its wind speed is $v_w \sim 30$ km s^{-1}, with a mass-loss rate of $\dot{m} \approx 0.11 M_\odot$ Myr^{-1}. The 1 $M_\odot$ companion star has not lost much mass yet and is still on the main sequence. Its radius is 0.0043au = 0.93 $R_\odot$.

We place the stars on a circular orbit at a separation of 10 au, and start the hydrodynamics solver. We adopt an SPH particle mass of $3 \times 10^{-11} M_\odot$ and integrate the hydrodynamics with a time step of one day, after which we launch new SPH particles using the simple wind module discussed in Section 6.3.1. Gas particles accreted by the 1 $M_\odot$ star, more than 200 au from the primary, are removed according to the method described in Section 6.3.2. Accretion is modeled by placing a sink particle (see Section 6.3.4) at the location of—and moving on the same orbit as—the 1 $M_\odot$ star. We include gravity by following the stars as dark matter particles in the hydro code.

Synchronization between wind particles and accreted particles is accomplished by

```
accreted = sink.accrete(interstellar_gas_particles)
interstellar_gas_particles.remove_particles(accreted)
interstellar_gas_particles.synchronize_to(hydro.gas_particles)
```

Here, `stellar_particles` and `sink_radius` are, respectively, the stellar particles added to the sink and the radius of the sink.

Figure 6.7 presents the accretion rate onto the 1 $M_\odot$ star as a function of time. The red curve is calculated using a sink particle that did not move and had a radius equal to the star's Bondi–Hoyle–Lyttleton (Bondi & Hoyle 1944; Hoyle & Lyttleton 1939) radius of ~2 au.[2] The blue curve is for a normal-sized star with self-gravity included. As can be seen, both approaches give the same results, within the statistical uncertainty, with a mean accretion rate of $\sim 0.0022 M_\odot$ Myr^{-1}. One might possibly argue that the Bondi–Hoyle–Lyttleton accretion starts a little earlier. This can be understood from the fact that particles have to travel less distance before being swallowed by the accreting star, simply because its accreting surface is closer to the companion.

The accretion rate from a steady wind can be approximated as (Theuns et al. 1996)

$$\dot{M} \propto \mu^2 k^4 (1 + k^2)^{-3/2} \dot{M}_{\rm donor}. \qquad (6.31)$$

Here, $\mu \equiv m_2/(m_1 + m_2)$ and $k \equiv v_{\rm orb}/v_{\rm wind}$. The mass ratio hardly changes during the integration and $\mu \simeq 0.34$, but the velocity in the wind at the location of the companion star is considerably lower than the terminal wind velocity used as input

[2] The Bondi–Hoyle–Lyttleton radius $R_{\rm BHL}$ is the effective accretion radius of a gravitating body. For an object of mass M and incoming material moving at speed v, $R_{\rm BHL} \simeq 2GM/v^2$.

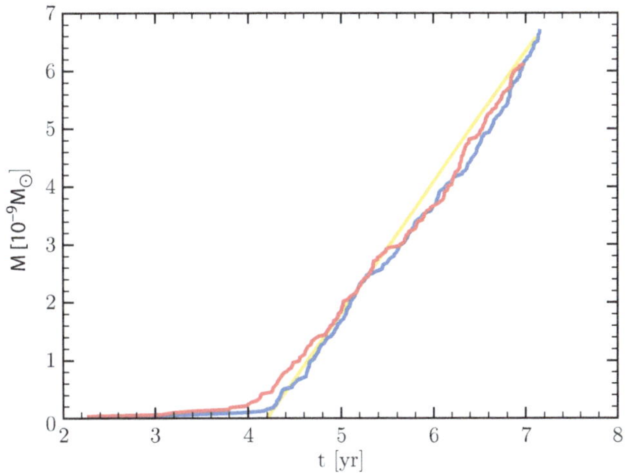

Figure 6.7. Accretion rate onto a 1 $M_\odot$ star embedded in the dense wind of a 2 $M_\odot$ companion on the asymptotic giant branch. The blue curve is calculated using the formal accretion radius from the stellar evolution code and the gravity solver in the hydrodynamics code. The red curve is calculated using the Bondi–Hoyle–Lyttleton accretion radius of the star (~2 au), without taking the primary's self-gravity into account. The yellow curve gives an estimate of the accretion rate using Equation (6.31) and adopting a wind velocity of 17.3 km s^{-1}. The hydrodynamics was resolved using `amuse.examples.hydro_give_or_take`. The plot was made with `amuse.examples.plot_accretion_from_windy_star`.

for generating the wind. The wind velocity measurement in the simulation that included self-gravity resulted in $\langle v_{\rm wind}\rangle \simeq 17.3$ km s^{-1}. With this wind velocity, the accretion rate calculated with Equation (6.31) is consistent with the simulations (the yellow line in Figure 6.7).

6.4.4 Continuing with Hydrodynamics after a Henyey Code Crash

In Section 6.2.5, we compared grid and SPH hydro simulations of a simplified exploding star. The initial conditions for these simulations were quite far from reality, if one is really interested in supernova explosions. Specifically, the density, temperature, and momentum structure of the two-component model in Section 6.2.5 could easily be improved if we adopted a Henyey stellar evolution code to generate the stellar structure at the moment of the explosion.

For our more realistic star, we will use the stellar structure from Section 8.3.3 and continue from the point when the stellar evolution model failed to converge, leading to a crash. As noted earlier, the termination of the stellar evolution code does not crash the AMUSE script, so we can pick up the last converged stellar model as input to a hydrodynamical model of the supernova explosion that crashed the Henyey code. For definiteness, we continue the calculation using an SPH code. (It would have been equally easy to adopt a grid-based code.)

We first convert the stellar model to an SPH particle realization, much as in Section 6.4.2. We then inject energy into the innermost SPH particles to start the

```
out_pickle_file = os.path.join(
    get_path_to_results(), "super_giant_stellar_structure.pkl"
)
if os.path.exists(out_pickle_file):
    return out_pickle_file
```

Code example 6.6: Routine to inject 10^{51} erg into the innermost regions of an SPH realization of an exploding star.

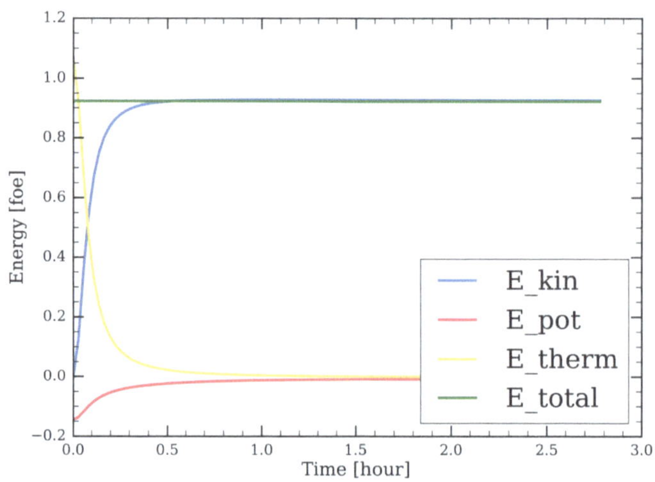

Figure 6.8. Evolution of the energy of an exploding 10 $M_\odot$ star. The star was first evolved using MESA to the tip of the asymptotic giant branch. After this, the one-dimensional Henyey model was converted to a particle (SPH) representation. At that point, we injected 10^{51} erg into the stellar core and continued the hydrodynamical evolution using `Gadget2`. The figure was produced with `amuse.examples.explode_evolved_star`, which should run in about 10 minutes on a laptop.

explosion. Code example 6.6 shows how we inject 10^{51} erg into the core of the SPH realization of the Henyey stellar model (see also Section 6.2.5). Note that selecting the right number of SPH particles is rather tricky and can easily lead to artificial numerical effects in the simulations. If too few are selected, individual SPH particles may penetrate the stellar envelope and leave the star without transferring their energy. On the other hand, if the supernova energy is divided among too many SPH particles, it may not be sufficient to drive the supernova explosion. Ideally, the energy should be divided among the particles corresponding to the degenerate core of the stellar model, and this may be achieved with a sufficiently high resolution in the hydrodynamical simulation. After the energy is injected, we start the hydro code, feed it the SPH particle set, and let it run. The resulting evolution of the thermal, kinetic, and potential energies is presented in Figure 6.8.

The function `get_path_to_results` referenced in Code example 6.6 is imported from `from amuse.test.amusetest`, to provide a standard directory in which temporary files can be stored.

6.5 Collapsing Molecular Cloud

In this section, we present a simple experiment related to hydrodynamics, and in particular on the collapse of a giant molecular cloud. Such simulations are one of the most exciting multiscale and multiphysics applications of AMUSE, and we continue this discussion with a broader range of codes in Section 10.5.1 by including sink particles and radiative feedback, which results in nice images. We are far from understanding the general collapse of molecular clouds, and much research is expected to be done in improving our knowledge.

Figure 6.9 presents the results of a simple experiment that can be used for pedagogical purposes or to study how a model molecular cloud collapses. Similar simulations have been performed by many authors, but new simulations, even at relatively low resolution, often give interesting results if novel initial conditions are tried. One might, for example, start with a more complex cloud structure, perhaps

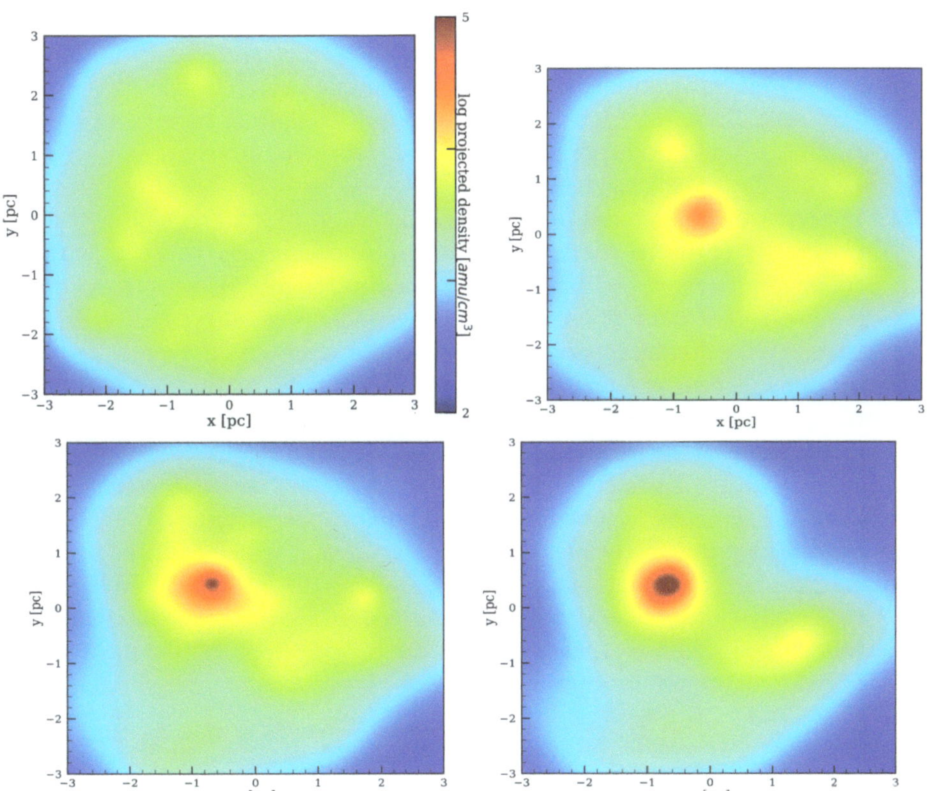

Figure 6.9. Collapse of a homogeneous $10^4\,M_\odot$ gas cloud with an initial radius of 3 pc and a turbulent (power law) velocity field. With these parameters, the free-fall time is $t_{ff} \approx 0.86$ Myr, and the central density is $6.0 \times 10^{-18}\,\mathrm{kg\,m^{-3}}$. The hydrodynamical evolution was carried out using the SPH code `Fi` using just 1000 SPH particles. The four panels show x–y projections of the cloud at 0.25 Myr 0.50, 0.75, and 1.0 Myr. The same color scheme is used for each panel ($\log_{10}$ projected density). The simulation was performed using the script `amuse.examples.molecular_cloud_collapse`.

based on an observed density distribution for which velocity, temperature, and/or composition information is available.

6.5.1 Initial Conditions

As a starting point, we adopt a homogeneous gas sphere with a power-law velocity power spectrum [$P(k) \propto k^{-4}$, where k is wavenumber] to mimic large-scale turbulence (Bate et al. 2003). This is a popular initial model for simulations of star-forming molecular clouds. At the start of the simulation, we scale the internal energy such that the cloud begins to collapse, resulting in a cloud with a mean temperature of about 23 K. We recognize that these initial conditions are not very realistic, but they are convenient and widely used in the community. The resolution of our simulation is low (only 1000 particles, i.e., 10 $M_{\odot}$ per SPH particle), and the softening is large (~0.1 pc). We can see the cloud form an interesting substructure as it evolves. The substructure in the collapsing cloud originates entirely from the initial velocity field, and while the initial model is idealized, the clumps and filaments that form are thought to be key ingredients in the collapse of real star-forming regions.

In the next few subsections, we extend this simple calculation in two key ways: adding a more realistic description of the equation of state of the gas, and allowing the possibility of star formation as the collapse proceeds. Both are implemented at the framework (`Python`) level.

1. Earlier, we encountered an adiabatic equation of state, which is basically a cheap way of approximating the result of the interplay between heating and cooling processes. More complex hydrodynamical problems generally require some form of explicit heating and cooling. Not all hydro codes have explicit treatments of heating or cooling; however, even if they do not, we can still heat and cool the gas via AMUSE. In Section 6.3.4, we describe how to introduce explicit cooling into the hydrodynamics solver to simulate the collapse of an isothermal molecular cloud.
2. When individual parcels of gas become sufficiently dense to collapse, we turn them into sinks (see Section 6.3.3). These sinks can be integrated using the internal hydrodynamics N-body integrator (usually a Leapfrog or Verlet code; see Section 4.2.2). They can continue to accrete, or can merge if two sinks approach sufficiently closely. The details are described in Section 6.5.1. In Chapter 9, we will demonstrate how a high-order direct N-body code can be incorporated to integrate the equations of motion of the sink particles more accurately.

In these cases, we will find generally it desirable to significantly increase the resolution of our calculations. We now consider these enhancements in turn.

A script that performs the entire simulation with cooling and sinks can be found in `amuse.examples.molecular_cloud_collapse_with_sinks`. Since this code is a bit more complicated than anything we have encountered so far, we have split some of the ingredients into `Python` classes. New to this script is a hydrodynamics class `hydro_model`, defined in `hydrodynamics_class`, and a rudimentary cooling class `cooling_model`, in `cooling_class`.

6.5.2 Star Formation in a Collapsing Molecular Cloud

Star formation is implemented by creating sink particles in regions of gas that exceed a specified critical density `density_threshold` (see Section 6.3.2). The stopping condition for detecting high-density particles or cells is set by

```
code.parameters.stopping_condition_maximum_density = density_threshold
density_limit_detection = self.code.stopping_conditions.density_limit_detection
density_limit_detection.enable()
```

The exception is raised when any high-density gas is detected. Loosely, "high-density" generally means something substantially above the average, although there is a more formal prescription based on the resolution of the simulation, related to smoothing length, internal energy, and particle mass (Federrath et al. 2010; Bleuler & Teyssier 2014; Sormani 2017).

In Section 6.3.2, we used a simple particle as a sink, but we could do better by importing the AMUSE sink particle module. This internal module is not much different from a simple particle set, except that it takes care of some rudimentary aspects of conservation following accretion. It is accessed by

```
from amuse.ext.sink import new_sink_particles
```

Sink particles can be initialized from a host particle set `host_particles`:

```
sink_particles = new_sink_particles(host_particles, sink_radius)
```

The `host_particles` set can be stars in a gravity code or dark matter particles in an SPH code. The parameter `sink_radius` here is the radius of the sink particle, which is used to determine when ambient gas is accreted and when sinks merge.

New sinks can be created after a hydrodynamics code stopping condition is reached, or (as here) by searching for high-density regions in the gas distribution. In either case, the `Python` script must search for high-density particles (or cells), and then create new sinks and add them to the hydro code:

```
highdens = gas_particles.select_array(lambda rho:rho>density_threshold, ["rho"])
candidate_sinks = highdens.copy()
gas_particles.remove_particles(highdens)
newsinks_in_code = hydro.dm_particles.add_particles(candidate_sinks)
newsinks = Particles()
for nsi in newsinks_in_code:
    if nsi not in self.sink_particles:
        newsinks.add_particle(nsi)
sink_particles.add_sinks(newsinks)
```

Here, the newly found sinks are added to the (previously created) particle set `sink_particles`.

When two sinks approach within `sink_radius` (typically the sum of the individual radii) of one another, they can be merged in much the same way as stellar mergers in the stellar evolution example in Code example 5.5 of Section 5.2.8. As soon as the mass of a sink exceeds some threshold, we could start a stellar evolution code and take stellar evolution into account in setting the mass and size of the growing star (see Chapter 5). However, increasing the stellar mass due to accretion onto the sink requires some more thought, which would take us beyond the scope of this chapter.

6.5.3 Stellar Winds and Spherical Outflows

To make the problem a little more realistic, we include stellar evolution using the parameterized stellar evolution code `SeBa` and take the winds generated by young stars (particle set `stars`) into account using the internal `amuse.ext.stellar_wind` module:

```
from amuse.ext.stellar_wind import new_stellar_wind
```

This module generates new SPH particles near the mass-losing stars, with wind parameters (position, velocity, density, temperature) derived from the stellar evolution code, and hands these particles back to the SPH code. The SPH code later automatically takes care of incorporating the effect of the stellar wind into the residual gas:

```
wind = new_stellar_wind(
    mass_gas,
    target_gas=gas,
    timestep=time_step,
    derive_from_evolution=True,
)
wind.particles.add_particles(stars)
channel_to_wind = stars.new_channel_to(wind.particles)
```

Here, `mass_gas` is the mass of an individual gas particle in the stellar wind generated over a time interval `time_step`. Then, the `stars`, which are also in the stellar evolution code, are added as particles to the wind module, and we make a channel from the stars to the wind module in order to ensure that changes to the stars due to stellar evolution are properly propagated to the wind module. In the main event loop, the stellar and wind modules are called and the newly generated wind particles are synchronized with the gas particles in the hydro code:

```
stellar.evolve_model(time)
channel_to_wind.copy()
wind.evolve_model(time)
gas_particles.synchronize_to(hydro_code.gas_particles)
```

Note that the routine `synchronize_to` copies particle properties, it does not copy particles from one set to another.

In Section 10.4.1, we demonstrate how the wind module can be used in conjunction with stellar evolution, gravity, and hydro codes to simulate the intricate coupling between strong stellar winds and the dynamics of the stars in orbit around the supermassive black hole in the Galactic center.

6.5.4 The Collapse Calculation

After implementing the above changes into the basic script discussed in Section 6.5, we can start to experiment with a more realistic collapsing gas cloud. (In Figure 9.8, we show another method to arrive an interesting initial conditions for the combined distributions of gas and stars.) Initially, the cloud mass was 10^3 $M_\odot$, distributed in 10^5 equal-mass particles, and the radius was 2 pc, with the same initial turbulent velocity field as before. Figure 6.10 presents a snapshot of the calculation 4.5 Myr after the start of the run. By this time, the age of the oldest star is almost 3.5 Myr, and the most massive star, born 1.5 Myr after the start of the simulation, is about 6.9 $M_\odot$.

Recently, Hacar et al. (2017) have measured the radio intensity of the molecular N_2H^+ emission in the embedded star cluster NGC 1333. This cluster lies at the center of a stellar clump with a diameter of ~ 2 pc. Star formation has been going on there for some 3 Myr. Hacar et al. (2017) estimate the total mass in gas in NGC 1333

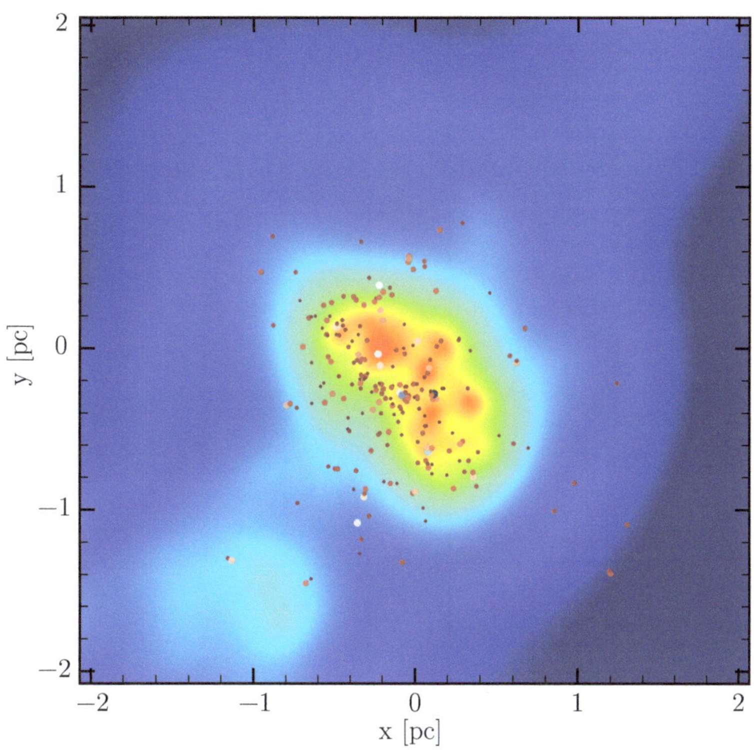

Figure 6.10. Projection of the central portion of a collapsing molecular cloud at an age of 4.5 Myr. The density scale is indicated to the right of the figure in $\log_{10}$ units of amu/cm^3. The colored bullet points are the stars formed from the densest gas (color coding from dark red for the low-mass stars to bright white for the most massive stars). The size of the points is proportional to the logarithm of the mass; the most massive has mass 6.9 $M_\odot$. The script for this simulation is `amuse.examples.molecular_cloud_collapse_with_sinks`; it uses the two support routines in `amuse.examples.hydrodynamics_class` and `amuse.examples.cooling_class`. This figure was generated using `amuse.examples.plot_molecular_cloud_collapse_with_sinks`.

to be ~580 $M_\odot$. The protostars are classified as type 0 to III (see Adams et al. 1987; Lada & Lada 2003), allowing an approximate age range to be associated with each. The radio intensity of the observed lines is a proxy for surface density of the dense gas, giving us a (limited) view of the changing cluster environment. We can convert between N_2H^+ radio intensity $I_{N_2H^+}$ (in units of $K\,km\,s^{-1}$) and projected molecular hydrogen density N (in cm^{-2}) using the approximate empirical prescription provided by Hacar et al. (2017):

$$N \simeq (7.9\ I_{N_2H^+} + 8.2) \times 10^{21}. \tag{6.32}$$

In computing N for our simulation, we first remove gas with density exceeding $2.64 \times 10^{-21}\,g\,cm^{-3}$, because such gas would be unresolved in the observations. We then convert the projected density of the remaining gas in the model to intensity, in order to compare this with the observations.

Figure 6.11 shows the projected gas density profile in our simulation, relative to the center of mass of the stellar distribution, at an age of 4.5 Myr, and compares it with the inferred projected density distribution in NGC 1333. The various lines indicate the densities in gas and stars. The squares are taken from Figure 3 of Hacar et al. (2017). The projected gas density in our simulation is somewhat smaller than the observed distribution, and our stellar density is somewhat higher. However, apart from the youngest stars, our model density distribution matches the observations reasonably well.

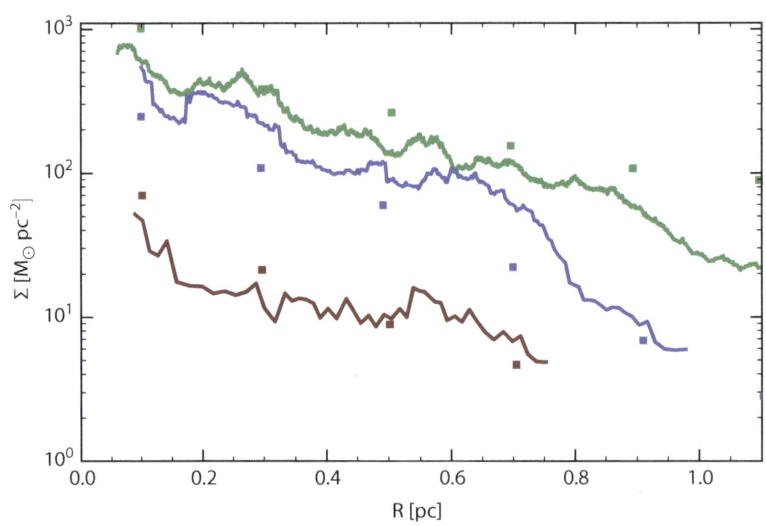

Figure 6.11. Projected density profile (on the x–y-plane) of the simulated molecular cloud and stars at an age of 4.5 Myr (lines) and the observed data for NGC 1333 from (Hacar et al. (2017); squares). The top curve (green) and squares represent the density distribution of the gas. The blue curve and squares are projected radial stellar density profiles; the red curve and squares show the density profile only of stars younger than 1.5 Myr (class 0 and class I objects). The observational data are taken directly from Figure 3 of Hacar et al. (2017). This figure was made using the simulation presented in Figure 6.10 and the script `amuse.examples.2017AandA...606A.123H_Fig3`.

Hacar et al. (2017) further argue that, on average, the younger stars (class 0 and I objects in the observations) are embedded in higher-density gas than the older stars (class II and III). They conclude that older and more massive stars blow bubbles around them, reducing the local gas density (as opposed to the alternative explanation that older stars have simply moved away from the sites where they formed). We test these assertions in Figure 6.12, where we plot the N_2H^+ intensity of gas near young stellar objects in our simulations. The shaded area in Figure 6.12 is constructed from Figure 10 of Hacar et al. (2017). It represents the spread in the intensity distribution for each protostellar class. In this figure, we have attributed tentative ages to each observed class of object, from class 0 ($\sim$0.5 Myr) to class III (2–3 Myr).

The red crosses plotted in Figure 6.12 are from our cloud-collapse simulation, including star formation, stellar evolution (using `SeBa`), and stellar winds. Because the simulation was performed with rather low-mass resolution (0.5 $M_\odot$ per particle), we adopted a lower mass (0.005 $M_\odot$) for the wind particles; otherwise, an insufficient amount of wind would be liberated by the stars to generate even a single SPH particle during the simulation. Performing SPH calculations with a mixed mass resolution of the SPH particles can, under some circumstances, result in problems, such as spurious mixing, if not handled carefully. However, when studying systems with small entropy gradients (Lombardi et al. 1999) or large density contrasts, there may be no other option than to run with different particle masses. Some SPH codes are, in fact, designed specifically with these objectives in mind (Gaburov et al. 2010).

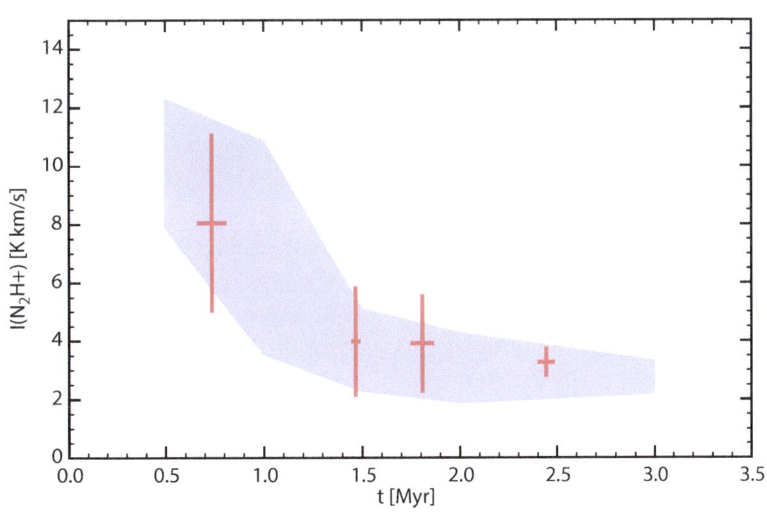

Figure 6.12. Intensity of emission from the molecule N_2H^+ in the direction of the newborn stars in our simulation. The horizontal axis is the age of the star; the vertical axis is (equivalent) intensity of N_2H^+ in units of K km s^{-1}. The shaded region shows the observed mean intensity of N_2H^+ in NGC 1333 as a function of the ages of the protostars studied by Hacar et al. (2017). The error bars give the 90% confidence intervals for the N_2H^+ intensity of the simulated young stellar objects. We adopt the following age ranges: 0–1 Myr for class 0, 1–1.5 Myr for class I, 1.5–2 Myr for class II, and 2–3 Myr for class III. This figure was made using the simulation presented in Figure 6.10 and the script `amuse.examples.2017AandA...606A.123H_Fig10`.

In Section 6.7.4, we encourage the reader to try this out by experimenting with multiple resolutions.

The comparisons of the radial distribution of the projected stellar and gas densities (Figure 6.11) and of the observable intensity as a function of stellar age (Figure 6.12) match the observations reasonably well, supporting Hacar et al.'s (2017) conclusions. This conclusion cannot be made if we start with a smaller ($\lesssim$1.4 pc) or larger ($\gtrsim$3 pc) molecular cloud, if we add more than 10% of break-up rotation to the cloud, or if we switch off the stellar-wind mass loss.

6.5.5 Collapsing Molecular Cloud in Grid-based Hydrodynamics

Simulating the collapse of a giant molecular cloud is really a multiscale and multiphysics effort; it probably takes several codes cooperating to realize such a simulation. For instance, one requires not only the hydrodynamics and radiative processes but also gravity, stellar evolution, and many other calculations to jointly lead to a satisfactory and reliable calculation. It is too early to dwell on the details of such a calculation, but in Section 10.2.1 we dwell on some of the more detailed efforts we have been pursuing in a larger collaborative effort in which AMUSE was pivotal.

The one aspect of such a calculation that is of interest here is the possibility of including specific features, such as asymmetric stellar outflow. One advantage of grid-based hydrodynamics is the natural ease with which specific hydrodynamical phenomena can be included in the calculations. One such example is the asymmetric outflow from low-mass young stars. These stellar jets can be described as the outflow of gas from the stellar surface in a cone with a specific opening angle. Since such processes tend to be implemented directly into the hydro code, it hardly counts as an AMUSE expansion, but rather an internal boundary condition on the hydrodynamics solver.

6.6 Validation

Validating gravity codes is relatively straightforward (see Section 4.6). Conservation of energy and angular momentum are generally the keys to reliable results. Hydrodynamics is a very different realm. The most common way to validate the results of a hydrodynamics solver is by means of convergence tests, where one repeats a calculation with the same initial conditions and code settings, but with different resolution, i.e., we vary the grid size or the number of SPH particles, depending on the type of code used. Convergence tests provide quantitative information about the behavior of the code, and (with luck) on the correct solution of the studied system. However, even if the results converge, there is still no guarantee that the converged solution (or, more often, the solution approaching convergence) is correct. This requires additional validation.

Explicit validation is only possible in a few very limited cases where an analytic or semianalytic solution is known (Landau & Lifshitz 1959). More often, one has to compare new simulation results with previously computed "standard" solutions (Klingenberg et al. 2007). Such comparisons have been undertaken for SPH

(Thacker et al. 2000) and for grid-based hydrodynamics, as in the Santa Barbara Cluster Comparison Project (Frenk et al. 1999). Others focus on specific aspects of physical processes, such as comparing the expansion of a shell of shocked neutral HII in an ionized medium using a variety of codes (Bisbas et al. 2015). Such comparisons reveal that both Eulerian mesh codes and standard SPH methods without artificial conductivity[3] (in addition to artificial viscosity) have significant problems even in rather idealized simulations (see also Saitoh & Makino 2016).

Thus, it is almost impossible to validate the results obtained by hydrodynamics codes, at least at the level of confidence typical in pure gravitational problems, and we advise caution in interpreting the results. In this section, we focus on validating codes via comparison with an exact solution (Section 6.6.1), and by comparing the results of a number of different codes. In the latter case, no analytic solution exists, but conventional wisdom states that, if several different codes give similar results, those results are probably qualitatively correct.

6.6.1 Riemann Shock Tube Problem

The one-dimensional Riemann shock tube test is one of the few nonlinear hydrodynamical problems for which an analytic solution exists (Landau & Lifshitz 1959). The test setup consists of two fluids of different densities and pressures separated by a membrane at the center of the grid. The simulation starts when the membrane is suddenly removed. The solution shows three distinct discontinuities in the fluid: (1) a contact discontinuity that marks the current interface between the two fluids, (2) a shock wave that moves from the high-pressure fluid (left) into the low-pressure fluid (right), and (3) an expansion wave that moves in the opposite direction (Tasker et al. 2008).

Figure 6.13 presents the exact Riemann solution at time 0.4 (in dimensionless code units; see Section 2.3.2), as well as the solutions for two hydro codes—one grid-based and one SPH. The exact solution is taken from Toro (1997). The numerical setup follows Tasker et al. (2008). All modern grid solvers are designed to pass this test (in fact, they are built around this solution as a means of computing fluxes between adjacent cells), so it is not surprising that the grid code performs extremely well. However, while SPH in one dimension also accurately reproduces the analytic solution, the three-dimensional SPH code used here clearly has far greater difficulty than the grid-based code at comparable resolution. Note also that the way the initial conditions were set up favors the grid code. We started with analytic density and pressure distributions, which naturally map onto a grid-based representation, whereas in the SPH representation, random fluctuations in the particle's positions and velocities lead to Poisson and shot noise.

6.6.2 Kelvin–Helmholtz Test

The Kelvin–Helmholtz instability was discussed in Section 6.1.4.2. Because this instability is fundamental to fluid dynamics, any solver must be tested to determine

[3] Artificial conductivity is an additional dissipative term introduced into the energy equation in order to handle discontinuities in the fluid's thermal energy (Monaghan 1997; Price 2008).

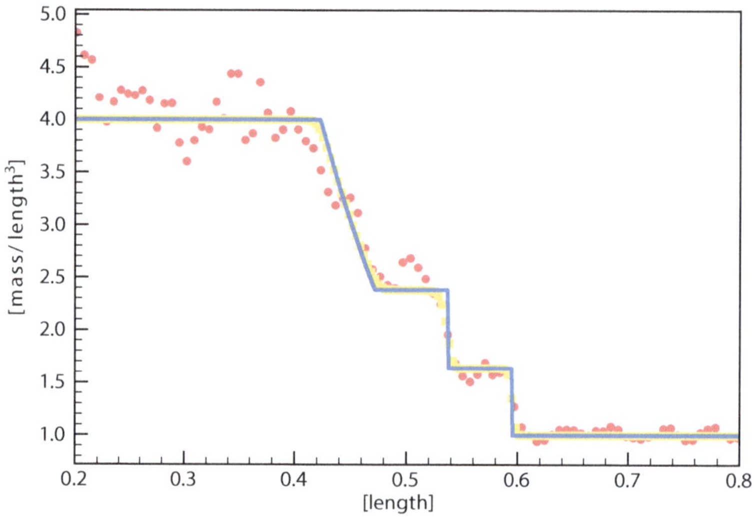

Figure 6.13. Riemann shock tube test using two fluids with a density ratio of 4 to 1, pressure ratio of 1 to 5.571, and with $\gamma = 5/3$. The solid blue line gives the analytic solution to this problem. The yellow squares are the results from the `athena` grid code in one-dimensional mode with 100 cells spanning the computational domain. The red bullets indicate the results using the SPH code `Fi` calculated in three dimensions using 10^7 particles. To be fair, we note that `athena` is working here in one dimension, whereas the SPH code `Fi` operates in three dimensions, using a rather small number of particles for this test. The apparently poor result for the SPH code then is really a demonstration of how low resolution in SPH can still give reasonable results. The source code for this test can be found in `amuse.examples.test_riemann_shock_tube_problem`. This figure was made using the script `amuse.examples/plot_riemann_shock_tube_rho`. Running the analytic code takes a few seconds, `athena` takes about 10 minutes, and `Fi` about 18 hours (single core) on a laptop.

whether the growth rate of the instability and its developing structure are consistent with earlier calculations (see also Section 6.6). However, it is not precisely known how the development of the instability is affected by the resolution—this is still an active area of research, involving both laboratory experiments and numerical simulations (Kajishima & Taira 2017). A small script to perform a Kelvin–Helmholtz test is part of the AMUSE test suite, and can be found at `amuse.examples.kelvin_helmholtz`.

Figure 6.14 presents the results of the Kelvin–Helmholtz instability test for three different grid-based hydrodynamics solvers. The red fluid moves to the left and has twice the density and half the speed of the blue fluid, which moves to the right. Even though the three images differ in detail, the characteristic structure of the Kelvin–Helmholtz instability is clearly evident. We did not perform this test for the SPH codes because they tend to perform very poorly unless we go to extremely high resolution and introduce a special treatment for mixing (Read et al. 2010).

6.6.3 Cloud-shock Test

Another hard test for hydro codes is the cloud-shock test, more popularly known as the blob test (Agertz et al. 2007). This test aims to study interacting multiphase

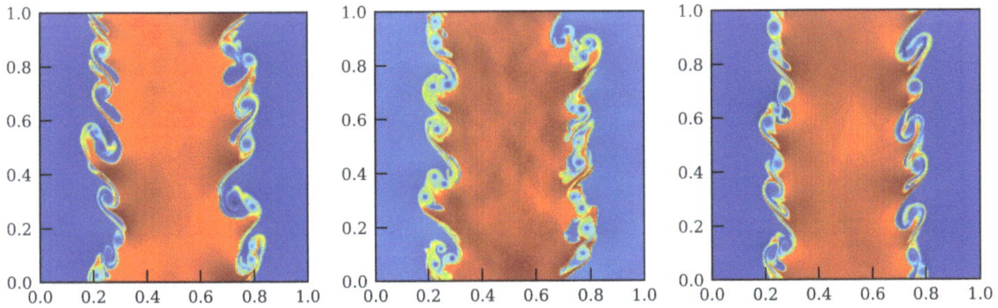

Figure 6.14. Result of the Kevin-Helmholtz test for three different grid-based hydrodynamics codes, `athena` (left), `capreole` (middle), and `mpiamrvac` (right), on a two-dimensional 211 × 211 mesh (see Pelupessy et al. (2013) for details). We seeded the initial system with random velocities of amplitude $0.01c_s$. Here, we show the density distribution at $t = 1$. The source code for this test took about a minute to run and can be found in `amuse.examples.kelvin_helmholtz`.

fluids, which are generally poorly resolved by SPH techniques, whereas Eulerian grid-based methods are able to resolve the most important dynamical instabilities. In this test, a cloud is compressed by a high Mach number inflow, and forms a contact discontinuity between the high-density and the low-density material. The experiment is meant to mimic the shock-induced star formation process, except that there is no gravity. We present the results of this test using three different grid-based hydrodynamics solvers in AMUSE in Figure 6.15.

SPH codes use the local density distribution to derive the spatial derivatives of other quantities, which therefore should always remain differentiable. It is exactly this assumption that breaks down in the blob test, causing SPH codes to perform poorly. By working instead with the internal energy density (or pressure), which changes smoothly across the discontinuity boundary, as the volume element used for the kernel integration, the contact discontinuity can be resolved without numerical problems (Saitoh & Makino 2013).

6.6.4 Boss–Bodenheimer Test

Finally, we perform the Boss–Bodenheimer test for fragmentation in a rotating protostar (Boss & Bodenheimer 1979). This test is often used to validate cloud collapse and the star formation process in hydrodynamical simulations. In this case, we use an SPH code, which provides a satisfactory solution. The experimental setup is a uniform spherical isothermal gas cloud of mass 1 $M_\odot$ and radius 0.01 pc, to which a bar-like density perturbation of amplitude 50% is added. The bar initially rotates as a rigid body with constant angular speed 1.56×10^{-12} rad s^{-1}. The calculation was performed using `Fi` with 50,000 SPH particles. The cloud collapses on a timescale of roughly the initial free-fall time ($t_{\rm ff} \approx 0.175$ Myr), and forms two blobs containing approximately 80% of the total mass. The remaining gas is ejected.

In the AMUSE test, illustrated in Figure 6.16, we can qualitatively see how the perturbation develops. A more useful validation is realized by making a comparison with other simulation codes. This test could, for example, be run relatively easily

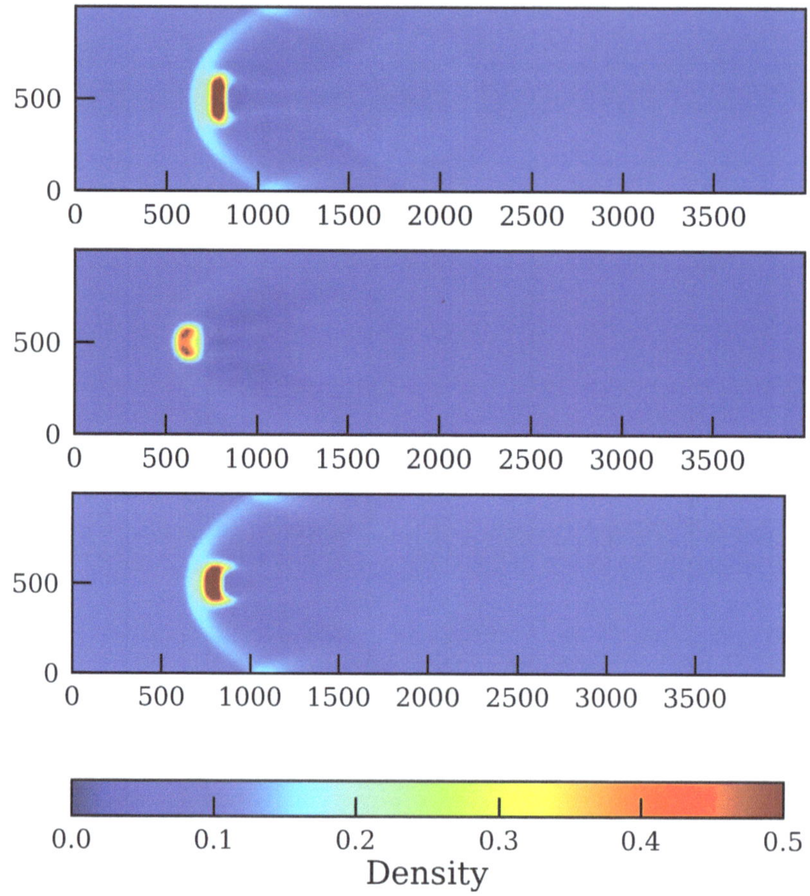

Figure 6.15. Comparison of the results of the cloud-shock test (Agertz et al. 2007) for three grid-based hydrodynamics codes: `capreole` (top), `mpiamrvac` (bottom), and `athena` (bottom). The color bar indicates the relative normalized density. Panels show slices through the density distribution at time $t = \tau_{\rm KH}$ (see Equation (6.25)). The resolution for `athena` and `capreole` is $320 \times 320 \times 1280$, for `mpiamrvac` it is $80 \times 80 \times 320$ with two levels of refinement for comparable effective resolution. The source code for this test is a standard application in AMUSE and can be found in `amuse.examples.cloudshock`. This image was first presented in Pelupessy et al. (2013). The various codes have a quite similar runtime of about 2 minutes.

using a grid-based hydrodynamics solver, and the reader may take up this challenge in Section 6.7.2.

There is no analytic solution to this problem, except that one can check conservation of mass, energy, and angular momentum throughout the test. We stopped the calculation after 1.5 $t_{\rm ff}$ because the local densities in the two blobs became rather high, dramatically slowing the SPH code. We could continue the calculation beyond this stage by introducing sink particles and allowing them to absorb some of the highest-density gas (see Section 6.3.4).

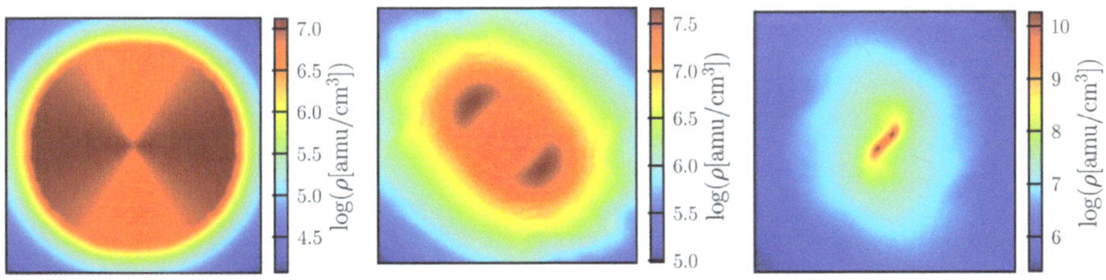

Figure 6.16. Three snapshots of the Boss–Bodenheimer test for a 1 $M_\odot$ uniform isothermal spherical cloud with radius 0.01 pc. Initially (left), a nonaxisymmetric 50% $m = 2$ perturbation is visible. It rotates as a rigid body with constant angular speed of 1.56×10^{-12} rad/s. The middle image shows the projected gas distribution at an age of $0.836t_{\rm ff}$. The right panel shows the state of the system at $1.5t_{\rm ff}$. The calculation was performed using `Fi` with 50,000 SPH particles. The source code for this test is a standard simple AMUSE example and can be found in `amuse.examples.boss_bodenheimer_test`. It runs in about five minutes on a laptop.

6.7 Assignments

6.7.1 Convergence Test

Ideally, aside from unphysical behavior of a code due to incorrect or absent physics, its behavior should be independent of the number of particles in the SPH code, the opening angle in a tree code, the softening length, or the number of cells in the grid code. In reality, the results generally depend on these resolution and accuracy parameters. In practice, you will probably want to perform production calculations with the minimum number of grid cells or SPH particles that produce acceptable results, in order to have as efficient a simulation as possible. (Here, efficiency can be expressed in terms of computer time, wall-clock time, disk storage, memory usage, etc., and is generally problem-dependent.)

For hydrodynamics codes, the number of SPH particles or the number of grid cells are the most common handles for measuring convergence. It is generally best to start a convergence test with as few SPH particles or as coarse a grid as one dares. In subsequent simulations, the resolution can be increased, typically by factors of two, until the outcome becomes independent of the resolution. The outcome might be a graph with some measure of the resolution (number of grid cells in each dimension, or the number of SPH particles) along the x-axis and the desired result on the y-axis.

In practice, simulations do not always reach convergence because the required resolution may be beyond the available resources, such as computer time or total memory. In these cases, it is still possible to validate the result so long as the behavior of the system asymptotically approaches a certain value, and the performance curve can be extrapolated to the desired resolution.

In Section 8.5.4, we discussed how a collision between two stars can be resolved using an SPH code. Here, we use this example to carry out a convergence test for the SPH simulations. We perform the test by increasing the number of SPH particles from 10/$M_\odot$ to $10^4/M_\odot$, and then increasing the number of particles by a factor of 10 for each subsequent simulation. The results are presented in Figure 6.17. We will

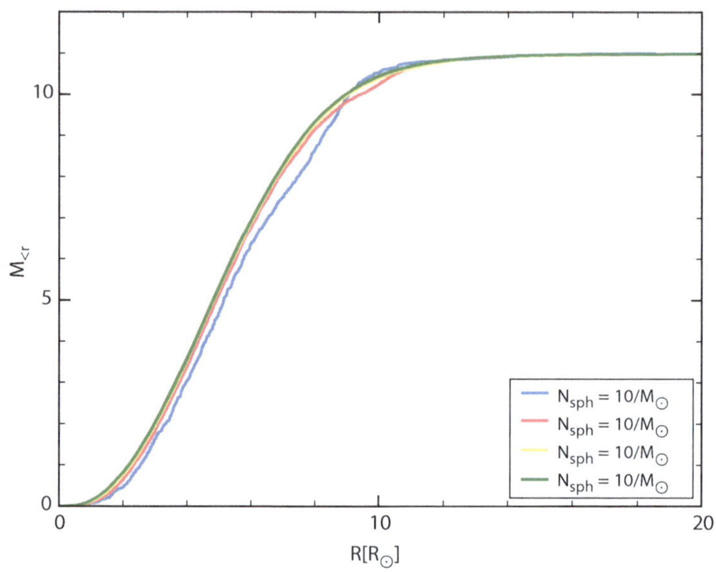

Figure 6.17. Convergence test for a hydrodynamical simulation of a merger of two stars of masses 10 $M_\odot$ and 1 $M_\odot$, with initial parameters otherwise identical to those discussed in Section 8.5.4. Here, we adapt the calculation to perform a convergence test using 10, 100, 10^3 and 10^4 equal-mass SPH particles per solar mass to resolve the stars. The source code for this convergence test is `amuse.examples.merge_two_stars_sph_convergence`.

run the simulation as described in Section 8.5.4.2 and plot the temperature and density profiles of the merger product with varying resolution.

- Perform your own resolution test, starting with two stars of masses 6 $M_\odot$ and 5 $M_\odot$, each evolved to an age of 100 Myr.
- For the test case in Figure 6.17, the results seem to converge rather quickly. Is this a general result? Do your simulations also converge for $N \gtrsim 1000$ SPH particles per $M_\odot$?
- Plot the wall-clock computer time as a function of resolution, and discuss how the cost in computer time grows with increasing resolution.
- Instead of using an SPH code, the same problem can also be done with a grid code (see Section 6.2.5). In that case, use the number of grid cells as a measure of resolution. Perform a similar analysis as for the SPH convergence test and address the same questions.

6.7.2 Testing the Boss–Bodenheimer Test

In Section 6.6.4, we performed the Boss–Bodenheimer test using the SPH code `Fi`. We continued the calculation up to 1.5 t_{ff}, after which two blobs of relatively high density started to appear. This is the classic result of the Boss–Bodenheimer test. We stopped the calculation because the high density forced the SPH code to run very slowly.

We can imagine at least two solutions that enable us to continue the calculation. One is to introduce sink particles in the highest-density regions of the gas distribution and integrate them as `dm_particles`. Another is to repeat the test using one of the available grid-based codes in AMUSE. Specifically, the AMR code `Athena` should do a reasonable job of following the formation of the binary system in this test calculation; `mpiamrvac` is less suitable because the test requires self-gravity. Running the SPH code with sink particles is straightforward, but converting the three-dimensional particle distribution including the velocities to a grid is somewhat more elaborate.

Do the grid and SPH codes give similar masses and orbital parameters for the two stars that form?

6.7.3 The Dissolving Bump

In Section 6.4.1 we discussed the dissolution of a massive perturbation in a protoplanetary disk around a 10 $M_\odot$ star. In this assignment, we will study the timescale on which such a bump dissolves. From the calculation in Section 6.4.1, we already know that the bump has dissolved after about 20 orbits, but we did not discuss the actual timescale for dissolution. The timescale can be derived from the calculations if we are able to recognize the bump at runtime and measure its mass.

The bump in the hydro particle set can be identified by using `hop` (see Section 8.5.4.3). In this assignment, you will repeat the calculation in Section 6.4.1 but identify and measure the bump's mass at runtime. Make a plot of the mass of the bump as a function of time and discuss the mass evolution. What happens to the bump-dissolution timescale if the mass of the central star and the mass of the disk and bump are all reduced by a factor of 10?

6.7.4 Collapsing Molecular Cloud with Sink Particles

In Section 6.5, we demonstrated the collapse of a giant molecular cloud. Even at this rather low resolution, the high densities reached in the simulation brought the code to a grinding halt. The solution was to introduce sink particles to reduce the computational load on the hydro solver. The size of a sink particle, and hence the limiting density that can be resolved, depends on the resolution of the simulation. This limiting density can be calculated by assuming that the volume containing $n_{\rm crit} = 64$ particles (a typical number of SPH neighbors) is barely Jeans unstable (see Section 6.1.4.3)

$$\rho_{\rm threshold} = \frac{\pi^5 e^3}{36 G^3 m^2 n_{\rm crit}^2}. \tag{6.33}$$

Here, $e \simeq c_s^2$ is the specific internal energy of the gas, which in our case (a cold molecular cloud of temperature $T = 15$ K, similar to what we used in Section 6.5) is $e \simeq 0.08$ km^2 s^{-2}, and $m \simeq 2 \times 10^{29}$ kg is the mass of a single SPH particle in a somewhat more refined simulation capable of resolving solar-mass objects. We convert between e and T using

$$T = \frac{2e\mu}{3k}, \tag{6.34}$$

where

$$\mu = 2m_p\left(X + \frac{1}{2}Y\right)^{-1} \simeq 2.4m_p, \tag{6.35}$$

for molecular hydrogen gas and solar composition. In this case, we find $\rho_{\rm threshold} \simeq 8.5 \times 10^{-20}$ g cm^{-3} and a sink radius of $\lambda_J = 0.13$ pc (see Equation (6.27)).

Run the hydro code and generate sink particles as soon as the local density (i.e., the density of a single SPH particle) exceeds the threshold density $\rho_{\rm threshold}$. Turn that particular particle into a sink particle by removing it from the gas and adding it to the hydro code as a `dm_particle`. We will have to do the mergers of the sink particles by hand, as we discussed in Section 4.5.1. Monitor the number of sink particles, and how their total mass increases with time. To validate the method, also plot the maximum gas density found in any of the SPH particles as a function of time. This maximum density should be limited by the adopted threshold density.

Measure, in several calculations, the moment when the first sink particle forms. Use the hydro code stopping condition to check for sink particle formation, then terminate the integration and print the time at which this occurs. This time is not always the same, but varies from run to run. Does this time also vary systematically for different resolutions? Why is it not the free-fall timescale?

6.7.5 A Star-forming Region

In Section 6.5.3, we discussed the correlation of the intensity of N_2H^+ emission with the positions of protostars and young stellar objects. Our calculation was realized with dual resolution using two different masses for the particles in the SPH code, in order to allow a relatively low-resolution model of the star formation process. This is usually a bad idea (but see Section 6.5.3), and can cause a whole raft of other problems, but in that example we wanted to have some quick results in order to obtain a first impression of the physical processes and their consequences. Here, we encourage the reader to further discuss this resolution issue.

Perform the same simulation as in Section 6.5.3, with a total of 200,000 SPH particles (corresponding to a mass resolution of 0.005 $M_\odot$), and reproduce Figures 6.11 and 6.12. Was the observed numerical correlation a consequence of the dual resolution in the SPH simulation?

6.7.6 Neutron Star Hits Companion

Construct a hydrodynamical realization of a main-sequence star of mass 10 $M_\odot$. Fire a neutron star at that star, from a distance equal to twice the stellar radius ($R_\star$) at a velocity of 1000 km s^{-1}. This is pretty fast, but during a supernova, the newly formed neutron star can acquire such a high speed due to asymmetries in the explosion. This problem can be run with a grid or an SPH code, although the latter makes it easier to include the neutron star as a point mass in the form of a dm_particle.

1. Perform a convergence test on the number of SPH particles for this problem where the neutron star is launched head-on into the star (see Section 6.7.1).

   Run a first test with a few (say, 100) SPH particles in the target star, and measure the behavior of the system. Does the neutron star stick to the companion star? How much mass is lost by the target star, and how much kinetic energy does the ejected mass carry? Run subsequent simulations with higher resolution, e.g., 500 SPH particles, 2000 articles, 4000, etc., until the results converge or your patience runs out.

   Make a figure of the resulting diagnostics of the convergence test as a function of resolution.
2. Repeat the convergence test for an impact parameter of $d = 0.5R_\star$. Was the resolution adopted after the initial convergence test (for $d = 0$) sufficient?
3. Finally, repeat the experiment with the same target star on the giant branch. Sadly, the convergence test will have to be redone because the different structure of the target star is likely to change the behavior of the adopted code.

   For this second experiment, the same code used to generate Figure 6.6 (`amuse.examples.star_to_sph`) can also be use used to make an SPH realization of a giant star. You may want to use a dm_particle to represent the stellar core, in order to avoid spending all the available computer time on the hydrodynamics of the giant's central region.

6.7.7 Supernova Explosion

In Section 6.2.5, we discussed the use of a simple two-component stellar model with constant density in each of the components to serve as a starting model for a hydrodynamical simulation of an exploding star. It is relatively straightforward to instead adopt the result of an actual Henyey code and convert the one-dimensional density structure of a star into a three-dimensional grid. In Section 6.2.5, we explored a similar approach but converted the stellar grid structure to a particle representation in order to resolve the hydrodynamics using an SPH code.

In this assignment, we will simulate a type Ic supernova using a grid code. We start with a 12 $M_\odot$ star, which we evolve to a radius of 100 $R_\odot$, which happens at an age of ~14.6 Myr. We assume that, at this stage, the star overfills its Roche lobe in the binary system to which it belongs, and loses its envelope. The stripping of the hydrogen-rich layers has to be done in the stellar structure model, using a Henyey code; see Section 5.2.8. Code example 6.7 demonstrates how this can be done.

Figure 6.18 shows the evolution of a 12 $M_\odot$ star in a Hertzsprung–Russell diagram, up to the time when it reaches a size of 100 $R_\odot$. The star is subsequently stripped of its hydrogen layers, using the procedure in Code example 6.7. This leaves us with a helium star of ~1.85 $M_\odot$, which we continue to evolve using the Henyey code for another 0.21 Myr.

```
    number_of_zones = se_star.get_number_of_zones()
    composition = se_star.get_chemical_abundance_profiles()
    # first zone with X > 1.0e-9
    index = (composition[0] > 1.0e-9).nonzero()[0][0]

    print(
        f"Creating helium star, from the inner {index} (out of {number_of_zones})"
        f"shells."
    )
    helium_star_in_code = stellar_evolution.new_particle_from_model(
        {
            "mass": (se_star.get_cumulative_mass_profile() * se_star.mass)[:index],
            "radius": se_star.get_radius_profile()[:index],
            "rho": se_star.get_density_profile()[:index],
            "temperature": se_star.get_temperature_profile()[:index],
            "luminosity": se_star.get_luminosity_profile()[:index],
            "X_H": composition[0][:index],
            "X_He": composition[1][:index] + composition[2][:index],
            "X_C": composition[3][:index],
            "X_N": composition[4][:index],
            "X_O": composition[5][:index],
            "X_Ne": composition[6][:index],
            "X_Mg": composition[7][:index],
            "X_Si": composition[7][:index] * 0.0,
            "X_Fe": composition[7][:index] * 0.0,
        },
        0.0 | units.Myr,
    )
```

Code example 6.7: Snippet demonstrating how to strip the outer helium-rich layers of an evolved star. This is a fragment from the script `amuse.examples.helium_star`

Ideally, we would like to continue evolving the star until just before the supernova occurs, but the AMUSE Henyey codes will likely encounter problems and crash at an earlier stage. Using a `stopping_condition` or exception handling to catch the right moment to ignite the supernova does not really work in these cases. Instead, we simply ignite the supernova at the moment at which the Henyey code fails to converge. In the example from `amuse.examples.helium_star`, we evolved the helium star for 10^3 internal time steps because this brings the star sufficiently close to exploding.

Once the stellar evolution is finished, convert the one-dimensional Henyey representation of the star into a three-dimensional grid. Use this grid as the initial conditions for exploding the star, using the "thermal bomb" approach adopted earlier. Inject a total energy of $\sim 10^{51}$ erg into the inner 1.4 $M_\odot$ of the star, and divide that explosive energy across this volume, assuming constant energy per unit mass.

As a second objective, use an SPH code to simulate the supernova explosion. In that case, you will have to convert the initial three-dimensional stellar realization into a representative particle set. There are two complementary ways to do this:

- Adopt the method described in Section 6.4.2, where the Henyey model is converted directly to a particle set, or from the grid generated from the Henyey code at the beginning of this exercise.

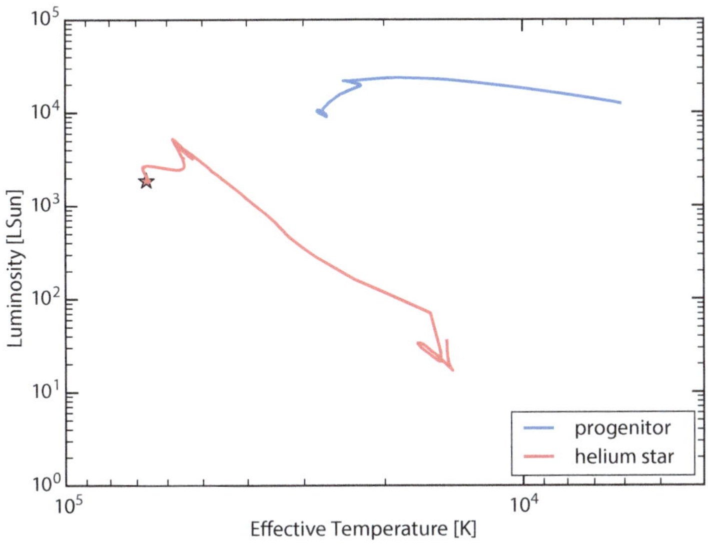

Figure 6.18. Blue curve: Evolutionary track of a 12 $M_{\odot}$ star for 14.6 Myr. The outer layers of the star are then stripped, as described in the text, and the helium core is evolved for a further 0.21 Myr (red track), eventually exploding as a supernova, marked by the star. The source code for this test is a standard simple AMUSE example and can be found in `amuse.examples.helium_star`.

- Use the separate routine `convert_grid_to_SPH` to convert the grid you generated from the Henyey code directly to a particle set.

For the grid code and the SPH simulations, plot the temperature and the density in the supernova shell, as functions of time:

- Compare the temperature profile of the stellar evolution model with the result after the stellar model is converted to the grid.
- Run both initial realizations with the same grid-based hydro code, and compare the results.

References

Adams, F. C., Lada, C. J., & Shu, F. H. 1987, ApJ, 312, 788

Adams, M., et al. 2014, Chombo Software Package for AMR Applications Design Document LBNL-6616E, *Lawrence Berkeley National Laboratory*

Agertz, O., Moore, B., Stadel, J., et al. 2007, MNRAS, 380, 963

Bate, M. R., Bonnell, I. A., & Bromm, V. 2003, MNRAS, 339, 577

Bedorf, J., & Portegies Zwart, S. 2020, ScPA, 1, 001

Benz, W., Slattery, W. L., & Cameron, A. G. W. 1986, Icar, 66, 515

Benz, W., & Hills, J. G. 1987, ApJ, 323, 614

Bernoulli, D. 1738, Papers Imp. Acad. Sci. St. Petersburg, 5, 175

Binney, J., & Tremaine, S. 1987, Galactic Dynamics (Princeton, NJ: Princeton Univ. Press)

Bisbas, T. G., Haworth, T. J., Williams, R. J. R., et al. 2015, MNRAS, 453, 1324

Bleuler, A., & Teyssier, R. 2014, MNRAS, 445, 4015

Bondi, , & Hoyle, 1944, MNRAS, 104, 273
Boss, A. P., & Bodenheimer, P. 1979, ApJ, 234, 289
Bradt, H. 2014, Astrophysics Processes (Cambridge: Cambridge Univ. Press)
Chandrasekhar, S. 1933, MNRAS, 93, 390
Coggeshall, S. V. 1991, PhFlA, 3, 757
Colella, P., & Woodward, P. R. 1984, JCoPh, 54, 174
Courant, R., Friedrichs, K., & Lewy, H. 1928, MatAn, 100, 32
Federrath, C., Banerjee, R., Clark, P. C., & Klessen, R. S. 2010, ApJ, 713, 269
Frenk, C. S., White, S. D. M., Bode, P., et al. 1999, ApJ, 525, 554
Fryxell, B., Olson, K., Ricker, P., et al. 2000, ApJS, 131, 273
Gaburov, E., Lombardi, J. C., & Portegies Zwart, S. 2010, MNRAS, 402, 105
Gardiner, T. A., & Stone, J. M. 2005, JCoPh, 205, 509
Gardiner, T. A., & Stone, J. M. 2008, JCoPh, 227, 4123
Gerritsen, J. P. E., & Icke, V. 1997, A&A, 325, 972
Godunov, S. K. 1959, SbMat, 47, 271
Godunov, S. K., & Kulikov, I. M. 2014, CMMPh, 54, 1012
Hacar, A., Tafalla, M., & Alves, J. 2017, A&A, 606, A123
Hernquist, L., & Katz, N. 1989, ApJS, 70, 419
Hills, J. G., & Day, C. A. 1976, ApJL, 17, 87
Hoyle, F., & Lyttleton, R. A. 1939, PCPS, 35, 405
Hugoniot, H. 1885, C. R. Hebd. Seances Acad. Sci., 110, 794
Inutsuka, S.-I. 2002, JCoPh, 179, 238
Iwasaki, K., & Inutsuka, S.-I. 2011, MNRAS, 418, 1668
Kajishima, T., & Taira, K. 2017, Numerical Simulation of Turbulent Flows (Cham: Springer) **207,**
Keppens, R., Meliani, Z., van Marle, A. J., et al. 2012, JCoPh, 231, 718
Klingenberg, C., Schmidt, W., & Waagan, K. 2007, JCoPh, 227, 12
Lada, C. J., & Lada, E. A. 2003, ARA&A, 41, 57
Landau, L. D., & Lifshitz, E. M. 1959, Fluid Mechanics (Oxford: Pergamon Press)
Lombardi, J. C., Sills, A., Rasio, F. A., & Shapiro, S. L. 1999, JCoPh, 152, 687
Longair, M. S. 2011, High Energy Astrophysics (Cambridge: Cambridge Univ. Press)
Lützgendorf, N., Helm, E. v. d., Pelupessy, F. I., & Portegies Zwart, S 2016, MNRAS, 456, 3645
MacNeice, P., Olson, K. M., Mobarry, C., de Fainchtein, R., & Packer, C. 2000, CoPhC, 126, 330
Meliani, Z., Keppens, R., Casse, F., & Giannios, D. 2007, MNRAS, 376, 1189
Mellema, G., Eulderink, F., & Icke, V. 1991, A&A, 252, 718
Monaghan, J. J. 1997, JCoPh, 136, 298
Monaghan, J. J., & Gingold, R. A. 1983, JCoPh, 52, 374
Murante, G., Borgani, S., Brunino, R., & Cha, S. H. 2011, MNRAS, 417, 136
Pakmor, R., Springel, V., Bauer, A., et al. 2016, MNRAS, 455, 1134
Pelupessy, F. I., & Portegies Zwart, S. 2012, MNRAS, 420, 1503
Pelupessy, F. I., van der Werf, P. P., & Icke, V. 2004, A&A, 422, 55
Pelupessy, F. I., van Elteren, A., de Vries, N., et al. 2013, A&A, 557, A84
Plummer, H. C. 1911, MNRAS, 71, 460
Porth, O., Xia, C., Hendrix, T., Moschou, S. P., & Keppens, R. 2014, ApJS, 214, 4
Price, D. J. 2008, JCoPh, 227, 10040

Price, D. J., Wurster, J., Tricco, T. S., et al. 2018, PASA, 35, e031
Rankine, W. J. M. 1870, Philos. Trans. R. Soc. London, 160, 277
Read, J. I., Hayfield, T., & Agertz, O. 2010, MNRAS, 405, 1513
Regev, O., Umurhan, O. M., & Yecko, P. 2016, Modern Fluid Dynamics for Physics and Astrophysics Graduate Texts in Physics (Berlin: Springer)
Robertson, B. E., Kravtsov, A. V., Gnedin, N. Y., Abel, T., & Rudd, D. H. 2010, MNRAS, 401, 2463
Safronov, V. S. 1960, AnAp, 23, 979
Saitoh, T. R., & Makino, J. 2013, ApJ, 768, 44
Saitoh, T. R., & Makino, J. 2016, ApJ, 823, 144
Shore, S. N. 2007, Astrophysical Hydrodynamics: An Introduction (New York: Wiley)
Sormani, M. C., Treß, R. G., Klessen, R. S., & Glover, S. C. O. 2017, MNRAS, 466, 407
Springel, V. 2000, GADGET-2: A Code for Cosmological Simulations of Structure Formation, Astrophysics Source Code Library, ascl:0003.001
Springel, V. 2005, MNRAS, 364, 1105
Springel, V. 2010, MNRAS, 401, 791
Stone, J. M., Gardiner, T. A., Teuben, P., Hawley, J. F., & Simon, J. B. 2008, ApJS, 178, 137
Stone, J. M., Gardiner, T. A., Teuben, P., Hawley, J. F., & Simon, J. B. 2010, Athena: Grid-based Code for Astrophysical Magnetohydrodynamics (MHD), Astrophysics Source Code Library, ascl:1010.014
Tasker, E. J., Brunino, R., Mitchell, N. L., et al. 2008, MNRAS, 390, 1267
Thacker, R. J., Tittley, E. R., Pearce, F. R., Couchman, H. M. P., & Thomas, P. A. 2000, MNRAS, 319, 619
Theuns, T., Boffin, H. M. J., & Jorissen, A. 1996, MNRAS, 280, 1264
Toomre, A. 1964, ApJ, 139, 1217
Toro, E. F. 1997, Riemann Solvers and Numerical Methods for Fluid Dynamics: A Practical Introduction (Berlin: Springer)
van der Holst, B., Keppens, R., Meliani, Z., et al. 2012, MPI-AMRVAC: MPI-Adaptive Mesh Refinement-Versatile Advection Code, Astrophysics Source Code Library, ascl:1208.014
van Leer, B. 1979, JCoPh, 32, 101
Von Neumann, J., & Richtmyer, R. D. 1950, JAP, 21, 232

Chapter 7

Radiative Transfer

Nel mezzo del cammin di nostra vita, mi ritrovai per una selva oscura, ché la diritta via era smarrita.

When halfway through our life's journey, I found myself in a gloomy wood, because the path which led straight had been lost.

—Dante Alighieri

Our human perception of the Universe is dominated by electromagnetic radiation. Radiation forms the basis for essentially all astronomical research, including all of the great images reported by observational astronomers. Radiative transport is the foundation for understanding how radiation interacts with matter. Solving the differential equations describing the interaction between radiation and matter is no trivial task. In AMUSE, radiative transfer is addressed numerically using two quite distinct techniques: by photon propagation through a grid, and by Monte Carlo ray-tracing. These approaches are complementary. In AMUSE, these codes are typically used to determine the temperature and ionization state of the gas. Radiative transfer is expensive in terms of computer time, particularly if the timescales are governed by the hydrodynamics or the gravitational dynamics of the environment.

7.1 In a Nutshell

Radiative transfer deals with the way in which electromagnetic radiation travels through space and interacts with matter. Radiation is mediated by photons, which have energy, direction, and polarization state. Because the number of photons is generally very large, we normally work with statistical quantities, such as mean intensity and flux, that describe the bulk properties of the radiation field. These quantities may be frequency-dependent, or integrated over some portion of the electromagnetic spectrum. Radiation travels freely through vacuum without loss of energy, but when a ray interacts with matter, energy can be injected or removed.

doi:10.1088/2514-3433/adfc3cch7

Radiative flux. The radiative flux F is defined as the total amount of energy passing through a unit of area of some surface per unit of time. Flux is measured in units of erg s^{-1} cm^{-2}, or the SI equivalent (Wm^{-2}). If (as is usually the case) the energy is distributed over a range of frequencies, it is convenient to work in terms of F_ν (unit erg s^{-1} cm^{-2} Hz^{-1}), where $F_\nu d\nu$ is the energy flux in the frequency range [ν, $\nu + d\nu$). In that case, the total (bolometric) flux is

$$F = \int_0^\infty F_\nu \, d\nu. \tag{7.1}$$

Radiative intensity. In general, a radiation field will consist of photons moving in all directions at any point in space. The specific intensity I is the radiative flux per steradian,

$$I \equiv \frac{F}{\Delta\Omega}, \tag{7.2}$$

where $\Delta\Omega$ is solid angle around some direction $\hat{\boldsymbol{s}}$. Here, we adopt the convention that the unit vector $\hat{\boldsymbol{s}}$ represents a direction at any point in space in which radiation moves. Later, we will use the scalar s to describe distance along the direction (ray) $\hat{\boldsymbol{s}}$. For a frequency-dependent field, the radiative intensity is $I_\nu(\boldsymbol{x}, \hat{\boldsymbol{s}})$ (units erg s^{-1} cm^{-2} Hz^{-1} $ster^{-1}$).

Now, imagine an element of area with normal unit vector $\hat{\boldsymbol{n}}$. The radiative flux crossing this area moving outward (i.e., in the direction $\hat{\boldsymbol{n}}$) is

$$F_\nu = \int I_\nu \cos\theta \, d\Omega, \tag{7.3}$$

where $\cos\theta = \hat{\boldsymbol{n}} \cdot \hat{\boldsymbol{s}} > 0$. For a locally isotropic field, $F_\nu = \pi I_\nu$. In vacuum, the intensity I_ν is constant. For radiation from a point source, the flux at distance r from a source scales as $1/r^2$. The constancy of the intensity along the path in vacuum plays a fundamental role in radiative transport and forms the basis for the radiative-transfer equation.

Blackbody radiation. The blackbody radiation is the radiation field emitted by a perfect emitter at temperature T. The intensity of radiation from that surface is

$$I_\nu = B_\nu(T) = \frac{2h\nu^3}{c^2}(e^{h\nu/kT} - 1)^{-1}, \tag{7.4}$$

independent of emission angle. Inverting this equation allows us to determine the temperature of a blackbody emitter by measuring the intensity I_ν in some direction. In practice, an emitting object is not a perfect blackbody, but has some thermal emissivity ε_ν. The thermal flux emitted from its surface [Equation 7.3] then is

$$F_\nu = \varepsilon_\nu \pi B_\nu(T). \tag{7.5}$$

7.1.1 Underlying Equations

In vacuum, the intensity I_ν is constant, but when radiation interacts with matter, I_ν changes. Matter along a ray may absorb radiation, decreasing I_ν, or emit energy, increasing I_ν. In addition, radiation can be scattered by the medium—changed in direction but not in energy. The resulting *transfer equation* describes these processes. Many excellent books have been written on this subject (Chandrasekhar 1960; Peraiah 2001), often developing the details in specific geometries, such as the spherically symmetric and locally planar atmospheres of stars. Here, we will avoid specific geometries because we are mostly interested in the general problem, where no simplifying assumptions exist.

7.1.1.1 Absorption

Consider first a beam of radiation passing through a purely absorbing medium. As the beam traverses distance δs, the probability that a photon of frequency ν will be absorbed is $\delta P = \alpha_\nu \delta s$. The absorption coefficient α_ν can be expressed in terms of the density ρ and opacity κ_ν of the medium: $\alpha_\nu = \rho\kappa_\nu$. Thus, the fractional change in intensity of the beam is $\delta I_\nu / I_\nu = -\alpha_\nu \delta s$, so

$$\frac{dI_\nu}{ds} = -\alpha_\nu I_\nu. \tag{7.6}$$

The solution to this equation is

$$I_\nu(s) = I_\nu(0)\, e^{-\tau_\nu(s)}, \tag{7.7}$$

where $\tau_\nu(s) = \int_0^s \alpha_\nu(s')\, ds'$ is the *optical depth* of the medium.

If $\tau_\nu \ll 1$, absorption is negligible and the medium is said to be *optically thin*. If $\tau_\nu \gtrsim 1$, the medium is *optically thick*. In terms of $d\tau_\nu = \alpha_\nu ds$, Equation 7.6 becomes

$$\frac{dI_\nu}{d\tau_\nu} = -I_\nu. \tag{7.8}$$

The absorption coefficient α_ν has dimensions of inverse length. The distance a photon of frequency ν is expected to travel before being absorbed is the mean free path, $l_{\nu,\mathrm{free}} = 1/\alpha_\nu$.

7.1.1.2 Emission

In an optically thin medium, absorption may be ignored and the change in the intensity I along some trajectory (ray) depends only on the emission j_ν (unit erg s^{-1} cm^{-3} Hz^{-1} ster^{-1}) by material along that line. For isotropic emission, the intensity changes according to

$$\frac{dI_\nu}{ds} = j_\nu(s). \tag{7.9}$$

If j_ν is independent of I_ν, integrating Equation 7.9 gives

$$I_\nu(s) = I_\nu(0) + \int_0^s j_\nu(s')\, ds'. \tag{7.10}$$

The first term on the right-hand side is the incoming (or background) intensity, which in astronomical situations is often either absent or due to a background star or galaxy. The second term describes the total emission between 0 and s.

For a medium with both absorption and emission, Equations (7.6) and (7.9) combine to give

$$\frac{dI_\nu}{ds} = j_\nu(s) - \alpha_\nu(s)I_\nu, \tag{7.11}$$

which is easily solved:

$$I_\nu(s) = I_\nu(0)e^{-\tau_\nu(0,\, s)} + \int_0^s j_\nu(s')e^{-\tau_\nu(s',\, s)}\, ds', \tag{7.12}$$

where $\tau(s_0, s_1) = \int_{s_0}^{s_1} \alpha_\nu ds$ is the optical depth between s_0 and s_1. The first term on the right-hand side of Equation 7.12 is the attenuated input field; the second may be identified as the (partly reabsorbed) radiation emitted by the medium. Often, the radiative-transfer equation is rewritten in terms of the *source function* $S_\nu \equiv j_\nu/\alpha_\nu$:

$$\frac{dI_\nu}{ds} = \alpha_\nu(s)[S_\nu(s) - I_\nu]. \tag{7.13}$$

A system is in local thermodynamic equilibrium (LTE) if the local gas temperature equals the local Planck temperature of the radiation field (see Section 5.1.2.1). For a medium in LTE at temperature T, it can be shown that $j_\nu = \alpha_\nu B_\nu(T)$, so $S_\nu = B_\nu[T(s)]$. However, many of the systems we will study are not in LTE. Indeed, the whole idea of performing radiative-transfer simulations is to be able to make qualitative and quantitative statements about systems that may be far from LTE. This does not pose a serious limitation to our numerical solution, but it does make it difficult to validate our results from first principles.

7.1.1.3 Scattering

Pure absorption heats the absorbing medium, while pure emission cools it. A medium may also scatter radiation. Scattering may be thought of as absorption followed immediately by re-emission of the absorbed energy: it neither heats nor cools the gas, so it is unimportant for the local temperature structure of the medium. However, it is often critical in determining the details of the radiation field—and hence the appearance of a region.

The removal of photons from a beam due to scattering can be described by a scattering absorption coefficient $\alpha_\nu^{\rm scat} = \kappa_\nu^{\rm scat}\rho$, where $\kappa_\nu^{\rm scat}$ is the scattering opacity. Thus, the total absorption coefficient has contributions from both pure absorption and from scattering,

$$\alpha_\nu = \alpha_\nu^{\rm abs} + \alpha_\nu^{\rm scat}. \tag{7.14}$$

Similarly, the emission j_ν has contributions from emission and from scattering,

$$j_\nu = j_\nu^{\rm emis} + j_\nu^{\rm scat}. \tag{7.15}$$

Here, $j_\nu^{\rm emis}$ corresponds to the absorption coefficient $\alpha_\nu^{\rm abs}$, so $j_\nu^{\rm emis} = \alpha_\nu^{\rm abs} B_\nu(T)$ in LTE.

The expression for $j_\nu^{\rm scat}$ introduces considerable complexity into the transfer equation because it couples beams moving in all directions at any point. We denote by $p_\nu(\hat{s}, \hat{s}')$ the probability that a photon initially moving in direction $\hat{s}'$ is scattered into direction $\hat{s}$ (i.e., into the direction of the beam under consideration in the transfer equation), where $\int p_\nu(\hat{s}, \hat{s}')\, d\Omega' = 1$. It follows that

$$j_\nu^{\rm scat} = \int \alpha_\nu^{\rm scat} I_\nu(\hat{s}')\, p_\nu(\hat{s}, \hat{s}')\, d\Omega'. \tag{7.16}$$

Scattering turns a relatively simple ordinary differential equation into an integro-differential equation that is much harder to solve.

The physics of the scattering interaction is contained entirely within the function p_ν. If the medium is composed of particles with radii a much smaller than the wavelength ($\lambda = c/\nu$) of the radiation, the interaction is in the so-called Rayleigh regime, which is well understood from a theoretical perspective (Bohren & Huffman 1983). If the particles have radii much larger than the wavelength ($a \gg \lambda$), as is generally the case with dust particles, then the scattering cross section is $\sigma \sim \pi a^2$ and we are in the geometric optics regime. Radiative transfer through a medium of large particles can be complex due to effects at the edges of the particles, or if the particles are partially transparent or (usually) not spherical. Radiative transfer through a medium composed of particles with sizes comparable to the wavelengths of interest ($a \simeq \lambda$) is most challenging, and we refer the interested reader to Chandrasekhar's textbook on the topic (Chandrasekhar 1960).

7.1.2 Physics of the Integrator

When numerically solving the radiative-transfer equation [Equation 7.11], the objective may be to construct an image of the simulation, compute the energy spectrum, or determine the temperature and ionization structure of a gaseous body. These are rather different objectives, and a wide range of numerical methods have been developed to address each.

Radiative transfer is difficult because the emission j_ν and the absorption α_ν are generally not known in advance—the radiation field $I_\nu(\mathbf{x}, \hat{s})$ we wish to compute affects both. Local absorption and emission of energy can change the temperature and ionization structure of the medium, in turn producing large changes in j_ν and/or α_ν. At the same time, all the rays from any direction that pass through location $\mathbf{x}$ contribute via scattering to the transfer equation at $\mathbf{x}$. Even if scattering can be neglected, in a system with multiple sources (e.g., a star-forming region), conditions at any location depend on the solution to the transfer equation along multiple directions. In other words, radiation from one region of a molecular cloud affects conditions in another part of the cloud, which in turn may affect conditions in the original region. Solving this problem is a major challenge in computational radiative transfer.

Photons travel at the speed of light. For the problems at hand, we assume that the solution to the radiative-transfer equation is established instantaneously each time the solver is called. The underlying assumption is that the effective travel time of a photon is much shorter than the timescale for local changes in the scattering medium though which the photons travel. We do not expect that this assumption will pose a problem or affect simulation results in any negative way (meaning that the consequences of the assumption are likely smaller than the uncertainties in the numerical scheme).

However, we know that the photon travel time can be long, in some cases, compared to the material response time. In such cases, delay times may be important. For example, it may be the case that the matter needs some characteristic time to respond by cooling down or heating up. In material with large optical depth, radiation may undergo many absorptions and re-emissions and the accumulation of short response times can become important. Although probably not very relevant in astrophysics, this aspect of radiative transfer has been included in the study of photonic crystals (Baba 2008).

7.2 Radiative Transfer in AMUSE

7.2.1 Radiative-transfer Modules

Table 7.1 lists the three radiative-transfer codes currently available as standard in AMUSE. Note that it does not include the radiative-transfer routines built into the `Flash` (Fryxell et al. 2000) hydro module; see Chapter 6. The codes listed differ in their description of both the radiation field and the medium with which it interacts, but all employ essentially the same iterative process to solve the transfer equation and determine the intensity field $I_\nu(\mathbf{x})$ at each spatial location.

A starting approximation to $I_\nu(\mathbf{x})$ is typically the solution obtained during the previous invocation of the module. Alternatively, we might simply set $I_\nu = 0$ everywhere. The emission $j_\nu(\mathbf{x})$ is then calculated at every location, taking into account scattering from the existing field, emission from the gas, and injection of energy from stars and other external sources. Finally, the transfer equation [Equation 7.11] is solved along each photon trajectory to give a new estimate for $I_\nu(\mathbf{x})$. The matter density is held static while the radiation field is updated, but its temperature and ionization structure (and hence its emission and absorption) change at each iteration in response to the changing radiative environment. As illustrated in Figure 7.1, this iterative process stops when changes to I_ν become acceptably small.

We now discuss some specifics of the AMUSE modules listed in Table 7.1. Although the AMUSE philosophy is to hide such details whenever possible, the radiative-transfer modules are more complex than gravitational (see Chapter 4) or even hydrodynamical (see Chapter 6) codes, and a little inside knowledge can be useful in deciding which radiative module to use for a given problem.

Data structure. The three modules (see Table 7.1) differ widely in their description of the medium through which the radiation propagates. `Mocassin` bins the gas onto a Cartesian grid, although there is a compile option to use a Voronoi (1908) tessellated grid (Hubber et al. 2016), while `SimpleX` uses an unstructured Delaunay-tessellated grid (Delaunay 1934), with more grid points in regions of higher opacity. The tessellated versions works best with SPH hydro simulations whereas regular Cartesian grid methods would be more suitable in combination with grid-based hydrodynamics codes. `SPHray` has no grid, but rather (as the name

Table 7.1. Radiative-transfer Codes Incorporated into AMUSE

Module	Mocassin (ref. 1)	SimpleX (2)	SPHRay (3)
Language	Fortran 95	C++	Fortran 95
Parallelization	MPI	OPENMP	OPENMP
Data Structure	Cartesian grid	Unstructured Delaunay grid	No grid; SPH particle distribution
Radiation Representation	Monte Carlo energy packets	Photon numbers and directions	Monte Carlo energy packets
Ray Propagation	Absorption computed from transfer equation; (re)emission creates new packets	Photons transported to adjacent cells; absorption computed from transfer equation; (re)emission/scattering adds/redirects photons	Absorption computed from transfer equation; (re)emission creates new packets
Time Dependence	Thermal and ionization equilibrium	Thermal and ionization equilibrium	Time-dependent temperature equation; nonequilibrium solution
Physics	Photoionization, photoelectric heating, and recombination; dust absorption	Photoionization, photoelectric heating, and recombination; dust absorption	Photoionization, photoelectric heating, and recombination

References: (1) Ercolano et al. (2003), (2) Altay et al. (2008); Paardekooper et al. (2010), (3) Ritzerveld & Icke (2006); Altay et al. (2008, 2011). In Section A.8, we present a more general overview of the various code application domains.

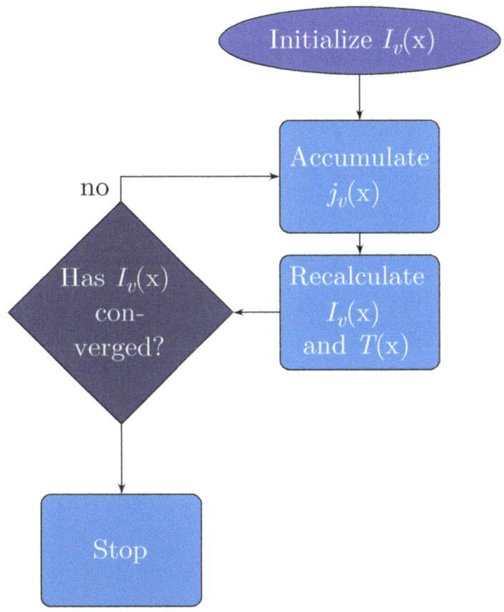

Figure 7.1. Workflow for a radiative-transfer code.

suggests) propagates photons through the interpolated density and temperature (as well as emission and absorption) fields defined by an SPH hydro simulation.

Radiation representation. `Mocassin` and `SPHray` are Monte Carlo methods, using a large number (typically millions) of energy packets to model the radiation field. The packets are created at sources and by emission/scattering events, and are removed by absorption as each packet propagates through the medium. `SimpleX` keeps track of the numbers of photons in each grid cell moving in the directions of the adjacent cell centers (a variable number, but averaging ~16 in three dimensions). Photons created by sources may be absorbed, emitted, or scattered as they move from cell to cell. Because only a limited number of directions is possible, the speed of the code is independent of the number of sources.

Ray propagation. In the Monte Carlo codes `Mocassin` and `SPHray`, each energy packet performs a random walk through the medium. Along each portion of a packet's trajectory, the absorption/scattering probability is

$$p(\tau) = e^{-\tau}, \tag{7.17}$$

where τ is optical depth measured along the path (Section 7.1.1.2). Operationally, the optical depth to the next absorption (or scattering) is determined by choosing a random number $\xi \in [0, 1)$ and setting $\tau = -ln(1 - \xi)$. The directions of emitted or scattered packets are chosen randomly (uniformly for emission or according to $p_\nu(\hat{s}, \hat{s}')$ for scattering; see Section 7.1.1.3). The process is repeated until the packet is absorbed or leaves the computational domain. `SimpleX` uses a similar optical depth formulation to determine the amount of absorption/scattering between one cell center and the next, it then modifies the photon distribution at each cell center to model emission and scattering. Because photons cannot move in straight lines on the irregular grid, care is taken to ensure that photons moving through optically thin regions on average maintain the correct direction.

Time dependence. Mocassin and SimpleX determine the physical conditions in the medium by assuming local radiative and ionization equilibrium at each iteration, and therefore return the steady-state solution to the problem after each invocation. SPHray solves an energy-balance equation for the temperature and ionization state of the medium as it computes the radiation field, and thus it generally returns an instantaneous nonequilibrium (time-dependent) solution.

Physics of the medium. All modules contain treatments of photoionization, photoelectric heating, recombination, and other directly related atomic/molecular cooling processes. Dust absorption is treated in all three radiative-transfer codes, although the implementations differ and not all internal possibilities are exposed to the current AMUSE interface. For SPHray, it is possible to set a stellar radiation spectrum, which is realized by means of a table that can be identified via the function

```
rt_code = Sphray()
rt_code.function get_spectra_file("spectra_file_name")
```

Normally, this file refers to a standard file in the AMUSE repository. It generally uses seven frequency channels to resolve a radiative source. Absorption in SPHray is computed from the transfer equation in each cell, after which re-emission creates new source. The transfer equation is solved using a semi-implicit iteration scheme (Iliev et al. 2006). Mocassin allows multiple sources with different spectra and can be used in conjunction with dust absorption, but this functionality is currently not supported by the AMUSE interface.

7.2.2 Ionization of a Molecular Cloud

We can run a radiative-transfer code in AMUSE in much the same way as running an N-body, stellar evolution, or hydrodynamics code. The procedure is a little more elaborate because radiative transfer requires two ingredients—a gaseous region through which the radiation transfer is to be calculated, and a list of ionizing sources. The typical application for radiative transfer in AMUSE will be star-forming regions, circumstellar disks, dense stellar winds, or supernova nebulae. The modules used to solve the radiative-transfer equations differ somewhat, and the user may have to adjust some of the fundamental code parameters before initiating the particular code.

Although the radiative-transfer codes in the AMUSE framework share essentially the same basic interface, they differ slightly in detail (see Table 7.1). Each is well suited to the problem of an embedded star cluster with radiative feedback. Code example 7.1 presents a simple example of how to start and run a radiative-transfer problem in AMUSE. It is not very elaborate, but it demonstrates the basic functionality and calling sequence for running a radiative-transfer problem. In this example, we opt for the SPHRay code, but we could have used either Simplex or Moccasin as well.

We start by creating a gaseous region—in this case, a homogeneous spherical distribution of $N = 10^4$ neutral gas particles slightly offset from the stellar source. This region is surrounded by a somewhat lower-mass disk with an outer radius of 5 pc (Code example 7.2). A specific internal energy of (9 km $s^{-1})^2$, but no ionizing radiative flux. We explicitly set the ionizing luminosity of the source particle to

```python
def rad_minimal(
    number_of_particles=10000,
    luminosity=100 | units.LSun,
    boxsize=100 | units.parsec,
    time_end=0.1 | units.Myr,
    seed=12345,
):
    np.random.seed(seed)
    ism, converter = generate_ism_initial_conditions(number_of_particles, boxsize)

    source = Particle()
    source.position = (1, 0, 0) | units.parsec
    source.luminosity = luminosity / (20.0 | units.eV)

    radiative = Sphray(convert=converter, mode="openmp")
    radiative.set_boundary(0)  # vacuum
    radiative.set_spectra_file("spectra/thermal6e4.cdf")
    radiative.set_raynumber(1.0e4 | units.Myr**-1)
    print(radiative.parameters)

    radiative.src_particles.add_particle(source)
    radiative.gas_particles.add_particles(ism)

    radiative.evolve_model(time_end)

    print(radiative.gas_particles)
    mux = mu()
    temperature = mux / constants.kB * radiative.gas_particles.u

    print(f"min ionization: {radiative.gas_particles.xion.min()}")
    print(f"average ionization: {radiative.gas_particles.xion.mean()}")
    print(f"max ionization: {radiative.gas_particles.xion.max()}")
    plot_ionization_fraction(
        radiative.gas_particles.position, radiative.gas_particles.xion, temperature
    )

    plot_gas_temperature(radiative)

    radiative.stop()
```

Code example 7.1: Minimal routine to determine the global degree of ionization of a Plummer (1911) distribution of hydrogen gas with a scale radius of 0.5 pc and a mean density of 83 atoms cm^3. The initial conditions are generated using the code in Code example 7.2. Each of the particles was irradiated by a star with luminosity 100 $L_\odot$ for a time interval of 10^6 yr. This snippet was taken from the source code in `amuse.examples.rad_minimal`.

100 $L_\odot$, load the specific energy distribution for the star's spectral type, and indicate to what resolution, in terms of photons, we wish to integrate. The gas in the radiative-transfer code requires an ionization fraction `xion`, which we set to 0.5 (half the hydrogen atoms are already ionized).

The cloud is generated using

```python
ism = new_plummer_gas_model(number_of_particles, converter)
ism.xion = 0.5
```

and the disk with

```
def generate_ism_initial_conditions(number_of_particles, boxsize):
    converter = nbody_system.nbody_to_si(1.0 | units.MSun, 0.5 | units.parsec)
    ism = new_plummer_gas_model(number_of_particles, converter)

    mass_disk = 0.2 * ism.mass.sum()
    n_disk = int(mass_disk / ism[0].mass)
    converter = nbody_system.nbody_to_si(mass_disk, 1 | units.parsec)

    from amuse.ext.protodisk import ProtoPlanetaryDisk

    disk = ProtoPlanetaryDisk(
        n_disk,
        convert_nbody=converter,
        densitypower=1.5,
        Rmin=1,
        Rmax=5,
        q_out=2.0,
        discfraction=0.1,
    ).result
    disk.x += 1 | units.pc
    ism.add_particles(disk)
    ism.xion = 0.5

    hydro = Fi(converter)
    hydro.gas_particles.add_particles(ism)
    hydro.evolve_model(1 | units.hour)
    hydro.gas_particles.new_channel_to(ism).copy()
    hydro.stop()
    ism = ism.select(lambda r: r.length() < 0.5 * boxsize, ["position"])
    print(
        f"Max density: {ism.rho.max().in_(units.MSun/units.parsec**3)} ",
        f"Mean density: {np.mean(ism.rho.value_in(units.g/units.cm**3))} g/cc ",
        f"{ism.rho.max().in_(units.amu/units.cm**3)}",
        f"{np.mean(ism.rho.value_in(units.amu/units.cm**3))} amu/cc",
    )
    return ism, converter
```

Code example 7.2: Routing for generating a gas distribution as initial conditions for the radiative-transfer calculation (see Code example 7.1). This snippet was taken from `amuse.examples.rad_minimal`.

```
from amuse.ext.protodisk import ProtoPlanetaryDisk
disk = ProtoPlanetaryDisk(
    number_of_particles,
    convert_nbody=converter,
    densitypower=1.5,
    Rmin=1,
    Rmax=10,
    q_out=1.0,
    discfraction=0.1,
).result
disk.h_smooth = 0.01 | units.pc
disk.x += 1 | units.pc
disk.xion = 0.5
```

which we center around the off-center source star at $x = 1\,\text{pc}$.

In order to allow the radiative-transfer code to use the gas as a scattering medium, the local density and smoothing length of the gas particles must be determined. In principle, this parameter could have been determined by the Plummer- and disk-generating functions, but here we use an SPH code (Hernquist & Katz 1989; Pelupessy et al. 2004; Gerritsen & Icke 1997) to calculate the density for each SPH particle. In the following snippet, we instantiate the SPH code `Fi` and run it for a short while (1 minute in this example) before copying the relevant information back to the local interstellar medium particles,

```
hydro = Fi(converter)
hydro.gas_particles.add_particles(ism)
hydro.evolve_model(1 | units.minute)
hydro.gas_particles.new_channel_to(ism).copy()
```

For `SimpleX`, it is important to ensure that there is no scattering nor any sources outside the box, and that there is not too much vacuum inside the box. It can be tricky to choose the right box size. In a coupled simulation where the sources and the gas particles are moving, this can make it difficult to keep the numerical solver stable. In the ray-tracing radiative-transfer codes, `SPHray` and `Mocassin`, this is less of an issue. In our small illustration, we need only remove the particles that are outside the box, which we do via an array operation

```
ism = ism[ism.position.lengths() < 0.5 * boxsize]
```

We now can initialize the star as a radiative source, which is a single particle at the origin of our coordinate system. The radiative-transfer code requires the ionizing luminosity of the radiative source to be expressed in photons per second. We make the conversion, assuming that each photon deposits 20 eV, so if the total ionizing luminosity is `Lstar` (W), then the quantity passed to the radiative-transfer code is `Lstar/(20. | units.eV)`, or 1.2×10^{46} photon/s in this case,

```
source = Particle()
source.position = (1, 0, 0) | units.parsec
source.luminosity = Lstar / (20. | units.eV)
```

The source, typically, is not a scattering object, in which case we do not give it a degree of ionization, nor a density or smoothing length.

We could add the source to the gas particles as well as to the source particles, in which case it would also behave as a scattering and absorbing object. This would require us to specify values for the particle's mass (`mass`), luminosity, density (`rho`), ionization state (`xion`), smoothing length `h_smooth`, and internal energy (`u`), in much the same way as we did for the gas. In this example, we do not allow the source particle to scatter or absorb its own radiation.

At this point, we start the radiative-transfer code. For most radiative-transfer codes, we simply specify how many rays are desired. This sets the resolution of the solution. For our example, we decided to use 10^5 rays per Myr, which solves the problem in about 1 minute. Be aware, though, that the calculation timescales quite steeply with the number of rays, and with the number of SPH particles.

For SimpleX, which is a Delaunay-tessellated radiative-transfer solver, we have to specify the size of the box within which we want it to solve radiative-transfer equations. This will pose some interesting challenges later, when we combine the radiative solver with gravitational dynamics and/or hydrodynamics.

```
rad = Simplex()
rad.parameters.box_size = boxsize
```

As noted earlier, the correct operation of the SimpleX code is quite sensitive to the size of the box, and may require some experimentation

```
radiative.particles.add_particle(source)
radiative.particles.add_particles(ism)
```

It is interesting to note that, in contrast to the hydrodynamical solvers discussed in Section 6.2.2, some radiative-transfer codes have only one set of particles, whether they are sources or are part of the scattering medium (but see Section 8.6.1 for a counterexample).

We are now ready to solve the radiative-transfer equation by evolving the model.

```
rad.evolve_model(time_end)
```

Here, we integrate the radiative transfer to time_end = 1 Myr. This calculation is fast because it is carried out for a relatively small number of scattering and absorbing particles and a single radiative source. However, we will see that, when multiple sources are introduced and/or the scattering medium is resolved at higher resolution, this operation can quickly become far more computationally intensive than the gravitational dynamics, stellar evolution, and even the hydrodynamics calculations combined. For some problems, the radiation integration time interval may be as short as a few years, which is usually small compared to the timescales on which other processes operate (see Figure 1.5), making radiative transfer one of the most expensive operations in a multiphysics environment.

In Figures 7.2 and 7.3, we present the result of the simulation at an age of 1 Myr using SPHRay. By that time, the disk on the star's right side is fully ionized, but the cloud (the star's left) blocks the radiation, leaving the left side of the disk neutral. The ionization front, presented in Figure 7.2 is much broader than expected based on the theory, but the morphology of the adopted cloud is rather complex, compared to a homogeneous sphere commonly used in such theoretical models; this example does not lead to a clear Strömgren (1939) sphere. Another reason for the lack of a Strömgren sphere originates from the two assumptions made in our model: the finite resolution and the limited number of rays in the ray-tracing.

7.3 Examples

7.3.1 Ionization Front in an H_2 Region

Most radiative-transfer codes require a large number of inputs, usually including an ionizing source and an absorbing medium. The former is often a star, and the latter a

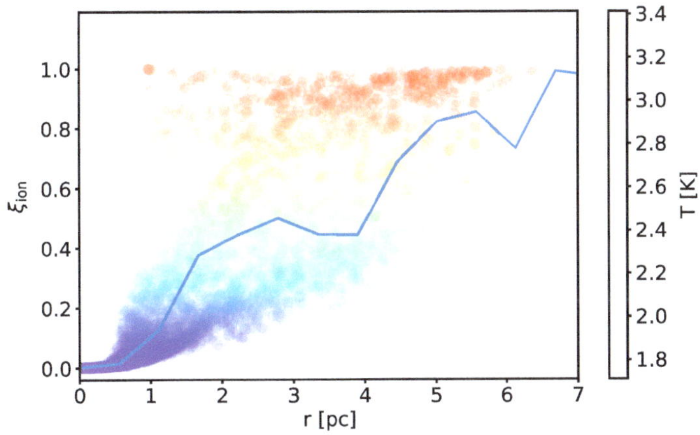

Figure 7.2. Degree of ionization ξ_{ion} due to a $100\,L_\odot$ star at the center of a 3 pc spherically symmetric Plummer distribution of 10^4 gas particles. The curve gives the average degree of ionization, binned in intervals of 0.1 pc. The figure was made using the script in Code example 7.1 and took 20 seconds on a laptop, see also `amuse.examples.rad_minimal`.

gaseous region. We addressed stellar evolution quite extensively in Chapter 5 and hydrodynamics of gaseous bodies in Chapter 6. If you skipped these chapters, this might be a good moment to revisit them before continuing.

In this example, we use stellar evolution and hydrodynamics together with radiative transfer to calculate the radial electron temperature structure of an initially homogeneous H_2 region with density 100 cm^{-3}, which is illuminated by a $120\,M_\odot$ (initial mass) star at an age of 3.3 Myr. At that age, such a star has, according to the `SeBa` stellar evolution code (Portegies Zwart & Verbunt 1996, 2012), a mass of 56.8 $M_\odot$, a luminosity of about $1.5 \times 10^6\,L_\odot$, and a temperature of 50,000 K. The stellar evolution was carried out using the simple script given in Code example 5.3. After the star is located at the cloud center, we initialize the gas distribution in a grid as in Code example 7.3. We then set up the radiative-transfer module; Code example 7.4 presents a snippet of code to do this using `Mocassin`. Of course, these initial conditions are a bit contrived. One might wonder how a 3.3 Myr old star of $120\,M_\odot$ could arrive in the middle of a cloud of molecular hydrogen, but anything is possible in the computer.

The number of settable parameters in a radiative-transfer code is quite large. Most of the fundamental code parameters used in radiative transfer can be queried from the helper function of the running code (see Section 2.2.1). In this example, we rely heavily on the code defaults and only set a few parameters explicitly to illustrate how this is done. Two important attributes are the grid resolution and the chemical abundances of the gas in the grid. In our example, we set these as follows:

```
setup_grid(outer_radius, Ngrid)
setup_abundancies(radiative_transfer)
```

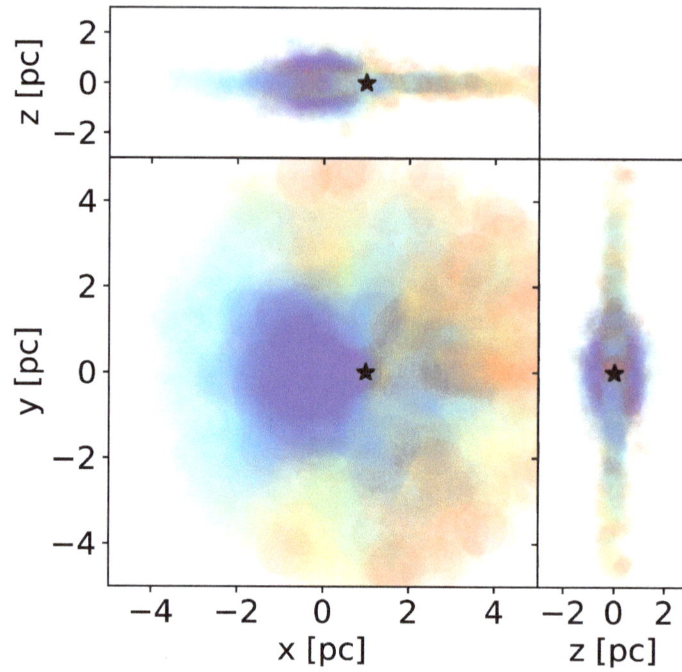

Figure 7.3. Top and side views of the gas distribution around the 100 $L_{\odot}$ star from the calculation presented in Figure 7.2. The bump to the left of the star shields the disk from radiation, and remains cold. The temperature distribution in the disk is radially structured, depending on clumpiness in the particle distribution in the disk. The color is coded according to the colorbar in Figure 7.2. `amuse.examples.rad_minimal`.

```
def make_grid(
    number_of_grid_cells, length, constant_hydrogen_density, inner_radius,
    outer_radius
):
    grid = new_regular_grid([number_of_grid_cells] * 3, length.as_vector_with_length
    (3))
    grid.radius = grid.position.lengths()
    grid.hydrogen_density = constant_hydrogen_density
    grid.hydrogen_density[grid.radius <= inner_radius] = 0 | units.cm**-3
    grid.hydrogen_density[grid.radius >= outer_radius] = 0 | units.cm**-3
    return grid
```

Code example 7.3: Initialization of a grid with a uniform-density spherical cloud.

Abundances of various elements are initialized as fractions relative to hydrogen. The following small snippet sets a few abundances, but printing the entire table will allow you to quickly develop an appreciation for the complexity of the chemical network available in these radiative-transfer codes.

```
def initiate_radiative_transfer_code(outer_radius, Ngrid):
    radiative_transfer = Mocassin(number_of_workers=4)

    radiative_transfer.set_input_directory(
        radiative_transfer.get_default_input_directory()
    )
    radiative_transfer.set_mocassin_output_directory(
        radiative_transfer.output_directory + os.sep
    )
    radiative_transfer.initialize_code()
    radiative_transfer.set_symmetricXYZ(True)

    setup_grid(radiative_transfer, outer_radius, Ngrid)
    setup_abundancies(radiative_transfer)

    radiative_transfer.parameters.initial_nebular_temperature = 6000.0 | units.K
    radiative_transfer.parameters.high_limit_of_the_frequency_mesh = (
        15 | mocassin_rydberg_unit
    )
    radiative_transfer.parameters.low_limit_of_the_frequency_mesh = (
        1.001e-5 | mocassin_rydberg_unit
    )

    radiative_transfer.parameters.total_number_of_photons = 10000000
    radiative_transfer.parameters.total_number_of_points_in_frequency_mesh = 600
    radiative_transfer.parameters.convergence_limit = 0.09
    radiative_transfer.parameters.number_of_ionisation_stages = 6
    radiative_transfer.commit_parameters()

    return radiative_transfer
```

Code example 7.4: Routine to initialize the Mocassin radiative-transfer code for an experiment with a single star as ionizing source in a homogeneous spherical cloud of molecular hydrogen gas.

```
def setup_abundancies(code):
    table = code.abundancies_table()
    for atom in table.keys():
        table[atom] = 0.0
    table["H"] = 1.0
    table["He"] = 0.1
    table["C"] = 2.2e-4
    table["N"] = 4.e-5
    table["O"] = 3.3e-4
    table["Ne"] = 5.e-5
    table["S"] = 9.e-6
```

After initializing the grid and the abundances, we set the rest of the initial conditions, such as the temperature of the ambient gas and the limits, to the frequency mesh,

```
radiative_transfer.parameters.initial_nebular_temperature = 6000.0 | units.K
radiative_transfer.parameters.high_limit_of_the_frequency_mesh = (
    15 | mocassin_rydberg_unit
)
radiative_transfer.parameters.low_limit_of_the_frequency_mesh = (
    1.001e-5 | mocassin_rydberg_unit
)
```

We then initialize the run-specific code parameters, such as the total number of photons, the convergence limit, and the number of ionization stages in the experiment,

```
radiative_transfer.parameters.total_number_of_photons = 10000000
radiative_transfer.parameters.total_number_of_points_in_frequency_mesh = 600
radiative_transfer.parameters.convergence_limit = 0.09
radiative_transfer.parameters.number_of_ionisation_stages = 6
```

Finally we initialize the grid in the radiative-transfer code and load the pre-evolved star into the code as the ionizing source,

```
radiative_transfer.grid.hydrogen_density = grid.hydrogen_density
radiative_transfer.particles.add_particle(star)
```

After all model- and code-specific parameters are set, the grid is initialized, and the ionizing source is defined, then we can start the event loop. The loop is controlled by setting a maximum value to the number of photons and tracking the degree to which convergence has been reached,

```
max_number_of_photons = (
    100 * radiative_transfer.parameters.total_number_of_photons
)
previous_percentage_converged = 0.0
```

The full event loop is shown in Code example 7.5. Figure 7.4 presents the radial temperature profile of the molecular cloud.

7.4 Validation

A number of standard tests can be used to validate the results of radiative-transfer codes in a steady state. Approximate analytic solutions exist for a few problems, and radiative-transfer codes are tested exhaustively against them. Every serious radiative-transfer code performs well on all of these tests.

These tests are important, and any code must pass them, but the real challenge lies in problems where the medium through which the radiation passes evolves dynamically in response to the radiation. Very few analytic (or even approximate analytic) solutions exist for such problems, so definitive tests in these cases are rare. There is, however, a large research community actively addressing these issues, and they have designed several standard tests to which every code should be expected to yield qualitatively similar solutions. Two excellent examples of large collaborative efforts in this area have been carried out by Iliev et al. (2006, 2009). These tests are limited to ionizing radiation passing through an optically thick medium, which is an important process and one of the hardest problems in radiation transfer, as discussed in Section 7.1.1.1. It is difficult, in this context, to compare methods that make different assumptions about the problem, but at least having a suite of tests allows a researcher to validate the results of a radiative-transfer code against earlier calculations.

```
    step = 0
    while percentage_converged < min_convergence:

        radiative_transfer.step()
        percentage_converged = radiative_transfer.get_percentage_converged()
        print(
            "percentage converged :",
            percentage_converged,
            ", step :",
            step,
            ", photons:",
            radiative_transfer.parameters.total_number_of_photons,
        )

        if previous_percentage_converged > 5 and percentage_converged < 95:
            convergence_increase = (
                percentage_converged - previous_percentage_converged
            ) / previous_percentage_converged
            if (
                convergence_increase < 0.2
                and radiative_transfer.parameters.total_number_of_photons
                < max_number_of_photons
            ):
                radiative_transfer.parameters.total_number_of_photons *= 2
        step += 1
        previous_percentage_converged = percentage_converged

```

Code example 7.5: Event loop for the radiative-transfer code.

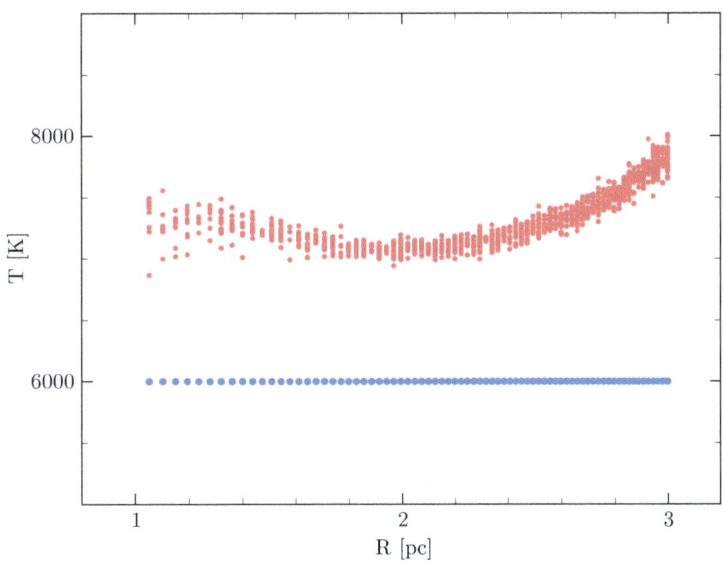

Figure 7.4. Radial electron temperature profile of a centrally illuminated H_2 cloud. The blue points indicate the initial temperature, the red points the final temperature. A script to generate this figure can be found at `amuse.examples.radiative_h2cloud`. It should take roughly 20 minutes to run.

Because testing codes against codes is the principal remaining validation strategy, we show here just one example. This is the last validation example in this book. Any further calculations using multiphysics solvers must be validated to the best degree possible, but we cannot offer any guarantee for the correctness of the results. The importance of radiative hydrodynamics across many theoretical and numerical astronomical problems has led to a wealth of papers on the cross-validation of methods and implementations. A few of the more recent ones include: Teyssier (2002); Leibrandt et al. (2005); Agertz et al. (2007); Iliev et al. (2009); Kolb et al. (2013); Ishii et al. (2017); Kim et al. (2017); Abe et al. (2018).

We run test suite number 5 of Iliev et al. (2009) and present the resulting temperature evolution in Figure 7.5. This test calculates the expansion of a HII region in a homogeneous medium using four different solvers (see also Pelupessy et al. (2013); for details). We simplify the problem here by adopting a solution with single-frequency radiative transfer to study the thermal evolution of the gas, but the same strategy should work with a full radiative solver. We carry out this test using `Fi` or `Gadget2` for the hydrodynamics solver and `SimpleX` or `SPHRay` for the radiative transfer. The coupling of the hydrodynamics code with the radiative-transfer solver is realized using the strategy discussed in Section 8.4.2.

Both radiative codes show different behavior in the fall-off of the temperature, mainly due to the different ways in which high-energy photons are treated. A similar divergence in the ionization-front profiles is observable in Iliev et al. (2009). The

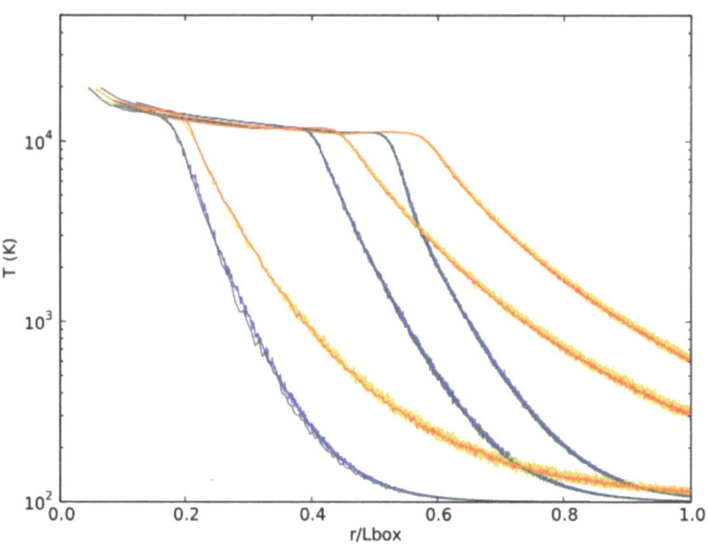

Figure 7.5. Expansion of an ionization front in an initially homogeneous medium (see test #5 of Iliev et al. 2009). We show here the temperature structure as a function of distance (normalized to the box size L_{box}) after 10, 30, and 200 Myr. The source has a ionizing luminosity $L = 5 \times 10^{48}$ photons s^{-1}. Calculations are performed using `Fi` with `SimpleX` (blue curve), `Gadget2` with SimpleX (green), `Fi` with `SPHRay` (yellow), and `Gadget2` with `SPHRay` (red). A more extensive version of this test can be found in Pelupessy et al. (2013), see their Figure 22. Using script `amuse.examples.iliev_test5`, this figure can be generated in about 10 minutes on a fast workstation.

choice of hydrodynamic code has a minor effect on the temperature profile. Other differences can be traced to the specifics of the numerical treatments and the parameter settings of individual methods. The coupling with SPHRay, for example, results in more scatter in the temperature in Figure 7.5 than does the calculation with `SimpleX`. This is caused by using only 2×10^6 rays, which is a rather small number compared to the number of rays used in the Monte Carlo solver.

Unresolved results are generally hard to spot in radiative-transfer calculations, and can only be validated by repeating the simulation at higher resolution. The calculation presented in Figure 8.10 was performed using from 10^6 up to a maximum of 10^9 rays per hour. The latter calculation is expensive in terms of computer time, but the difference between the runs with 10^8 and 10^9 rays was smaller than the thickness of the line. One might then argue that 10^8 rays is sufficient. Note, however, that changing the resolution of the underlying SPH model would again change the results, and a new convergence test would be needed for validation. Therefore, the response of the results to the resolution depends not only on the resolution of the radiative-transfer calculation but also on the resolution of the underlying gas code. This poses interesting challenges for validating coupled radiative-hydrodynamics problems.

7.5 Assignments

7.5.1 Habitability of a Protoplanetary Disk

The habitability of a protoplanetary disk depends on the local temperature. We can calculate the disk's temperature structure using a radiative-transfer code acting on a gaseous disk. This assignment therefore consists of two parts: (1) obtaining a static hydrodynamical realization of a protoplanetary disk, and (2) allowing the disk to evolve. The radiative-transfer codes available in AMUSE, however, are dedicated to ionizing radiation, and calculating the temperature structure in the disk around a low-mass star will therefore not give the correct answer. In this example, we consider a disk around a high-mass star, which we can address using the available radiative-transfer codes.

1. Generate the initial conditions for a protoplanetary disk of 1% of the central star's mass, extending from 1 au to 100 au. Adopt a Toomre Q parameter of 2 (see Section 8.6.1), which will result in a relatively thin disk. In Section 6.4.1, we discussed how to generate such a disk model. Perform the simulation for a star of 10 $M_\odot$ at an age of 10 Myr (see Section 5.2.2). You should run a stellar evolution code to obtain the appropriate luminosity of the star.
2. Measure the temperature in the disk for a duration of 1 Myr, and make a plot of local temperature as a function of radius. Determine the extent of the so-called habitable zone in the disk—the region where the temperature lies between 273 K and 373 K (a broad range, but one in which Earth organisms are known to survive and thrive; see Miroshnichenko et al. 2001). Life also seems to be able to flourish in more extreme environments (Rothschild & Mancinelli 2001), and you may want to extend the range of the habitable zone to include psychrophilic organisms at temperatures below the freezing point of water (Cavicchioli et al. 2002).

3. For a more massive (brighter) star, the habitable zone moves outward. Plot the range of the habitable zone as a function of the luminosity of the central star. What can you say about the radial temperature structure within the habitable zone as the mass of the star increases, other than that it gets bigger and moves outward? Specifically, make a plot of the mean temperature in the habitable zone as a function of the distance to the star and as a function of the mass of the star.

7.5.2 Bumpy Disks

When running the disk with a bump problem in Section 6.4.1, it was clear that the bump shields part of the disk from the central irradiating source. As a consequence, the disk behind the bump remains cooler compared to its surroundings. One might wonder whether it is possible to create a habitable zone behind such a bump, or is the lifetime of the bump too short for life to develop? In order to prevent the bump from dispersing, we put a massive planet or a secondary star in the bump.

In this assignment, you will simulate a protoplanetary disk with a bump, in order to study the temperature evolution of the region behind the bump and compare this with the temperature elsewhere.

1. Generate the initial conditions for a protoplanetary disk with a bump around a 10 $M_\odot$ star. The mass of the disk should be 1% of the central star's mass, ranging from 1 au to 100 au. Adopt a Toomre Q parameter of 10, which will result in a relatively thick disk. Generate a Plummer density distribution with a mass of 10% of the disk and a radius of 10 au, and put it in a circular orbit at a distance of 50 au around the central star (see Section 6.4.1). While running, use `hop` (see Section 8.3.1) to keep track of the position, velocity, size, and mass of the bump.
2. Measure the temperature in the disk for as long as the bump has at least 30% of its initial mass. Plot the mean temperature in the bump as a function of time, and of the point diagonally across the disk at the other side and at the same distance from the central star. This will give you a sort of differential measurement of the temperature in the disk with and without bump.
3. Repeat the calculation for a 10 $M_\odot$ star at the terminal-age main sequence, at the beginning of the giant branch, and when it ascends the asymptotic giant branch.
4. Is there any moment in time at which the bump would be a better place to live than at the other side of the disk?

References

Abe, M., Suzuki, H., Hasegawa, K., et al. 2018, MNRAS, 476, 2664

Agertz, O., Moore, B., Stadel, J., et al. 2007, MNRAS, 380, 963

Altay, G., Croft, R. A. C., & Pelupessy, I. 2008, MNRAS, 386, 1931

Altay, G., Croft, R. A. C., & Pelupessy, I. 2011, SPHRAY: A Smoothed Particle Hydrodynamics Ray Tracer for Radiative Transfer, Astrophysics Source Code Library, ascl:1103.009

Baba, T. 2008, NaPho, 2, 465

Bohren, C. F., & Huffman, D. R. 1983, Absorption and Scattering of Light by Small Particles (New York: Wiley)
Cavicchioli, R., Siddiqui, K. S., Andrews, D., & Sowers, K. R. 2002, COBt, 13, 253
Chandrasekhar, S. 1960, Radiative Transfer (New York: Dover)
Delaunay, B. 1934, Bull. Acad. Sci. USSR(Vll), Classe Sci. Mat. Nat., 7, 793
Ercolano, B., Barlow, M. J., Storey, P. J., & Liu, X. W. 2003, MNRAS, 340, 1136
Fryxell, B., Olson, K., Ricker, P., et al. 2000, ApJS, 131, 273
Gerritsen, J. P. E., & Icke, V. 1997, A&A, 325, 972
Hernquist, L., & Katz, N. 1989, ApJS, 70, 419
Hubber, D. A., Ercolano, B., & Dale, J. 2016, MNRAS, 456, 756
Iliev, I. T., Ciardi, B., Alvarez, M. A., et al. 2006, MNRAS, 371, 1057
Iliev, I. T., Whalen, D., Mellema, G., et al. 2009, MNRAS, 400, 1283
Ishii, I. T., Ohnishi, D., Nagakura, G., Ito, G., Yamada, G., et al. 2017, JCoPh, 348, 612
Kim, J.-G., Kim, W.-T., Ostriker, E. C., & Skinner, M. A. 2017, ApJ, 851, 93
Kolb, S. M., Stute, M., Kley, W., & Mignone, A. 2013, A&A, 559, A80
Leibrandt, D. R., Drake, R. P., Reighard, A. B., & Glendinning, S. G. 2005, ApJ, 626, 616
Miroshnichenko, M. L., Hippe, H., Stackebrandt, E., et al. 2001, Extmo, 5, 85
Paardekooper, J. P., Kruip, C. J. H., & Icke, V. 2010, A&A, 515, A79
Pelupessy, F. I., van der Werf, P. P., & Icke, V. 2004, A&A, 422, 55
Pelupessy, F. I., van Elteren, A., de Vries, N., et al. 2013, A&A, 557, A84
Peraiah, A. 2001, An Introduction to Radiative Transfer: Methods and Applications in Astrophysics (Cambridge: Cambridge Univ. Press)
Plummer, H. C. 1911, MNRAS, 71, 460
Portegies Zwart, S. F., & Verbunt, F. 1996, A&A, 309, 179
Portegies Zwart, S. F., & Verbunt, F. 2012, SeBa: Stellar and Binary Evolution, Astrophysics Source Code Library, ascl:1201.003
Ritzerveld, J., & Icke, V. 2006, PhRvE, 74, 026704
Rothschild, L. J., & Mancinelli, R. L. 2001, Natur, 409, 1092
Strömgren, B. 1939, ApJ, 89, 526
Teyssier, R. 2002, A&A, 385, 337
Voronoi, G. 1908, JRAM, 134, 198

Part III

Advanced Strategies

Chapter 8

Elementary Coupling Strategies

> I would rather have questions that can't be answered than answers that can't be questioned.
>
> —R. Feynman

The simplest multiscale or multiphysics coupling strategy is the linear propagation of information from one code to another. In some sense, one code produces the conditions that another code uses to continue the calculation. These couplings are often unidirectional; one code informs another, but there is no back propagation of information. So long as the response from one code to the other is linear, we can refrain from complex code-coupling strategies, as described in the next chapter (Chapter 9). Linear coupling strategies lead to new AMUSE applications without increasing your script's complexity, while increasing its applicability.

8.1 Multicode Coupling Problems

In this chapter, we discuss how one code can produce self-consistent input or boundary conditions for another code without using some intermediate state, such as a file. We encountered such coupled problems in Section 4.5.1, where a gravitational dynamical simulation required interrupts in order to resolve collisions, in Section 5.2.10 for treating stellar explosions, and in Section 6.3.5 where we discussed how a hydrodynamics solver can be stopped to allow a special treatment of high-density clumps to be addressed nonhydrodynamically.

In these examples the physical solvers were not coupled, but instead we adopted a simple linear information propagation strategy (`A`→`B`), in which one code (`A`) generated the initial conditions for another (`B`), but there was no back coupling from the other code (`B`) to the first (`A`).

In this chapter, we focus on weakly coupled problems in which one code generates the data for another code to proceed, after which a third code (or the first one)

doi:10.1088/2514-3433/adfc3cch8

continues or is (re)started. Towards the end of this chapter, in Section 8.7.2, we will discuss two different strategies of multiscale and multiphysics coupling. These couplings are relatively simple solutions to the very complex but weakly coupled problem of multiscale and multiphysics simulations. The numerics become considerably more complex once a code responds nonlinearly to small perturbations induced by another code. This more advanced topic of code coupling is discussed in Chapter 9.

A word of caution is in order here. There is probably no unique way to combine numerical simulations of different physical domains. Although many researchers claim self-consistency in their numerical solutions, this rarely is actually the case. The AMUSE approach described here lends itself well to problems in which the physics is weakly coupled, but even in those cases validation is a dark art rather than a well-documented science.

8.2 Advanced Data Handling

8.2.1 Dedicated Channels and Copy Operations

Key to coupling codes is the ability to transfer data from one code to another, taking the proper unit conversions into account. In AMUSE we achieve this objective by the use of channels. Channels are one of the cornerstones of AMUSE, allowing one code to modify the data on which another code operates, as we now discuss.

Channels are used to communicate data between modules and the driving `Python` script. We encountered channels previously in Section 2.2.3, where we learned how to copy data from an N-body code to the AMUSE framework, and vice versa. Here, we extend that discussion. Recall that channels are dedicated MPI (or sockets) communication links that do not need to be reestablished on every use. While it is possible to access code data directly from the running module (e.g. as `module.particles.mass`), we strongly encourage the use of channels because they are faster, more efficient, more transparent, and result in cleaner code compared to directly accessing the module's particle set. Perhaps most important, they are independent of specific hardware and software implementations of interprocess communication. For monophysics solvers, channels may seem like overkill, but with multiphysics solvers they are extremely useful, if not essential.

To run stellar evolution and gravity concurrently, we require three channels: one to copy data from the gravity solver to the local `Python` framework, one to copy data from the stellar-evolution code to the framework, and one to copy data from the framework to the gravity code. This can be done as follows:

```
channel_from_gravity = gravity.particles.new_channel_to(bodies)
channel_from_stellar = stellar.particles.new_channel_to(bodies)
channel_to_gravity = bodies.new_channel_to(gravity.particles)
```

The local particles (called `bodies` in this example) can contain more information than either of the running codes provide. The first and second channels are used to copy data from the specific modules to the framework, whereas the third channel allows us to write to the gravitational dynamics module (see Section 2.2.3). In this case, we do not require a channel from the framework to the stellar-evolution code because the gravitational dynamics do not affect the stellar evolution. That would change if we were to incorporate collisions, in which case the mass of a merger product, and therefore the evolution of that star, would be affected by the merger process.

Code example 8.1 presents an initialization sequence for combining stellar evolution with gravitational dynamics, demonstrating that communication through the `Python` script is often not necessary because codes can communicate directly with each other through channels. We can arrange for the stellar-evolution code to communicate directly with the gravity solver by declaring a direct channel between the two modules and copying the relevant information,

```
channel_from_stellar_to_gravity = stellar.particles.new_channel_to(
    gravity.particles
)
```

The data stream still passes through the AMUSE framework, to ensure proper unit conversion, but in this case the local `bodies` particle set is unaffected. This is a bit tricky because now the copies of the data in the various running codes are not synchronized with the data in the local `Python` script.

After the particle set is constructed, the codes initialized, and the channels created, these channels can now be used to communicate between the running modules and the framework. For example, the gravity solver affects stellar positions and velocities, whereas the stellar-evolution module affects stellar masses. After taking steps in `stellar`, the new stellar masses must be copied to the `gravity` module before the next gravity step occurs.

Code example 8.2 presents an example of the sequence of calling the codes and the associated copy commands. Here, we assume that the channels `channel_stellar_to_gravity` and `channel_gravity_to_framework` are constructed to copy data from `stellar` to `gravity` and back from `gravity` to the framework. Between the calls to `gravity.evolve_model` and `stellar.evolve_model`, there is no need to synchronize the data because the gravity module does not affect the parameters used by the stellar-evolution code.

In Code example 8.2, we adopted an interlaced time-stepping strategy, which is a slight improvement over the linear temporal interstepping discussed so far. The more elaborate event-driven approach will be discussed in Section 8.4.1, and illustrated in Code example 8.7.

At the end of the snippet shown here, the `bodies` in the script are up to date and can be used to write a snapshot for later analysis, as we do in the last line of the script, where the `bodies` array is saved to a file called `filename`. Note that the order in which channels are called is important for those parameters that share the same name but may conceivably have different meanings in different codes. For example, both codes here use the particle attribute `mass`, which means essentially the same thing in each code. However, the attribute `radius` can and does mean different things in different modules, requiring some care when managing data transfer (see, e.g., Section 8.7.1).

```
masses = new_salpeter_mass_distribution(number_of_stars, mass_min, mass_max)
mass_total_init = masses.sum()
converter = nbody_system.nbody_to_si(mass_total_init, radius_virial)
bodies = new_king_model(number_of_stars, W0, convert_nbody=converter)
bodies.mass = masses
bodies.scale_to_standard(convert_nbody=converter)

stellar = Sse()
stellar.parameters.metallicity = metallicity
stellar.particles.add_particle(bodies)
bodies.radius = stellar.particles.radius

gravity = Ph4(converter)
gravity.parameters.timestep_parameter = 0.01
gravity.particles.add_particles(bodies)

channel_from_stellar_to_gravity = stellar.particles.new_channel_to(
    gravity.particles
)
channel_from_gravity_to_framework = gravity.particles.new_channel_to(bodies)
```

Code example 8.1: Initializing a simple solver combining stellar evolution with gravity.

```
energy_total_init = gravity.kinetic_energy + gravity.potential_energy
energy_difference_gr = 0 | energy_total_init.unit

time = 0.0 | time_end.unit
while time < time_end:
    time_step = min(stellar.particles.time_step.amin(), time_end - time)

    stellar.evolve_model(time + time_step / 2)
    channel_from_stellar_to_gravity.copy()

    energy_total_gr = gravity.kinetic_energy + gravity.potential_energy
    gravity.evolve_model(time + time_step)
    energy_difference_gr += (
        gravity.kinetic_energy + gravity.potential_energy - energy_total_gr
    )

    stellar.evolve_model(time + time_step)
    channel_from_stellar_to_gravity.copy()
    channel_from_gravity_to_framework.copy()

    time += time_step
    write_set_to_file(bodies, filename, append_to_file=True)
```

Code example 8.2: Event loop combining stellar evolution with gravity.

8.2.1.1 Nonsymmetric Cross Synchronization of Specific Attributes

A certain number of attribute names have specific purposes for different modules in AMUSE. The `radius`, used in particle sets, for example, is used in an N-body code as the collision radius that triggers the associated stopping conditions (see Section

```
    masses = new_salpeter_mass_distribution(N, Mmin, Mmax)
    Mtot_init = masses.sum()
    converter=nbody_system.nbody_to_si(Mtot_init,Rvir)
    bodies = new_king_model(N, W0,convert_nbody=converter)
    bodies.mass = masses
    bodies.scale_to_standard(convert_nbody=converter)

    gravity = ph4(converter)
    gravity.parameters.timestep_parameter = 0.01
    gravity.particles.add_particles(bodies)
    collision_detection = gravity.stopping_conditions.collision_detection
    collision_detection.enable()

    stellar = SeBa()
    stellar.parameters.metallicity = z
    stellar.particles.add_particle(bodies)
    stellar.evolve_model(0|units.Myr)

    supernova_detection = stellar.stopping_conditions.supernova_detection
    supernova_detection.enable()

    channel_from_se = stellar.particles.new_channel_to(bodies)
    channel_from_gd = gravity.particles.new_channel_to(bodies)
    channel_to_gd = bodies.new_channel_to(gravity.particles)
    channel_from_se.copy_attributes(["mass","radius", "age",
                                     "temperature", "luminosity"])
```

Code example 8.7: Declaration of the solvers and data structures used in Code example 8.8.

4.5.1). In the stellar-evolution modules, the same attribute refers to the photospheric radius of the star. This is confusing, and at times may even cause the script to behave inappropriately.

To prevent such name confusion, we would like to be able to copy a specific attribute from a running code to another local attribute name in the script. This can be realized through channels, in case the receiver's name does not match that of the sender.

The physics modules tend to use generic terms such as `mass`, `radius`, `position`, etc. In a complicated production script, one may wish to copy the `mass` from the particle set of a stellar-evolution code, `stellar.particles`, to a parameter with another name. This might happen, for example, when the gravity solver treats the combined mass of a star and a protoplanetary disk as a single particle, whereas the stellar-evolution code uses only the stellar mass to calculate the evolution of the star. In the local particle-set `bodies`, the attributes might then be called `bodies.mass_of_star_and_disk` for the total mass of the star and the disk, and `bodies.radius_of_star_and_disk` for the equivalent radius, whereas the mass of the star alone could be called `bodies.stellar_mass`. In the main script we declare the parameter `bodies.mass_star_and_disk= bodies.stellar_mass+bodies.disk_mass`.

We can tune the various channels to copy the right parameters. In this case we can, for example, say

```
channel_to_stellar = bodies.new_channel_to(
    stellar.particles,
    attributes=["radius"],
    target_names=["radius"]
)
channel_to_gravity = bodies.new_channel_to(
    gravity.particles,
    attributes=["mass_of_star_and_disk",
                "radius_of_star_and_disk"],
    target_names=["mass", "radius"]
)
```

When the copy function is called, only the specific named attributes are copied. Similar channels can be declared for the reverse operations, such as copying the mass from the stellar-evolution module back to the local particle set

```
channel_from_stellar = stellar.particles.new_channel_to(
    bodies,
    attributes=["mass"],
    target_names=["stellar_mass"]
)
```

In the main event loop we would write

```
    channel_to_stellar.copy()
    channel_to_gravity.copy()
```

to copy just the masses from the local particle set into the stellar-evolution code, and the masses and radii from the local set into the gravity code. Similarly, we would use

```
    channel_from_stellar.copy()
```

after the stellar-evolution step to copy the updated stellar masses back into local memory. We present an example of how this works in practice in Section 8.4.4.

8.2.2 Particle Subsets and Supersets

Sometimes, it is desirable to perform certain tasks on a subset of the particles in a set. This can be accomplished by generating new particle sets, but managing multiple related particle sets can unnecessarily complicate the bookkeeping, increase the chance of mistakes, and reduce readability. We can achieve the same result using channels.

We demonstrate this approach by simulating a system of stars with planetesimal disks around them. All particles should be evolved using the gravitational dynamics module, but only the stars should be affected by stellar evolution. To do this, we define two particle sets, called `planetesimals` and `stars`. Both sets contain masses, positions, and velocities, but the `stars` set also includes stellar data, such as radius, luminosity, and temperature.

We send `stars` to the stellar-evolution code:

```
stellar.particles.add_particle(stars)
```

after which we can call `evolve_model` to evolve them. But, we send both `stars` and `planetesimals` to the gravity module:

```
gravity.particles.add_particle(stars)
gravity.particles.add_particle(planetesimals)
```

We subsequently declare the appropriate channels to copy data from the physics modules to the framework:

```
channel_from_stellar_to_stars = stellar.particles.new_channel_to(stars)
channel_from_gravity_to_stars = gravity.particles.new_channel_to(stars)
channel_from_gravity_to_planetesimals \
    = gravity.particles.new_channel_to(planetesimals)
```

and a similar set of channels from the framework to the physics modules.

In many cases, however, it may be easier to declare a *superset* of particles

```
gravitating_bodies = ParticlesSuperset([planetesimals, stars])
```

After this, we can treat `gravitating_bodies` as the particles affected by gravity, keeping the `planetesimals` separate from the `stars` for the stellar evolution. Sending the superset to the gravity module can be accomplished with

```
gravity.particles.add_particles(gravitating_bodies)
```

Only the stars are sent to the stellar-evolution module.

In order to guarantee consistency among the various parameters in the particle subsets, we create channels to and from the gravity module, and one channel from the stellar-evolution code:

```
channel_to_gravity = gravitating_bodies.new_channel_to(gravity.particles)
channel_from_gravity = gravity.particles.new_channel_to(gravitating_bodies)
channel_from_stellar = stellar.particles.new_channel_to(stars)
```

Now, we can easily copy particle attributes after the gravity or stellar-evolution steps, e.g.,

```
channel_from_stellar.copy_attributes(["mass", "radius"])
```

We can select subset of particles within a particle set by using standard `Python` tools. For example, suppose we have a superset of stars and planetesimals, as in `gravitating_bodies` just discussed. Filtering out the planetesimals can be realized by selecting on their masses—since the planetesimals have low masses, we can write

```
mass_min = 1 | units.MEarth
low_mass = gravitating_bodies[gravitating_bodies.mass < mass_min]
```

to select objects with masses less than 1 Earth mass. In principle, we could use the more arcane but versatile `Pythonlambda` function. Regrettably, however, as of 27 November 2023 Lambda functions are no longer supported in `Python` in tandem with the AMUSE `select` operation.

8.2.3 Interrupting Codes and Stopping Conditions

Sometimes it is necessary to have a community code stop and hand control back to the framework, leaving subsequent decisions to the user. As discussed in Section 4.5.1, Section 5.2.10, and Section 6.3.5 (interestingly enough the radiative-transfer module has no stopping conditions) we call these events *stopping conditions*. In Section A.5.2, we present a more complete overview of the various stopping conditions. There are many reasons why a community code should stop and return control to the user script, and the circumstances may be quite different for each physical domain. Stopping conditions are intrusive to the community code, and therefore are inconsistent with the AMUSE philosophy of "keep community codes pristine." Nevertheless, we have taken the step of implementing stopping conditions in most AMUSE community codes. The details vary from one code to another. Barnes–Hut tree codes, for example, are generally not built to detect physical collisions between stars, while in regularized *N*-body codes, a "collision" may have an entirely different meaning. In all cases, checks for physical (or other) collisions can usually be incorporated into the interface code or at a high level in the community code without entailing sweeping changes in the source.

In an AMUSE script, one can initiate stopping conditions after initializing the worker code. In a gravity module this is realized by

```
detect_collision = gravity.stopping_conditions.collision_detection
detect_collision.enable()
```

Many, but not all, AMUSE *N*-body codes have collision detection enabled. Those codes that do not will produce a warning message when the user code attempts to set a stopping condition, but subsequently they simply continue without the ability to stop, offering no other indication that stopping is disabled.

In the time-evolution loop, you can check for a stopping condition on return from `evolve_model()` and take appropriate action:

```
if detect_collision.is_set():
    for cp in range(len(detect_collision.particles(0))):
        particles_in_code = Particles(
                particles=[detect_collision.particles(0)[cp],
                           detect_collision.particles(1)[cp]])
                local_colliding_particles = \
                    particles_in_code.get_intersecting_subset_in(bodies)
        merge_two_particles(gravity, local_colliding_particles)
        gravity.particles.synchronize_to(bodies)
```

The *N*-body integrator checks for the exception after each internal step and returns to the calling function when the event (in this case, a collision) is detected. The

N-body system is synchronized before returning to the framework to allow the calling function to perform further tasks on the datasets without corrupting the integrity of the N-body data.

The flag `detect_collision.is_set()` returns `True` if the stopping condition is met, after which the code for resolving stopping conditions is invoked. Inside this code, the bodies that caused the community code to stop are returned to the `Python` framework and the user script must deal with the exception. In this example, the gravity solver stops when two bodies approach within a distance smaller than the sum of their radii. The resolution of the exception in this case is to delete the colliding particles and insert a new particle at their center of mass. The following routine accomplishes this as follows:

```
def merge_two_particles(gravity, particles_in_encounter):
    new_particle = Particles(1)
    new_particle.mass = particles_in_encounter.total_mass()
    new_particle.position = particles_in_encounter.center_of_mass()
    new_particle.velocity = particles_in_encounter.center_of_mass_velocity()
    new_particle.radius = particles_in_encounter.radius.sum()
    gravity.particles.add_particles(new_particle)
    gravity.particles.remove_particles(particles_in_encounter)
```

In this routine, a new particle is created and the center-of-mass data of the merging particles are copied to the new particle and added to the `gravity` instantiation. The two merging stars are not needed anymore, so they are removed from `gravity`. Best practice probably requires that we also flag the event, writing the merging particles to a file before removing them from the system (not shown here). Removal is achieved by explicitly removing the particles from the local `bodies`, adding the center of mass, and synchronizing `gravity.particles` with the local data:

```
        gravity.particles.synchronize_to(bodies)
```

Of course, for all this to happen, one must set the radii of the particles in advance. For example, you could write

```
    bodies.radius = 0.001 | nbody_system.length
```

just after the particles have been initialized, but before they are sent to the community module. When working with physical units, that line might look as follows

```
    bodies.radius = 10 | units.RSun
```

8.2.4 Collision Handling

We decide when the point-mass approximation fails based on the sizes of two stars (as determined by the stellar-evolution code) and the distance between them (from the N-body code). A collisional stopping condition is triggered when the distance between two stars becomes smaller than the sum of their radii.

Because a stopping condition is an exception, it must be checked for and addressed as soon as possible after it is raised (see Section A.5.2). In the case of a collision, once the gravity module reports the exception, we should next evolve the stars to the moment the collision was detected. It is possible (although not common) that the collision may be avoided by a drastic change in stellar radius due to stellar evolution. For example, the gravity module may decide that two giant stars have collided, but when stellar evolution is applied these stars may turn into (non-colliding) compact objects before the collision is resolved. In such a case, the collision should probably not occur, and the dynamical integration can simply continue.

Synchronization between the modules (here gravity and stellar evolution) is very important. Recall that there are usually two copies of the data for each module—one in local framework (Python) memory, the other in the module itself. The module data is up to date at the end of a step, and we must synchronize the modules with the local framework before resolving the stopping condition. To manage a collision between two stars, the framework needs to know the stellar ages, masses, radii, luminosities, and temperatures. Some of this information comes from the gravity module, the rest from the stellar-evolution module.

Note that we must keep track of two different "radii"—the natural stellar radius in the stellar-evolution module, and the collision radius used by the gravity code to identify close dynamical encounters (see Section 8.7.1). Probably the safest way to manage this is by initializing two separate local variables in the framework particle set. Selective channels can be declared to ensure proper communication between the modules and the framework script:

```
channel_from_stellar = stellar.particles.new_channel_to(bodies,
    attributes=["mass", "radius", "age"],
    target_names=["mass", "radius", "age"])
channel_from_gravity = gravity.particles.new_channel_to(bodies,
    attributes=["x", "y", "z", "vx", "vy", "vz", "mass", "radius"],
    target_names=["x", "y", "z", "vx", "vy", "vz", "mass", "collision_radius"])
channel_to_gravity = bodies.new_channel_to(gravity.particles,
    attributes=["mass", "collision_radius"], target_names=["mass", "radius"])
```

In channel_from_gravity, we explicitly copy all the positions and velocities to ensure that the particle set in the framework is consistent with the gravity module. Later in the event loop (see Code example 8.3) we will simply call the copy operator for the various channels. We copy data from the stellar evolution to the framework with

```
channel_from_stellar.copy()
```

and copy data from the N-body code to the framework with

```
channel_from_gravity.copy()
```

These lines ensure that the masses and radii used in the gravitational module are consistent with those in the stellar-evolution module and in the framework.

```
    while time < t_end:
        time += dt

        bodies.radius *= 1.e+5
        channel_from_framework_to_gd.copy_attributes(["mass", "radius"])
        E_dyn = gravity.kinetic_energy  + gravity.potential_energy
        gravity.evolve_model(time)
        dE_dyn = E_dyn - (gravity.kinetic_energy  + gravity.potential_energy)

        Nenc += resolve_stellar_encounter(time, stellar, gravity, bodies,
    stopping_condition)

        E_stellar = gravity.kinetic_energy + gravity.potential_energy
        stellar.evolve_model(time)
        dE_stellar = E_stellar - (gravity.kinetic_energy + gravity.potential_energy)

        channel_from_gd_to_framework.copy()
        channel_from_se_to_framework.copy_attributes(["age", "mass", "radius"])

        write_set_to_file(bodies.savepoint(time), filename, 'hdf5', append_to_file=
    True)
        print_diagnostics(time, bodies.mass.sum(), E_dyn, dE_dyn, dE_coll,
    dE_stellar)
```

Code example 8.3: Main code declaration and event loop for simulating a star cluster with collisions.

Stars are small objects, and collide infrequently. In order to make more stars collide (for debugging or illustrative purposes), we can increase the stellar radii used by the gravitational dynamics, multiplying the radii provided by the stellar-evolution code by some factor. This has to be done after the stellar-evolution information is copied back into the local particle set. The magnified particle-set data must then be copied into the gravity module's particle set to ensure that the dynamics is aware of the new stellar radii:

```
stellar.evolve_model(time)
channel_from_stellar.copy()
bodies.collision_radius = 1.0e5 * bodies.radius
channel_to_gravity.copy()
gravity.evolve_model(time)
```

This is a pretty dirty trick, but it will guarantee many interesting collisions!

Code example 8.3 presents part of the main routine for simulating the dynamical evolution of a cluster of stars, including stopping conditions for collisions. The function that resolves the collisions is presented in Code example 8.4. The same routine is also used in Section 8.5.1, where we discuss an event-driven simulation of a star cluster with collisions.

How do we model the merger product? If we are using a parameterized stellar-evolution code, the structure of the merger product cannot easily be determined. We can mimic the mixing of fresh hydrogen into the merger product's core by reducing the stellar age in some mass-dependent way, resulting in an effective rejuvenation of the star (see Section 5.3.12). Afterwards, we place the merged star on a new

```
def resolve_stellar_encounter(time, stellar, gravity, bodies, stopping_condition):
    Nenc = 0
    if stopping_condition.is_set():
        E_coll = gravity.kinetic_energy + gravity.potential_energy
        print("At time=", gravity.model_time.in_(units.Myr), "number of encounters=", len(stopping_condition.particles(0)))
        for ci in range(len(stopping_condition.particles(0))):
            particles_in_encounter = Particles(particles=[stopping_condition.particles(0)[ci], stopping_condition.particles(1)[ci]])
            particles_in_encounter = particles_in_encounter.get_intersecting_subset_in(bodies)

            merge_two_stars(bodies, particles_in_encounter)
            bodies.synchronize_to(gravity.particles)
            bodies.synchronize_to(stellar.particles)
            Nenc+=1
            print("Resolve encounter Number:", Nenc)
            gravity.evolve_model(time)
        dE_coll = E_coll - (gravity.kinetic_energy + gravity.potential_energy)
    return Nenc
```

Code example 8.4: Simple function to resolve stellar collisions.

```
def merge_two_stars(bodies, particles_in_encounter):
    com_pos = particles_in_encounter.center_of_mass()
    com_vel = particles_in_encounter.center_of_mass_velocity()
    new_particle=Particles(1)
    new_particle.mass = particles_in_encounter.total_mass()
    new_particle.age = min(particles_in_encounter.age) * max(particles_in_encounter.mass)/new_particle.mass
    new_particle.position = com_pos
    new_particle.velocity = com_vel
    new_particle.radius = 0 | units.RSun
    bodies.add_particles(new_particle)
    bodies.remove_particles(particles_in_encounter)
```

Code example 8.5: Simple function to merge and rejuvenate two stars while conserving linear momentum.

evolutionary track at the new age and continue the calculation. This can be realized simply by adding a few lines to the linear momentum- (and mass-)conserving routine `merge_two_particles()` discussed in Section A.5.2, which then becomes Code example 8.5. In the next subsection, we discuss a more sophisticated approach to the problem.

8.3 Runtime Analysis and Task Offloading

AMUSE contains many analysis tools designed to operate on simulation data. Here, we describe two modules, hop and `Kepler`, that are commonly used in tandem with running simulation codes. There are other codes that are often associated with AMUSE, and their number is growing continuously, but this book can only present a snapshot in time of the development.

8.3.1 The Hop Package

Hop (Eisenstein & Hut 1998) is a module designed to identify and quantify structure in a distribution of points.[1] Although it was originally developed to identify clumps in a three-dimensional spatial distribution of objects (stars in a cluster or galaxies in the Universe), we can easily apply hop to any set of particle data. For example, to distinguish planetesimals in a protoplanetary disk from other debris, we might look for clumps in the space of orbital velocity, inclination, and distance from the central star. Figure 8.1 shows a small example of how hop can be used to characterize clumpy structure in a distribution of stars.

The code snippet in Code example 8.6 gives the basic calling sequence to run hop. Several free parameters guide hop in finding structure and substructure; they include

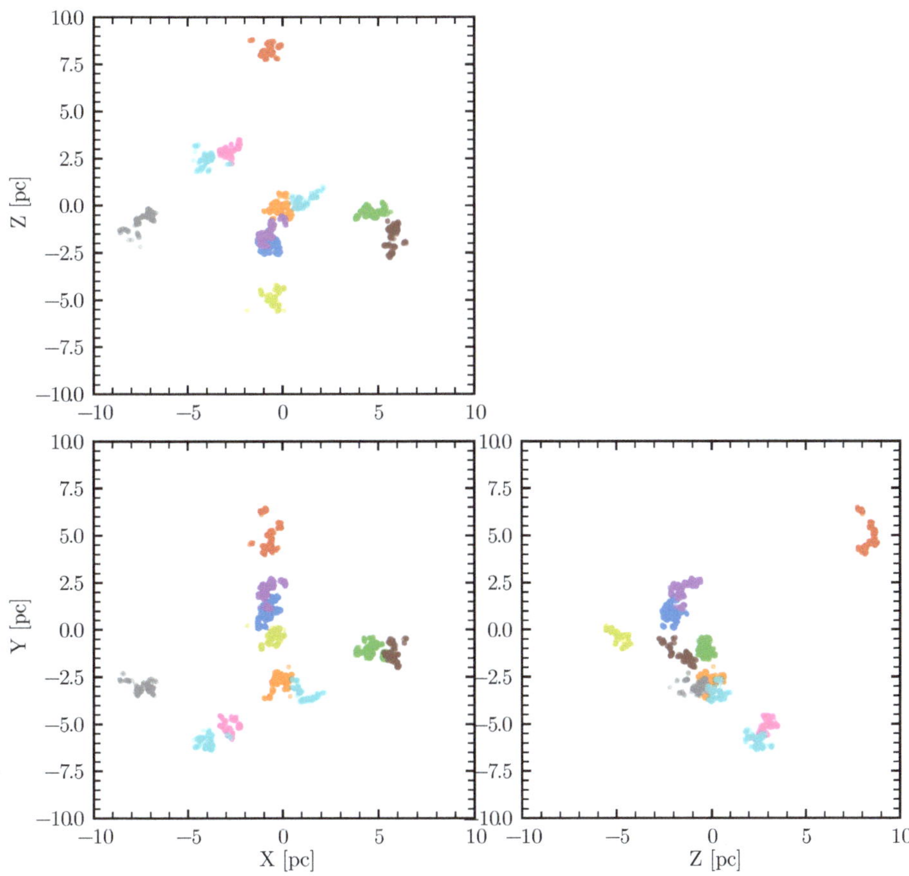

Figure 8.1. Cluster of 1000 stars initially distributed with fractal dimension 1.6 and a half-mass radius 0.5 pc. The 11 different clumps identified by hop are indicated by color. The figure was produced using the script `amuse.examples.plot_fractal_clumpy_cluster`

[1] The name derives from the algorithm used to divide the distribution into equivalence classes representing substructure by "hopping" from each particle to its neighbor with the highest local density.

```
hop = Hop(unit_converter)
hop.particles.add_particles(particles)
hop.calculate_densities()

mean_density = hop.particles.density.mean()
hop.parameters.peak_density_threshold = mean_density
hop.parameters.saddle_density_threshold = 0.99 * mean_density
hop.parameters.outer_density_threshold = 0.01 * mean_density

hop.do_hop()
result = [x.get_intersecting_subset_in(particles) for x in hop.groups()]
hop.stop()
```

Code example 8.6: Using hop to identify substructure in a particle distribution.

peak_density_threshold, the lower limit for recognizing a peak, saddle_density_threshold, the phase-space density of a saddle point, and outer_density_threshold, which is the minimum density threshold for recognizing a clump. The result in this example is a list of particle sets corresponding to the clumps identified by hop. The algorithm is not particularly fast, but hop is reliable and works intuitively.

8.3.2 The Kepler Package

Plotting positions and animating *N*-body systems is illustrative and often educational, but not always the most useful for astrophysical analysis. When studying the orbits of planets or multiple systems, the Keplerian orbital elements provide much more quantitative information. These are straightforward to compute from the masses, positions, and velocities of the orbiting bodies, but this is such a common operation that AMUSE contains built-in modules designed to calculate, manipulate, and use Keplerian orbits, and we can simply use one of those to do the work for us. One such module, a legacy of the older starlab package,[2] is simply called Kepler.

We begin by initializing the Kepler package

```
johannes = Kepler(convert_nbody)
```

where convert_nbody is a converter (see Section 2.3). We also generate a particle set containing just the two particles for which we want the orbital elements. For example, in the earlier example on solar-system orbits (Section 3.4.1 and Section 4.4.1), we could call this local particle set sun_earth and have its two particles point to the Sun (particle 0) and to Earth (particle 3).

[2] See github.com/amusecode/Starlab.

```
sun_earth = Particles()
sun_earth.add_particle(particles[0])
sun_earth.add_particle(particles[3])
channel_from_to_sun_earth = gravity.particles.new_channel_to(sun_earth)
```

We also created a local channel to these two particles, which we can use in the event loop to synchronize the gravity code data with the local particles in `sun_earth`.

Within the loop, we reinitialize `Kepler` with the two particles of interest—Sun and Earth. To print Earth's semimajor axis, this looks like:

```
johannes.initialize_from_particles(sun_earth)
print(f"Earth's semimajor axis={johannes.get_elements()}")
```

Note that we reinitialize an existing `Kepler` in the event loop, rather than creating a new one each time. This is done for reasons of efficiency. We could simply create and delete a new Kepler code every time we require these services, but this can incur a considerable overhead in large runs.

Using the `Kepler` package only for computing simple orbital elements is probably overkill. In AMUSE, the same result can be achieved by calling the routine `orbital_elements. orbital_elements_from_binary()`. This function takes two arguments, a particle set (here containing the two stars in a binary) and the gravitational constant `nbody_system.G`, and returns the six orbital elements (see Section 4.4.1). However, the `Kepler` package can also manipulate orbits—returning particle sets corresponding to a specified time, orbital separation, apocenter, or pericenter—making it a useful tool in many applications. For example, the `multiples` module (Section 8.7.1) uses `Kepler` to set up scattering experiments, perform two-body orbit integrations, classify scattering outcomes, and rescale particle data.

8.3.3 Recovering from a Code Crash

As discussed in Section 4.6, physics modules may crash if they are placed in regimes for which they were not designed. When a physics module in AMUSE crashes, it is undesirable for the framework also to crash. Using `Python` exception handling, we can detect when a code has terminated and take appropriate action. We demonstrate this by using a Henyey code to model the evolution of a $10\,M_\odot$ star until the code crashes. Another example, where we switch from one stellar-evolution code to another, is presented in Section 8.4.3.

Stellar-evolution codes tend to crash at the end of the nuclear fusion as the star approaches its death as a supernova explosion. (It may happen sooner, depending on the code!) Evolving a star to this late stage can be expensive because the stellar-evolution code has a built-in safety valve—if an evolution step fails to converge, the code iteratively reduces the time step and restarts from the previous self-consistent stellar model. This process is repeated until the code has reached a built-in limiting number of restarts, at which point it simply gives up.

```
try:
    while True:
        stellar.evolve_model()
except AmuseException as ex:
    print(f"Star evolved to t = {stellar.particles[0].age}")
```

Here, we call `evolve_model()` without an argument, which evolves the star using the internal stellar-evolution time step, rather than to a specified time. Once we reach the `print` statement the Henyey code has terminated, but the AMUSE interface has not. We can then obtain the last converged stellar-evolution model, which is as close to an exploding star as the stellar-evolution code can get. In Section 6.4.4, we showed how the postevolution product can converted into an SPH particle representation to continue the evolution through the supernova explosion.

Rather than using `MMAMS`, it can be desirable to use a hydrodynamical solver to resolve the collision. A first step in this process would be to convert the one-dimensional stellar-evolution structure from a Henyey code to a realization suitable for continuing the calculation with a hydrodynamics solver. One such routine is `convert_stellar_model_to_SPH` from AMUSE's `ext.star_to_sph` package.

In Figure 8.2, we present the density structure for a $1\,M_\odot$ a $3\,M_\odot$ and a $10\,M_\odot$ star, and the associated SPH realization of the same star using 10^4 SPH particles.

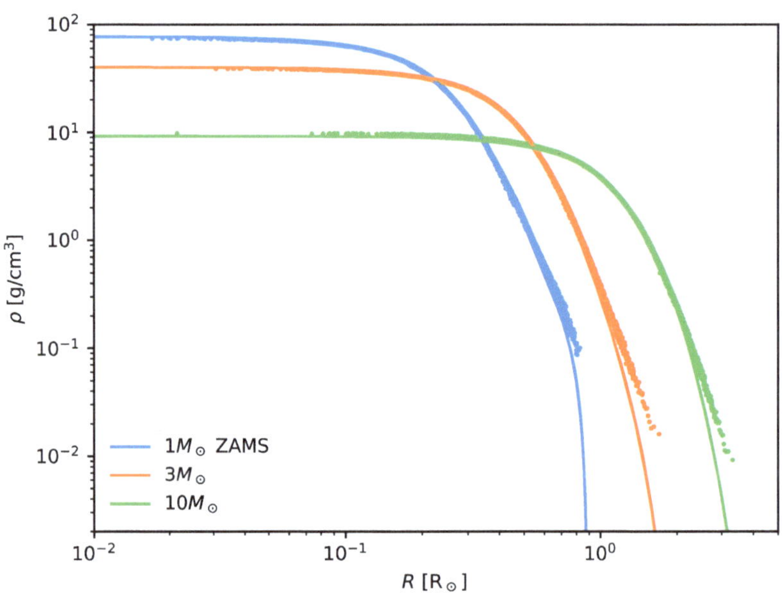

Figure 8.2. Density as a function of distance to the stellar center for three stars of $1\,M_\odot$ $3\,M_\odot$ and $10\,M_\odot$ each at an age of 1 yr. The solid lines show the results from the stellar-evolution model (`MESA` in this case), the dots give the densities attributed to the SPH particles using the `convert_stellar_model_to_SPH` routine. This figure was produced using the script `amuse.examples.single_star_to_sph`, which runs in about an hour. For convenience, the hydro data is also available in `amuse.examples.data.fig_star_in_sph_N10000M1.0MSun.npy` for the $1.0\,M_\odot$ star, and equivalently for the $3.0\,M_\odot$ and the $10.0\,M_\odot$ stars, in which case the plot can be generated using `amuse.examples.plot_single_star_to_sph`.

The SPH realizations nicely follow the density profile calculated using the stellar-evolution code for most of the stellar structure. Near the stellar surface, the density of individual particles exceeds the results from the stellar-evolution code. This is a result of the hydrodynamics solver placing the star in dynamic equilibrium, whereas the stellar-evolution code also requires thermal equilibrium in conjunction with the internal energy source, which is absent in the hydro code. Eventually, near the stellar surface, the hydro code is unable to resolve the density structure in the outermost layers. This latter discrepancy is mainly a resolution issue, with only 10^4 SPH particles, we cannot resolve densities below $\sim\rho_c/\sqrt{10^4}$.

8.4 Elementary Code-coupling Strategies

A straightforward but very important way to couple codes is by directly propagating information from one code to another at the user script level. In this approach, two or more codes work on portions of the data (typically contained in a particle set), which may also be used by other codes. A classic example can be found in star cluster evolution. Stars move due to gravity and at the same time they evolve due to internal nuclear processes. Such a system can be simulated by a straightforward combination of the methods discussed in Chapters 4 and 5, as we now describe.

8.4.1 Combining Gravity with Stellar Evolution

N-body codes have incorporated treatments of stellar evolution for almost four decades. As a consequence, we have by now a fairly good understanding of the numerical subtleties in coupling these pieces of physics. The most common way of solving the combination of stellar evolution and stellar dynamics is the strategy adopted in the seminal work of Terlevich (1987), which updates and synchronizes the two codes at regular time intervals (see also Aarseth 2003)—update the dynamics to some time, then update the stellar evolution to that time, then update the dynamics to the next time using the new stellar masses, etc. We call this *synchronized temporal discretization.*

In general, both codes will have their own internal time steps, which may be different for each object and may change dynamically with time, but in this prescription the time interval for the exchange of information is generally set by the dynamics—typically some fraction of the crossing time of the self-gravitating system. As discussed in Sections 4.1.2 and 5.1.1, this is appropriate in many cases, but it may not be so for systems in resonance, chaotic dynamics, or young massive star clusters, where stellar-evolution time scales may actually be shorter than the relevant dynamical time.

The flexibility afforded by AMUSE allows us to experiment with other time-stepping strategies. One way to improve on the synchronized temporal discretization strategy is to advance the stellar evolution for half a time step, then advance the dynamics for a full step using stellar masses defined at the midpoint, and finally update the stellar evolution to the end of the step. We saw in Section 4.2.3 how evaluating key quantities in the middle of the step leads to improved accuracy. In this case, the stellar evolution cares nothing for the dynamics, but using centered stellar masses improves the accuracy of the dynamical integration. We call this approach *interlaced temporal discretization*, and we explore its use in the next example. A more complete and elaborate approach to this general problem is presented in Chapter 9.

There are (at least) two complications in coupling stellar evolution with gravitational dynamics: (1) unexpected events in the stellar evolution, and (2) unexpected events in the gravitational dynamics. An example of an unexpected stellar event is a supernova explosion;[3] an unexpected dynamical event is a collision between two stars. Neither event can easily be predicted, but they can be handled straightforwardly using interrupts, as discussed in Section A.5.2.

8.4.2 Coupling Radiative Transfer with Hydrodynamics

Combining radiative transport with hydrodynamics is not trivial. In principle, we could adopt a linear coupling strategy between two modules, but a bridge-like approach (see Chapter 9) gives superior results in terms of both better performance and smaller errors. However, since we have not yet discussed a bridge, we adopt a simpler but still effective approach similar to the interlaced temporal discretization discussed in Section 8.4.1.

We have defined a small `RadHydro` class to simplify this function, to be incorporated later into a bridge with (for example) a gravity solver. In order for this class to comply with bridge specifications, it must have a few extra functions: the most important is `evolve_model()`. In addition, to make life easier, we include a method to copy data from the two worker codes to the framework, and one to clean up after we are done. For safety (and convenience), we start by synchronizing all relevant data between the two worker codes and the framework, giving us a self-consistent dataset to work with. For additional safety, albeit probably unnecessary in this example, we perform a similar synchronization after the two worker codes have reached their objectives.

Between the two synchronization operations, we run the hydrodynamical simulation for half a time step, followed by the radiative-transfer simulation for a full time step, and end with a second half time step for the hydrodynamics (see also Section 8.4.1):

```
while t < (tend-timestep/2):
    self.hydro.evolve_model(t+timestep/2.)
    update_rad_from_hydro_channel.copy()
    self.rad.evolve_model(t+timestep)
    update_hydro_from_rad_channel.copy()
    self.hydro.evolve_model(t+timestep)
    t = self.model_time
```

The rest of the radiation–hydrodynamics class is fairly straightforward. In Section 8.6.1, we present an example calculation using this `RadHydro` class using the code example in `amuse.examples.supernova_irradiation`.

[3] In practice, since the stellar-evolution codes actually operate one step ahead of the current time, to permit interpolation of their results on demand, a supernova is not completely unexpected. We ignore this operational reality here.

8.4.3 Switching from One Stellar-evolution Code to Another

Stellar-evolution code, in particularly the Henyey codes, can be rather fragile. If such a code becomes impossibly slow, or simply crashes, one would not necessarily stop the entire calculation, but escape to a safer (secondary) stellar-evolution code or prescription. Ideally, the new stellar-evolution code should pick up the failed star from the crashed code and continue its evolution from that same point in its evolutionary stage.

This turns out to be nontrivial because each stellar-evolution code will give a different radius, luminosity, and temperature for the same mass, age, and metallicity. These differences are often small, but can be quite substantial, see for example Section 5.6.1.

To make the transition process simpler, we designed a special module for stellar-evolution codes, allowing us to start with one code and then switch to another at some arbitrary moment. The module is briefly described in Appendix A. The moment to switch codes can be arbitrary but the evolutionary state of where the new stellar-evolution code starts is chosen to be as close as possible to the end-point of the previous code.

The handover algorithm performs an rms search on the relative differences of mass, radius, and luminosity between the last point known state in the primary code and the entire stellar history of the secondary code (here `SSE` or `SeBa`) until the star turns into a remnant.

In the following example, we show how a star starts to evolve using `Evtwin`, for 1 Myr, and then we switch (for no particular reason) over to `SSE`. The interface function `EVtwin2SSE()`, is really a small class with two stellar-evolution codes and a collection of other parameters. One of these parameters is `EVtwinAgeAtSwitch`, indicating at what time to switch from one code to the other. Another parameter is `EVtwinException`, which is used if the code-switch in nonvoluntary (i.e., the primary stellar-evolution code crashed).

```
from amuse.lab import Particle
from amuse.lab import units
from amuse.community.evtwin2sse.interface import EVtwin2SSE

stellar = EVtwin2SSE()
stellar.EVtwinAgeAtSwitch = 1 |units.Myr
star.mass = 10|units.MSun
stellar.particles.add_particle(star)

print(stellar.particles)
stellar.evolve_model(10|units.Myr)
print(stellar.particles)
```

8.4.4 Event-driven Simulations

So far, our simulations of star clusters have alternately called the gravitational dynamical evolution and stellar-evolution solvers at regular time intervals. This gives a somewhat limited perspective on reality—a massive star does not wait until a dynamical encounter is over before blowing up, and a stellar collision does not wait until the end of a stellar-evolution step. Instead, each may interrupt the otherwise orderly joint evolution process.

In conventional simulation languages, taking these interrupts into account at the proper times is difficult—the language semantics simply are not flexible enough to deal with these events. In AMUSE, an event-driven simulation can be easily realized with the proper use of timing, channels, declarations, and calls. In Code example 8.7, we create data structures and declare two codes, a stellar dynamics solver and a stellar-evolution code, together with communication channels. In principle, there is nothing really new about these, except that both codes now have stopping conditions enabled.

Code example 8.8 shows the event loop, which starts by querying the stellar-evolution code about the shortest safe time step, the minimum interval during which no star will experience a supernova or significant changes to its `stellar_type` (see Appendix A). The collision handling for the event-driven simulation is handled by a separate function (see the example script, Code example 8.4). The supernova handling is illustrated in Code example 8.9. Strictly speaking, this simulation is only event-driven in terms of the collision handling. The stellar evolution is

```
    while time < t_end:

        dt_min = min(dt, stellar.particles.time_step.min())
        print("Time steps:", dt.in_(units.Myr), dt_min.in_(units.Myr))
        time += dt_min

        bodies.radius *= 1.e+5
        channel_to_gd.copy_attributes(["mass", "radius", "vx", "vy", "vz"])
        E_dyn = gravity.kinetic_energy  + gravity.potential_energy
        gravity.evolve_model(time)
        dE_dyn = E_dyn - (gravity.kinetic_energy  + gravity.potential_energy)

        resolve_collision(collision_detection, gravity, stellar, bodies)
        channel_from_gd.copy()

        time = gravity.model_time

        E_stellar = gravity.kinetic_energy + gravity.potential_energy
        stellar.evolve_model(time)
        dE_stellar = E_stellar - (gravity.kinetic_energy
                                  + gravity.potential_energy)

        resolve_supernova(supernova_detection, bodies, time)
        channel_from_se.copy_attributes(["mass", "radius", "age",
                                         "temperature", "luminosity"])

        write_set_to_file(bodies.savepoint(time), filename, 'hdf5')
        print_diagnostics(time, bodies.mass.sum(), E_dyn,
                          dE_dyn, dE_coll, dE_stellar)
```

Code example 8.8: Event-driven simulation with collisions and supernovae.

```
def resolve_supernova(supernova_detection, bodies, time):
    if supernova_detection.is_set():
        print("At time=", time.in_(units.Myr), \
              len(supernova_detection.particles(0)), 'supernova(e) detected')

        Nsn = 0
        for ci in range(len(supernova_detection.particles(0))):
            print(supernova_detection.particles(0))
            particles_in_supernova \
                = Particles(particles=supernova_detection.particles(0))
            natal_kick_x = particles_in_supernova.natal_kick_x
            natal_kick_y = particles_in_supernova.natal_kick_y
            natal_kick_z = particles_in_supernova.natal_kick_z

            particles_in_supernova \
                = particles_in_supernova.get_intersecting_subset_in(bodies)
            particles_in_supernova.vx += natal_kick_x
            particles_in_supernova.vy += natal_kick_y
            particles_in_supernova.vz += natal_kick_z
            Nsn += 1

        print('Resolved', Nsn, 'supernova(e)')
```

Code example 8.9: Function for resolving a supernova event.

predictable so long as there are no additional complications; for example, due to mass transfer in a close binary system.

We run the gravitational dynamics for that safe time frame. If a collision occurs during this time, the *N*-body code is interrupted and the encounter is resolved, ending the dynamics step. In Code example 8.8, we ensure that the stars are evolved to the same time as the *N*-body integrator. For technical reasons, individual time step codes do not normally guarantee to evolve the system to exactly the time given in the call to the community code. This can be overruled by forcing the *N*-body code to synchronize to the target time with

```
gravity.parameters.force_sync = True
```

We normally would not recommend synchronizing an *N*-body code in this way, but it is often necessary in multiscale and multiphysics problems. After the dynamics step, the stellar evolution is updated to the current age of the system, which is the moment of a collision, the moment of a supernova, or some predetermined output time. In Section 10.1.3, we give an example of `force_sync` in the `venice` module in AMUSE.

The results of such an event-driven simulation can be distinctly different from a regular calculation in which both modules are called at fixed time intervals. This is demonstrated in Figure 8.3, where we compare three simulations, one without stellar evolution, and one with constant time steps between stellar dynamics and stellar evolution. The third simulation (green curve in Figure 8.3) is event-driven. In this simulation there was no real chance of stellar collisions because the initial cluster was quite large, so we followed a slightly different event-driven strategy: we evolved

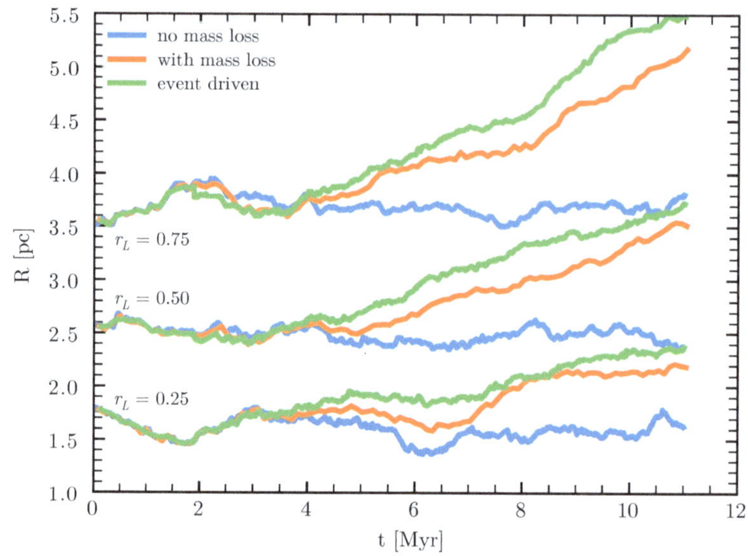

Figure 8.3. Lagrangian radii containing 25%, 50%, and 75% of the total cluster mass for three simulations of a cluster of 1000 stars. Each simulation is performed using exactly the same initial realization. In one of these runs (blue curve), stellar mass loss was ignored. In the second (orange curve), mass loss was applied at constant time intervals of 0.1 Myr after the dynamics step (note that in Figure 8.4 we adopted a very small time step of only 10% of the inner orbital period). The green curve is the result of an event-driven simulation (see Section 8.4.4), and should, therefore, be considered most reliable. The figure was produced using the script `amuse.examples.gravity_stellar_comparison`, which should run in 5–10 minutes on a laptop.

all stars for a minimum stellar-evolution step, followed by the dynamics for the same amount of time.

Figure 8.3 shows the evolution of the 25%, 50%, and 75% Lagrangian radii (see Section 4.1.3.1) of three initially identical clusters of 1000 stars over a time interval of 10 Myr. The stars were initially distributed as a $W_0 = 3$ King model (King 1966, see Section 3.2), with stellar masses drawn from a Salpeter mass function between $1\,M_\odot$ and $100\,M_\odot$ (see Section 3.3). All runs were performed using the same dynamical integrator. The differences in the Lagrangian radii are therefore a consequence of how stellar mass loss propagates into the dynamical evolution of the cluster.

In the cluster without stellar mass loss, the Lagrangian radii shrink slightly over the 10 Myr period studied, since pure dynamics dictates the evolution. During the first ~3 Myr, the orange and blue (and green) clusters show basically the same dynamical evolution. The differences only become evident later, when the most massive stars start to lose significant mass. After ~4 Myr, the Lagrangian radii of the runs with stellar mass loss start to expand.

In the cases with mass loss, about 18% of the initial cluster mass of $3200\,M_\odot$ is lost, causing the 25%, 50%, and 75% Lagrangian radii to expand by 30–50%. The outermost Lagrangian radii expand slightly faster than the more inner Lagrangian radii. These expansion factors are significantly larger than would be naively expected based on an adiabatic response of the cluster, which would predict an 18% increase in size (Carathéodory 1909). The reason is that much of the mass lost is ejected impulsively by the strong winds of the massive Wolf–Rayet stars and (after about 3 Myr) supernova. In addition, the orbits of the mass-losing stars are not circular, as is assumed in the naive adiabatic argument, but are generally quite eccentric.

At the start of these particular simulations, none of the 10 most massive stars ($\gtrsim 20\,M_\odot$) lay within the inner 25th percentile. By the end of the simulation, at 10 Myr, this number increased to four in the run without stellar evolution, and two in the other cases. Three massive stars were initialized outside the 75th percentile and had insufficient time to complete an orbit around the cluster center. The more rapid expansion of the outermost Lagrangian radii compared with the innermost radii is therefore in part due to the mass lost by these more massive stars, and not by the segregation of the most massive stars to the cluster center.

Event-driven simulations are generally somewhat slower than fixed-time interaction simulations, but they are probably worth the cost because the physics is better represented by the numerical model.

8.4.5 Evolution of a Hierarchical Triple System

In this section, we demonstrate an interlaced coupling between stellar evolution and gravitational dynamics by integrating a triple composed of three massive stars, each of them losing mass copiously. The orbital evolution of such a system cannot be described analytically, so the problem must be addressed numerically. This problem turns out to be rather hard because the various time scales of the problem are comparable. This example is not much different than the one presented in Section 8.4.4, but here we apply it to a tightly coupled triple system. Of course, one could consider performing this simulation with the more dedicated triple evolution code `Tres`, discussed in Section B.2.1.

The triple system θ Muscae comprises the WC5/6 Wolf–Rayet star WR48 on an inner orbit with period $P_{\rm in} \simeq 19$ days around an O6/7V star, with an O9.5/B0Iab companion on a much wider outer orbit (Hill et al. 2002; Sugawara et al. 2008). The other orbital parameters are not known, but each of the three stars has a strong stellar wind. We can easily integrate the equations of motion of such a triple system within AMUSE, taking the mass loss of each star into account, by coupling a stellar-evolution code with an N-body integrator.

We adopt masses at birth of $60\,M_\odot$ (for WR48) and $30\,M_\odot$ for the inner two binary stars, and $20\,M_\odot$ for the outer companion. The nature of the Wolf–Rayet star suggests that the system is at least a few megayears old. We therefore began the simulation by evolving the three stars to an age of 4.5 Myr using `SeBa`. By that time, the masses of the three stars had fallen to $12.48\,M_\odot$, $28.52\,M_\odot$, and $19.81\,M_\odot$. It is difficult to judge whether or not these masses are consistent with the observations, but at least their spectral types are. The mass-loss rates at that age are $-1.6 \times 10^{-5}\,M_\odot\,{\rm yr}^{-1}$, $-4.2 \times 10^{-7}\,M_\odot\,{\rm yr}^{-1}$, and $-5.2 \times 10^{-8}\,M_\odot\,{\rm yr}^{-1}$. Further details on the stellar parameters are presented in Table 8.1.

After this initial period of stellar evolution, we initialized the N-body code (in this case `Huayno`) in a configuration consistent with the observed orbit. We assumed

Table 8.1. Adopted Parameters for the Stellar Wind Binary θ Muscae

Parameter	Unit	Star 1	Star 2	Star 3
Spectral type		WR5	O6V	O9.5III
Mass	$M_\odot$	12	20	30
Radius	$R_\odot$	2.2	10	50
Luminosity	$L_\odot$	2.2×10^5	1.4×10^5	0.55×10^5
Temperature	K	7.1×10^4	4.5×10^4	3.9×10^4
Mass-loss rate	$M_\odot\ yr^{-1}$	1.8×10^{-5}	1.0×10^{-6}	1.0×10^{-6}
Wind velocity	$km\ s^{-1}$	2000	1000	1000

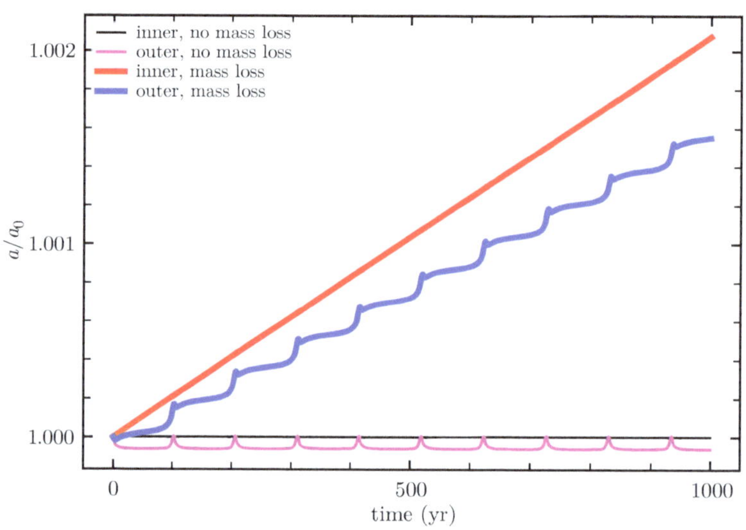

Figure 8.4. Evolution of the orbital semimajor axes of the inner (blue) and outer (red) orbits of the hierarchical triple system θ Muscae. All quantities are expressed relative to their initial values (at $t = 4.5$ Gyr). At the start of the integration, at 4.5 Gyr, the masses were 12.48, 28.52, and 19.81 $M_\odot$, respectively. The integration spanned a time interval of 10^3 yr. The script to generate the figure, `amuse.examples.plot_triple_with_wind`, takes a few minutes to run. The code that generated the data is available in `amuse.examples.evolve_triple_with_wind`.

that the outer orbit has an inclination of 30° relative to the inner one, with an inner eccentricity of 0.2, and an outer eccentricity of 0.6, although these are not known from observations. For definiteness, we chose an initial mean anomaly and argument of periastron of 180°, and an outer orbital separation of 100 au.

Figure 8.4 presents the orbital semimajor axes and eccentricities of the inner and outer binaries during an integration from 4.0 Myr to 4.55 Myr using the interlaced scheme just described. The Wolf–Rayet star exploded as a supernova shortly after this, at an age of 4.6 Myr. During the integration, we continued to evolve the stars using `SeBa`, and the mass loss was propagated to the N-body integrator after every

0.1 $P_{in} = 1.9$ days. The evolution of θ Muscae over the relatively small time frame presented in Figure 8.4 barely requires the use of a stellar-evolution code, but it allows us to illustrate how stellar evolution can be combined with the dynamics of this complicated triple system.

The orbits of both binaries expanded due to mass loss. On the timescale of the integration, the inner two stars lost 0.18 $M_\odot$, whereas the outer star lost 0.026 $M_\odot$. From Figure 8.4 we can see that the outer binary expanded by about 0.8%, which is about a factor of two more than the fractional mass loss from all stars combined. The inner binary was much more strongly affected by mass loss, and expanded by about 2.7% The variation in the eccentricity of the inner binary is caused by its interaction with the outer star. The classical Lidov–Kozai (von Zeipel 1910; Lidov 1962; Kozai 1962) period is about 0.44 Myr, and corresponds to the large excursions in the eccentricity of the inner binary. The eccentricity of the outer binary remained almost constant.

In this example, and in fact in all other current approaches, mass is lost in small chunks at fixed time intervals. This causes the system to receive a small shock after every mass-loss step, and this may have undesirable side effects. In fact, in constructing Figure 8.4, we found that the outcome of the simulation is remarkably sensitive to the choice of integration parameters, especially the time interval on which stellar mass is applied. The script used to perform this calculation can be found in `amuse.examples.evolve_triple_with_wind`, and we invite readers to explore its outcome with different choices for various parameters. It will take between a few minutes and many days to run, depending on the parameter values chosen; the parameters leading to Figure 8.4 took a few hours on a laptop. This sensitivity is a topic of ongoing research, and AMUSE modules capable of incorporating mass loss directly into the dynamical integration are currently under development.[4] While these are clearly counter to the AMUSE philosophy of not making major changes to the internals of community modules, sometimes code validation requires more intrusive measures.

The relatively poor performance of the code in this particular strongly coupled problem motivated us to develop a new symplectic N-body code with the possibility of including a mass-loss term in the integrator. We call this code `symple` for its simple structure, its simple implementation, and the symplecticity of the adopted integrator. In fact, `symple` can be run with a wide range of integrators of different order ranging from first order to 10th order, each of which is symplectic. For more details about `symple`, we refer to Section A.3.4.

Although it goes well beyond the scope of this chapter, as a preview of AMUSE capabilities introduced in Chapter 6 and soon in Chapter 9, Figure 8.5 shows an image of the wind produced in θ Muscae, where now we integrate the equations of motion of the individual stars and simultaneously resolve the winds using a hydrodynamical simulation (Jänes et al. 2014). Specifically, the simulation shows

[4] See `AMUSE_updates.pdf` for ongoing developments.

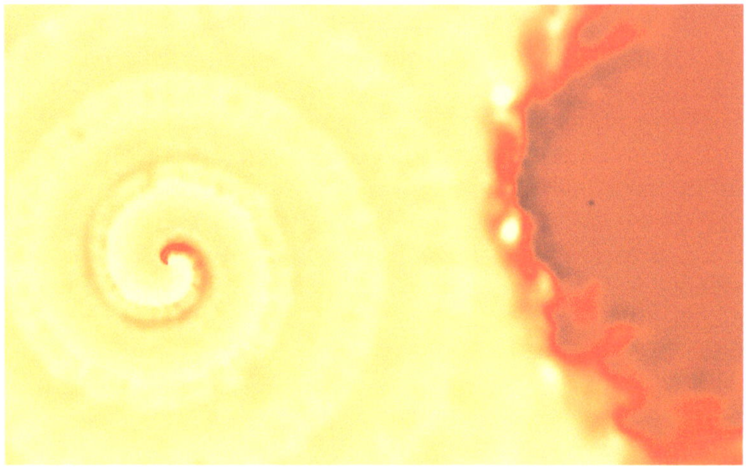

Figure 8.5. Evolution of the Wolf–Rayet triple-star system θ Muscae using `Huayno` (Jänes et al. 2014) as the gravity solver and `fi` for the hydrodynamics (see Chapter 6). The hydrodynamics was resolved using the stellar wind routines in AMUSE (see van der Helm et al. 2018). The script used to produce this figure is too long (and takes too long to run) to reproduce in the text, but its basic principles are described in Chapters 6 and 9.

the gas density in the orbital plane of the hierarchical triple system. While the winds collide, the three stars orbit each other, resulting in the tight spiral (left) from the inner binary that interacts with the massive wind of the tertiary outermost star. The simulation was performed using the SPH code `Fi` using a mass of $10^{-11}\,M_\odot$ for each SPH particle. Both the inner and the outer binaries are initially circular and in the same plane with orbital period of $P_{\rm in} = 19.1$ days and $P_{\rm out} = 130$ years.

The wind was resolved in the hydrodynamics solver using the `AcceleratingWind` routines from `amuse.ext.stellar_wind`, which produces a wind acceleration inversely proportional to the distance squared (or $\propto 1/r^2$). The results of the calculations are presented in Figure 8.5.

Evidently, our "simple" example of integrating a triple-star system has already proved quite challenging. What about an entire star cluster? Generating initial conditions for a star cluster in which we integrate the equations of motion and simultaneously evolve the stars is basically the same issue as discussed in Chapter 4 and 5. Once the initial realization has been created, we can start the stellar-evolution code and the gravity solver in much the same way as we did here for θ Muscae.

8.5 Stellar Cluster with Collisions

8.5.1 Sticky Stars

To demonstrate the working of stopping conditions and collision handling in direct N-body codes, we present an example of a simple cluster in which stars are allowed to collide. We opted for an $N = 1000$ body system initially distributed as a $W_0 = 7$ King model with a Salpeter mass distribution between 0.1 $M_\odot$ and $10\,M_\odot$ (see Section 3.3). Each star was given a radius of 0.001 N-body length units, which

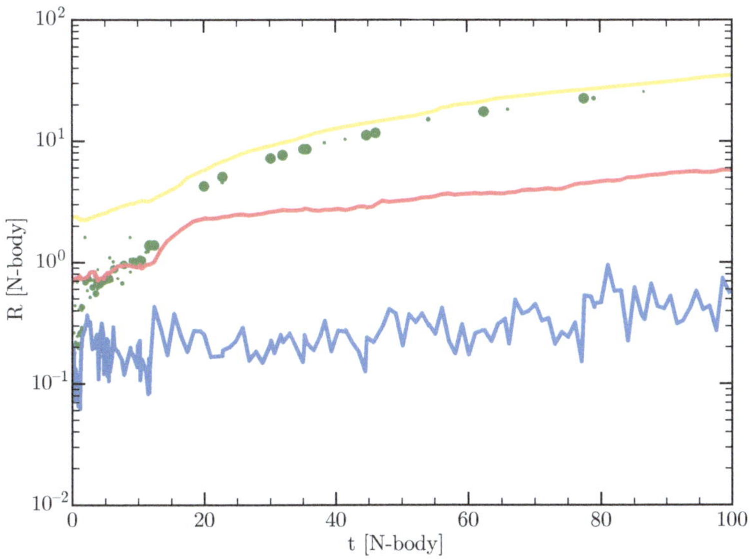

Figure 8.6. Direct N-body simulation (using ph4) of a small star cluster over 100 N-body time units. Initially, the cluster contained $N = 1000$ stars with a Salpeter mass distribution spanning two orders of magnitude in mass, distributed in a virialized King model with $W_0 = 7$. The initial radii of all stars were taken to be 0.001 N-body length unit. The bullets indicate the times and locations of stellar collisions; the size of a bullet indicates the mass of the collision product. We assume that no mass is lost and the mean stellar density remains constant. The three curves (bottom to top) illustrate the core radius (blue), the half-mass radius (red), and the 90% Lagrangian radius (yellow) of the cluster. The script that produced this figure is `amuse.examples.gravity_collision`. It should take a few minutes to run on a laptop.

resulted in many interesting collisions. The code example is too long to reproduce here, but it is available in `amuse.examples.gravity_collision`.

We ran one simulation with virialized initial conditions. Collisions were resolved using the stopping conditions described in Section A.5.2. For collision products, we assumed no mass loss and calculated a new radius such that the mean density remained constant. The integration was performed using the direct integrator `ph4`. Stellar evolution was not included (see Section 5.1). Figure 8.6 presents the evolution of the core and the half-mass radii of the N-body system; the moments at which collisions occur are indicated by bullets. The size of a bullet is proportional to the mass of the collision product.

8.5.2 Using a Separate Code to Manage a Collision

If instead of running a parameterized stellar-evolution code we are using a full Henyey code, such as MESA or EVtwin, we could in principle use the detailed structural information from the stellar-evolution code to construct three-dimensional models of the colliding stars and model the collision as a hydrodynamical simulation, as we will discuss later in Section 8.5.4. This should, in principle, give the most accurate estimate of mass loss, energy released, etc., but it is typically a computationally expensive calculation—perhaps too expensive to be done routinely.

A useful intermediate solution to the merger problem is provided by the `MMAMS` module (Lombardi et al. 2003; Gaburov et al. 2008), as described in Section 5.2.9. The script presented earlier in Code example 5.5 can be used to determine the structure of the merger product. The only additional actions needed are to reposition the merger product in the center-of-mass frame of the two colliding stars, and to conserve momentum, as in Code example 8.5.

8.5.3 Restarting Stellar Evolution after a Merger

After the collision has been detected and managed, the evolution of the merger product can be continued. For a parameterized code, we simply use a rejuvenation formula to place the merged star on a new track at some evolutionary time (see also Section 5.3.12). We do have a script to transfer the stellar evolution from a parametrized code to a full Henyey code, which is an example in the AMUSE repository called `hr_diagram_cluster`.

However, if we are using a Henyey code, we can continue the detailed evolution of the merger product. Code example 8.10 shows how this can be done. In this example, we evolve two stars to the same time using `MESA`, merge them using MMAMS, and then continue the evolution of the merger product using `MESA`.

Figure 8.7 presents (yellow dashed line) the evolutionary track of the collision product resulting from the merger of stars of masses $3\,M_{\odot}$ and $1\,M_{\odot}$. The stars collided at 150 Myr, which is half the main-sequence lifetime of the more massive star (determined by first evolving the primary to the end of its nuclear burning lifetime using `EVTwin`). For comparison, we overplot the evolutionary tracks of single stars of masses $3\,M_{\odot}$ and $4\,M_{\odot}$, as well as the track for a main-sequence star of mass $3.70\,M_{\odot}$ (the value returned by MMAMS).

The evolutionary track of the merger product makes several rapid excursions in temperature and luminosity before settling after $\sim 10^5$ yr (comparable to its Kelvin–Helmholtz timescale, see Equation (5.6)) onto a new "rejuvenated" track, after which its evolution continues along the main sequence at a slightly lower luminosity than the $4\,M_{\odot}$ star. The merger evolutionary track differs significantly from that of a canonical $3.70\,M_{\odot}$ star, mainly due to its nonstandard abundance profile. Subsequently, on the subgiant and giant branches, the track is quite similar to that of the $4\,M_{\odot}$ star.

In many cases, a simplified parameterized prescription for the consequences of a stellar merger may be quite satisfactory, although not very accurate. One can ask whether spending significantly more computer time on this complex problem is worth the effort, but many studies have demonstrated that detailed modeling of stellar collisions can give results quite different from the simplified semianalytic prescriptions described in Section 5.3.12 (Lombardi et al. 1995, 2002, 2003).

8.5.4 Using a Hydro Code to Simulate a Stellar Merger

In Section 8.5.2, we demonstrated how to use a separate code to resolve a stellar collision using a parameterization of the collision process. Instead of a

```python
def merge_two_stars_and_evolve(
    mass_primary, mass_secondary, time_collision, time_end
):
    stars = Particles(2)
    stars.mass = [
        mass_primary.value_in(units.MSun),
        mass_secondary.value_in(units.MSun)
    ] | units.MSun
    stellar_evolution = Mesa(**MESA_OPTIONS)
    stellar_evolution.parameters.metallicity = 0.0142
    stellar_evolution.particles.add_particles(stars)
    while stellar_evolution.model_time < time_collision:
        stellar_evolution.evolve_model()
        print_stars(stellar_evolution)
    n_shell = min(
        stellar_evolution.particles[0].get_number_of_zones(),
        stellar_evolution.particles[1].get_number_of_zones(),
    )
    merger_code = Mmams(**MMAMS_OPTIONS)
    merger_code.parameters.target_n_shells = n_shell
    merger_code.parameters.dump_mixed_flag = True
    merger_code.parameters.do_shock_heating_flag = True
    merger_code.commit_parameters()
    handler = CollisionHandler(
        merger_code,
        stellar_evolution_code=stellar_evolution,
    )
    merger_product = handler.handle_collision(
        stellar_evolution.particles[0],
        stellar_evolution.particles[1],
    )
    merged = stellar_evolution.particles[0]
    stellar_evolution.evolve_model(keep_synchronous=True)
    p = stellar_evolution.particles[0]
    while stellar_evolution.model_time < time_end:
        stellar_evolution.evolve_model()
        print(
            f"Time={stellar_evolution.model_time} {p.stellar_type} "
            f"{p.mass} {p.radius} {p.temperature} {p.luminosity}"
        )
        if p.stellar_type >= 4 | units.stellar_type:
            break
    merger_code.stop()
    stellar_evolution.stop()
```

Code example 8.10: Merge two stars and evolve the merger product. The code example is available in `amuse.examples.merge_two_stars_and_evolve`. Several lines related to plotting Figure 8.7 have been omitted.

parameterized code, we could also use a hydrodynamics solver for the collision. Here, we do just that, using an SPH code to resolve the merger.

8.5.4.1 Generating Stellar Structure in a Particle Set

To start the merger simulation, we first need the structure of two stars. We saw in Chapter 5 how this can be realized. Code example 8.11 demonstrates the process again.

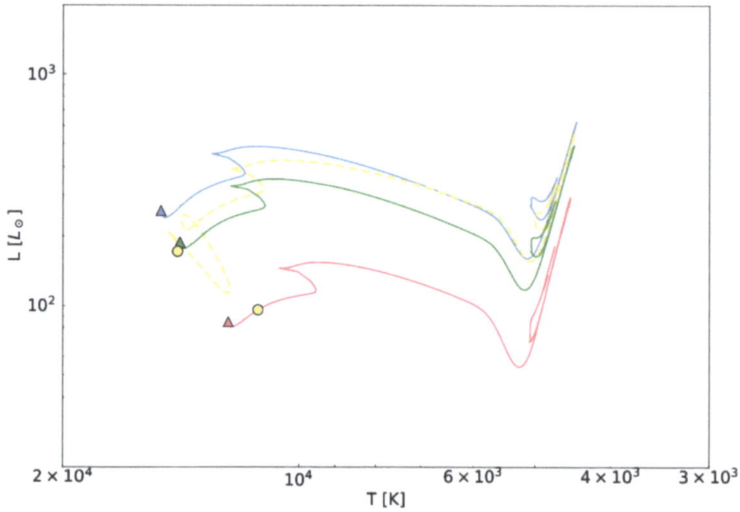

Figure 8.7. Evolutionary tracks for a 3 $M_\odot$ star (red curve) a 4 $M_\odot$ star (blue), and a 3.70 $M_\odot$ star (green) from the main sequence (filled triangles) to the bottom of the AGB. The yellow dashed curve shows the evolutionary track of the merger product of a 3 $M_\odot$ star and a 1 $M_\odot$ star that collide halfway along the primary's main-sequence track, with the merger structure determined by MMAMS. The yellow circles represent the evolutionary states of the primary immediately before the collision and of the merger product immediately afterward. This figure was produced using the script `amuse.examples.merge_two_stars_and_evolve`, which should run in 10–15 minutes on a laptop.

```
def return_evolved_star_hydro(mass, time, number_of_sph_particles):
    star = Particle(mass=mass)
    stellar = Evtwin()
    star = stellar.particles.add_particle(star)
    stellar.evolve_model(time)
    number_of_sph_particles = number_of_sph_particles * int(mass.value_in(units.MSun
    ))
    star_in_sph = convert_stellar_model_to_sph(
        star, number_of_sph_particles
    ).gas_particles
    stellar.stop()
    return star_in_sph
```

Code example 8.11: Using a Henyey code to construct an SPH realization of a single star of mass `mass` evolved to time `time`. The complete code example is in `amuse.examples.merge_two_stars_sph`.

In addition to evolving the star to the desired time, this code also converts the stellar structure provided by the Henyey code to an SPH particle representation:

```
star_in_sph = convert_stellar_model_to_SPH(star, N_sph).gas_particles
```

The method `convert_stellar_model_to_SPH` returns a `Python` class having `gas_particles` as an attribute. We run the same procedure for the primary and the secondary stars, then add the two SPH realizations together after offsetting the secondary with respect to the primary and giving them some relative velocity. This is

```python
def merge_two_stars_sph(
    mass_primary, mass_secondary, time_collision, number_of_sph_particles
):
    primary_in_sph = return_evolved_star_hydro(
        mass_primary, time_collision, number_of_sph_particles
    )
    primary_in_sph = relax_sph_realization(primary_in_sph)
    secondary_in_sph = return_evolved_star_hydro(
        mass_secondary, time_collision, number_of_sph_particles
    )
    secondary_in_sph = relax_sph_realization(secondary_in_sph)
    radius = primary_in_sph.x.max() + secondary_in_sph.x.max()
    total_mass = primary_in_sph.mass.sum() + secondary_in_sph.mass.sum()
    secondary_in_sph.x += 0.8 * radius
    secondary_in_sph.y += 0.6 * radius
    secondary_in_sph.vx -= (constants.G * total_mass / radius).sqrt()

    converter = nbody_system.nbody_to_si(mass_primary, 1.0 | units.au)
    hydro = Gadget2(converter)
    hydro.gas_particles.add_particles(primary_in_sph)
    hydro.gas_particles.add_particles(secondary_in_sph)
    hydro.evolve_model(2.0 | units.hour)
    hydro.gas_particles.new_channel_to(primary_in_sph).copy()
    hydro.gas_particles.new_channel_to(secondary_in_sph).copy()
    hydro.stop()
    return primary_in_sph, secondary_in_sph
```

Code example 8.12: Using an SPH code to merge two stars. For the full code example see `amuse.examples.merge_two_stars_sph`.

illustrated in Code example 8.12. The script looks similar to Code example 5.5, except that here we convert the evolved stars into N smoothed gas particles.

The conversion from Henyey to SPH is done on a grid, which introduces artifacts and generally leaves the SPH star slightly out of equilibrium. We address this problem by relaxing the SPH models (in this case, using a separate SPH code), as shown in Code example 8.13. Here, we slowly reduce the velocities of the SPH particles while running the hydrodynamics code with short steps. This is the same process as for creating a "glass" solution for initial conditions as is done in many simulations (see, e.g., Section 3.3 on page 40 in White 1994). In this example, we have opted to relax the star for 2.5 dynamical times in 250 steps per dynamical timescale, which is rather short. It is better to have the star relax for far longer: J. Lombardi (2010, private communication) recommends at least 25 dynamical time scales.

By default, the output of `convert_stellar_model_to_SPH` represents a star at rest at the origin. Hence, before combining the two SPH realizations of the stars in a hydro solver, their positions and velocities must be changed to reflect the initial condition of the collision. This can be done by shifting each of the stars to its desired center-of-mass position, with all particles within each star being given the same velocity. In Code example 8.12, we position the secondary star such that it is almost touching the primary star, with a transverse velocity of roughly the escape speed in a direction that results in a grazing collision.

```
def relax_sph_realization(sph_star):
    dynamical_timescale = sph_star.dynamical_timescale()
    converter = nbody_system.nbody_to_si(dynamical_timescale, 1 | units.RSun)
    hydro = Gadget2(converter, number_of_workers=2)
    hydro.gas_particles.add_particles(sph_star)

    to_framework = hydro.gas_particles.new_channel_to(sph_star)

    ts_factor = 2.5
    time_end = ts_factor * sph_star.dynamical_timescale(mass_fraction=0.9)
    n_steps = ts_factor * 100
    velocity_damp_factor = 1.0 - (ts_factor * 2 * np.pi) / n_steps
    time_step = time_end / float(n_steps)
    time = 0 | units.day
    while time < time_end:
        time += time_step
        hydro.evolve_model(time)
        hydro.gas_particles.velocity = (
            velocity_damp_factor * hydro.gas_particles.velocity
        )
    to_framework.copy()
    hydro.stop()
    return sph_star
```

Code example 8.13: Relaxing the SPH realization of a star in order to prepare it for a hydrodynamical collision simulation. This code fragment is from `amuse.examples.merge_two_stars_sph`. Here, we introduced a scaling factor `ts_factor` = 2.5. This factor indicates how slowly (or how quickly) we reduce the velocities of the particles.

In order to prevent the two stars from moving to the left due to the momentum added to the system, one could move all the SPH particles to their center-of-mass frame after merging,

```
all_sph_particles = ParticlesSuperset([primary_in_sph, secondary_in_sph])
all_sph_particles.move_to_center()
```

Given that SPH is Galilean-invariant, the shift to the center-of-mass frame is unnecessary, but it makes graphical presentations of the encounter much more convenient.

8.5.4.2 Resolving the Collision with the Hydrodynamics Solver

With the initial conditions established, we instantiate the hydrodynamics solver `Gadget2`

```
hydro = Gadget2()
```

and add the SPH particle sets of the two stars to the community code:

```
hydro.gas_particles.add_particles(primary_in_sph)
hydro.gas_particles.add_particles(secondary_in_sph)
```

Here, we add the particle sets for each of the two stars separately, in order to keep track of which star each SPH particle initially belonged to.

In Code example 8.12, we run the hydrodynamics code for a simulation time of only two hours without saving any diagnostic data. For production simulations, this is not desirable. In many cases, it is better to run the specific code in a loop over time, testing conservation of energy and angular momentum, printing some diagnostics, and writing the raw data to a file at the end of each step. Such a loop might look like

```
time_end = 10 | units.hour
time_step = 1 | units.hour
while hydro.model_time < time_end:
    hydro.evolve_model(hydro.model_time + time_step)
    channel_from_hydro_to_framework.copy()
    print_diagnostics(hydro)
    write_set_to_file(primary_in_sph, "hydro.amuse")
    write_set_to_file(secondary_in_sph, "hydro.amuse")
hydro.stop()
```

Here, `print_diagnostics(hydro)` might print time, energy, angular momentum, and other useful diagnostic information reflecting the progress of the simulation.

8.5.4.3 Recognizing a Star in the Collision Product

It is not completely trivial to recognize the stellar merger product in the messy hydrodynamical blob that results from a collision. Such collisions may be highly asymmetric, rotate rapidly, and have a giant envelope of barely bound low-density gas. Figure 8.8 demonstrates the asymmetries one can expect from a rather violent collision, in this case between a 80 $M_\odot$ and a 20 $M_\odot$ star, both at an age of about 0.49 Myr. In this calculation, we adopted Gadget2 to perform the hydrodynamical simulation using 10^5 equal-mass SPH particles. The amount of mixing seen in Figure 8.8 is somewhat surprising when compared to earlier results, particularly because such collisions tend to lead to relatively little mixing (see the discussion in Lombardi et al. 1996). The use of more realistic stellar structure models cannot be the reason for the discrepancy (see Sills & Lombardi 1997), but it is possible that the very high stellar masses adopted, compared to the initial conditions in Lombardi et al. (1996), could lead to more mixing.

Here, we adopt a rather simple approach to the problem of identifying the SPH particles making up the merger product. We analyze the results of the SPH simulation using the hop clump finder (see Section 8.3.1). The three control parameters for peak density threshold, saddle point density cutoff, and the outer density threshold are specified to help hop decide which particles belong to a coherent structure. Each of the groups in the list of particle sets (hop.groups) identified by hop belongs to an identified structure. For our stellar merger calculation, we are interested in the largest one.

Figure 8.8. Projected density distribution of the gas ejected shortly after the merger between two massive stars. The results of the hydrodynamical simulation are shown 8.8 days after the merger between a $M_{\rm in} = 80\,{\rm M}_\odot$ star and a $m_{\rm in} = 20\,{\rm M}_\odot$ star in an $a_{\rm in} = 77\,{\rm R}_\odot$ orbit with an eccentricity of $e_{in}^{\rm B} = 0.87$. Projected density is shown in blue, while the nitrogen/hydrogen ratio is presented in mustard. The origin is at the center of mass of the original two-body system. The two massive loops of ejected gas (at left) are characteristic of a merger. We do not present the plotting routine for this figure, but instead refer to Portegies Zwart & van den Heuvel (2016) for more details. Adapted with permission from (Portegies Zwart & van den Heuvel, 2016). © 2016 The Authors. Published by Oxford University Press on behalf of the Royal Astronomical Society.

```
def find_clumps_with_hop(sph_particles, unit_converter):
    hop = Hop(unit_converter)
    hop.particles.add_particles(sph_particles)
    hop.calculate_densities()

    mean_density = hop.particles.density.mean()
    hop.parameters.peak_density_threshold = mean_density
    hop.parameters.saddle_density_threshold = 0.99*mean_density
    hop.parameters.outer_density_threshold = 0.05*mean_density
    hop.do_hop()

    blob = [x.get_intersecting_subset_in(sph_particles) for x in hop.groups()]
    hop.stop()
    return blob
```

Finding clumps in an SPH simulation is commonly done in postprocessing, but it can also be used to monitor the progress of the merger simulation. In the simulation presented in Figure 8.8, we monitored the clumpiness of the particle distribution at run time, and continued until only a single blob of particles remained. After the detection of this final blob, we ran the hydrodynamics for a further 5.5 years to ensure that it had come into dynamical equilibrium. This final blob is considered to be the merger product, and the residual material is identified as ejecta from the merger process. In this simulation, the merger resulted in a single $83\,{\rm M}_\odot$ star.

Once coherent structures have been identified, we often are interested in the original properties of particles that are part of a particular group. For performance and memory optimization, the particles in hop are given only the attributes necessary to identify individual clumps, whereas the original particle set (sph_particles in the above example) may contain additional information, such as chemical composition. We therefore want to find the intersecting subset of particles with the same key among the original sph_particles and the particles in hop. This is done in the line

```
blob = [x.get_intersecting_subset_in(sph_particles) for x in hop.groups()]
```

Finally, we stop hop and return the appropriate subset of particles.

8.5.4.4 Converting from Hydro Back to Stars

In Section 8.5.2, we discussed how a separate code can be used to manage a head-on collision between a 3 $M_{\odot}$ and a 1 $M_{\odot}$ star, each evolved to an age of 150 Myr, continuing the evolution of the resulting merged star afterwards. There we adopted the output of the MMAMS (Gaburov et al. 2008; Lombardi & Sawyer Warren 2014) package to resolve the collision, but now we can adopt the methods discussed in the previous subsection, using SPH rather than MMAMS to determine the structure of the merger product.

As before, we first evolve the stars for 150 Myr, which is half of the primary's main sequence lifetime. Both stars are then converted to SPH realizations and relaxed (see Section 8.5.4.1). The resulting particle sets are combined, with the stellar surfaces just touching and zero relative velocity at infinity. We then use the SPH code Fi to resolve the hydrodynamics of the encounter. The free-fall timescale at this moment is just a few minutes, but to ensure that the gas distribution is in equilibrium, we continue the hydrodynamical calculation for 10^3 times longer. After the SPH calculation, we identify the merged star by using hop to search for the most massive blob (see Section 8.5.4.3), and then we strip away any low-density unbound splash material. The leftover merger product has a mass of 3.73 $M_{\odot}$, very close to the MMAMS result for the same encounter (see Section 8.5.2).

After stripping away the envelope and relaxing the SPH system again, we put the star back into the Henyey stellar-evolution code. This latter step is a rather delicate process because the stellar-evolution code expects a stellar structure that is in thermal equilibrium. The hydrodynamics blob will be in dynamical equilibrium, but thermal equilibrium cannot be guaranteed. The procedure for converting the SPH star into a Henyey model is

```
from amuse.ext.sph_to_star import convert_SPH_to_stellar_model
new_stellar_model = convert_SPH_to_stellar_model(star_in_SPH)
```

This function call can be quite expensive in terms of computer time, and may even fail after an hour or so of crunching. If it succeeds, however, you have a good chance of being able to evolve the star using MESA or EVtwin. After this conversion, we continue the evolution of the collision product to study its track in the Hertzsprung–Russell (H–R) diagram.

Figure 8.9 displays the same information for this SPH/hop merger as was presented in Figure 8.7 for the MMAMS merger (see Section 8.5.2). The red, blue, and green tracks show, respectively, the evolution of main-sequence stars of masses 3 $M_{\odot}$, 4 $M_{\odot}$, and 3.73$M_{\odot}$ (the value returned by SPH/hop). The yellow dashed line is the evolutionary track of the merged object. As with the merger discussed in Section 8.5.2, the evolutionary track of the merger product makes several rapid excursions in temperature and luminosity before settling (after $\sim 10^5$ yr, again comparable to the Kelvin–Helmholtz timescale; see Equation (5.6)) onto a new stable track. Its subsequent evolution takes it along a significantly more luminous track than either the MMAMS equivalent or the 4 $M_{\odot}$ star.

Some cautionary notes are in order. First, recognizing the bound blob after the hydrodynamical simulation depends on several parameters in the hop algorithm (see Section 8.5.4.3), and may require some experimentation. Second, we have used a relatively small number of SPH particles (100 equal-mass particles per solar mass) here to model the merging stars. In part, this is to reduce the cost of the SPH simulation, but it also reflects the fact, already noted, that converting from an SPH to a Henyey representation is a very fragile process that may easily fail to produce a workable Henyey model, and whose outcome depends sensitively on the details of how SPH particles are combined into Henyey shells. In particular, the initial

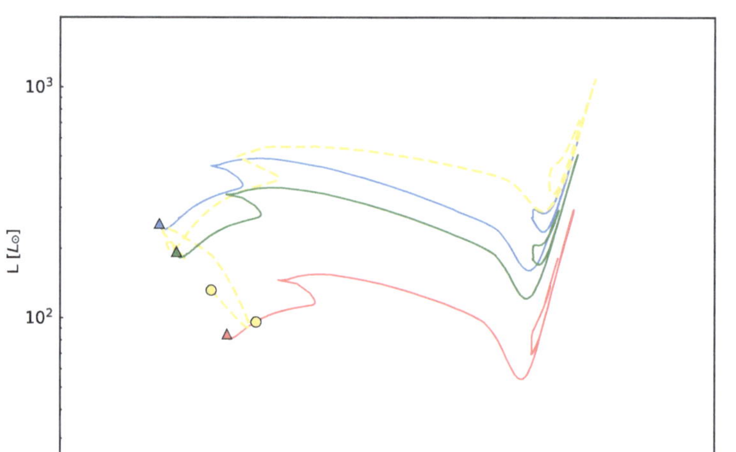

Figure 8.9. Evolutionary tracks for a 3 $M_{\odot}$ star (red curve), a 4 $M_{\odot}$ star (blue), and a 3.73 $M_{\odot}$ star (green) from the main sequence (filled triangles) to the bottom of the AGB. The yellow dashed curve shows the evolutionary track of the merger product of a 3 $M_{\odot}$ star and a 1 $M_{\odot}$ star that collide halfway along the primary's main-sequence track, with the merger structure determined using SPH and hop. The yellow circles represent the evolutionary states of the primary immediately before the collision and of the merger product immediately afterward. This figure was produced using the script amuse.examples.merge_two_stars_and_evolve, which should run in 10–15 minutes on a laptop.

location of the merger product on the H–R diagram can vary dramatically with these details, and probably should not be taken too seriously. As with many advanced topics in AMUSE, these details are still under active development: see the updates page `AMUSE_updates.pdf` for new material.

It is hard to validate the evolution of the merger product. Understanding the causes of these deviations from the standard evolutionary track would require a detailed analysis of the stellar structure, which could easily expand into a PhD project. We will not dwell further on the dark art of stellar astrophysics. It is clear, however, that it is worth exploring the possibility of resolving stellar collisions using hydrodynamical models. At this point, however, we consider `MMAMS` a viable option that at least leads to predictable behavior (see Section 8.5.2).

8.6 Complex Linear Code-coupling Strategies Applied

8.6.1 Heating of a Protoplanetary Disk

The abundance of several species of short-lived radionuclides (SLRs), such as ^{26}Al and ^{60}Fe, in the solar system has puzzled astronomers for some time (MacPherson et al. 1995). Key questions include: How did these volatile elements, particularly ^{26}Al, enter the solar system? How did they come to be part of meteorites? And, how did chondrules (the metallic grains containing these elements) form before chondrites (the meteorites that contain them)? There seems to be a lot we do not really understand.

One proposed explanation involves a nearby supernova when the Sun was still part of its parent star cluster. This is a two-stage process. First, the radiation from a bright supernova illuminates and heats the disk. Second, a few decades later, the hot ionized gas from the explosion plows through it. The idea is that it might be possible for the first burst of radiation to heat the disk enough to vaporize the dust, such that the subsequent wave of radioactive material could deposit the SLRs. Later cooling of the disk might subsequently have caused these SLRs to become part of the newly formed dust. This is all speculative, but worthy of further exploration.

Here, we study the first stage of the process: the heating and ionization of a protoplanetary circumstellar disk by the radiation from the supernova. For the gaseous disk, we adopt the protoplanetary disk model discussed in Section 6.4.1. To mimic the thick protoplanetary disk, we adopt a Safronov–Toomre Q parameter of $Q = 25$, rather than $Q = 2$ as in the earlier example of a thin disk (see Section 6.4.1). The disk was constructed as in the earlier example, with 10^4 equal-mass SPH particles at distances between 1 au and 100 au from the Sun, a total mass of 0.01 $M_\odot$, and a power-law density profile with an exponent of -2. We arbitrarily assume that the disk is oriented at an angle of 15° (almost face-on) to the line of sight of the exploding star. The hydrodynamical simulation was run using `Fi` with an isothermal equation of state and an artificial viscosity parameter of $\alpha = 0.1$ (see Section 6.2.4).

For the evolution of the supernova luminosity, we follow Sanders et al. (2015) and adopt a simple broken power-law luminosity function with characteristic luminosity and time scales. The script `amuse.examples.supernova_IIp_Lightcurve` presents code from Sanders et al. (2015) to reproduce those light curves. We consider two distinct supernovae, with peak luminosities of a few times $\sim 10^9\,L_\odot$, reached about 10 days after the onset of the explosion. For the first, we adopt the best-fit parameters for type-IIp supernova SN PS1-11aof, which has a peak luminosity of 2×10^{43} erg/s. The second supernova, PS1-10a, is somewhat dimmer. The light curves for the two supernovae are shown as the blue curves in Figure 8.10.

We used the radiative-transfer code `SPHRay`, which is well suited to such simulations, because it does not impose any limits on the spatial geometry or the computational domain. The hot supernova radiation was mimicked by setting some additional parameters in the radiative-transfer code.

```
radiative = SphRay(redirection="file", number_of_workers=24)
radiative.parameters.number_of_rays = number_of_rays / time_step
radiative.parameters.default_spectral_type = -3.0
```

Here, `radiative.parameters.number_of_rays = number_of_rays / time_step` defines the number of rays per time step used in the radiative-transfer

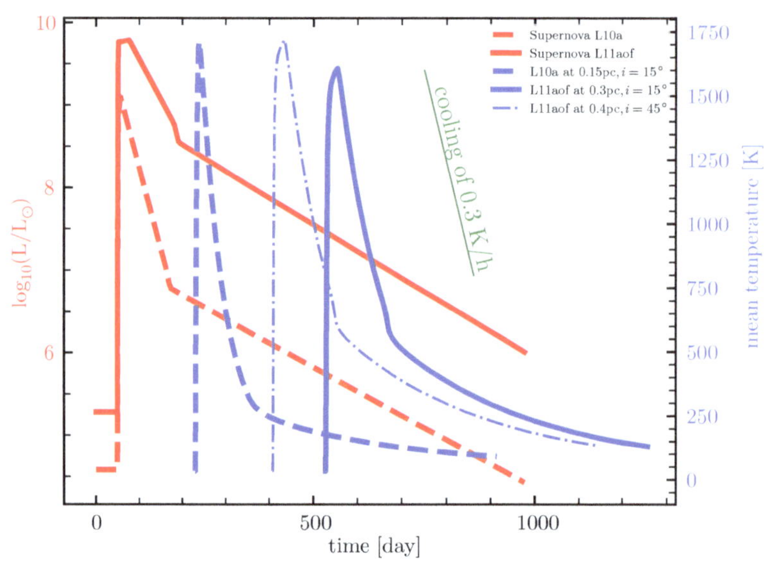

Figure 8.10. Light curves for the type-IIp supernovae PS1-10a and PS1-11aof (dashed and solid blue curves, respectively, scale in blue on the left, Sanders et al. 2015). The red curves give the temperature of our model disk at distances of 0.15, 0.3, and 0.4 pc (from left to right). The mean disk temperature (axis scale in red on the right) gives the disk temperature calculated using supernova PS1-10a at $d = 0.15$ pc and $\Theta = 15°$ (the dashed curve) and PS1-11aof at a distance of 0.3 pc with $\Theta = 45°$ and at $d = 0.4$ pc with $\Theta = 15°$ (the solid curves). The thin green line (upper right) indicates a cooling of 0.3 K/hour, which is consistent with cooling after the supernova radiation has heated the disk for SN PS1-11aof, and which is sufficient for the formation of vitreous structures in the chondrules of the circumstellar disk. The calculations are performed at a resolution in the radiative-transfer code of `Nray/dt=`10^7`/(1|units.hour)`. The figure was made using the script `amuse.examples.supernova_irradiation`. The calculation took several days on a laptop, and for this reason we include the data files and the routine to plot those in `amuse.examples.plot_supernova_IIp_and_disk_temperature`.

solver. In our case, we adopted `number_of_rays / time_step`=10^7/`(1|units.hour)`. The parameter `default_spectral_type` is used to indicate the slope of the source's spectral energy distribution. For a supernova, a value of -3 is closer to the actual observed spectral energy distribution than the default value of -1.

In the radiation transport code, the irradiating source is represented by a particle, called `src_particle`. We can update this particle's parameters at run time by using a separate channel.

```
channel_from _src_to_rad = \
        supernova.particles.new_channel_to(radiative.src_particles)
supernova.evolve_model(time)
channel_from_src_to_rad.copy_attributes(["luminosity"])
```

Note that there is only one source particle in our calculation, even though the code's denotative is plural.

The calling sequence for the radiative-transfer and hydrodynamics modules is similar in structure to Code example 8.2 in Section 8.2.1, where we combined stellar evolution with gravitational dynamics. The two codes are called `radiative` and `hydro`. Both codes have appropriate channels defined to and from the local particle sets: `disk` for the SPH particles, `star` for the stellar particle in the hydro code, and `supernova.particles` for the local source particle to the radiative-transfer code. The event loop then becomes:

```
while time < time_end:
    time += time_step / 2
    supernova.evolve_model(time)
    channel_from_src_to_rad.copy_attributes(["luminosity"])
    radiative.evolve_model(time)
    channel_rad_to_hydro.copy()

    time += time_step / 2
    hydro.evolve_model(time)
    hydro_to_disk.copy()

    supernova.evolve_model(time)
    channel_from_src_to_rad.copy_attributes(["luminosity"])
    update_source_particle(time)
    radiative.evolve_model(time)
    rad_to_disk.copy()
```

First, we update the supernova luminosity, perform an evolution step with a time step of `time_step / 2` in the radiative-transfer code, and copy the irradiated SPH particle data to the hydrodynamics code. Next, we update the time to a full step, evolve the hydrodynamics of the disk, and copy the data back to the framework. Finally, we repeat the first snippet, updating the supernova luminosity and performing another half radiative time step. In the last line, we copy the data from the radiative-transfer code back to the local particle set.

The red curves in Figure 8.10 show the temperature evolution of three identical protoplanetary disks around a 1 $M_\odot$ star at a distance of 0.15 pc from SN PS1-10a (left dashed curve), and at 0.3 and 0.4 pc from the considerably brighter SN PS1-

11aof (right two solid curves). The mean temperature in the disk exceeds 1500 K for each of these supernovae and distances, which is high enough to melt agglomerates with the highest melting points. Carrying out any of these simulations with a lower resolution, using less than 10^7 rays per hour, in the radiative transport calculation would lead to lower temperatures, whereas adopting a larger number of rays hardly affect the results, except that the script takes longer to run.

The shock wave reaches the disk a few tens of years after supernova radiation. This is a more hydrodynamical process, which we will discuss further in Section 10.3.

8.6.2 Integrating a Star Cluster with Interrupts

The close proximity of the Trapezium cluster allows detailed observations of circumstellar disk sizes using HST/WFPC2 (Vicente & Alves 2005). This size distribution is well described by a power law, but the origin of this distribution is not well understood. Vicente & Alves (2005) argue that "[Despite] the young age of the Trapezium, given that disk destruction is well underway, it is perhaps too late to tell if the present-day disk size distribution is primordial or if it is a consequence of the massive star formation environment."

We can write a small AMUSE script to test this hypothesis, and to study the evolution of the size distribution of circumstellar disks. Such a script should integrate the equations of motions of the stars and keep track of close encounters in order to truncate disks depending on the distance, mass, and velocity of a close encounter. Any direct N-body code will be sufficiently accurate for such a study. And since the number of stars is relatively small, these calculations can be performed on a laptop. Since the vast majority of close encounters are simple two-body interactions, we can integrate the cluster for a small time frame (smaller than a typical crossing time of ~2.4 Myr, see Equation (4.5)), evaluate all close encounters (using the `Kepler` package to find the distances of closest approach), resolve the effect of the encounters on the circumstellar disks, and compare the numerically obtained distribution of disk sizes with the observed distribution.

Code examples 8.14 to 8.16 present the main routine for these runs. The complete script is available in `amuse.examples.gravity_kepler_disks`. It generates the data for Figure 8.11, and should take 5–10 minutes to run on a laptop. This simulation started with 2000 stars drawn from a Kroupa (2001) distribution with masses between $0.01\,M_{\odot}$, and $100\,M_{\odot}$, with positions drawn from a fractal distribution (Goodwin & Whitworth 2004) of dimension $F = 1.6$, and velocities scaled so that the cluster was in overall virial equilibrium. At run time, we check for close encounters. If such an event is identified, we calculate the pericenter distance for the two-body encounter and truncate the disks of both stars.

The disk truncation radius, $r_{\rm trunc}$, is calculated from the distance of closest approach between two stars, q, and the masses of the two stars m and M by (Breslau et al. 2014)

$$r_{\rm trunc} = 0.28q\left(\frac{m}{M}\right)^{0.32}. \tag{8.1}$$

```
    while time < t_end:
        time += dt

        gravity.evolve_model(time)
        Etot_prev_se = gravity.kinetic_energy + gravity.potential_energy
        while stopping_condition.is_set():
            channel_from_gd_to_framework.copy()
            Ek_enc = gravity.kinetic_energy
            Ep_enc = gravity.potential_energy
            for ci in range(len(stopping_condition.particles(0))):
                particles_in_encounter = Particles(
                    particles=[stopping_condition.particles(0)[ci],
                               stopping_condition.particles(1)[ci]])
                local_particles_in_encounter = particles_in_encounter.
                                            get_intersecting_subset_in(bodies)

                resolve_close_encounter(gravity.model_time,
                                        local_particles_in_encounter, Johannes)
                Nenc+=1
                print(f"At time= {gravity.model_time} Nenc= {Nenc} ",
                      (f"Rdisk= {local_particles_in_encounter.disk_radius}")
                channel_from_framework_to_gd.copy_attributes(["radius"])
            dEk_enc += Ek_enc - gravity.kinetic_energy
            dEp_enc += Ep_enc - gravity.potential_energy
            gravity.evolve_model(time)

        channel_from_framework_to_gd.copy_attributes(["mass"])
```

Code example 8.14: Main routine of a script to perform an N-body simulation with interrupts to resolve star–disk encounters. The main event loop is presented in Code example 8.15. The star–disk interaction routine is given in Code example 8.16.

```
def resolve_close_encounter(time, bodies, Johannes):
    Johannes.initialize_from_particles(bodies)
    rcom = bodies.center_of_mass()
    vcom = bodies.center_of_mass_velocity()

    a, e = Johannes.get_elements()
    p = Johannes.get_periastron()
    print("Close encounter at t=",  time.in_(units.Myr), "a=", a.in_(units.AU), "e="
    , e, "p=", p.in_(units.AU), "M=", bodies.mass.max().in_(units.MSun), bodies.mass
    .min().in_(units.MSun), "at d=", rcom.in_(units.parsec), "with v=", vcom.in_(
    units.kms))

    truncate_disks_due_to_encounter(bodies, p)
```

Code example 8.15: Continuation of the script to run an N-body simulation with interrupts to resolve star–disk encounters. The star–disk interaction routine is given in Code example 8.16.

If r_{trunc} is smaller than the current disk size, the disk is truncated at $r_{\rm trunc}$. The truncation routine is presented in Code example 8.16.

Figure 8.11 compares the size distributions of the circumstellar disks with the observed distribution in the Orion Trapezium cluster. The observed distribution is

```
def truncate_disks_due_to_encounter(bodies, p):
    q = bodies[1].mass/bodies[0].mass
    rtr_prim = 0.28*p / q**(0.32)
    rtr_sec = 0.28*p * q**(0.32)
    dm0 = truncate_disk_due_to_encounter(bodies[0], rtr_prim)
    dm1 = truncate_disk_due_to_encounter(bodies[1], rtr_sec)

    mtot = bodies.mass.sum()
    bodies[0].accreted_mass += dm1 * bodies[0].mass/mtot
    bodies[1].accreted_mass += dm0 * bodies[1].mass/mtot

    bodies[0].radius = min(bodies[0].radius, 0.5*p)
    bodies[1].radius = min(bodies[1].radius, 0.5*p)

    bodies[0].mass = bodies[0].stellar_mass + \
        bodies[0].disk_mass + bodies[0].accreted_mass
    bodies[1].mass = bodies[1].stellar_mass + \
        bodies[1].disk_mass + bodies[1].accreted_mass
```

Code example 8.16: Routine to truncate a circumstellar disk during a pericenter passage with another star.

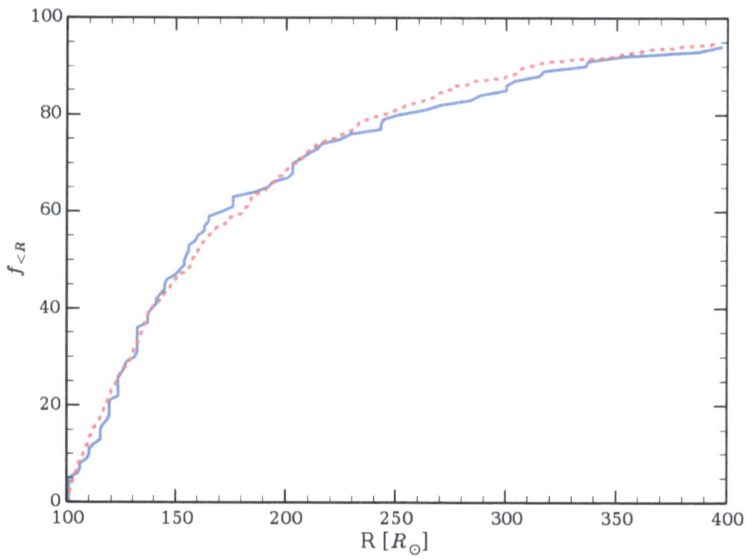

Figure 8.11. Cumulative size distribution of circumstellar disks in the Trapezium cluster. The observations are complete for disk radii exceeding 100 au. The red dotted line gives the 95 observed disk sizes from the observations. The black solid curve gives the result of our simulations at 1 Myr. A Kolmogorov–Smirnov (KS) comparison (Kolmogorov 1933; Smirnov 1948) of the two cumulative distributions results in a KS-statistic of 0.058 with a two-tailed p-value of 0.94, indicating that we cannot reject the hypothesis that both distributions are indistinguishable. The complete script that produced this figure is in `amuse.examples.compare_disk_size_Trapezium`. It simply plots data read from a file and runs in a few seconds on a laptop.

limited by the telescope's resolution (Vicente & Alves 2005). The pixel size in these observations is 45 AU, and disks as small as 1.5 pixels (or 67.5 au) can be resolved. The authors argue that their sample is complete down to a minimum radius of

100–150 au. Of the 162 protoplanetary disks, 95 are larger than 100 au (Vicente & Alves 2005). Although the observations are limited and the dynamical model is oversimplified, the similarity between the two datasets is striking.

8.7 Multiscale Modules in AMUSE

8.7.1 The `Multiples` Module: Combining Few-body with Many-body Dynamics

The N-body integrators described in Chapter 4 are designed to move large numbers of stars around a stellar system, and they are extremely efficient in accomplishing that task. However, in reality many stars are born with a companion, in tightly bound binary systems, or they may acquire companions through dynamical interactions. In these circumstances, while many of the direct N-body codes in AMUSE can still follow the motion with reasonable accuracy, the short time steps associated with the tight orbits mean that the integration effectively grinds to a halt once close binaries form.

In traditional N-body codes, such as NBODY6/7 and `starlab`, a substantial fraction of the total code base is devoted to following the internal motion of binaries and multiples simultaneously with the large-scale motion of the rest of the system. The algorithms involved have varying degrees of efficiency and accuracy, but all gradually relax the accuracy of the perturbed binary integration until "unperturbed" binaries (usually at a perturbation level of 1 part in 10^{6-7}) are treated as isolated unperturbed objects. The alternative approach, widely used in Monte Carlo schemes, treats multiples (where for simplicity the term will henceforth include binaries) as completely unperturbed until they interact strongly with another object. (A pioneering N-body variant on this approach was the `GORILLA` code presented by Tanikawa & Fukushige 2009.) Many papers have been devoted to the relative merits of these approaches, but, in our opinion, there is no convincing evidence that the improved treatment of lightly perturbed binaries inherent in the N-body approach has any significant advantage over the much simpler Monte Carlo treatment.

Accordingly, in the AMUSE `multiples` module, we depart from the N-body norm and adopt the Monte Carlo approach to multiple interactions. The module is designed to handle close encounters between stars and multiples in N-body systems. The basic idea is that the N-body modules manage multiple centers of mass, but are completely ignorant of the internal dynamics of these systems. In operation, the `multiples` module maintains a database of all stable multiples in the system. A stable multiple is a binary, a hierarchical triple deemed stable by the Mardling (2008) criterion, or a higher-order multiple system in which the Mardling criterion applied to the outermost orbits indicates approximate stability. As illustrated in Figure 8.12, when two centers of mass approach within the sum of their (dynamical) radii, a stopping condition is thrown and control is passed to `multiples`. The radii are typically chosen to be the 90° turnaround distance ($\sim G\langle m\rangle/\langle v^2\rangle$) for a single star, and a few outer binary semimajor axes for a multiple system. In virial equilibrium and N-body units, this distance is $1/2N$. (All critical `multiples` parameters have reasonable defaults and can be modified by the user script, if desired.)

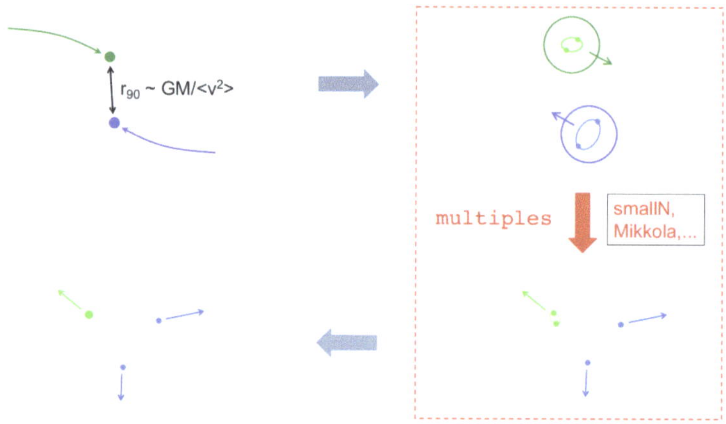

Figure 8.12. Two interacting stars (top left) are identified by the N-body code, expanded into their component stars (top right), integrated to completion by a standalone small-N code (bottom right), and reinstated into the N-body code (bottom left).

Once a dynamical interaction is detected and passes some basic tests (e.g. that the centers of mass are actually approaching), the internal structures of the interacting bodies are reinstated from the `multiples` database and the dynamics of the resulting small-N system followed to completion using a dedicated code. The AMUSE `smallN` module, a robust shared-step time-symmetrized code with analytic extensions (Hut et al. 1995; McMillan & Hut 1996) is often used for this task; the `Mikkola` module, which implements Mikkola's (Mikkola & Merritt 2008) algorithmic regularization code is valuable for problems where post-Newtonian gravitational effects must be included. "Completion" in this context means that the system has resolved itself into a collection of mutually receding stable systems (as just defined). We always expect this to be the eventual outcome of a small-N interaction. In principle, the tidal field of the N-body system could be included in the small-N calculation, but this is not currently implemented, although an attempt is made to absorb tidal errors by reorienting the final top-level system. To enforce a degree of isolation, in the current implementation a close perturber (lying within $\sim$3 times the initial separation of the interacting centers of mass) will veto the small-N calculation.

A major approximation in this approach is that interactions occur instantaneously —two center-of-mass particles in the N-body code are replaced by some number of new center-of-mass particles with no change in the system time. For most interactions, which take only a two-body encounter time ($\sim 1/N$ dynamical times) to complete, this causes a negligible error, but in some long-lived "hierarchical" resonances, as defined by Hut (1993), this may not be acceptable. We address this shortcoming as follows. In a pure scattering experiment, we would require that all top-level centers of mass of stable multiples be mutually receding and unbound at the end of the simulation. However, if we also accept centers of mass that form wide bound pairs, hierarchical resonances resulting in long bound intermediate orbits can be returned to the N-body code, even though the criteria for completion are not fully met. The internal motion is frozen until the N-body code again selects these bodies to continue the interaction.

Thus, long resonances are not treated as instantaneous, providing a rudimentary means for the N-body code to interact with the small-N system.

Allowing wide bound systems as an acceptable outcome of a scattering has another important consequence. In practice, it is very convenient to use `multiples` to also manage two-body encounters of single stars, even though most could easily be managed by the N-body integrator. This allows `multiples` to detect wide binaries formed by the N-body code, allowing us to model binary formation as well as binary interactions.

Code example 8.17 illustrates a simple `multiples` integration. Basically, we set up the system and tell multiples what N-body, `kepler`, and small-N codes to use and (since `multiples` is written in N-body units) optionally specify a converter.

```
def integrate_system(number_of_stars, t_end, seed=None):

    total_mass = number_of_stars | units.MSun
    length = 1 | units.parsec
    converter = nbody_system.nbody_to_si(total_mass, length)
    gravity = ph4(convert_nbody=converter)
    gravity.initialize_code()
    gravity.parameters.set_defaults()
    gravity.parameters.epsilon_squared = (0.0 | units.parsec)**2

    if seed is not None:
        numpy.random.seed(seed)
    stars = new_plummer_model(
        number_of_stars, convert_nbody=converter)
    stars.mass = total_mass / number_of_stars
    stars.scale_to_standard(
        convert_nbody=converter,
        smoothing_length_squared=gravity.parameters.epsilon_squared
    )
    index = numpy.arange(number_of_stars)
    stars.id = index+1
    stars.radius = (0.5 / number_of_stars) | units.parsec

    gravity.particles.add_particles(stars)

    stopping_condition = gravity.stopping_conditions.collision_detection
    stopping_condition.enable()

    init_smalln(converter)
    kep = Kepler(unit_converter=converter)
    kep.initialize_code()
    multiples_code = multiples.Multiples(gravity, new_smalln, kep,
                                         constants.G)
    multiples_code.neighbor_perturbation_limit = 0.05
    multiples_code.global_debug = 1

    energy_initial = print_diagnostics(multiples_code)
    multiples_code.evolve_model(t_end)
    print_diagnostics(multiples_code, energy_initial)
```

Code example 8.17: A simple N-body solver with close-encounter management. The full script can be found in `amuse.examples.basic_multiples`. It should run in a few seconds on a laptop.

```
SMALLN = None
def init_smalln(converter):
    global SMALLN
    SMALLN = SmallN(convert_nbody=converter)
def new_smalln():
    SMALLN.reset()
    return SMALLN
```

Code example 8.18: Initializing the small-N part of the `multiples` solver.

From that point on, we use `multiples` in exactly the same way as any other integrator—its `evolve_model` method invokes the integrator and takes appropriate action if encounters occur. The full script, including diagnostic and other functionality, can be found in `amuse.examples.basic_multiples`.

Note that the stellar ID defined in the script is important because it is used for internal bookkeeping within the module. The only slightly unusual aspect of the `multiples` setup is that, instead of providing the small-N module as an argument, the current module expects a function that resets and returns a small-N integrator (see Code example 8.18). This was part of the original design of the module, and is currently under review; it may be eliminated in future versions of AMUSE.

Finally, the `multiples` database enables the implementation of treatments of binary evolution (Section 5.3), secular internal evolution of multiple structure (Section 4.4.4), e.g., by the von Zeipel–Lidov–Kozai (von Zeipel 1910; Lidov 1962; Kozai 1962) mechanism, and external perturbations due to wide encounters in the N-body code. These are not yet fully implemented, and represent work in progress.

8.7.2 The `Kira` Module: Combining Stellar and Binary Evolution with Stellar Dynamics

In 1992, with Piet Hut and Jun Makino, we began work on the `starlab` project (Portegies Zwart et al. 1999). Central to this effort were two codes, the stellar-evolution code `SeBa` and a fourth-order accelerated Hermite integrator not dissimilar to `ph4`. The combined code was known as `kira`. Many of the fundamental ideas we learned in creating `kira` have percolated into the AMUSE framework.

One aspect in which `kira` was ahead of AMUSE, as described thus far, was in its handling of binary evolution in conjunction with the dynamics. So far, we have briefly discussed the evolution of binary stars in Section 5.3, but that was for binaries in isolation and not in a dynamical environment. Here, we adopt the same fundamental structure for single stars and binary stars to simulate a cluster with primordial binaries. Binaries and higher-order multiple encounters are managed by the `encounter` module in `amuse`, and it is possible to ionize binaries or make new binaries by dynamical encounters. We call this script `kira.py` in recognition of the original package. It is available in `amuse.examples.kira`. We use the `multiples` module to treat close multiple stellar encounters (see Section 8.7.1). Like many of the more complex AMUSE scripts described in this book, this is a work in progress and likely to change.

There are a few additional details to consider when modeling a star cluster containing binaries (either primordial or dynamically formed). The small orbital periods of binaries tend to dramatically slow down the underlying N-body code, and the many small time steps required to resolve the binaries lead to a steady accumulation of round-off and integration errors. These factors mandate special methods to manage the binaries. A further complication is that binaries are affected both by the stellar evolution of the components and by occasional encounters with other stars (and binaries) in the system. The evolution of single stars can be handled by any of the stellar-evolution codes available in AMUSE. Binary evolution, however, is a different matter.

The initial conditions for the simple simulation described here begin with a Plummer sphere of 100 particles with masses randomly selected from a Salpeter mass function between 0.1 $M_\odot$ and 100 $M_\odot$. We then select `Nbin=50` stars randomly using

```
binaries = center_of_masses.random_sample(Nbin)
```

These stars are then provided with a secondary companion, with orbital parameters chosen from standard distributions. The mass of the secondary is selected randomly from a uniform distribution between the minimum stellar mass and that of the primary. The orbital eccentricity is selected between 0 and 1, assuming a thermal distribution ($p(e) = 2e$; Duquennoy & Mayor 1991), and the semimajor axis is selected using the prescription in Equation 1.6.3 of Eggleton (2006), with a period between 10 days and 100 years. Other binary orbital parameters are initialized in much the same way as described in Section 5.3.

With stars and binaries initialized, we can start the stellar-evolution code and the global gravity solver. We use `SeBa` to resolve the stellar evolution because it has a natural way to handle the evolution of single stars as well as binaries. For the global dynamical evolution of the stars, we use `Hermite`. Stable binaries and close two-body encounters are not integrated, but we follow their dynamical evolution via the `Kepler` package. The global N-body code integrates the center of mass of the close two-body encounters as well as of the center of masses of the stable binary systems. Once another star perturbs a binary (or higher-order system), we resolve its interaction using the `multiples` module (Section 8.7.1). The `small` Ncode it relies on is based on the old `Kira_smallN` package from `starlab`.

We start by importing the `encounters` module, which is a general-purpose module designed to handle stellar encounters of all types (collisions, multiple encounters, etc.) and is still under development

```
from amuse.couple import encounters
```

We then realize `encounters_code` using the `kepler` package to handle stable two-body systems, and `smallN` for multibody systems

```
encounter_code = encounters.HandleEncounter(
    kepler_code = new_kepler(converter),
    resolve_collision_code = new_smalln(converter)
)
```

Here, we use helper functions new_kepler and new_smalln, which return instantiations of the two codes with common parameters (such as timestep_-parameter in smallN) set to reasonable values. We now start multiples handling using Multiples, which is part of the encounters module. We give it a global gravity code and the encounter_code we just created using the kepler and smallN modules:

```
multiples_code = encounters.Multiples(
    gravity_code = Hermite(converter),
    handle_encounter_code = encounter_code
)
```

Multiple encounters will now be resolved using smallN, whereas two-body stable binaries are addressed using the kepler solver, as described in Section 8.7.1.

Having set up all code dependencies, it is time to copy all stellar and binary data to the multiples module and declare channels to the code:

```
multiples_code.particles.add_particles((stars-binary_stars).copy())
multiples_code.singles_in_binaries.add_particles(singles_in_binaries)
multiples_code.binaries.add_particles(binary_stars)
multiples_code.commit_particles()
channel_from_stars_to_particles = (
    stellar.particles.new_channel_to(multiples_code.particles)
)
```

The multiples module now takes care of all two-body and higher-order interactions in the star cluster simulation.

After an encounter, binary parameters may have changed. This affects the stellar and binary evolution of these systems, and the stellar-evolution code must be informed of these changes. We do this via a stopping condition, in this case binaries_change_detection, which stops the multiples module when the orbital separation or eccentricity of a binary is significantly affected by the dynamical encounter:

```
stopping_condition \
        = multiples_code.stopping_conditions.binaries_change_detection
stopping_condition.enable()
```

We run the multiples module in a (by now) familiar loop:

```
while time < time_end:
    time += time_step
    multiples_code.evolve_model(time)

    if stopping_condition.is_set():
        resolve_changed_binaries(stopping_condition, stellar)

    stellar.evolve_model(ti)
    channel_from_stars_to_particles.copy_attributes(["mass", "radius"])
    update_dynamical_binaries_from_stellar()
```

Newly formed binaries, binaries with orbital elements affected by a dynamical encounter, and binaries that broke up (and must be removed) are handled by the code in Code example 8.19. (Note that we encountered the routine `calculate_orbital_elements` previously in Section 4.4.1.)

Similarly, after the stellar-evolution step we must update the dynamical information in the binaries. This is realized by reading the binary parameters in the stellar-evolution code, calculating the new coordinates of the binary components, and updating the dynamics. Special care is required for binaries that merged during the evolution process, or that were disrupted in a supernova explosion, but those can be handled with separate stopping conditions for `supernova_detection` and `merger_detection` in the stellar-evolution module.

```
def resolve_supernova(stellar, stars, multiples_code):
    singles = stars.as_particle_in_set(multiples_code.singles)
    singles_in_stellar = singles.as_particle_in_set(stars)
    singles.velocity += singles_in_stellar.supernova_kick_velocity
```

Figure 8.13 presents the results of a small test simulation using the script `kira` from the `amuse.examples` module. The initial conditions for these calculations are a standard star cluster with 100 stars drawn from a Salpeter (1955) mass function with a minimum mass of $10\,M_\odot$, distributed in a Plummer (1911) sphere with a characteristic radius of 1 pc. Half of the primary stars are given a secondary companion

```


def resolve_changed_binaries(stopping_condition, stellar, converter):
    new_binaries = stopping_condition.particles(0)
    for bi in new_binaries:
        print(("add new binary:", bi))
        a, e = calculate_orbital_elementss(bi, converter)
        bi.semi_major_axis = a
        bi.eccentricity = e
        stellar.binaries.add_particle(bi)
        print(("new binary parameters", a, e))
        print(bi)

    lost_binaries = stopping_condition.particles(1)
    for bi in lost_binaries:
        print(("remove old binary:", bi.key))
        stellar.binaries.remove_particle(bi)

    changed_binaries = stopping_condition.particles(2)
    for bi in changed_binaries:
        bs = bi.as_particle_in_set(stellar.binaries)
        a, e = calculate_orbital_elementss(bi, converter)
        bs.semi_major_axis = a
        bs.eccentricity = e
        print(("Modified binary parameters", a, e))
        print(bs)
```

Code example 8.19: Morphing binaries in kira.py.

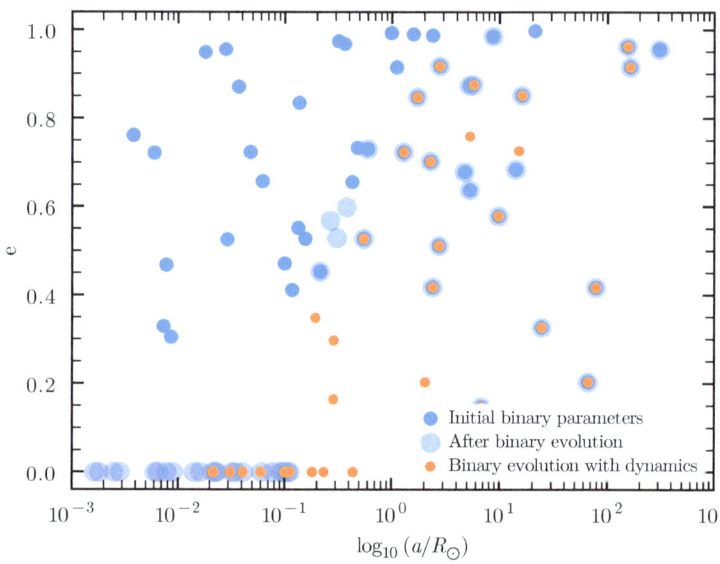

Figure 8.13. Distribution in semimajor axis and eccentricity of a population of binary stars in an N-body system at $t = 0$ Myr (dark blue bullets) and at an age of 10 Myr (big blue and orange bullets). The big blue bullets indicate what happens in stellar evolution but without taking the dynamical evolution into account. The orange bullets show the same binary population but they are affected by stellar evolution, binary evolution, and by stellar dynamics. We deliberately plot the blue dots first and larger than the red dots to illustrate how some of the binaries disappear or move within the parameter space. The script that produced this figure is `amuse.examples.kira`. It should run in about a minute on a laptop.

with mass chosen randomly between 10 $M_{\odot}$ and the primary mass, in an orbit with eccentricity taken from a thermal distribution and semimajor axis drawn from the distribution proposed by Eggleton (2006). The other binary parameters and the binaries' orientations in space are random. We integrate the equations of motion using the `Hermite` integrator—see Section 4.3.2; binary evolution is managed using `SeBa` (Portegies Zwart & Verbunt 1996; Portegies Zwart & Yungelson 1998; Toonen et al. 2012; Portegies Zwart & Verbunt 2012).

Figure 8.13 shows the distributions of orbital semimajor axis and eccentricity of the binaries after integrating for 10 Myr. We integrate for such a short time because that is already sufficient to see differences compared to a calculation without stellar and binary evolution.

The blue bullets in Figure 8.13 indicate the initial conditions (orbital separation and eccentricity) of the primordial binaries in the simulation. The red dots show the orbital parameters after 10 Myr of integration using `kira.py`. For several of the closest binaries the orbital parameters have already been affected by stellar evolution and tidal interactions within the binary. Most of these are affected by tidal circularization, particularly those with short orbital periods and eccentricities $\gtrsim 0.6$. Some of the tightest ($a \lesssim 0.1\,R_{\odot}$) circular binaries have disappeared because they were Roche-lobe overflowing at birth and have merged to form a single star.

8.7.3 The `LonelyPlanets` Module: Combining Planetary with Stellar Dynamics

Instead of integrating multiple stars in a cluster, one might be interested in the multiscale problem of integrating planetary systems. Integrating a single isolated planetary system is usually done with a regular N-body code, but a planetary system in a star cluster is quite challenging because of the wide range of time scales involved. One way to perform such a calculation is with a divide-and-conquer strategy.

For this purpose, we developed `LonelyPlanets` (not to be confused with the popular travel guide books, which deal with one specific rather peculiar planet). We mean to integrate multiplanet systems, possibly with moons, asteroids, and comets together with the dynamical evolution of the stellar cluster.

`LonelyPlanets` works through a two-stage process. In a first simulation, we select a star (in our example randomly), and monitor its most prominent (i.e. strongest) perturbers during the integration of the cluster. The event loop for `LonelyPlanets` is listed in Code example 8.20, and should look quite familiar by now, with the exception that we keep track of the perturbers. In this example, the perturber is any of the five nearest stars. In practice, one probably wants to keep track of the strongest perturber, rather than the nearest star, since a massive far away star can perturb a planetary system more severely than a low mass nearby star.

We find the nearest neighbors simply by sorting stars on the distance to our selected star,

```
def nearest_neighbors(bodies, star, Nnn=3):
    r = ((bodies-star).position-star.position).lengths().value_in(units.pc)
    r, p = (list(t) for t in zip(*sorted(zip(r, bodies-star))))
    b = Particles()
    for pi in p[:Nnn]:
        b.add_particle(pi)
    return b
```

```
    while gravity.model_time < time_end:
        gravity.evolve_model(gravity.model_time + dt)
        channel_from_gd.copy()

        star.age = gravity.model_time

        target_star.append(star.copy())
        model_time.append(gravity.model_time)
        perturbers = find_perturbers(star, bodies, Nnn)
        nn_distance[str(model_time[-1])] = (
            star.position - perturbers.position
        ).lengths()
        nn_bodies.append(perturbers)
        print(f"time={model_time[-1].in_(units.Myr)}, x={star.x.in_(units.pc)}")

```

Code example 8.20: Snippet of the stage 1 `LonelyPlanets` event loop.

In the second stage, we generate a planetary system around the selected star and integrate it together with the earlier (in stage 1) generated database of perturbers. We further optimize the stage 2 calculation (as illustrated in Code example 8.21) by including the perturbing stars only if their perturbations on the planetary system exceed some predetermined limit. In our case, we adopt a simple distance criterion of 3000 au. This saves a little computer time, but with only five perturbers and eight planets the gain is rather small.

```
    for ti in range(len(model_time)):
        star = planetary_system[planetary_system.type == "star"][0]
        planetary_system.position -= star.position
        planetary_system.velocity -= star.velocity
        planetary_system.position += ss_star[ti].position
        planetary_system.velocity += ss_star[ti].velocity
        channel_to_gd.copy()

        perturbed = nearest_encounter(
            ss_star[ti], nn_bodies[ti], maximum_perturber_distance)
        if perturbed:
            npert += 1
            print(f"Integrate with stars: N={len(nn_bodies[ti])} "
                  f"at time={model_time[ti].value_in(units.Myr)}")
            gravity.particles.add_particles(nn_bodies[ti])
        else:
            nunpert += 1
        print(f"Time={model_time[ti].in_(units.Myr)}: "
              f"N= {len(gravity.particles)} = "
              f"{len(planetary_system)}+{len(nn_bodies[ti])}")
        gravity.evolve_model(model_time[ti])

        if perturbed:
            gravity.particles.remove_particles(nn_bodies[ti])
        channel_from_gd.copy()

        star = planetary_system[planetary_system.type == "star"][0]
        planets = planetary_system - star
        get_orbital_elements_of_planetary_system(star, planets)
        amin = (np.min(planets.semimajor_axis.value_in(units.au)) | units.au,)
        sma.append(planets.semimajor_axis)
        ecc.append(planets.eccentricity)

        lost_planets = planets[planets.eccentricity > 1]
        if len(lost_planets) > 0:
            lp = lost_planets.get_intersecting_subset_in(gravity.particles)
            gravity.particles.remove_particles(lp)
        lost_planets = planets[np.isnan(planets.eccentricity)]
        if len(lost_planets) > 0:
            lp = lost_planets.get_intersecting_subset_in(gravity.particles)
            gravity.particles.remove_particles(lp)
```

Code example 8.21: Stage 2 `LonelyPlanets` event loop, in which the planetary system is being integrated with nearby stars.

This is a typical example of optimizing for a rather minor speed-up while taking the risk of creating a complex bug. Fixing the bug may take more time than just integrating for all neighboring stars.

In Figure 8.14, we present the result of the stage 1 calculation of `LonleyPlanets`, where we initialize 1000 stars from a Kroupa mass function (between $0.08\,M_\odot$ and $30\,M_\odot$) in a virialized King model (with $W_0 = 5$) 1 pc radius. We randomly selected one star, which happened to have a mass of $\sim 0.11\ M_\odot$ and was born in the cluster central region (at $(-0.026, -0.072, -0.14)$ pc from the center). This star acquired planets between 1 au and 100 au in circular orbits in the x–y-plane.

In the second stage of the calculation, we integrate the planetary system together with the most strongly perturbing stars. We select the three nearest neighbors and the three strongest perturbers. Sometimes these are the same stars, but sometimes a low-mass star happens to be near the planetary system, while other, more massive, stars are stronger perturbers. In Figure 8.15, we present the orbital separation and eccentricity of the 15 planets for the first 3 Myr of the evolution of the cluster. Although it is something of a jumble of lines, it is clear that the planet eccentricities

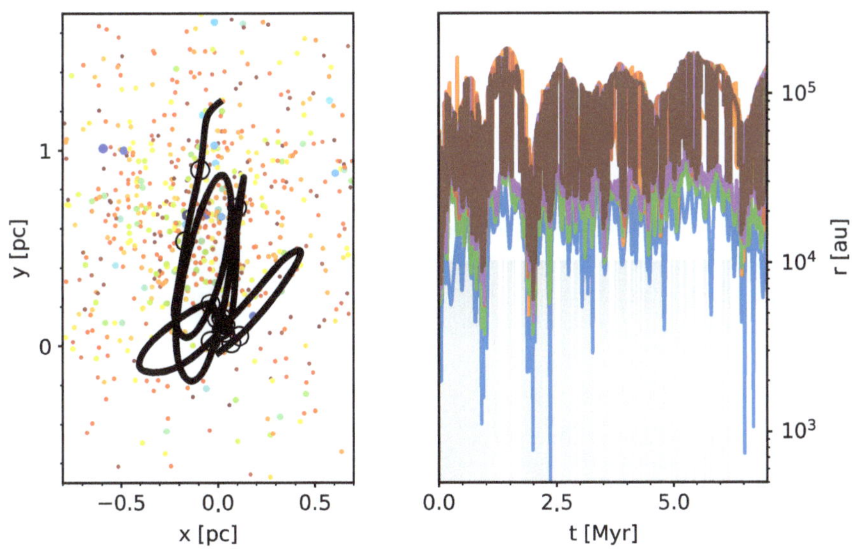

Figure 8.14. First stage calculation for `LonelyPlanets`. Strategy to reduce computing cost without loss of accuracy. This is a two-stage process. In the first step (left), we calculate the evolution of the cluster, storing all the close-encounter information. The points indicate the initial stellar positions (color and size represent mass). The orbit of a randomly selected star is presented as the solid black curve. The right panel shows, over the 3Myr of the evolution in this cluster, the distance to the three nearest stars from the selected star. In the second step of the calculation, we repeat the close encounters of our selected star with the nearest other stars, but then including planets and asteroids. Left: A representation of the cluster at zero age, with the selected star's orbit presented as a solid black curve. While integrating, we keep track of the strongest perturbers, and a small circle is drawn each time such a perturbation exceeds the equivalent of an encounter at around 5000 au. Right: The distance from our selected stars to the three strongest perturbers, as a function of time. The identity of these perturbers changes regularly, once their perturbation is overtaken by that of another (more strongly perturbing) star. The vertical dotted lines indicate the moments of strongest perturbation (and identify to the circles in the left-hand panel).

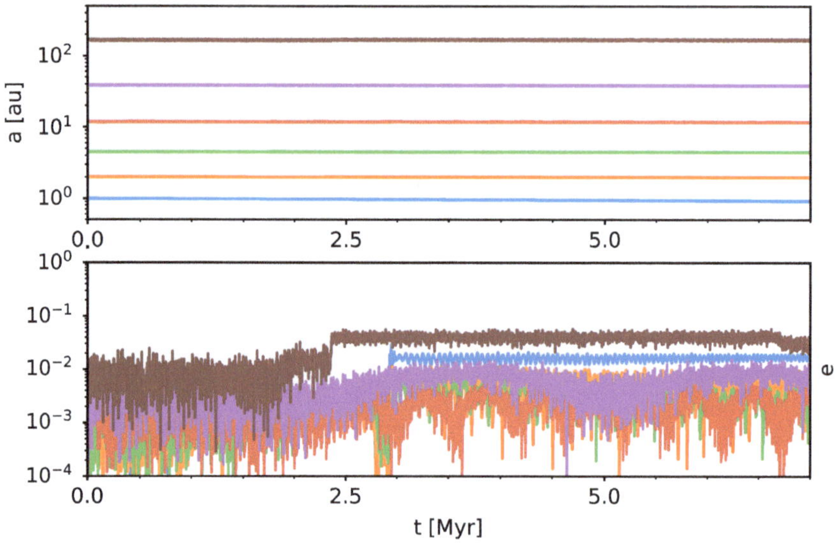

Figure 8.15. Second stage calculation for `LonelyPlanets`.

rapidly increase with time, and only four planets survive until an age of 3 Myr, when we stop the calculation.

It is relatively straightforward to include other effects in to the `LonelyPlanets` scripts, such as test particles (representing asteroids or comets), stellar evolution (representing not only the cluster stars but also the host stars of the planetary systems), etc. The method outlined here is a workable test version meant for integrating a single planetary system (composed of some 30 planets) in a cluster composed of 1000 stars among which is the planetary system.

Integrating the planetary system requires $\mathcal{O}[(1 + N_p + N_{cc})^2]$ operations per time step. Here, $f \leqslant 1$ indicates the fraction of time steps during which the planetary system is perturbed by the neighboring stars (the gray shaded areas in Figure 8.15). In the simplest case, $f \equiv 1$, indicating that the planetary system is always perturbed, irrespective of the mass or distance to the nearest perturber. This may be the case so long the planetary system is a cluster member, but as soon as the star escapes the cluster, $f \to 0$. This is nicely illustrated in Figure 8.15, where the last strong encounter with the planetary system, at an age of about 5.7 Myr, leads to a perturbation in the outermost planet (see the hiccup in the brown curve in Figure 8.15 at the same time) and causes the star to escape the cluster. The further integration of the planetary system is achieved by a standalone symplectic integrator. We can still include the galactic potential, in case we are interested in the evolution of the Oort cloud or asteroids along in the star's galactic orbit.

8.8 Validation

Compared to other coupling strategies, problems with linear coupling are relatively easy to validate, because each of the individual parts can to some degree be validated separately, as described in previous chapters. Energy is generally not strictly conserved in coupled problems, and neither is momentum, but wherever conservation laws do apply it is important to check them. Mass, number of objects, energy, and angular momentum are generally conserved in specific subproblems, but not throughout the simulation.

A simple validity check for conserving mass during stellar evolution and energy during stellar dynamics is

```
mass = stellar.particles.mass.sum()
stellar.evolve_model(time)
channel_stellar_to_framework.copy()
lost_mass = mass - stars.mass.sum()

channel_framework_to_gravity.copy()
Etot = gravity.kinetic_energy + gravity.potential_energy
gravity.evolve_model(time)
dE = Etot - (gravity.kinetic_energy + gravity.potential_energy)
```

Because mass is lost in stellar evolution, the kinetic and potential energies of the gravitational system are also affected. These can be checked independently, for example with

```
E_stellar = gravity.kinetic_energy + gravity.potential_energy
stellar.evolve_model(time)
channel_stellar_to_gravity.copy()
Etot = gravity.kinetic_energy + gravity.potential_energy
dE_stellar = E_stellar - Etot

gravity.evolve_model(time)
dE_gravity = Etot - (gravity.kinetic_energy + gravity.potential_energy)
```

Printing `dE_gravity` then provides a diagnostic on the quality of the simulation results. Similar strategies can be applied in hydrodynamics, and radiative transfer, as we will see in the next two chapters. We will experience similar issues in Chapter 9, and we postpone further discussion until there.

8.9 Assignments

8.9.1 Interlaced Timestepping

In Section 8.4.5, we discussed the evolution of the Wolf–Rayet colliding wind triple system θ Muscae, in which the WC5/6 star WR48 is in a 19 day orbit with an O6 companion, and both are orbited by a distant and relatively low-mass O9.5/B0Iab star. In the example presented in Section 8.4.5, stellar evolution and the dynamical evolution of the triple were applied at very short and constant time intervals of only one-tenth of the inner orbital period. It is not clear if such a small time step is actually needed to recover the correct physics. In this assignment we will study this.

In Section 8.4.4, we discussed some alternative time-stepping strategies, but there are other ways to do this. In this assignment, implement an alternative time-stepping scheme by allowing the time step to depend on the rate of mass loss, not allowing the total mass loss during the step to exceed a given value. If the time step becomes too large, the integrator will produce incorrect results, but the integration will take too long for time steps that are too small. Use the interlaced temporal discretization method described in Section 8.4.4 to study if this is really a desirable way of integrating two numerical solvers.

Run a series of simulations, systematically halving the time step from one to the next, and record the semimajor axis and eccentricity of the inner and the outer binary and their relative inclination. Make a plot of the orbital parameters and the wall-clock computer time as a function of the time step size. Based on these curves, can you decide what is the best time step size for simulation of a triple system such as θ Muscae?

- How does the optimum time step depend on the mass-loss rate in the star with the strongest wind?
- How does optimum time step depend on the initial relative inclination between the inner and the outer orbit?

8.9.2 Stellar Evolution and Dynamics

Generate a Plummer sphere of $N = 1000$ stars with a Salpeter stellar mass function between $M_{\min} = 0.1$ $M_\odot$ and $M_{max} = 100$ $M_\odot$ (power-law slope $\alpha = -2.3$). Evolve the system using an N-body integrator of your choice, including stellar evolution, and measure the moment when the cluster has lost half of its initial mass. Take, for example, the simple N-body code in Code example 4.1 as a starting point, and expand it with the stellar-evolution recipes from Code example 8.1.

Now change the power-law slope from $\alpha = -2.3$ (Salpeter) to $\alpha = -2.0$ and repeat the measurement. Continue this process with increasing slope until the initial mass function is flat. What can you conclude when you plot the time for the cluster to lose half of its initial mass as a function of the slope of the initial mass function?

- Why do clusters with relatively flat mass functions dissolve more quickly than those with a steep mass function?
- What changes when stellar evolution is switched off?
- Why is it not possible to continue the simulation to complete dissolution with a Salpeter mass function without stellar evolution?

8.9.3 Multiple Stellar Populations

Globular clusters had long been thought to contain single stellar populations—all stars formed at the same time with the same metallicity—but recent discoveries challenge this view. In particular, the discovery of numerous globular clusters, e.g., ω Centauri (Lee et al. 1999) and NGC2808 (Piotto et al. 2007), with two or more distinct main-sequence populations has been puzzling astronomers for at least two decades. One question that keeps cropping up is the rate at which two (or more) distinct stellar populations mix dynamically when their birth distributions differ. In this experiment, we will study this phenomenon by initializing a star cluster with two distinct stellar populations. These might be two populations with different metallicities, or with different ages.

Generate a first population of 500 stars by selecting masses from a Salpeter (or any other) mass distribution. Evolve the population to an age of 1 Gyr. Select positions and velocities for these stars from a Plummer model with a virial radius of 1 pc, and scale the star cluster to virial equilibrium. Now generate a second stellar population at zero age and using the same (Salpeter) initial mass function, but now using a virialized Plummer sphere with a virial radius of 3 pc for the stars. Before starting the N-body integrator, you will need to merge the two stellar distributions and scale the whole system to virial equilibrium.

Evolving the two populations dynamically can be done in the same N-body integrator. Use two stellar-evolution codes for the two populations. Evolve the entire cluster for 1 Gyr and study how the individual Lagrangian radii of the two populations, and of the cluster as a whole, evolve with time from the start of the N-body integration.

References

Aarseth, S. A. 2003, Gravitational N-body Simulations (Cambridge: Cambridge Univ. Press)

Breslau, A., Steinhausen, M., Vincke, K., & Pfalzner, S. 2014, A&A, 565, A130

Carathéodory, C. 1909, MatAn, 67, 355

Duquennoy, A., & Mayor, M. 1991, A&A, 248, 485

Eggleton, P. 2006, Evolutionary Processes in Binary and Multiple Stars (Cambridge: Cambridge Univ. Press)

Eisenstein, D. J., & Hut, P. 1998, ApJ, 498, 137

Gaburov, E., Lombardi, J. C., & Portegies Zwart, S. 2008, MNRAS, 383, L5

Goodwin, S. P., & Whitworth, A. P. 2004, A&A, 413, 929

Hill, G. M., Moffat, A. F. J., & St-Louis, N. 2002, MNRAS, 335, 1069

Hut, P. 1993, ApJ, 403, 256

Hut, P., Makino, J., & McMillan, S. 1995, ApJL, 443, L93

Jänes, J., Pelupessy, I., & Portegies Zwart, S. 2014, A&A, 570, A20

King, I. R. 1966, AJ, 71, 64

Kolmogorov, A. N. 1933, Giornale dellaIstituto Italiano degli Attuari, 4, 83

Kozai, Y. 1962, AJ, 67, 591

Kroupa, P. 2001, MNRAS, 322, 231

Lee, Y. W., Joo, J. M., Sohn, Y. J., et al. 1999, Natur, 402, 55

Lidov, M. L. 1962, Planet. Space Sci., 9, 719

Lombardi, J., Rasio, C., & Shapiro, S. L., F. A. 1995, ApJL, 445, L117

Lombardi, J. C., Thrall, A. P., Deneva, J. S., Fleming, S. W., & Grabowski, P. E. 2003, MNRAS, 345, 762

Lombardi, J., & Sawyer Warren, J. 2014, MMAS: Make Me A Star, ascl:1412.010

Lombardi, J. C., Rasio, F. A., & Shapiro, S. L. 1996, ApJ, 468, 797

Lombardi, J. C., Warren, J. S., Rasio, F. A., Sills, A., & Warren, A. R. 2002, ApJ, 568, 939

MacPherson, G. J., Davis, A. M., & Zinner, E. K. 1995, Metic, 30, 365

Mardling, R. A. 2008, IAU Symp. 246, Dynamical Evolution of Dense Stellar Systems, ed. E. Vesperini, M. Giersz, & S. Sills (Cambridge: Cambridge Univ. Press) 246

McMillan, S. L. W., & Hut, P. 1996, ApJ, 467, 348

Mikkola, S., & Merritt, D. 2008, AJ, 135, 2398

Piotto, G., Bedin, L. R., Anderson, J., et al. 2007, ApJL, 661, L53

Plummer, H. C. 1911, MNRAS, 71, 460

Portegies Zwart, S. F., & van den Heuvel, E. P. J. 2016, MNRAS, 456, 3401

Portegies Zwart, S. F., & Verbunt, F. 1996, A&A, 309, 179
Portegies Zwart, S. F., & Verbunt, F. 2012, SeBa: Stellar and Binary Evolution, Astrophysics Source Code Library, ascl:1201.003
Portegies Zwart, S. F., & Yungelson, L. R. 1998, A&A, 332, 173
Portegies Zwart, S. F., Makino, J., McMillan, S. L. W., & Hut, P. 1999, A&A, 348, 117
Salpeter, E. E. 1955, ApJ, 121, 161
Sanders, N. E., Soderberg, A. M., Gezari, S., et al. 2015, ApJ, 799, 208
Sills, A., & Lombardi, J. C. 1997, ApJL, 484, L51
Smirnov, N. N. 1948, Ann. Math. Stat., 19, 279
Sugawara, Y., Tsuboi, Y., & Maeda, Y. 2008, A&A, 490, 259
Tanikawa, A., & Fukushige, T. 2009, PASJ, 61, 721
Terlevich, E. 1987, MNRAS, 224, 193
Toonen, S., Nelemans, G., & Portegies Zwart, S. 2012, A&A, 546, A70
van der Helm, M. I., Saladino, T. P. G., Pelupessy, F. I., Portegies Zwart, S., & Pols, O. R. 2019, A&A, 625, 85
Vicente, S. M., & Alves, J. 2005, A&A, 441, 195
von Zeipel, H. 1910, AN, 183, 345
White, S. D. M. 1994, arXiv:astro-ph/9410043

Chapter 9

Advanced Coupling Strategies

> All you really need to know for the moment is that the universe is a lot more complicated than you might think, even if you start from a position of thinking it's pretty damn complicated in the first place.
>
> —Douglas Adams

The full potential of AMUSE becomes clear when we couple codes to solve complex multiphysics or multiscale problems. In this chapter, we discuss coupling issues and strategies, and present examples of coupled solutions. Computing requirements increase drastically when running multiphysics problems. As a result, some of the examples presented in this chapter may be quite challenging for a laptop or workstation.

Nonlinear coupling strategies lead to new applications and greatly increase the power of the framework. We are only at the brink of discovering the large number of possibilities and the complexities of the coupling strategies it enables. Many of these, however, may be challenging to validate. We discuss some basic and a few more advanced hierarchical coupling strategies.

9.1 Code-coupling Strategies

Often, two or more processes or physical phenomena are coupled in a single problem. Nature seems to have a good grasp on how to do this, but in the computer it is often far less clear how to achieve an accurate or efficient solution.

Many simulations entail combining physical domains with very different length and timescales. We have seen numerous examples in the preceding chapters. In Section 8.4.1, we introduced an interlaced temporal discretization scheme as a way to couple dynamics and stellar evolution. So far, we have primarily symmetrized the

doi:10.1088/2514-3433/adfc3cch9

interaction between two codes, on the general principle that resolving the effect of code A on the time step for code B, and vice versa, should produce more accurate results. That approach naturally led us to a drift–kick–drift scheme equivalent to the second-order predictor–corrector or Verlet method (Verlet 1967; see Sections 4.2.3 and 9.1.1). We will see below that this approach represents much more than just a convenient symmetric coupling; in particular, it provides a symplectic coupling when applied to symplectic codes (Section 4.2.3.2). We then generalize it to coupling an arbitrary number of codes, even when the modules involved are not symplectic nor even widely separated in scale.

Typically, when two systems interact, they either have comparable scales but different physics, or the same physics but different scales. For clarity, we start with a multiscale system, in which we identify one or more macroscopic (parent) systems, and one or more microscopic (child) systems. We prefer the parent–child terminology because it is less confusing in discussions.

A child system is characterized by smaller length and timescales compared to the parent system. The child system then dictates the time resolution (i.e., time step size) and the smallest resolvable distance in the problem. Child systems are numerically expensive because of their short timescales.

The parent system typically has larger length and longer timescales. The longer timescale generally allows for larger numerical time steps, making the parent systems relatively cheap compared to integrating the children. The cost per step, however, can be many times more expensive because the large extent of the parent and the large number of objects. An example might be a stellar cluster with thousands of stars and a size scale on the order of a parsec. The typical timescale in such a system would be about a megayear. If the cluster is part of the Galaxy, with 100 billion stars, and a multi kpc size, the typical timescale is easily 100 times larger than the child systems.

In Figure 9.1, we illustrate the hierarchy of scales by presenting an artist's conception of a planetary system, within a young star cluster in a spiral galaxy. Although the images are real observed systems (the planetary system is an artist's impression, though), they are not actually associated. The small bars indicated in the image, represent 10 kpc, 10 pc, and 10 au, for the galaxy, the star cluster and the planetary system, respectively. The associated orbital timescales for these systems will be on the order of 100 Myr, 1 Myr, and 100 yr The galaxy may contain some 10^{11} stars, the cluster 10^4 stars, and the planetary system may contain several planets.

One scale often dominates in computational expense, memory requirements, communication bandwidth, and storage space. The key to a successful and rewarding multiscale simulation is to balance these requirements. This scale discrepancy is often addressed by approximating either the parent or the child system, e.g., using a semianalytic model. Replacing the child system with an approximate method is often referred to as subgrid physics (whether or not a grid actually exists). Replacing the parent system with an approximate method is often referred to as a background (field) method or external boundary conditions. An example of child subgrid physics is the incorporation of a `multiples` solver

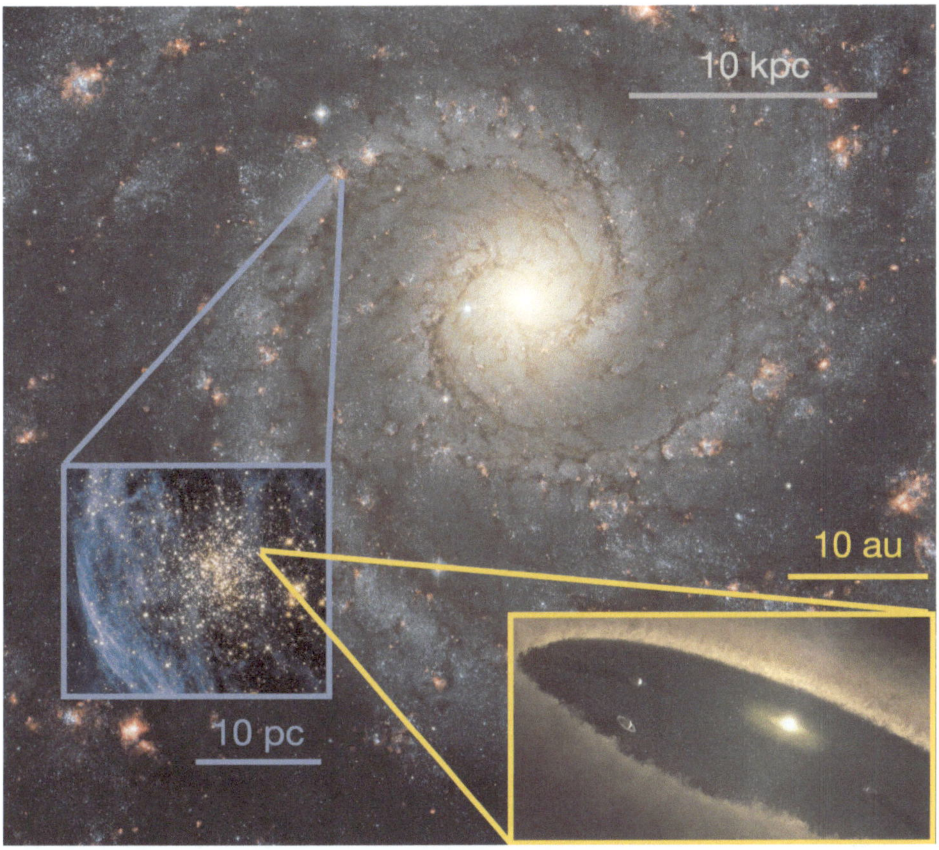

Figure 9.1. Hierarchy of scales illustrated with a hypothetical young planetary system (lower right: artist impression of HD 95086 by Spitzer Space Telescope, JPL, NASA; image credit: NASA/JPL–Caltech) around a star in a young stellar cluster (illustrated with NGC 1860; image credit: NASA, ESA and P. Goudfrooij (STScI); Processing: M. H. Özsaraç (Türkiye Astronomi Derneği)), in the galaxy M74 (image credit: ESA/Hubble & NASA, R. Chandar; CC BY 4.0). Images were taken from APOD https://apod.nasa.gov/apod.

(see Section 8.7.1) or a parameterized stellar evolution solver (see Section 5.2) into a gravitational N-body code. An example of a parent background method is approximating the galactic tidal field by an analytic potential within which a star cluster evolves. Such approximations are inflexible and prevent feedback between child and parent. Our goal in this chapter is to provide a flexible and self-consistent way to realize feedback between the child and parent systems.

In many dynamical cases of interest, the spatial and temporal scales of the child and parent systems are well separated, and we can split the underlying Hamiltonian for the combined system into rapidly and slowly changing parts. Child and parent evolution can then be followed simultaneously by adopting a kick–drift–kick (KDK) algorithm, allowing both scales to be integrated separately and the results combined into a consistent solution. If we now choose an accurate but expensive method for the child system and a cheap but less accurate method for the parent system, we may gain the ability to save computer time and tune the accuracy, depending on the scale at which we couple the solvers. The resulting *bridge method* is flexible and dynamic, in the sense that we can control the degree of coupling between solvers, and, if necessary, we can introduce a cascade of bridges to address hierarchical coupling, as we will demonstrate in the matrix of coupled codes in `Venice` (see Section 10.1.3).

This method can be generalized, both to higher order and to a variety of physical solvers. In principle, the method can have any even order (the odd orders cancel out), but the second-order coupling scheme is sufficiently accurate for most applications.

9.1.1 The Babylonian Integrator

In computational astrophysics, probably the most widely used symplectic integrator is the Verlet (leapfrog) method or Babylonian scheme described in Section 4.2.3. Although only of second order, its simplicity and long-term stability make it the method of choice for many applications. It is often implemented in "standard" symplectic form as a KDK scheme:

$$\begin{aligned} v' &= v_n + \frac{1}{2}a(x_n)\delta t \\ x_{n+1} &= x_n + v'\delta t \\ v_{n+1} &= v' + \frac{1}{2}a(x_{n+1})\delta t. \end{aligned} \tag{9.1}$$

A moment's reflection reveals that this is equivalent to the second-order predictor–corrector implementation described earlier. We note that, in practice, the acceleration used in the first kick of the next step is the same as that used in the last kick of the current step (see Equation (9.1)), meaning that only one acceleration need be computed per step, a fact of great importance in large N-body calculations.

An alternative (also symplectic) drift–kick–drift scheme is

$$\begin{aligned} x' &= x_n + \frac{1}{2}v_n\delta t \\ v_{n+1} &= v_n + a(x')\delta t \\ x_{n+1} &= x' + \frac{1}{2}v_{n+1}\delta t. \end{aligned} \tag{9.2}$$

This approach is much less widely used than its KDK cousin, mainly because of its poorer performance in applications where variable time steps are required (Springel 2005).

Higher-order symplectic schemes also exist (e.g., Yoshida 1990; Tricarico 2012; Rein & Spiegel 2015), and are sometimes employed in astrophysical contexts. The math for these methods goes beyond the scope of this book, but to be complete we explain the higher-order symplectic bridge in Section A.4. Finally, we note that time-reversibility in even a nonsymplectic scheme is sufficient to ensure that the obtained solution stays close to the true solution and that the error in the energy does not drift. Hut et al. (1995) describe an iterative algorithm to make any integrator time-reversible; it is implemented in the AMUSE `smallN` module.

9.1.2 The Bridge Method

In the derivation of the original Bridge scheme, Fujii et al. (2007) consider a star cluster (subscript c, indices i and j, total number N_c) orbiting a parent galaxy (subscript g, indices k and l, total number N_g). The internal orbits of stars in the cluster are integrated using a fourth-order Hermite scheme (Makino 1991) with direct summation of the interstellar gravitational forces. Interactions among the particles comprising the galaxy, and between galaxy particles and cluster stars, are computed using a hierarchical tree force evaluation method (Barnes & Hut 1986).

To proceed, we need a little dynamics (and in this context, we admit that we exaggerated a little in Section 1.1.3). The Hamiltonian H of the full system can be divided into two parts:

$$H = H_A + H_B, \tag{9.3}$$

where

$$H_A = -\sum_{k<l}\frac{Gm_{gk}m_{gl}}{r_{kl}} - \sum_{k=1}^{N_g}\sum_{i=1}^{N_c}\frac{Gm_{gk}m_{ci}}{r_{ki}} \tag{9.4}$$

is the total gravitational potential energy of the galaxy particles, including their mutual potential energy with the cluster stars, and H_B is the total kinetic energy of the system plus the total internal potential energy of the cluster stars,

$$H_B = \sum_{k=1}^{N_g}\frac{p_{gk}^2}{2m_{gk}} + \sum_{i=1}^{N_c}\frac{p_{ci}^2}{2m_{ci}} - \sum_{i<j}\frac{Gm_{ci}m_{cj}}{r_{ij}}. \tag{9.5}$$

The quantity p is momentum, and r_{ij} is the distance between cluster stars (or galaxy particles) i and j.

Hamilton's equations for a particle can be written as

$$\frac{df}{dt} = D_H f, \tag{9.6}$$

where $f = (\mathbf{x}, \mathbf{p})$ is a six-dimensional phase-space vector representing the particle's dynamical state (position $\mathbf{x}$ and momentum $\mathbf{p} = m\mathbf{v}$), and the differential operator D_H is defined by

$$D_H g \equiv \{g, H\} = \frac{\partial g}{\partial x}\frac{\partial H}{\partial p} - \frac{\partial g}{\partial p}\frac{\partial H}{\partial x} \tag{9.7}$$

for any time-dependent quantity $g(t)$. Here, the syntax $\{.,\ .\}$ denotes a Poisson bracket (see Landau & Lifshitz 1969; Section 42). The formal solution to Equation (9.6) is

$$f(t) = e^{tD_H} f(0), \tag{9.8}$$

or, for a single time step,

$$f(t + \delta t) = e^{\delta t D_H}\, f(t). \tag{9.9}$$

This solution is of little use by itself because we have no explicit way to evaluate it. However, writing $H = H_A + H_B$, and $D_H = D_A + D_B$, it is generally possible to derive an approximation to the exponential,

$$e^{\delta t\,(D_A + D_B)} = \prod_{i=1}^{K} \ e^{a_i \delta t\ D_A} e^{b_i \delta t\ D_B} + O(\delta t^{K+1}). \tag{9.10}$$

The approximate expression on the right side of Equation (9.10) is symplectic—that is, it preserves phase-space volume and hence is time-reversible (see Section 4.2.3.2 and Section 9.1.1)—and formulae for the coefficients a_i and b_i exist for all orders K, giving rise to a family of practical integration schemes known generically as symplectic integrators.

This all looks like rather scary mathematics, but the applications are really quite simple. For example, the second-order ($K = 2$) version of Equation (9.9) is

$$f(t + \delta t) = e^{\frac{1}{2}\delta t\ D_A}\ e^{\delta t\ D_B}\ e^{\frac{1}{2}\delta t\ D_A} f(t) \tag{9.11}$$

(i.e., $a_1 = \frac{1}{2}$, $b_1 = 1$, $a_2 = \frac{1}{2}$, $b_2 = 0$). In a system in which H_A is the total potential energy and H_B is the total kinetic energy of some dynamical system, it is easily shown that $e^{a\tau D_A}$ advances only the velocity by time $a\tau$,

$$\mathbf{v}(t + a\tau) = \mathbf{v}(t) + a\tau\ \mathbf{a}(t) \tag{9.12}$$

(a "kick"), where $\mathbf{a}$ is acceleration, while $e^{b\tau D_B}$ advances only the position,

$$\mathbf{x}(t + b\tau) = \mathbf{x}(t) + b\tau\ \mathbf{v}(t) \tag{9.13}$$

(a "drift"). Thus, the combination of operations in this case is just the KDK Verlet–leapfrog scheme discussed in Section 4.2.3.

Applied to the galaxy–cluster system, as discussed in more detail by Fujii et al. (2007), this scheme translates to the following algorithm (advancing from time "0" to time "1"):

1. Kick cluster star velocities with time step $\frac{1}{2}\delta t$ using accelerations due to the galaxy; kick galaxy particle velocities with time step $\frac{1}{2}\delta t$ using accelerations due to both galaxy and cluster:

$$\begin{aligned}
\mathbf{v}_c' &= \mathbf{v}_{c0} + \frac{1}{2}\delta t\ \mathbf{a}_{g\to c0} \\
\mathbf{v}_g' &= \mathbf{v}_{g0} + \frac{1}{2}\delta t\ \mathbf{a}_{\mathrm{all}\to g0}.
\end{aligned}$$

2. Update (drift) cluster positions and kicked velocities with time step δt, using the internal cluster integration scheme and kicked velocities; update galaxy particle positions with time step δt using kicked velocities:

$$
\begin{aligned}
\mathbf{x}_{c0} &\rightarrow (\text{high-order scheme}) \rightarrow \mathbf{x}_{c1} \\
\mathbf{v}'_{c} &\rightarrow (\text{high-order scheme}) \rightarrow \mathbf{v}'_{c1} \\
\mathbf{x}_{g1} &= \mathbf{x}_{g0} + \delta t\ \mathbf{v}'_{g}.
\end{aligned}
$$

3. Recompute accelerations and complete the step by updating (kicking) all cluster and galaxy velocities:

$$
\begin{aligned}
\mathbf{v}_{c1} &= \mathbf{v}'_{c1} + \frac{1}{2}\delta t\ \mathbf{a}_{g\rightarrow c1} \\
\mathbf{v}_{g1} &= \mathbf{v}'_{g} + \frac{1}{2}\delta t\ \mathbf{a}_{all\rightarrow g1}.
\end{aligned}
$$

This basic scheme carries over to the generalized Bridge class in AMUSE. Note that the overall scheme will be symplectic if and only if the higher-order scheme is symplectic, but, even if it is not, the interlaced symmetry of the algorithm still guarantees second-order accurate results.

9.1.3 Implementation of Bridge

The Bridge scheme can be written in `Python`, as presented in the code example below. The main `evolve_model` method has a loop over time in which three tasks are performed; the system is kicked for half a time step, drifted for the full-time step, and then kicked again (we saw a similar approach in Section 8.4.2). The source code is shown below:

```
def evolve_model(time, time_end, time_step):
    first = True
    while self.time < (time_end-time_step/2):
        if first:
            kick(system, partners, time_step/2)
            first = False
        else:
            kick(system, partners, time_step)
        drift(system, time+time_step)
        time = time+time_step
    if not first:
        kick(system, partners, time_step/2)
    return 0
```

The kick operator is composed of two steps. In the first loop, each system is synchronized, and then kicked with the function `get_gravity_at_point()` function in the second loop:

```
class GalacticCenterGravityCode(object):
    def __init__(self,R, M, alpha):
        self.radius=R
        self.mass=M
        self.alpha=alpha

    def get_gravity_at_point(self,eps,x,y,z):
        r2=x**2+y**2+z**2
        r=r2**0.5
        m=self.mass*(r/self.radius)**self.alpha
        fr=constants.G*m/r2
        ax=-fr*x/r
        ay=-fr*y/r
        az=-fr*z/r
        return ax,ay,az

    def circular_velocity(self,r):
        m=self.mass*(r/self.radius)**self.alpha
        vc=(constants.G*m/r)**0.5
        return vc
```

Code example 9.1: Semianalytic power-law description of the gravitational field in the inner few hundred parsecs of the Galactic center. The quantity `mass` is the mass within distance `radius` of the center, and `alpha` is the slope of the power-law mass distribution.

```
def kick(systems, partners, time_step):
    for subsystem in systems:
        if do_sync[subsystem]:
            subsystem.synchronize_model()
    for subsystem in systems:
        for y in partners[subsystem]:
            if subsystem is not y:
                kick(subsystem, y.get_gravity_at_point, time_step)
    return 0
```

Code example 9.1 gives an example of `get_gravity_at_point` for a static background potential.

The drift operator evolves each subsystem for the appropriate time step `time_step`:

```
def drift(system, time_end):
    for subsystem in systems:
        time_step = time_end - subsystem.model_time
        subsystem.evolve_model(time_step)
    return 0
```

9.1.4 Determining the Bridge Time Steps

One of the key elements and uncertainties in a bridge is the time step. Considerable research has gone into methods to automatically determine the bridge time step. It is not some parameter easily set and forgotten about; it is important to carefully choose its proper value.

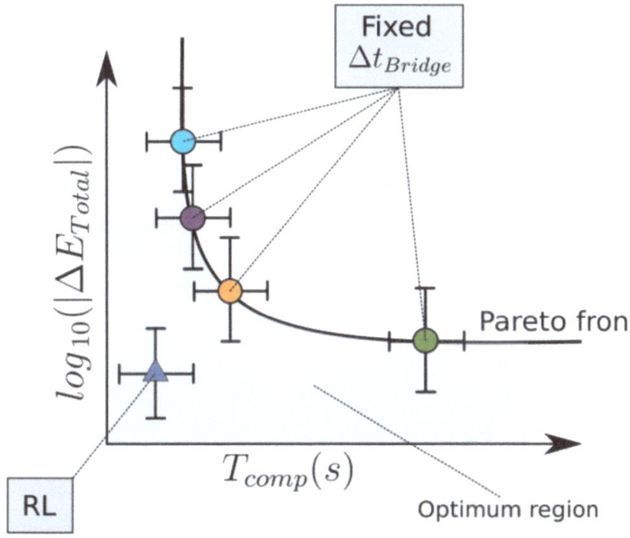

Figure 9.2. Perato curve for the bridge time step when tuned using reinforcement learning. The bullet points represent fixed bridge time steps, whereas the reinforcement algorithm allows bridge time steps to change with time. Figure from Saz Ulibarrena and Portegies Zwart (2025).

Over the last several years, we have experimented with autotuning bridge time steps, but the successes have been marginal. One method entails reinforcement learning, in which a trainable algorithm makes its own decisions on the time step size, considering the computational expense (wall clock time) and energy error produced by the integration (Saz Ulibarrena & Portegies Zwart 2025).

In Figure 9.2, we present part of the analysis as explored in Saz Ulibarrena & Portegies Zwart (2025), in which a reinforcement algorithm was used to find the optimum bridge time step. The two axes represent integration (wall clock) time and energy error. The curve shows the trade-off between the two if we reduce the bridge time step.

The automated routine to determine the bridge time step, using reinforcement learning, is interesting and in principle powerful. If fully operational, it would have two major advantages: 1) it would not require any expert knowledge to choose a bridge time step, and 2) the calculation would be faster with a smaller overall energy error. In practice, however, it still turns out to be hard to find an optimal choice. For now, we are not really encouraging using reinforcement learning to determine your bridge time step, but in the near future we hope that this method can be made into a workable module.

9.2 Using Bridge

Fortunately, you do not have to code all the intricate details of the Bridge method yourself. AMUSE has a `bridge` class that takes care of the details of implementing the classic and higher-order Bridge schemes just described. We illustrate this by developing a simple model of a star cluster moving in an analytic model of the potential near the Galactic center (McMillan & Portegies Zwart 2003; Harfst et al. 2010). Then, we show how this model can be extended easily to reproduce the seminal results of Fujii et al. (2007), which represented the first fully self-consistent treatment of star clusters near the Galactic center.

9.2.1 Star Cluster in a Static Galactic Potential

Code example 9.1 presents a simple class that provides a power-law analytic approximation to the mass distribution and gravitational field within a few hundred parsecs of the Galactic center,

$$\begin{aligned} M(r) &= M_{\rm gal}\left(\frac{r}{R_{\rm gal}}\right)^{\alpha} \\ a(r) &= -\frac{GM(r)}{r^2}, \end{aligned} \tag{9.14}$$

where $M(r)$ is mass interior to radius r and we take $M_{gal} = 1.6 \times 10^{10}\,{\rm M}_{\odot}$ within radius $R_{gal} = 1\,{\rm kpc}$, and $\alpha = 1.2$ (Mezger et al. 1996). We initialize the class with

```
galactic_center = GalacticCenterGravityCode(Rgal, Mgal, alpha)
```

Key to this (and any) bridge class is the function `get_gravity_at_point()`, which takes a position in space and a softening length as arguments, and then returns the gravitational acceleration at the specified location. This function allows any gravity code to be bridged with any other in AMUSE. (In this particular case, the softening length is ignored, but the argument is required by bridge because it is used by other codes for which softening may be necessary.)

Next, we initialize the star cluster by starting an N-body code and generating a realization of particle masses, positions, and velocities. In the example in Code example 9.2, we initialize a code called `nbodycode`, which could be either a direct N-body code, a tree code, or a gravity-enabled SPH code. The cluster is generated from a King (1966) model with `N` particles, total mass `Mcluster`, virial radius `Rcluster`, and a dimensionless potential depth `W0` (see also Section 4.1.3.1). This is illustrated in Code example 9.3. Additional parameters for the N-body code can be passed using the argument `parameters`. In this case, we provide the softening parameter as an argument because we will run this script with a tree code, which requires softening.

The bridge is created by constructing a new `gravity` solver that combines the N-body code `cluster_code` for the star cluster and the analytic `galaxy_code` for the galactic background. The bridge is declared by

```
gravity = bridge.Bridge()
```

```
Rgal = 1. | units.kpc
Mgal = 1.6e10 | units.MSun
alpha = 1.2
galaxy_code = GalacticCenterGravityCode(Rgal, Mgal, alpha)

m=galaxy_code.mass*((100|units.pc)/galaxy_code.radius)**galaxy_code.alpha
cluster_code = make_king_model_cluster(BHTree, N, W0, M, R,
                                       parameters=[("epsilon_squared",
                                                    (0.01 | units.parsec)**2)])

stars = cluster_code.particles.copy()
stars.x += Rinit
stars.vy = 0.8*galaxy_code.circular_velocity(Rinit)
channel = {"from_framework": stars.new_channel_to(cluster_code.particles),
           "to_framework": cluster_code.particles.new_channel_to(stars)}
channel["from_framework"].copy_attributes(["x","y","z","vx","vy","vz"])

plot_cluster(f, stars.x.value_in(units.pc), stars.y.value_in(units.pc), c='r')

system = bridge(verbose=False)
system.add_system(cluster_code, (galaxy_code,))

times = quantities.arange(0|units.Myr, tend, timestep)
xcom = [] | units.pc
ycom = [] | units.pc
for i,t in enumerate(times):
    system.evolve_model(t,timestep=timestep)
    channel["to_framework"].copy_attributes(["x","y","z","vx","vy","vz"])
    com = stars.center_of_mass()
    xcom.append(com[0])
    ycom.append(com[1])

plt.plot(xcom.value_in(units.pc), ycom.value_in(units.pc), lw=1, c='k')

x = system.particles.x.value_in(units.parsec)
y = system.particles.y.value_in(units.parsec)
cluster_code.stop()
```

Code example 9.2: Initializing the gravity code to be coupled via bridge to a background potential.

and subsequently we identify the codes to be bridged with

```
gravity.add_system(cluster_code, (galaxy_code,))
```

This line adds `cluster_code` as a code to be bridged, and indicates that it interacts with `galaxy_code`. In general, the second argument is a list of codes with which the first code interacts.[1]

In this example, the bridge is unidirectional: the dynamical evolution of the particles in `cluster_code` code is affected by `galaxy_code`, but not the other way around. Multiple codes can be added to a bridge, and their interactions tailored

[1] A `Python` aside: The trailing comma after `galaxy_code` turns the single argument into a tuple that can be evaluated. Forgetting the trailing comma will lead to an error with the inscrutable and somewhat intimidating message "`object is not iterable`". It is only needed when there is only one item in the list.

```
def make_king_model_cluster(nbodycode, N, W0, Mcluster,
                            Rcluster, parameters = []):

    converter=nbody_system.nbody_to_si(Mcluster,Rcluster)
    bodies=new_king_model(N,W0,convert_nbody=converter)

    code=nbodycode(converter)
    for name,value in parameters:
        setattr(code.parameters, name, value)
    code.particles.add_particles(bodies)
    return code
```

Code example 9.3: Coupling a gravity code to a background potential using a bridge. The last three lines store the stars x and y positions and terminate the code. In the example, we adopted the following parameters: `number_of_particles = 1024`, `king_w0 = 3`, `distance_initial = 50.0 | units.pc`, `mass_cluster = 5.0e4 | units.MSun`, and `radius_cluster = 0.8 | units.pc`.

```
class CodeWithFriction(bridge.GravityCodeInField):

    def kick_with_field_code(self, particles, field_code, dt):
        self.LnL = 3.7

        R = particles.position.length()
        vx = particles.vx.mean()
        vy = particles.vy.mean()
        vz = particles.vz.mean()
        rho = field_code.density(R)
        vc = field_code.circular_velocity(R)
        X = 0.34
        m = particles.mass.sum()
        ax = -4*np.pi*self.LnL*constants.G**2 * rho*m*(vx/vc**3)*X
        ay = -4*np.pi*self.LnL*constants.G**2 * rho*m*(vy/vc**3)*X
        az = -4*np.pi*self.LnL*constants.G**2 * rho*m*(vz/vc**3)*X
        self.update_velocities(particles, dt, ax, ay, az)

    def drift(self, tend):
        pass

def get_system_state_relative_to_SMBH(time, black_holes):
    t = [] | units.Myr
    r = [] | units.pc
    SMBH = black_holes[black_holes.name=="SMBH"]
    for bi in black_holes-SMBH:
        t.append(time)
        r.append((bi.position-SMBH.position).length())
    return t, r

def get_system_state(time, black_holes):
    t = [] | units.Myr
    r = [] | units.pc
    for bi in black_holes:
        t.append(time)
        r.append(bi.position.length())
    return t, r
```

Code example 9.4: Example class of a code for adding an extra functionality to a bridge. In this case, for calculating the dynamical friction term. The source code is available in `amuse.examples.imbhs_with_friction`.

to the problem at hand. This can be very effective if, for example, our galaxy code is composed of several separate parts—for the bulge, bar, and disk, say—and we wanted to model several clusters in the bridge:

```
gravity.add_system(cluster1, (cluster2, cluster3, bulge, bar, disk))
gravity.add_system(cluster2, (cluster1, cluster3, bulge, bar, disk))
gravity.add_system(cluster3, (cluster1, cluster2, bulge, bar, disk))
```

and so on.

Once the bridge is initialized and the integrators added, we set the bridge time step:

```
gravity.timestep = 0.1 | units.Myr
```

It often requires some experimentation to determine what this time step should be. Too long a time step results in inaccurate inclusion of the coupling, whereas too short a time step may make the calculation too expensive. It is possible to change the bridge time step at runtime (see Section 10.1.3), but there is currently no ready algorithm to determine its optimum value.

In the event loop over `time`, we now call `evolve_model` from the bridged `gravity` solver

```
gravity.evolve_model(time)
```

Synchronization between the internal code particles and those in the `Python` script is realized via channels. The event loop is presented in Code example 9.1.

Figure 9.3 shows the results of a simulation of the orbital evolution of the stars in the Arches cluster in the field of the Galactic center. The 50,000 $M_\odot$ star cluster was represented by 1024 equal-mass particles (not a good approximation to the observed mass function of the Arches!) initially represented by a virialized King model (King 1966) with a radius of 0.8 pc. The cluster had a noncircular orbit in the Galactic field, starting at a distance of 50 pc. We adopted a Barnes–Hut tree code with a rather large softening of 0.1 pc, to allow us to quickly integrate the motion of the stars in the cluster without getting too distracted by the details of the internal cluster dynamics.

9.2.2 Adding Extra Functionality with a Kick to a Bridge

Sometimes one would like to add a functionality to the bridge operator in order to kick the system code. A simple example can be considered with the dynamical friction on a black hole of intermediate mass due to its orbit through a galactic nucleus. An example is presented in Code example 9.4. We discussed a similar issue in Section 4.1.2.3, but without addressing how to add such a term to the integration of an N-body system.

In principle, this process does not work much more differently than the bridging of two codes, except that in this case one of the codes has no `particle` attribute and does not need to be kicked nor drifting itself. In the example, below, we initiate the `Hermite` integrator two intermediate mass black holes that orbit in a background potential. This example is not very different than what we have already

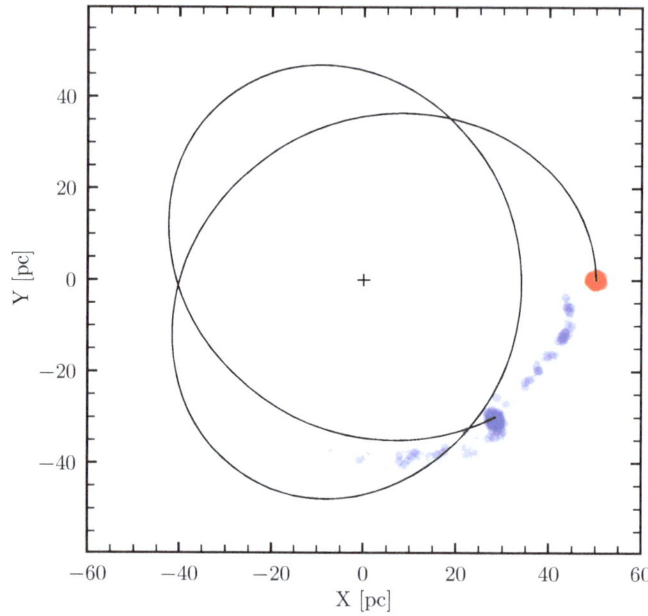

Figure 9.3. Snapshots of an Arches-like cluster orbiting near the Galactic center. Initially (in red) the cluster was composed of 5000 particles taken from a Salpeter mass function over two orders of magnitude with a mean mass of 10 $M_\odot$ leading to a total mass of 50,000 $M_\odot$, distributed as a King model in virial equilibrium with $W_0 = 3$ and a virial radius of 0.8 pc. The center of mass of the cluster was placed 50 pc from the Galactic center (along the x-direction) with 80% of the circular velocity (in the y-direction). A softening of 0.1 pc was introduced for all particles in the cluster. The integration was performed using the background galactic potential presented in Code example 9.1. The initialization and main event loop are shown in Code example 9.3 and Code example 9.2, respectively. The equations of motion of the cluster are integrated using the `BHtree` code. The black curve follows the cluster's center of mass, ending in the final snapshot (in blue). The source code for the figure is `amuse.examples.gravity_potential`.

discussed in Section 9.2.1. The difference is that we would like to exert a force on the black holes due to the effect of dynamical friction exerted on the black holes. We simply start the N-body code (here `Hermite`), and add the black holes. Then, we start the `DynamicalFriction` code, and bridge them together,

```
code1 = Hermite()
code1.particles.add_particles(cluster)
code2 = DynamicalFriction(Coulomb_logarithm=3.7)
system = Bridge(timestep=0.1 | units.Myr)
system.add_system(code1)
system.add_code(code2)
system.evolve_model( 10 | units.Myr )
```

The dynamical friction solver comprises of two parts: a Plummer potential and a dynamical friction term. We have created a small class to achieve this, see Figure 9.4.

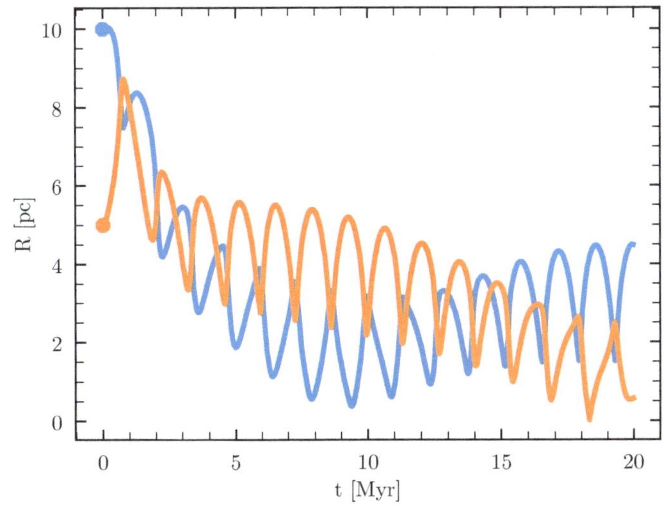

Figure 9.4. Time evolution of Galactocentric distance R of two massive compact objects of 50 000 $M_\odot$ each from a distance of 10 pc (blue) and 5 pc (orange). Both curves were integrated using a bridge between `ph4` for integrating the black holes and a background potential with dynamical friction using $\ln(\Lambda) = 3.7$. The sinusoidal variations along the curves are the result of the two black holes not being in a perfect circular orbit. The interference between the two black holes is a consequence of their mutual gravitational interactions. The source code takes a few seconds to run and can be found at `amuse.examples.imbhs_with_friction`

Since the `Hermite` code is fourth order, we use the symbol S_4 (for stars) for the subsystem. The potential is indicated with the letter G (for Galaxy). The bridge is indicated with the left-pointed arrow, and the augmentation is indicated with the letter f (for function) below the bridge arrow. The notation for the augmented bridge in which a fourth-order direct N-body code is bridged with a potential including an additional function for (in this case for the dynamical friction term) then has the following notation: $[S_4 \underset{f}{\leftarrow} G]$. this may look somewhat arcane, at first, but once augmented, bridge become hierarchical, multilayers or more complex in other means, this notation captures these complexities nicely.

A star cluster in the Galactic center will feel a frictional force due to the accumulation of stars in its wake. McMillan & Portegies Zwart (2003) used a direct integration technique to study the timescale on which the Arches star cluster sinks from its birth location at about 30 pc from the Galactic center towards the middle of the Milky Way. It turns out that the timescale depends quite sensitively on the Coulomb parameter $\ln(\Lambda)$. In their Figure 6, they present the results of several calculations for the distance to the Galactic center as a function of time for a black hole (or star cluster) of 50 000 $M_\odot$. In Figure 9.4, we repeat their calculation but using an external function included in the bridge step

$$\left[S_4 \underset{f}{\leftarrow} G\right]. \tag{9.15}$$

The equation to solve is $f = f(S, G)$ where we calculate the acceleration of the individual particles in the N-body code due to dynamical friction (see Equation 7 of McMillan & Portegies Zwart 2003),

$$\overrightarrow{a_f} = -4\pi \ln(\Lambda) G^2 \rho m \frac{\overrightarrow{v_c}}{v_c^3} \chi. \tag{9.16}$$

Here, G is the gravitational constant, ρ the local stellar density, with mass m, $\chi \simeq v_c/\sqrt{2}\,\sigma$ (with σ the local one-dimensional velocity dispersion), and $\ln(\Lambda) \simeq \ln(r)/\ln(R)$ is the Coulomb logarithm. Here, r is the distance of the stars to the potential center (which has size R) and $\overrightarrow{v_c}$ is the particle's velocity vector. The acceleration ($\overrightarrow{a_f}$) is negative because dynamical friction results in a drag force leading to deceleration.

The class performing this calculation together with hovering the stars on the background potential is presented in Figure 9.4.

In this example, we included dynamical friction as an additional drag force in the kick of the bridging step. Alternatively, we could include a term in the drift-part of the bridge step. Including additional forces to a bridge can be realized in a similar way by adding an extra term to the kick operator. In this way, constructing an N-body solver that includes the YORP- and Yarkovski-effects,[2] dynamical friction, tidal evolution, gas-drag, or post-Newtonian dynamics becomes relatively straightforward.

```
friction_code = CodeWithFriction(
    cluster_gravity,
    (potential,),
    do_sync=True,
    verbose=False,
    radius_is_eps=False,
    h_smooth_is_eps=False,
    zero_smoothing=False,
)
gravity = bridge.Bridge(use_threading=False)
gravity.add_system(cluster_gravity, (potential,) )
gravity.add_code(friction_code)
gravity.timestep = dt_bridge
```

9.2.3 The Classic Bridge

Now that we have seen some of the issues involved in setting up and using a simple Bridge system, we can quite easily create a bridge to reproduce the results of Fujii et al. (2007) on the dynamical evolution of a star cluster orbiting the Galactic center. We start by creating a classic Bridge combining a fourth-order Hermite predictor–corrector scheme for the cluster and a Barnes–Hut tree code for the Galaxy. After declaring the codes to be bridged—`ph4` and `bhtree`—we create the bridge solver and add the codes:

[2] Named after Ivan Osipovich Yarkovsky (1844–1902), see also Whipple (1950) and Bottke et al. (2002).

```
code1 = Ph4()
code2 = Bhtree()
combined_gravity = bridge.Bridge()
combined_gravity.add_system(code1, (code2,))
combined_gravity.add_system(code2, (code1,))
```

In this case, we have created a bidirectional solver, in which each code interacts with the other.

The simulation is then performed much as before:

```
combined_gravity.timestep = 0.1 | units.Myr
combined_gravity.evolve_model(10 | units.Myr)
```

Figure 9.5 presents an example calculation of the Arches cluster. It is similar to the calculation in Figure 9.3, but now we have adopted a particle representation for the Galactic center region and solved the system using the tree code `BHtree`, in much the same way as presented in Fujii et al. (2007). The Arches cluster was initialized with a King model with a dimensionless depth of $W_0 = 3$ and a virial radius of 0.8 pc, as in Figure 9.3. The cluster was placed in an almost circular orbit at a distance of 50 pc from the Galactic center. The latter was represented by a Plummer (1911) sphere with a total mass of 1.6×10^{10} $M_\odot$ and a scale radius of 100 pc (see Section 3.2).

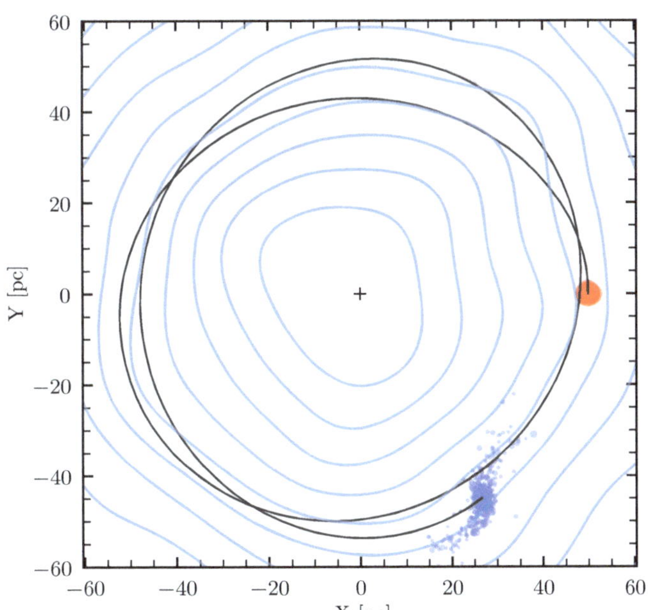

Figure 9.5. Calculation of a star cluster (using `ph4`) in a galaxy (using `BHTree`) for initial conditions comparable to those in Figure 9.3, but now we actively evolve the galaxy using 10^4 equal-mass particles. The star cluster was represented by 1000 point masses drawn from a Salpeter mass function between 0.1 and 100 $M_\odot$. This star cluster, as in Figure 9.3, was born at a distance of 50 pc from the galactic center (along the X-axis) and with a velocity of 80% of the circular velocity (in the Y-direction). Evolving to an age of 2.5 Myr took about 3 minutes. The source code for the figure is available in `amuse.examples.gravity_gravity`.

9.2.3.1 Bridging with a Drifter Code

In many cases, it is overkill to run a direct N-body code or tree code in a background potential, particularly if the particles hardly interact, as is the case with planetesimals in a planetary system or individual stars in a background galactic potential. In such cases, it is preferable to use a small `Python` script that just drifts the particles in the background potential, rather than calculating the (negligible) interparticle forces and integrating the orbits. In Code example 9.5, we present an example of such a simple drift code, which we use here to integrate a low-mass star cluster in a semianalytic Galactic potential. The drift code itself can be used in the same way as any other gravity code. Note that we use the `Python @property` built-in function which allows our small class to remain compatible with past and future versions (see Section C.3).

The example in Code example 9.5 is far from optimal, and the drifting can easily be done in parallel (for an example, see Section A.7.4) or using a graphical processing unit. This would require somewhat more elaborate functions, and you might want to explore the possibility of using

```
from amuse.community.fastkick import Fastkick
```

which is a more efficient and optimized implementation of the kick operation using a GPU. An example of the use of `FastKick` is given in Section 9.3.2.1.

Although the script (which is referenced in Code example 9.5) uses a different model for the background Galactic center potential adopted for producing

```
class drift_without_gravity:
    def __init__(self, convert_nbody, time=0 | units.Myr):
        self.model_time = time
        self.convert_nbody = convert_nbody
        self.particles = Particles()

    def evolve_model(self, t_end):
        dt = t_end - self.model_time
        self.particles.position += self.particles.velocity * dt
        self.model_time = t_end

    @property
    def potential_energy(self):
        return quantities.zero

    @property
    def kinetic_energy(self):
        return (
            0.5 * self.particles.mass * self.particles.velocity.lengths() ** 2
        ).sum()

    def stop(self):
        pass
```

Code example 9.5: Simple drift code to replace the N-body code in a background potential where interactions between particles are unimportant. The source code is available in `amuse.examples.gravity_drifter`.

Figure 9.3, the results are also different due to the lack of self-gravity among the stars. As a consequence, the stars stay together longer and the cluster disperses more slowly. We do not recommend using this script for simulating star clusters in a background potential, but it is a handy tool for integrating the orbits of individual (non-self-interacting) particles in the galactic potential.

9.3 Bridging Other Codes

Bridging high-precision gravity with low-precision gravity, or gravity with a background field, is a very effective coupling strategy. Bridge is guaranteed to work in these cases because we can write down a common Hamiltonian for the coupled systems, but Bridge is much more versatile than this. Coupled systems for which we cannot write down a coupled (or any) Hamiltonian can also be integrated using the classic bridge scheme. We could just as easily have replaced the analytic calculation by another N-body code or a hydrodynamical solver, or bridged one hydrodynamical solver with another, or included radiative transfer. One needs a little courage to do this because we are unaware of any formal proof that this procedure results in a unique or "correct" solution to the coupled problem. On the other hand, resolution and convergence tests indicate that the solutions obtained in this way do appear to be reliable and accurate, and therefore suitable for scientific interpretation. We leave a complete proof to the brave reader, and should one be forthcoming we will be happy to include it in future editions!

The range of applications of Bridge is up to the imagination of the researcher. In the next section, we present a few bridge topologies you might want to experiment with.

9.3.1 Bridge Hierarchies and Hierarchical Bridges

The simplest possible bridge can be realized by integrating two single particles, such as the Sun and the Earth, and following each—not with a regular integrator, but using Bridge directly. Classic Bridge works as a second-order Verlet–leapfrog integrator, which is perfectly adequate for integrating the solar system, or (almost) any N-body system. An example of how to integrate such a two-body system is given in Code example 9.6.

```
    star_gravity = Ph4(converter)
    star_gravity.particles.add_particle(ss[0])

    planet_gravity = Ph4(converter)
    planet_gravity.particles.add_particle(ss[1:])

    channel_from_star_to_framework = star_gravity.particles.new_channel_to(ss)
    channel_from_planet_to_framework = planet_gravity.particles.new_channel_to(ss)

    gravity = bridge.Bridge(use_threading=False)
    gravity.add_system(star_gravity, (planet_gravity,))
    gravity.add_system(planet_gravity, (star_gravity,))
```

Code example 9.6: Bridge construction for integrating the Sun–Earth system.

In this example, we declare a bridge with two components: one for the Sun, which is affected by Earth's gravity, and one for Earth, which is affected by the gravity of the Sun. The eventual integration is carried out in a loop over time

```
while time < time_end:
    time += time_step
    gravity.evolve_system(time)
```

in exactly the same way as a more usual integrator or N-body code would be used within the framework. It may seem strange to start an integrator (in this case `ph4`) and have it integrate a single particle (and indeed, some integrators will not even let you do this), but here the integrator simply performs the drift step and defines `get_gravity_at_point()`, letting the bridge complete the integration of the equations of motion. It probably makes far more sense to use the drifter code in Code example 9.5 instead of the direct N-body code.

We could make this example even stranger by adding the Moon as a third particle and constructing a bridge for all three elements (Sun, Earth, and Moon) as in Code example 9.7, a simple code to integrate a three-body system using second-order bridge. As you might imagine, this is not a very efficient way to integrate the Sun–Earth–Moon system. In fact, for comparable overall accuracy, our three-body bridge is almost 600 times slower than a direct integration using `ph4`, mainly because it is all done in `Python`, whereas `ph4` is compiled C++ (Stroustrup 2013).

We could extend the list of arguments indefinitely, even hierarchically, making the system quite complex. In Code example 9.8, we demonstrate how a hierarchy of

```
    star_gravity = Ph4(converter)
    star_gravity.particles.add_particle(star)

    planet_gravity = Ph4(converter)
    planet_gravity.particles.add_particle(planet)

    moon_gravity = Ph4(converter)
    moon_gravity.particles.add_particle(moon)

    channel_from_star_to_framework = star_gravity.particles.new_channel_to(ss)
    channel_from_planet_to_framework = planet_gravity.particles.new_channel_to(
        ss)
    channel_from_moon_to_framework = moon_gravity.particles.new_channel_to(ss)

    time = 0 | units.yr
    write_set_to_file(ss, filename,
                      timestamp=time, overwrite_file=True)

    gravity = bridge.Bridge()
    gravity.add_system(star_gravity, (planet_gravity, moon_gravity))
    gravity.add_system(planet_gravity, (star_gravity, moon_gravity))
    gravity.add_system(moon_gravity, (star_gravity, planet_gravity))
```

Code example 9.7: Bridge construction for integrating the Sun–Earth–Moon system. We call this method the multibridge; its topology is at the bottom left of Figure 9.6.

```
sp_gravity = bridge.Bridge()
sp_gravity.add_system(moon_gravity, (planet_gravity,))
sp_gravity.add_system(planet_gravity, (moon_gravity,))

gravity = bridge.Bridge()
gravity.add_system(sp_gravity, (star_gravity,))
gravity.add_system(star_gravity, (sp_gravity,))
```

Code example 9.8: Hierarchical bridge for integrating the Sun–Earth–Moon system. The majority of the code is identical to Code example 9.7 except that the last four lines there should be replaced with these lines to initialize the hierarchical bridge. This hierarchical-bridge topology is depicted on the bottom right of Figure 9.6.

Table 9.1. Performance of Various Bridge Implementations of the Sun–Earth–Moon Problem

	Code Example	100 yr Integration		
Code		$t_{\rm wall}$ [s]	$d\Delta E_{\max}$	$\Delta E_{\rm tot}$
ph4		4.40	4.21×10^{-5}	1.61×10^{-6}
ph4 normalized	4.1	1	1	1
Bridge with ph4	9.6	4.0	0.52	40
Three-body bridge	9.7	8.4	0.57	36
Hierarchical bridge	9.8	12.5	0.57	36
TBB fourth order	A.4	13	1.13	62
TBB sixth order	A.4	4.4	0.52	31
TBB eighth order	A.4	40	0.78	47

See also Figure 9.7. Values are relative to `ph4` as a standard, using a `timestep_parameter` of 0.15 (first line), which gives 4.40 s for the wall clock time, $\Delta E_{\max} = 4.21 \times 10^{-5}$ for the energy error, and $\Delta E_{\rm tot} = -1.61 \times 10^{-6}$ for the maximum energy error per output time step (0.1 yr) and the total energy error for integration over 20 years, respectively. In the following lines, we normalize the performance of various bridged integrators (using `ph4`), including higher-order bridges. The latter are generally more expensive, but have a favorable energy error per bridge time step. The various scripts are in `amuse.examples` and are called `no_bridge_ph4_integrator`, `three_body_bridge` (in the table named TBB), `three_body_bridge_hierachical`, and `three_body_bridge_n_order`.

level two can be achieved. Various topologies of bridge constructs are presented in Figure 9.6. The bridge for realizing the interaction between Sun and Earth is again bridged with the Moon, leading to a single composite bridge that integrates the entire system. In this case, the `ph4` instantiations take care of the `get_gravity_at_point()` used in the various bridges. This implementation is even slower than the example in Code example 9.6, but no matter—its purpose is to illustrate the method.

The relative performance of the various Bridge implementations just described is presented in Table 9.1. The three bridged methods use `ph4` in various ways, in combination with a bridge. The hierarchical-bridge method uses three nested bridges. The various bridge topologies are further illustrated in Figure 9.1.

Performance may be much better for other bridge topologies or different problems, but the key message here is that Bridge allows considerable flexibility that may compensate for (some) loss of performance in many circumstances.

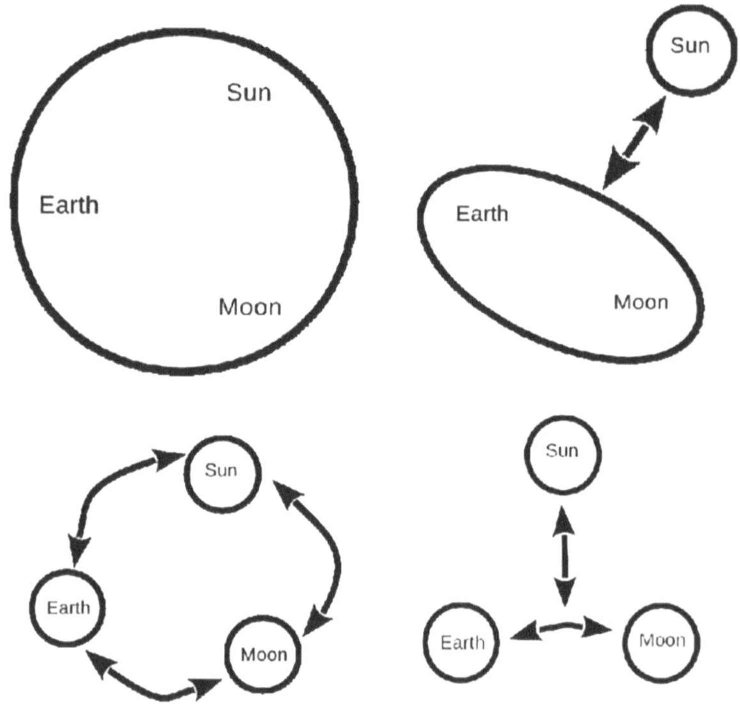

Figure 9.6. The various `Bridge` and `ph4` instantiations in Table 9.1. Top left indicates the use of a single N-body integrator for all three particles. At top right we present a two-way bridge (indicated by the two-pointed arrow) between an N-body realization for only the Sun, and separately for Earth and the Moon. At bottom left we show a three-body bridge, in which each particle is separately handled by an N-body integrator, and each is separately added to a bridge, see Code example 9.7. The bottom right depicts the hierarchical bridge, see Code example 9.8.

Of course, we do not want performance to drop by a factor of 600, but in practical situations, when running computer-intensive production-quality simulations, the overhead of such complex bridging strategies is actually rather small—generally, not more than a few percent.

Figure 9.7 compares the evolution of the energy error $(E - E_0)/E_0$ of `ph4` with those of the three bridge topologies listed in Table 9.1. To make the comparison more honest, we degraded the `ph4` time step by using a value of 0.6 for `timestep_parameter`. (This parameter is normally set to 0.1, which would have given a relative energy error better than 10^{-12} per step.) It is interesting to note that the energy error in the nonsymplectic fourth-order Hermite integrator `ph4` accumulates, whereas the energy errors in the symplectic bridged systems show no secular evolution. The performance of the fourth- and sixth-order bridge methods is quite poor. At present, these are implemented in `Python`, which is rather inefficient. Porting these to compiled C++ or FORTRAN has not been a priority because, normally, little time is spent in the bridge, except for these somewhat contrived examples.

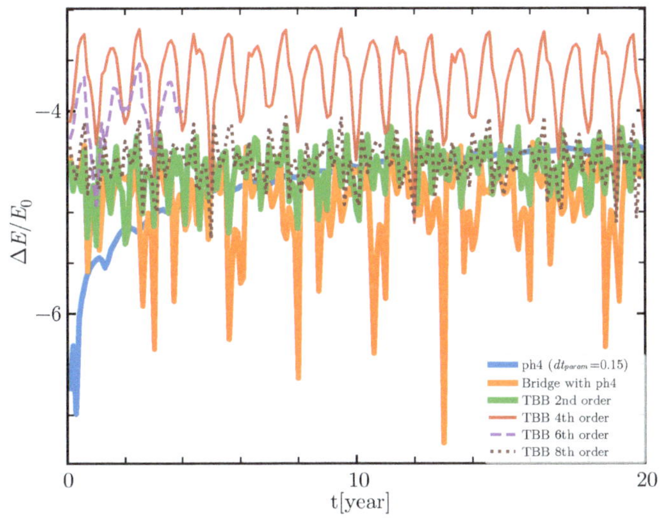

Figure 9.7. Evolution of the energy error while integrating the Sun–Earth–Moon system using a variety of bridge integrators. We use ph4 with a time step parameter $\eta = 0.15$ (blue curve) as a reference, which clearly shows a secular increase of the energy error in the calculation. The orange curve gives Earth–Moon in ph4 but subsequently bridged with the Sun. The other curves are calculated with all particles individually bridged with a second, fourth, sixth and eighth order bridge. Interestingly, the second-order bridge performs best for this case. The source code for reproducing this figure can be found in `amuse.examples.plot_sun_and_earth_and_moon`, but the worker codes are presented in Figure 9.6.

The example of a bridge coupling or a hierarchical bridge may appear a bit artificial for integrating the Sun–Earth–Moon system, but for other applications, the strict separation of codes, while combining them either sequentially or hierarchically, creates many interesting possibilities. In that sense, the timings listed in Table 9.1 are not representative of the ultimate performance of a more complex simulation in which each of the individual modules is more compute intensive. We encourage the reader to experiment with bridge until bridging is as familiar an operation as eating chicken pie (Cooder 1972).

If a higher-order bridge is needed, it can be easily implemented by

```
from amuse.couple import bridge
import amuse.ext.composition_methods as CM
gravity = bridge.Bridge(use_threading=False, method=CM.SPLIT_4TH_S_M4)
```

Here, `composition_methods` provides a library for different bridge-coupling strategies (see Section A.4.2). The first free parameter in such complex bridge couplings is the number of evaluations of the same bridge step (m). For an example, see Equation (A.10), where we describe a sixth-order bridge with $m = 11$ force evaluations per step. The other parameter is the number of symplectic operations (S, see also Section A.4). The argument `method` in the `Bridge` declaration indicates the specific bridge scheme to use, in this case a fourth-order method with $m = 4$; for a sixth-order method one might instead adopt SPLIT_6TH_SS_M13 (see Table A.16 in Section A.4.2).

An example of a higher-order bridge is given in amuse.examples.three_-body_bridge_order4m4. The default method is set to LEAPFROG, the classic second-order bridge. Note that, as discussed in Section A.4, for codes to be bridgeable to an order higher than second, they must be able to run with negative time steps (stepping backward in time). Not all codes in AMUSE comply with this requirement: those that cannot accommodate negative time steps, such as the Barnes–Hut tree code, can only be bridged to second order.

9.3.2 Bridging Gravity with Hydrodynamics

Many hydrodynamics solvers allow the possibility of adding dark matter particles (called dm_particles in AMUSE, as supposed to gas_particles; see Section 6.3.3). The dm_particles have a gravitational effect on the rest of the system, but no hydrodynamical interactions. But the numerical solver used to integrate the equations of motion for dm_particles is generally identical to the hydrodynamics solver, which typically is a second-order Verlet integrator with a hierarchical tree to evaluate the forces. These integrators are not very accurate, and they are generally not used for collisional N-body systems. Collisional gravitational problems need more accurate N-body integrators, of which many are available in AMUSE. In principle, if we could couple the high-order gravity modules with the hydrodynamics modules, we could have the best of both worlds: the high accuracy of a fourth-order predictor–corrector Hermite integrator (e.g., ph4), along with the hydrodynamics solver of choice (e.g., Fi). In AMUSE we can do this using Bridge.

Bridging gravity with hydrodynamics is not much different than bridging gravity with gravity, as discussed in Section 9.1. Essentially the same bridge technique can be adopted, and the script will not look much different from one in which two or more gravitational dynamics codes are coupled. We could replace ph4 in Code example 9.1 with any of the SPH codes in AMUSE without any further changes. Code example 9.9 presents an example of a coupling between an SPH code and a high-order direct N-body code. Here, we generalize the initialization of the hydro and the gravity codes by making them into classes. (For a brief overview of Python classes, we refer to Section C.3.7.) This requires a base class, as presented in Code example 9.10, and two specific classes for the gravity and hydrodynamics solvers. These derived classes are presented in Code examples 9.11 and 9.12. Section 9.4.3 presents an example in which we integrate the equations of motion of three stars using an N-body code, while we address the gas dynamics with an SPH code.

9.3.2.1 Using Bridge to Relax Multicomponent Systems

In Section 8.5.4.1, we discussed how to use a Henyey code to generate hydrodynamical (SPH) representations of two stars, which we then smashed into one another. After generating hydrodynamical blobs for each star, we relaxed both to a

```
def gravity_hydro_bridge(Mprim, Msec, a, ecc, t_end, n_steps,
                         Rgas, Mgas, Ngas):

    stars = new_binary_from_orbital_elements(Mprim, Msec, a, ecc,
                                             G=constants.G)
    eps = 1 | units.RSun
    gravity = Gravity(ph4, stars, eps)

    converter = nbody_system.nbody_to_si(1.0|units.MSun, Rgas)
    ism = new_plummer_gas_model(Ngas, convert_nbody=converter)
    ism.move_to_center()
    ism = ism.select(lambda r: r.length()<2*a,["position"])
    hydro = Hydro(Fi, ism, eps)
    model_time = 0 | units.Myr
    filename = "gravhydro.hdf5"
    write_set_to_file(stars.savepoint(model_time), filename, 'amuse')
    write_set_to_file(ism, filename, 'amuse', append_to_file=True)

    gravhydro = bridge.Bridge(use_threading=False)
    gravhydro.add_system(gravity, (hydro,))
    gravhydro.add_system(hydro, (gravity,))
    gravhydro.timestep = 2*hydro.get_timestep()

    while model_time < t_end:
        orbit = orbital_elements_from_binary(stars, G=constants.G)
        dE_gravity = gravity.initial_total_energy/gravity.total_energy
        dE_hydro = hydro.initial_total_energy/hydro.total_energy
        print(("Time:", model_time.in_(units.yr), \
              "ae=", orbit[2].in_(units.AU), orbit[3], \
              "dE=", dE_gravity, dE_hydro))

        model_time += 10*gravhydro.timestep
        gravhydro.evolve_model(model_time)
        gravity.copy_to_framework()
        hydro.copy_to_framework()
        write_set_to_file(stars.savepoint(model_time), filename, 'amuse',
    append_to_file=True)
        write_set_to_file(ism, filename, 'amuse', append_to_file=True)
        print("P=", model_time.in_(units.yr), gravity.particles.x.in_(units.au))
    gravity.stop()
    hydro.stop()
```

Code example 9.9: Combined solver for gravity and hydrodynamics using the generic class structure presented in Code example 9.10 (see `amuse.examples.gravity_hydro` for the full code example).

glass-like solution to reduce random noise before we started to resolve the collision using an SPH code. Such a relaxation process can be rather elaborate, particularly when a gaseous body is not just self-gravitating but is also affected by the presence of another potential, such as a semianalytic background or another self-gravitating system of particles.

In Code example 9.13, we demonstrate the use of `bridge` to relax a body of gas in the background potential of a stellar distribution. Bridge is used implicitly here by importing the `relax` routine

```
class BaseCode:
    def __init__(self, code, particles, eps=0|units.RSun):

        self.local_particles = particles
        m = self.local_particles.mass.sum()
        l = self.local_particles.position.length()
        self.converter = nbody_system.nbody_to_si(m, l)
        self.code = code(self.converter)
        self.code.parameters.epsilon_squared = eps**2

    def evolve_model(self, time):
        self.code.evolve_model(time)
    def copy_to_framework(self):
        self.channel_to_framework.copy()
    def get_gravity_at_point(self, r, x, y, z):
        return self.code.get_gravity_at_point(r, x, y, z)
    def get_potential_at_point(self, r, x, y, z):
        return self.code.get_potential_at_point(r, x, y, z)
    def get_timestep(self):
        return self.code.parameters.timestep
    @property
    def model_time(self):
        return self.code.model_time
    @property
    def particles(self):
        return self.code.particles
    @property
    def total_energy(self):
        return self.code.kinetic_energy + self.code.potential_energy
    @property
    def stop(self):
        return self.code.stop
```

Code example 9.10: Generic base class for a single code.

```
class Gravity(BaseCode):
    def __init__(self, code, particles, eps=0|units.RSun):
        BaseCode.__init__(self, code, particles, eps)
        self.code.particles.add_particles(self.local_particles)
        self.channel_to_framework \
            = self.code.particles.new_channel_to(self.local_particles)
        self.channel_from_framework \
            = self.local_particles.new_channel_to(self.code.particles)
        self.initial_total_energy = self.total_energy
```

Code example 9.11: Derived class for gravity includes channels.

```
from amuse.ext.relax_sph import relax
```

which is designed to generate a glass-like solution for a set of SPH particles by evolving them in a background field while imposing critical damping on the particle velocities. The routine `relax` takes several arguments, including the particle set for

```
class Hydro(BaseCode):
    def __init__(self, code, particles, eps=0|units.RSun,
                 dt=None, Rbound=None):
        BaseCode.__init__(self, code, particles, eps)
        self.channel_to_framework \
            = self.code.gas_particles.new_channel_to(self.local_particles)
        self.channel_from_framework \
            = self.local_particles.new_channel_to(self.code.gas_particles)
        self.code.gas_particles.add_particles(particles)
        m = self.local_particles.mass.sum()
        l = self.code.gas_particles.position.length()
        if Rbound is None:
            Rbound = 10*l
        self.code.parameters.periodic_box_size = Rbound
        if dt is None:
            dt = 0.01*numpy.sqrt(l**3/(constants.G*m))
        self.code.parameters.timestep = dt/8.
        self.initial_total_energy = self.total_energy
    @property
    def total_energy(self):
        return self.code.kinetic_energy \
            + self.code.potential_energy \
            + self.code.thermal_energy
```

Code example 9.12: Derived class for hydrodynamics. Hydro codes often require additional parameters, which can be handled in the derived class.

```


def relax_gas_and_stars(stars, gas):
    dynamical_timescale = gas.dynamical_timescale()
    converter = nbody_system.nbody_to_si(dynamical_timescale, 1 | units.parsec)
    hydro = Fi(converter, mode="openmp", redirection="file", redirect_file="fi.log")
    hydro.parameters.timestep = dynamical_timescale / 100
    hydro.parameters.eps_is_h_flag = True

    gravity_field_code = Fastkick(converter, mode="cpu", number_of_workers=2)
    gravity_field_code.parameters.epsilon_squared = (0.01 | units.parsec) ** 2
    gravity_field_code.particles.add_particles(stars)
    relaxed_gas = relax(
        gas,
        hydro,
        gravity_field=gravity_field_code,
        monitor_func=check_energy_conservation,
        bridge_options=dict(verbose=True, use_threading=False),
    )
    gravity_field_code.stop()
    return hydro
```

Code example 9.13: Using `bridge` to relax a body of gas in the background potential of a stellar distribution. The source code takes a few a minutes to run and is available in `amuse.examples.relax_gas_and_stars`.

the gas, the hydrodynamics code, and the code that generates the background potential.

The `relax` method uses bridge to couple the hydrodynamics with the background potential, and may require a few tuning parameters. These are specified as `bridge_options`, using a dictionary in the instantiation of `relax`. In Code example 9.13, we choose to increase the amount of diagnostic output from the bridge and to switch off threading. In order to monitor the progress of the relaxation process, we also include a user-defined call-back function that will be called after each bridge step. In the current example, we keep track of energy conservation while relaxing the gas; the monitoring code is in Code example 9.14. The monitoring function takes at least four arguments: the bridge system (or other code which is used to relax the gas), the current step, the current time, and the total number of steps. The last three arguments are only used for diagnostic output.

Code example 9.13 uses another interesting and handy utility in AMUSE, `FastKick`. This routine is actually a rather simple function that computes the kick in the bridge system, in this case for the gravity of the stars. Bridge requires such a code, but in this example it would be overkill to use a full N-body code to calculate the kicks between the stars and the gas particles because we do not need to move the stars, and therefore can omit the N^2 operation in the gravity module. Using a direct N-body code, or in some cases even a tree code, to resolve the kicks would in fact make the code considerably slower. (You can easily verify this by replacing `FastKick` by any gravity solver from Chapter 4.) `FastKick` is parallel (using the keyword `number_of_workers`) and has a GPU implementation, which is invoked by setting `mode = "gpu"`:

```
gravity_field_code = FastKick(converter, mode="gpu", number_of_workers=4)
```

The result is a fast and consistent kick implementation.

Figure 9.8 presents the results of generating initial conditions for 100 stars in a Plummer distribution with a virial radius of 0.33 pc with masses from a Salpeter (1955) function between 1 $M_\odot$ and 100 $M_\odot$. The gas was initially distributed identically to the stars with 10^5 equal-mass particles but with a mass 10 times higher than the total mass in stars. After generating the gas distribution, we iteratively

```
def check_energy_conservation(system, i_step, time, n_steps):
    unit = units.J
    U = system.potential_energy.value_in(unit)
    Q = system.thermal_energy.value_in(unit)
    K = system.kinetic_energy.value_in(unit)
    print(
        f"Step {i_step} of {n_steps}, t={time.as_quantity_in(units.yr)}: "
        f"U={U:.2e}, Q={Q:.2e}, K={K:.2e} {unit}"
    )
```

Code example 9.14: Call-back function to monitor energy conservation during the `relax` process. The source code is available in `amuse.examples.relax_gas_and_stars`.

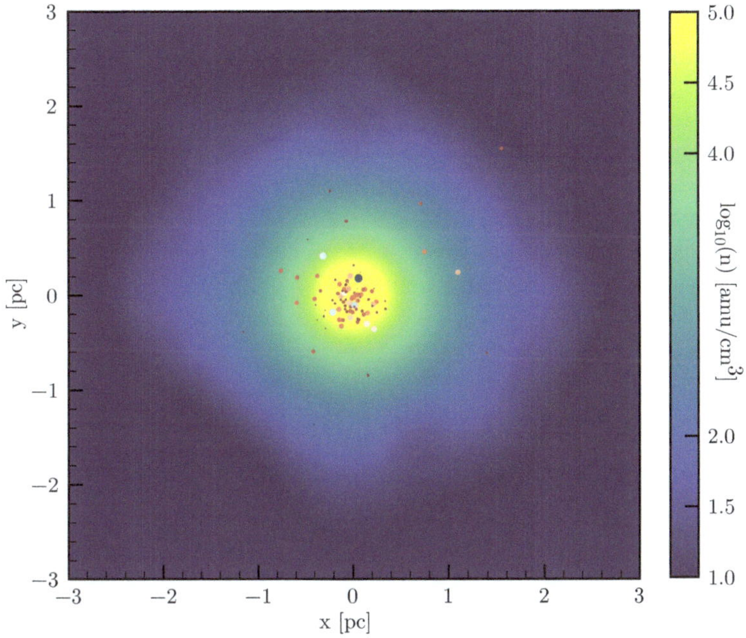

Figure 9.8. Initial conditions for stars and gas in equilibrium. The stars are distributed according to a Plummer (1911) sphere. The gas was initially also generated from a Plummer sphere, but subsequently the gas around each star up to the total mass of that star was removed. The resulting three-dimensional distribution then resembles Tynjetaler cheese. After the gas around the stars was removed, we brought the gas distribution into equilibrium with the stars using the `relax` routine in Code example 9.13. The source code to reproduce this figure can be found in `amuse.examples.relax_gas_and_stars`, which takes about 2 hours to run.

removed the gas particles around each star with a mass equal to the mass of that particular star. These 88941 remaining gas particles were then relaxed. This method was used by Pelupessy & Portegies Zwart (2012) to generate the initial conditions for their simulations. In Figure 6.10, we present the results of a hydrodynamical calculation in which we allowed the gas to clump and form stars through sink particles. The two distributions are quite different, and it is not clear what gives the most satisfactory initial conditions for further simulations.

Generating initial conditions in this way can be expensive in terms of computer time, but that is sometimes the price we have to pay for a realistic or at least a desirable realization.

9.4 Examples

9.4.1 Dissolving Star Cluster in the Galactic Potential

The Sun was probably born in a star cluster with a mass of 500–10,000 $M_{\odot}$ and a virial radius of about 1 pc (Portegies Zwart 2009). In the next few subsections, we explore the orbital evolution of the Sun and its parental cluster over the past 4.6 Gyr.

This work is motivated by the idea that some of the Sun's siblings may still be around in the solar neighborhood and may be identifiable (Portegies Zwart 2009). As has been pointed out by Mishurov & Acharova (2011), this may not be so easy due to the lumpiness of the Galaxy. In addition, the Galaxy may have changed considerably since the Sun's birth and reconstruction of the comoving Galactic potential over the last 4.6 Gyr is not trivial. Nevertheless, let us explore this interesting question for ourselves.

We perform a series of simulations of a small cluster with 1000 stars orbiting in the Galactic disk, for a variety of representations of the Galactic potential. The basic approach is to choose a starting point 4.6 Gyr ago, and then follow the orbits of the Sun and its siblings forward in time to assess their current locations with respect to the Sun.

9.4.1.1 Dissolution of a Star Cluster in a Smooth Galaxy Potential

We first adopt a three-component model of the Galactic potential from Bovy (2015), consisting of a bulge, modeled as a power-law density profile with an exponential cut-off, a Miyamoto–Nagai disk (Miyamoto & Nagai 1975), and an NFW dark matter halo (Navarro et al. 1996). The calling sequence to create this model is

```
from amuse.ext.galactic_potentials import MWpotentialBovy2015
galaxy_model = MWpotentialBovy2015()
```

We initialize the protosolar cluster by selecting a plausible starting point and investigating how the stars spread out during a 4.6 Gyr integration, without pretending that the starting point represents the actual birthplace of the Sun.

We initialize the cluster with 1000 stars with masses drawn from a Salpeter distribution between 0.1 $M_\odot$ and 10 $M_\odot$, distributed in a King (1966) model with a dimensionless depth parameter $W_0 = 3$ and a virial radius of 100 pc. We did not add stellar evolution because with the limited number of stars only a few will be sufficiently massive to show any noticeable effects. We evolve the cluster to an age of 4.6 Gyr using the `BHtree` direct N-body code, which we bridge with the static potential using a bridge time step of 10 Myr. The resulting stellar distribution in the Galaxy is shown in Figure 9.9.

After having orbited the Galactic center some 27 times in 4.6 Gyr, the solar cluster has dispersed along its orbit. Based on these results, Portegies Zwart (2009) concluded that 10 to 100 siblings of the Sun could still be within a distance of 1 kpc of the current location of the Sun in the Galaxy.

9.4.1.2 Use of a Rotating Bridge

In order to say something substantive about the Sun's past, we must adopt a more sophisticated model for the Galactic potential. A rather detailed model of the Galactic potential is available in AMUSE in the `Galaxia` package (Martínez-Barbosa et al. 2016). This model consists of semianalytic axisymmetric prescriptions for the bulge, disk, and halo (Allen & Santillan 1991), and a parameterization for the bar (Romero-Gómez et al. 2007; Romero-Gómez et al. 2011) that includes transient spiral arms (Cox and Gómez 2002). The package has a lot of free parameters for

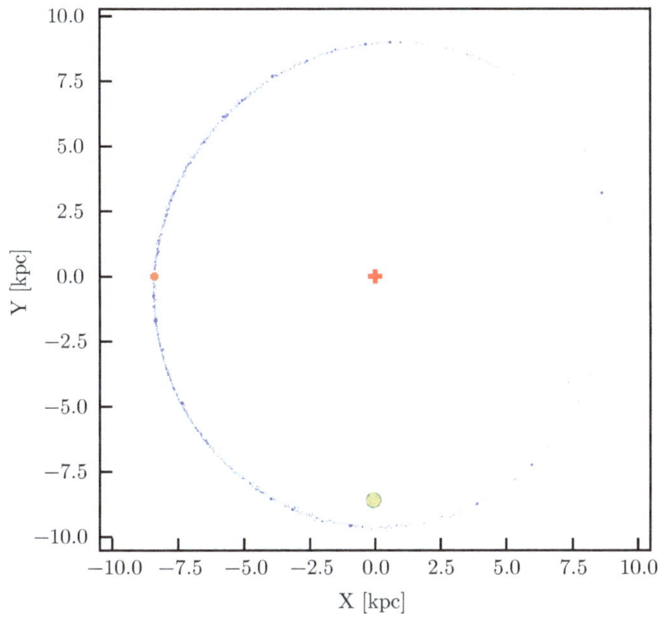

Figure 9.9. Distribution of the solar siblings (blue points) and the Sun (green dot, at left) in the Galaxy, projected in Galactic X and Y coordinates. The Galactic center is indicated by the large red cross, but no Galactic particles are shown because the Galaxy is represented here by a smooth density distribution (in contrast to Figure 9.10 where part of the galaxy is represented by giant molecular clouds and star clusters). The birth location of the Sun is indicated by the green cross at the bottom of the figure, and the initial cluster of 1000 sibling particles by the yellow points. The orbit of the Sun over the last 4.6 Gyr is indicated by the light blue line. The simulation was performed using the script `amuse.examples.solar_cluster_in_galaxy_potential`.

which reasonable choices are needed; some of them will be tuned using data from the *Gaia astrometric satellite* (Gaia Collaboration et al. 2016) as new data become available. Here, we adopt the `Galaxia` potential for the Milky Way, for which parameters are summarized in Table A.15. The current location of the Sun is taken to be $\mathbf{x} = (-8400, 0.0, 17.0)$pc, $\mathbf{v} = (-11.4, -232, -7.41)\,\mathrm{km\,s^{-1}}$ (Karachentsev & Makarov 1996). This is a little farther away than the current best estimates of $\sim$8.0 $\pm$ 0.3 kpc (Camarillo et al. 2018).

We can use the `Galaxia` model in combination with bridge to integrate the Sun's orbit backward in time for 4.6 Gyr to study where it originated. The underlying `Galaxia` galaxy model is rotating, so we take Coriolis forces into account in our simulation by adopting a rotating bridge module in AMUSE (see Martínez-Barbosa et al. 2016). This model is rather complex, and a simple second-order scheme for the standard bridge is insufficient to integrate the equations of motion. In the code example below, we adopt a sixth-order bridge with $m = 13$ symmetric evaluations per step by setting the method option in bridge to `SPLIT_6TH_SS_M13` (see Section A.4). This integration scheme does an excellent job and allows us to take rather large time steps without significant loss of accuracy.

The Galaxia module and bridge are initialized as follows

```
from amuse.community.galaxia.interface import BarAndSpirals3D
from amuse.ext.rotating_bridge import Rotating_Bridge
from amuse.ext.composition_methods import SPLIT_6TH_SS_M13
galaxy = BarAndSpirals3D()
system = Rotating_Bridge(omega, timestep=dt_bridge, method=SPLIT_6TH_SS_M13)
```

We then initialize the Sun at its current position and velocity, and build the bridge with two components

```
drifter = drift_without_gravity(sun_in_rotating_frame)
system.add_system(drifter, (galaxy,))
system.add_system(galaxy, ())
```

The function `drift_without_gravity` is the drifter code discussed earlier in Section 9.2.3.1 and `Sun_in_rotating_frame` is the particle set that contains the Sun, corrected for the rotating frame of the `Galaxia` model. An example of how this approach works is available in `amuse.examples.integrate_sun_in_galaxy_potential`.

Adopting the current solar coordinates and integrating backward in time for 4.6 Gyr, we find the birthplace of the Sun to be $\mathbf{x} = (6641, -4953, 77.2)$pc, with velocity $\mathbf{v} = (131.9, 197.4, 2.737)\,\mathrm{km\,s^{-1}}$ (right at the edge of Borg space in the Delta quadrant of the Galaxy[3]). This example should really be viewed as an illustration of how to use Bridge in conjunction with a more complex galactic potential, rather than as the ultimate method of finding the Sun's birth location. Still, the question in itself remains interesting.

9.4.1.3 Dissolution of a Star Cluster in a Lumpy Galaxy

The previous two experiments neglected the fact that the Galaxy is lumpy, in the sense that stars and gas tend to form clusters and clouds. In the next study, we add a population of clumpy objects to the Galactic potential, in the form of a number of softened point masses representing star clusters and molecular clouds, and have those objects also orbit in the Galactic potential. These objects scatter the protosolar cluster, causing it to disperse more quickly, spreading the siblings out over a larger volume of space.

We could add clusters and clouds to our Galactic model using any of a number of published references (see e.g., Heyer & Dame 2015), but the `GalactICs` module described earlier (see Section 4.4.5) already contains detailed prescriptions for clusters and molecular clouds, so we choose to create these objects using the `GalactICs` model. However, since `GalactICs` does not produce the same disk potential as our adopted Bovy (2015) analytic galaxy potential model (Section 9.4.1.1), we use positions from `GalactICs` but give each molecular cloud a circular velocity in the Bovy (2015) potential. Our adopted `GalactICs` model is generated with a total of 40,000 particles, 3157 of which represent the disk population, the rest

[3] See http://www.startrek.com/

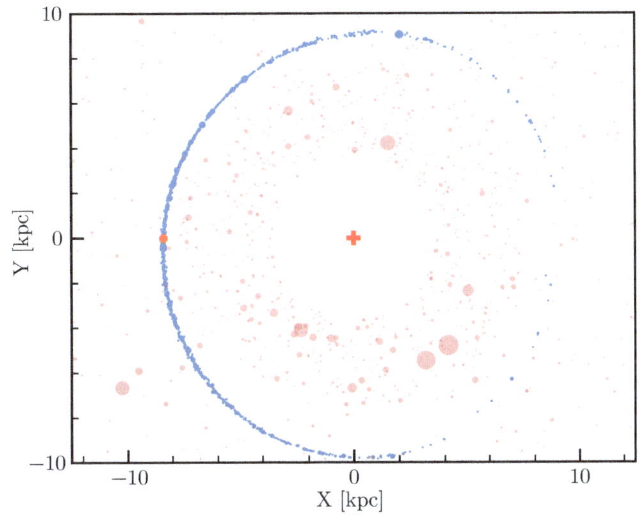

Figure 9.10. Current distribution of the solar siblings (blue bullets) relative to the Sun (green dot). The Galactic center is indicated by the red cross. No galactic particles are shown because the Galaxy is represented as a smooth potential, but we do show the giant molecular clouds included in the calculations (pink circles). The simulation was performed using the script `amuse.examples.solar_cluster_in_semilive_galaxy`, which should take a few minutes to run on a laptop.

follow the distribution for the Galactic halo. We use the former population of particles for the star clusters and molecular clouds, and ignore the rest. We illustrate this operation in `make_giant_molecular_clouds` in the routine `solar_cluster_in_semilive_galaxy`, which is used to generate Figure 9.10. To make the simulation somewhat more realistic, we give each of these clusters a mass from a power-law mass distribution between $10^3\,M_\odot$ and $10^7\,M_\odot$ with a power-law slope of -2.3 (close to Salpeter). The massive objects are softened with a softening length of 100 pc, and are integrated using the `BHtree` tree code (Barnes & Hut 1986).

The cluster itself has the same initial conditions and the same Galactic potential as the one adopted for Figure 9.9. Figure 9.10 presents the distribution of the stars for this second calculations at an age of 4.6 Gyr. In comparison with Figure 9.9, the siblings in the Galaxy with the massive objects are considerably more spread out due to encounters than in the case with the smooth galaxy potential. In a realistic Galaxy, the spiral arms should also be taken into account, as was done more properly in Martínez-Barbosa et al. (2016), although in that study (using `Galaxia`) the massive objects representing star clusters and molecular clouds were not taken into account.

9.4.1.4 Dissolution of a Star Cluster in a Live Galaxy

We can easily replace the analytic potential used in the previous examples with a live evolving Galaxy by bridging with another gravity solver. For the Galaxy, we probably do not want to use a direct N-body solver; a tree code is more appropriate.

We start by generating an initial galaxy model with $N = 10000$ particles using Halogen (Zemp et al. 2008; Zemp 2014) as in the code example below.

```
def initialize_galaxy_model(N, converter):
    instance = Halogen(converter)
    instance.initialize_code()
    instance.parameters.alpha = 2.0
    instance.parameters.beta = 5.0
    instance.parameters.gamma = 0.0
    instance.parameters.number_of_particles = N
    instance.commit_parameters()
    instance.generate_particles()
    galaxy = instance.particles.copy()
    instance.cleanup_code()
    instance.stop()
    return galaxy
```

Note that the parameters used here are comparable, but not identical, to the settings used to generate the halogen panel of Figure 3.1.

We create the cluster using a $W_0 = 3$ King model with a virial radius of 3 pc, which we place at the same location as in the previous experiment. In this case, we adopted $N = 1000$ stars randomly selected from a Salpeter mass function between 0.1 $M_\odot$ and 10 $M_\odot$, resulting in a total mass of $\sim$100 $M_\odot$. We evolved the galaxy using BHTree and the cluster with ph4. The results were bridged so that the cluster affects the Galaxy and the Galaxy the cluster:

```
gravity = bridge.Bridge(use_threading=False)
gravity.add_system(cluster_gravity, (galaxy_gravity,) )
gravity.add_system(galaxy_gravity, (cluster_gravity,) )
```

In this example, we have switched off the internal threading of the bridged codes by sending use_threading=False as an argument in Bridge. Under usual circumstances, threading (or multiprocessing) is enabled in bridge because the two bridge steps are intrinsically parallel and most computers allow the use of multiple cores simultaneously. On some older computers, however, threading produces memory conflicts, which can be prevented by explicitly turning it off.

We set the bridge time step to 1/32 of the initial orbital period of the cluster around the Galactic center

```
gravity.timestep = time_orbit/32.
```

resulting in a bridge time step of $\sim$12 Myr. The results of the calculation, after evolving for $\sim$4.6 Gyr, are presented in Figure 9.11. We see that the siblings have been dispersed considerably compared to the smooth galaxy (Figure 9.9) or the galaxy in which we take some molecular clouds into account (Figure 9.10). Of course, the granularity of the Galactic potential, which is responsible for the large sibling dispersion evident here, depends on the details of the Galactic model and is likely far smaller than in the simulation described here. The dispersion of the solar cluster is an active field of research, and our lack of understanding of the past evolution of the Galaxy seriously hinders this analysis. The consequences of the

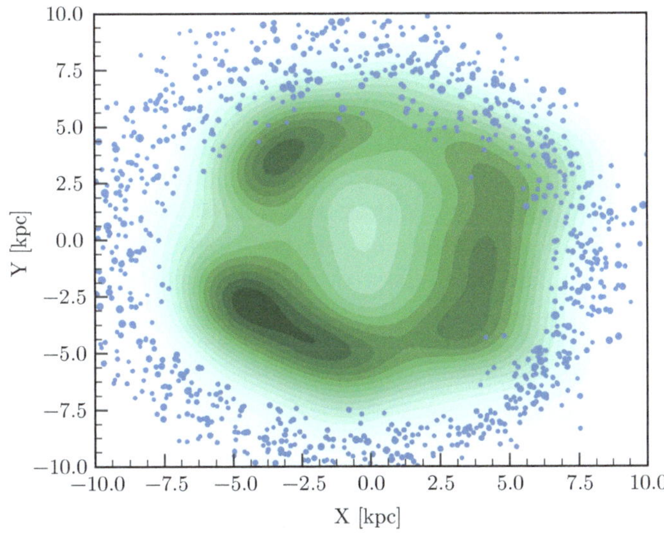

Figure 9.11. Distribution of the Solar siblings (blue bullets) in the Milky Way (green shaded contours). We cut out the central 2 kpc from the Galaxy to keep contrast in the disk. The Sun in this simulation could be anywhere where you see a blue bullet point. The simulation was performed using the script `amuse.examples.solar_cluster_in_live_galaxy`; the figure was made with `amuse.examples.plot_solar_cluster_in_live_galaxy`.

degree by which the solar cluster dispersed may have profound consequences for our ability to identify and find the lost siblings of the Sun.

9.4.2 Did the Sun Originate in M67?

In Section 5.4.2, we discussed the blue stragglers in the star cluster M67. We now return to this cluster in relation to the birthplace of the Sun. In recent years, the argument that the Sun originated from M67 has been made, and even though in the previous section we argued that the Sun originated from a lower-mass cluster, we investigate here the arguments for M67 as the parental cluster.

We begin this discussion with the star M67-1194, which is remarkably similar to the Sun in terms of mass and composition (Önehag et al. 2011). Further dynamical analysis led Gustafsson et al. (2016) to conclude that the Sun might have originated in this cluster, although according to Pichardo et al. (2012) this conclusion is hampered by differences in the orbits of the Sun and the cluster. In addition, 982 of the stars near the cluster's turn-off have a mean age of 3.9 ± 1.2 Gyr Xiang et al. 2017), which is somewhat young compared to the Sun, and M67 is rather massive, making it unlikely that the solar system would have survived in its current form (Portegies Zwart 2009). Nevertheless, it is interesting to test the hypothesis that the Sun originated from this cluster, and therefore we further analyze this possibility, simply because M67 is the closest massive cluster today with metallicity comparable to the Sun (Pichardo et al. 2012).

The hypothesis that the Sun originated in M67 raises a number of interesting questions. When did the Sun escape from the cluster? And, does the solar system still bear any signature of this escape? This question may be linked to the formation and morphology of the Oort cloud and other trans-Neptunian objects.

We perform a simple experiment to study when the Sun and M67 had closest approaches. Possibly a recent close approach might be associated with the moment the Sun escaped the cluster. We integrate the orbits of the Sun and the center of mass of M67 backward in time in a potential representing the Milky Way. The last stellar encounter prior to the Sun's escaping the cluster might be a rather violent event, which could easily affect the orbits in the outer parts of the current solar system. It has therefore been argued that a single encounter, which may also have caused some long-lasting effect on the outer parts of the solar system, initiated the Sun's escape from the cluster, and the moment of escape may be identified with the closest approach between the Sun and the cluster.

We realize, of course, that the orbit of the Sun or the cluster may have been affected by a passing molecular cloud, but we ignore that in this example (although we have already demonstrated in Sections 9.4.1.3 and 9.4.1.4 that this makes a big difference!). We encourage the reader to experiment with a more realistic Galaxy model and study its consequences in relation to this question in Section 9.5.2. In Section 9.4.1, we present a method by which we might include this effect in the calculation, at least statistically.

We use the potential from `Galaxia` discussed in Section 9.4.1.2 to carry out this calculation. We initialize the main script with two point masses, one for M67 and one for the Sun. For the Sun, we adopt the current position and velocity used earlier in Section 9.4.1.2. M67 is located at position (0.77, 0.0, 0.49) kpc and velocity (31.92, −21.66, −8.87) km s^{-1} relative to the Sun (Schönrich et al. 2010). Using a bridge containing both the Sun and M67 in the Galactic potential, we integrate the equations of motion backward in time for the age of the Sun, 4.56 Gyr. The age of M67 seems to be somewhat smaller; but if the Sun originated from the cluster, the ages must of course be about the same (we consider it unlikely that the Sun is a blue straggler). The result of this backward integration is presented in Figure 9.12. According to this calculation, the Sun and M67 passed each other at a distance of ~92 pc about 259 Myr ago.

Our assumed Galactic background potential did not evolve over the last ~259 Myr, which is probably not realistic. Including the effects of transient spiral arms, massive molecular clouds, and star clusters surely affected the orbits over that period quite substantially (see for example Section 9.4.1.4). In Section 9.5.2, we explore when and with what velocity the Sun could have been ejected from the cluster and still end up at its current location.

9.4.3 Inspiral of a Binary Star into a Common Envelope

When two stars orbit each other, mass may flow from one to the other in a phase of Roche-lobe overflow (RLOF). Binaries are particularly vulnerable to RLOF when the more massive star ascends the giant branch. The stability of mass transfer, in

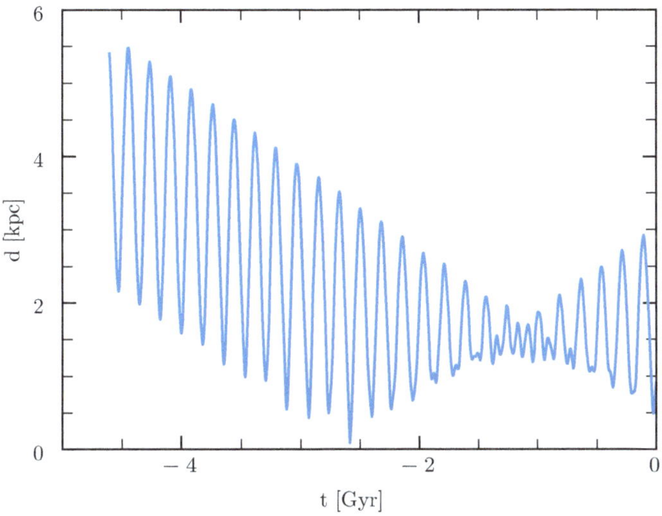

Figure 9.12. Distance between the Sun and the star cluster M67 as a function of time (negative numbers indicate the past). According to this calculation, the Sun and M67 had a closest approach of ~92 pc some 2.59 Gyr ago. The simulation was performed using the routine in `amuse.examples.Sun_and_m67_in_the_galaxy`, and took a few minutes to run.

terms of how rapidly mass is transferred to the companion, depends on the structure of the Roche-lobe-filling star, the response of the accreting star, and the resultant change in the orbital parameters of the binary system. RLOF in binary systems has been studied quite exhaustively, both theoretically and numerically (Webbink 1984; Taam & Sandquist 2000; Ivanova et al. 2013).

In most triple systems, RLOF probably does not proceed much differently than in a binary. For the majority of such systems, the most massive star, which evolves most rapidly, is a member of the inner binary. As a consequence, mass transfer is likely to begin in the inner binary system. The outer orbit is generally relatively wide, so mass lost from the inner binary is felt by the outer star as a slow wind. This affects the orbit of the outer star, and possibly the star itself, but probably not in a way much different than if the inner binary simply were a single star with an excess wind.

In some triples, however, this simple picture is inadequate. In those cases, the outer star is the most massive, and in a subset of these the outer orbital period is sufficiently small that mass transfer will start when the outer star ascends the giant branch. Table 9.2 presents an overview of such triples. The systems listed in Table 9.2 are indicated in Figure 9.13, where we also demonstrate that the most massive outer star experiences RLOF at some moment during the evolution.

RLOF in these triples is considerably different than mass transfer in an ordinary binary system. One reason is that the mass cannot simply be accreted by the inner object, but may be ejected via a slingshot effect by the rapidly orbiting inner binary. During mass transfer, the inner binary also evolves: both stars may accrete some of

Table 9.2. Known Triples for Which the Outer (Tertiary) Star Is More Massive Than Either of the Two Inner Stars, and the Roche Radius Is Smaller Than the Maximum Size of the Outer Star at the Tip of the Asymptotic Giant Branch

Name	m_1 [M$_\odot$]	m_2	m_3	$a_{\rm in}$ [au]	$e_{\rm in}$	a_{out} [au]	$e_{\rm out}$	$t_{\rm RLOF}$ [Myr]	$m_{\rm RLOF}$ [M$_\odot$]	$r_{\rm Roche}$ [au]
τ CMa	17.8	17.8	50.0	0.076	0.0	2.49	0.285	4.3	41.5	0.95
V* CQ Dra	0.8	0.26	6.3	0.006	0.0	5.42	0.0	70.1	6.02	2.64
ξ Tau	3.2	3.1	5.5	0.133	0.0	1.23	0.15	83.7	5.50	0.42
V* V1334 Cyg	1.62	1.83	4.4	4.62	0.0	10.7	0.0	168	1.67	3.99
V* DL Vir	2.5	1.4	3.68	0.037	0.0	6.68	0.46	269	3.44	2.34
V* d Ser	2.02	1.94	2.5	0.047	0.0	1.94	0.47	799	2.46	0.62
HD 97131	1.29	0.9	1.5	0.037	0.0	0.80	0.191	2960	1.48	0.26
KIC002856960	0.24	0.22	0.76	0.008	0.0064	0.73	0.612	$>H_0$	0.72	0.12

Column 1 gives the common name of the triple system; columns 2–4 give the masses of the triple components; columns 5–8 give the semimajor axis and eccentricity of the inner and outer orbits; columns 9–11 give the estimated moment at which the outer (most massive) star starts to overfill its Roche lobe, the mass of that star, and its size.

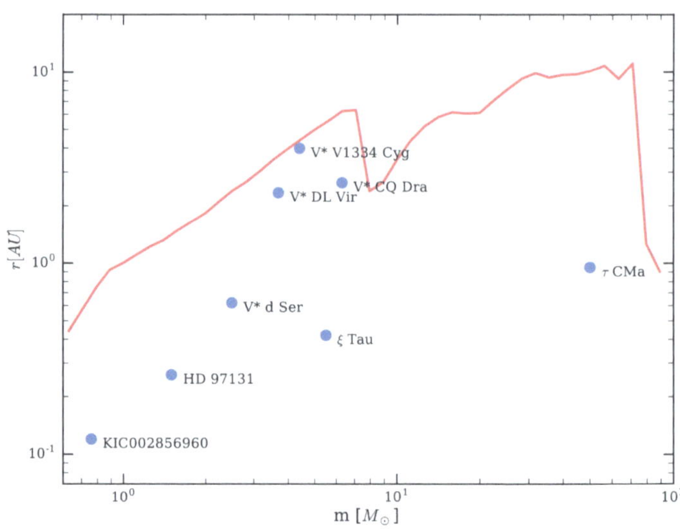

Figure 9.13. Maximum stellar radius at the tip of the asymptotic giant branch as a function of mass (red curve), calculated using `SeBa`. Blue bullets indicate the radius of the (most massive) outer component at the moment of RLOF, for the triple systems listed in Table 9.2. This figure was made using the script `amuse.examples.plot_rlof_triples`.

the material delivered by the outer star, and the leftover material either forms a circumbinary disk or is ejected from the system. At the same time, orbital angular momentum exchange may affect the eccentricity of the inner binary via the von Zeipel–Lidov–Kozai (Lidov 1962; Kozai 1962) effect.

Simulating all these effects requires both an accurate gravitational solver and some way of taking the hydrodynamics of the system properly into account. The subtleties of von Zeipel–Lidov–Kozai are not well handled by the rather low-accuracy gravity solvers normally used in SPH codes. Instead, we adopt here a fourth-order Hermite solver to integrate the equations of motion of the two inner binary components and the core of the outer most star (see also Section 8.4.5). The giant envelope is modeled using an SPH code, and its core is represented as a dark matter particle in the hydro code. The coupling between the gravity solver and the hydrodynamics code is realized using Bridge.

We adopt the observed parameters of the triple system ξ Tau as initial conditions for our simulation. We start by evolving the $5.5\,M_{\odot}$ primary star using EVTwin (Eggleton et al. 2011; Eggleton 1971, 2006), up to the moment when it fills its Roche lobe, at 0.42 au at an age of 83.7 Myr. The star is then converted to a hydro realization (see Section 8.4.2) with a core mass of $1.4\,M_{\odot}$ and placed in the triple system with the orbital parameters listed in Table 9.2. Since the stars have lost virtually no mass, we do not adjust the orbital parameters to take pre-RLOF mass loss into account. The orbit of the outer star is placed at the relatively low inclination of 9°; because of the von Zeipel–Lidov–Kozai effect, an inclination initially higher than ~45° would drive the inner binary into a state of mass transfer before the outer star had a chance to undergo RLOF.

Figure 9.14 presents the evolution of the semimajor axis of the outer orbit of the ξ Tau triple system. Since the Roche-lobe-filling giant hardly loses any mass during its evolution, we simply adopt the sum of the total gas mass plus the core mass as the

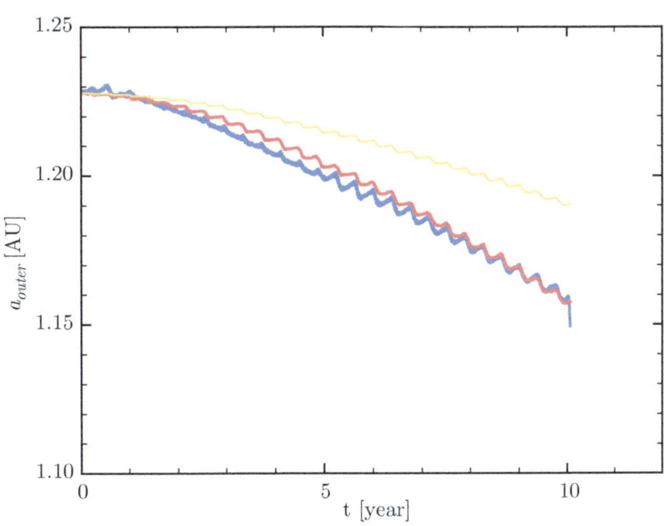

Figure 9.14. Evolution of the semimajor axis of the outer orbit in ξ Tau. In blue, we plot the result of the SPH simulations. The semianalytic solution (Equation (9.17)) using $\eta = 6$ is overplotted in red. Two additional curves are plotted for $\eta = 4$ (yellow) and $\eta = 8$ (green). The simulation was performed using the script `amuse.examples.xi_tau_rlof`, the plotting was done with `amuse.examples.plot_xi_tau_orbital_evolution`.

giant's total mass. The outer orbit shrinks within ~10 years from the initial 1.23 au to ~1.15 au after the onset of RLOF. During the same period, the eccentricity of the outer orbit decays from 0.15 to ~0.1.

We can derive an expression for the effect of mass transfer in the system, starting with a binary of primary mass m_3^0 (note here that the primary is the outer component) and secondary mass $m_{\rm in}^0 = m_1 + m_2$, and ending after mass transfer with masses m_3 and $m_{\rm in}$ for the primary and secondary stars, respectively. The degree of mass conservation is expressed by the parameter $f = (m_3 + m_{\rm in})/(m_3^0 + m_{\rm in}^0)$. Ignoring the internal angular momentum of the outer giant star and the inner binary system, we can express the evolution of the semimajor axis according to the redistribution of mass (Huang & Taam 1990):

$$\frac{\dot{a}}{a} = -2\frac{\dot{m}_3}{m_3}\left(1 - \eta\frac{m_3}{m_{\rm in}} - \frac{1}{2}\frac{(1-\eta)m_3}{m_3 + m_{\rm in}} - f\frac{(1-\eta)m_3}{m_3 + m_{\rm in}}\right). \tag{9.17}$$

Here, η is the specific angular momentum of mass leaving the system, relative to the specific angular momentum of the accreting system (in this case the center of mass of the inner binary system). Interestingly, this is essentially the same equation (Equation (5.18)) we already saw in relation to the nonconservative evolution of semidetached binaries in Chapter 5. Figure 9.14 includes an overplot of the evolution of the orbital separation of the outer binary in this semianalytic model (Equation (9.17)) with $\eta = 6$, a value consistent with losing mass through the second Lagrangian point (L_2) of the outer binary system.

9.4.4 Budding Planets in a Protoplanetary Disk

In an attempt to model the early evolution of a planetary system, we simulate a young star surrounded by a gaseous disk and a number of young planets. This is a typical example in which hydrodynamics and gravitational dynamics are both relevant and operate on similar timescales, presenting an outstanding example of the power of AMUSE. The script used to perform these calculations should by now look rather familiar, except perhaps in the technique used to write composite data files (see Section A.2.2).

We initialize the protoplanetary system using a central star of mass $m_\star = 1.73\,{\rm M}_\odot$, which happens to be consistent with the mass of β Pic. This system is thought to harbor at least one rather massive planet and a gas disk, but its true composition may be considerably more complex (Winn & Fabrycky 2015). For this simulation we limit ourselves to a collection of planets with total mass $0.01m_\star$ and a gaseous disk of the same mass. The planets were arranged according to the Oligarchic growth model proposed by Kokubo & Ida (2002), and for which Tremaine (2015) proposed a more general scenario (see also Section 3.4.2). The disk was initialized with a Safronov–Toomre Q-parameter $Q = 1$ (see Section 6.4.1),

Our calculation had seven planets in circular orbits ranging from a ~0.4 $M_{\rm Jup}$ planet at a separation of 1.0 au to a ~8.1 $M_{\rm Jup}$ planet at ~53 au. The gaseous disk was generated using routine `ProtoPlanetaryDisk`, as already discussed in Section 6.4.1, between 1 au and 100 au in the same plane as the planets. The equations of motions were integrated using `mercury` (Chambers & Migliorini 1997; Chambers 2012) and the hydrodynamics was resolved using `Fi` (Hernquist & Katz 1989; Gerritsen & Icke 1997; Pelupessy et al. 2004). We neglected the evolution of the parent star, since we evolved the system for only 10^4 yr.

Figure 9.15 presents a snapshot of the simulation about 280 yr after the start of the simulation. The planets are represented by various colored bullets with sizes proportional to the logarithm of the planet mass. The thin disk ($Q = 1$) is perturbed by the planets, creating spiral waves. This is particularly visible at middle left, where a slightly inclined planet drives a large spiral arm in the disk. The leading arm (spiraling inwards) has its low-density wake on the outside, whereas the trailing arm

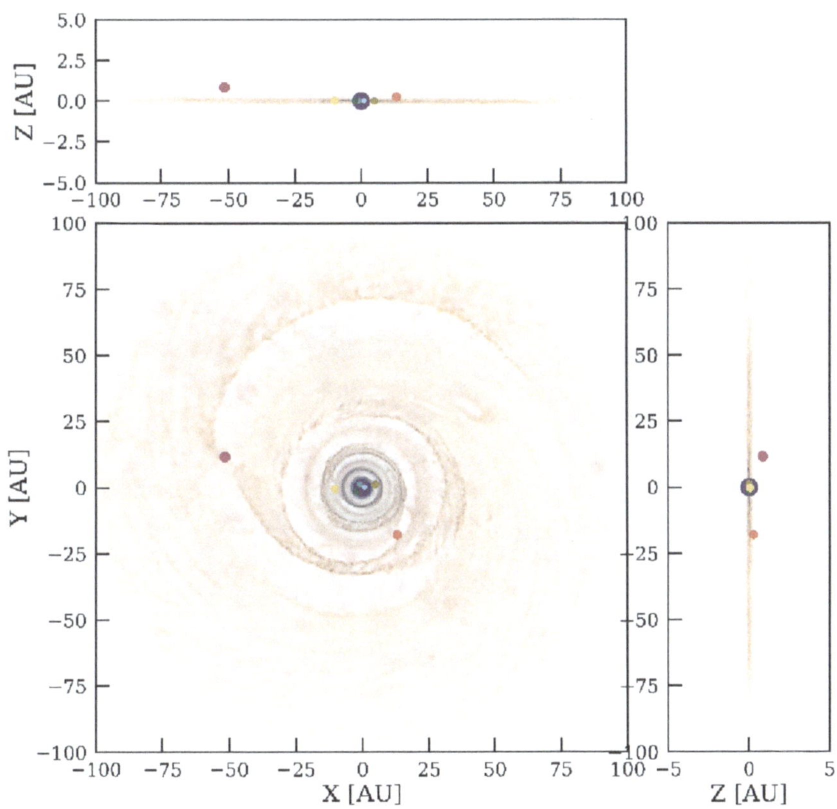

Figure 9.15. Young star with planets (bullets) and a rather flat gaseous disk. This image was made about 280 years after the start of the simulation. The planetary system was generated according to the Oligarchic growth model (see `amuse.examples.make_planets_oligarch`) with a total mass of 1% of the central star's (1.73 $M_\odot$) mass, with the same mass in the gaseous disk (see `ProtoPlanetaryDisk`). The simulation was performed using the routine in `amuse.examples.planetary_system` (which took about a week to run); plotting was done with `amuse.examples.plot_planetary_system`.

(going outwards) has the low-density wake on the inside. The other planets are barely visible in the image, but their presence can been seen by the structure they induce in the disk.

Although this is a rather simple demonstration of the capabilities of bridge, much more elaborate experiments including moons and other stars can easily be implemented by adjusting the example script in `amuse.examples.planetary_system`.

9.4.5 Planetary Systems in Star Clusters

Stars in star clusters are often found in pairs, or even triples or higher-order multiples. "Democratic" multiples, in which no pair of objects dominates the binding energy, tend to be unstable and quickly decay into simpler components—ultimately single stars and binaries, the fundamental building blocks of star clusters. Some higher-order hierarchies can survive for a very long time, although eventually most of those also dissolve. The realization that many stars are born with multiple planets (and smaller objects) demands a numerical treatment of hierarchical systems. (We note in passing that a stellar distribution around a supermassive black hole is in many ways a similar numerical problem.)

So far we have only discussed binaries in our discussion of stellar evolution, and how to manage encounters in star clusters in Sections 8.7.1 and 8.7.2, Here, we briefly discuss the ongoing development of how to handle rich hierarchical systems of stars, planets, and maybe even planetesimals. The module discussed in Section 8.7.2 is intended for simulating star clusters with a rich population of binaries, whereas the `Nemesis` module described here is meant for integrating planetary systems inside a star cluster.

The relatively short orbital periods of planetary systems with respect to the crossing time of a star cluster give rise to a number of interesting numerical challenges. One of these is a wide range in timescales, which can be days for a planet orbiting a star compared to a million years for a star orbiting the center of mass of a cluster. This wide separation in scales allows us to treat the planetary system differently than the star cluster, often using a different code. The bridge method is ideally suited for such a hierarchical structure.

In `Nemesis` (see Section 10.1.2), each planetary system is integrated with its own individual integrator. When several stars or planetary systems approach each other within a predefined limit, both systems are merged and the integration is continued using a single integrator. If a multibody system dissolves, single objects are picked up by the global N-body integrator, whereas multibody subsystems continue to be integrated with individual solvers. Integrators in this method are dynamically created and deleted at runtime. `Nemesis` is fast, efficient, parallel, and effective in addressing multiplanetary systems in a star cluster. At the time of writing, the `Nemesis` module is not yet available in AMUSE because we are still in the testing phase (see Figure 9.16). We can say, however, that the method seems to work as expected, and that `Nemesis` will soon become part of the AMUSE code base.

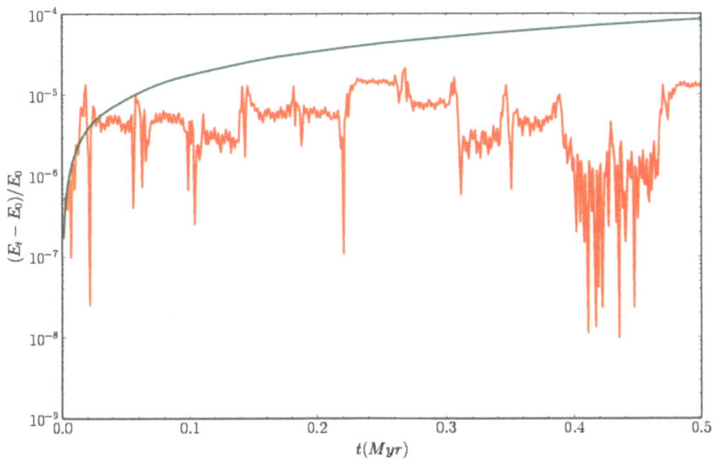

Figure 9.16. Evolution of the numerical integration error in the energy of the solar system when orbited by second 1 $M_\odot$ star. In this example, the other star (allegedly called Nemesis) orbits the solar system in the ecliptic plane with a semimajor axis of 1 pc and an eccentricity of 0.6. The green curve gives the energy error when integrated using `Hermite`; the red curve gives the result of the Nemesis module.

9.5 Assignments

9.5.1 Drift with Gravity Code

In Section 9.3.1, we discussed a small experiment in which we integrated the equations of motion for the Sun, Earth, and Moon using `Bridge`. In that example we employed three incarnations of `ph4` to integrate a single Sun, a single Earth, and a single Moon, and coupled these three codes using multiple bridges. This was a little excessive, and it would have been much better to use the drifter code in Code example 9.5 instead of the direct N-body codes. Rewrite Code example 9.6, Code example 9.7, and Code example 9.8 using the drifter code from Code example 9.5, and compare them with a small script in which all the particles are integrated using the direct N-body code. Which of the various methods would you prefer, based on reliability, speed, and accuracy?

9.5.2 How Did the Sun Escape from M67?

If the Sun escaped from the star cluster M67, as discussed in Section 9.4.2, this might have happened quite a while ago, when the cluster and the Sun had a closest approach. In the small calculation performed earlier, we adopted a simple analytic potential for the Galactic background, launched the Sun and M67 from their current locations, and integrated backwards in time for 4.56 Gyr. According to this calculation, at closest approach, the distance between the Sun and the M67 was 92 pc. However, it was not mentioned in Section 9.4.2 that the relative velocity between Sun and the cluster at that moment was $\sim$48 km s^{-1}, which is rather high. An encounter with another cluster member could have ejected the Sun, but this would probably have led to a velocity of the order of the cluster's escape speed, which is an order of magnitude smaller. An encounter that propelled the solar system to a speed of almost 50 km s^{-1} would have profound consequences for the further evolution of most of the planets orbiting the Sun.

In this assignment, we recognize that the current coordinates of M67 are not precisely known, and ask: if we were allowed some freedom, would it be possible to find a better match between the trajectories of the Sun and the cluster? First repeat the calculation in Section 9.4.2 and confirm the earlier result. Then, allow the current position and velocity of M67 relative to the solar system to vary within the observed uncertainty, and search for the closest historical distance and relative velocity between the two point masses in the Galactic potential. The underlying assumption is that the Sun was ejected from M67 and that its orbit has not changed appreciably since then.

We could search parameter space by hand, but this is a rather painful process. A better method would be to adopt an MCMC approach using the Metropolis–Hastings algorithm (Metropolis et al. 1953; Hastings 1970) to find the optimum solution. This algorithm works quite well for searching a large parameter space. There are more efficient alternatives (Feroz et al. 2009), but MCMC tends to find solutions close to the optimum quite effectively and it is rewarding to implement by yourself.

The Metropolis–Hastings algorithm works as follows:

1. Perform a calculation with some predetermined initial set of parameters, such as the Sun's current location and velocity in the Galaxy and that of M67.
2. Integrate both objects backwards in time in the Galactic potential for 4.6 Gyr and measure the minimum "distance" (in position r and velocity v) between the Sun and M67. Let us call this phase-space distance d. In order to be able to add distance and velocity, we have to normalize them, in which case one could define d by $d^2 = (r/r_{\rm min})^2 + (v/v_{\rm min})^2$, where $r_{\rm min}$ and $v_{\rm min}$ are the minimum joint values found for r and v. After the first calculations $d^2 = 2$.
3. Introduce a small change to the initial position and velocity of M67 within the observational uncertainty, or simply adopt a variation of ~100 pc in position and ~1 km s^{-1} in velocity. You could use a more formal approach from the literature, but this procedure will give a reasonable solution; we can later return to this assumption to see if the final solution is reasonable.
4. Repeat the calculation with the introduced variation in the position and velocity of M67, and measure the new minimum phase-space distance, which we call d'. If $d' > d$, the Sun passed the orbit of M67 at a larger phase-space distance; but if $d' < d$, the new solution is better.
5. Now draw a random number X on the interval [0, 1]. If $X < \exp(-(d' - d)/d)$, we accept the new initial conditions as the best coordinates for M67, and we update $r_{\rm min}$ and $v_{\rm min}$ accordingly. Otherwise, we keep the previous best initial conditions.
6. Repeat this procedure (from step 3 to step 5) until you are happy with the results. While performing the iterative procedure you may want to reduce the uncertainty. It is generally best to do this slowly and preferably only when you adopt the new initial realization. You will find that the quality of the calculation improves with the number of iterations. The rate at which it improves depends on your choices for the variations in position and velocity introduced in each new calculation. It may be rewarding to experiment with this to find the most efficiently converging Markov chain.

Instead of writing your own MCMC code, you could adopt a package, such as `emcee` (Foreman-Mackey et al. 2013), which, due to its `Python` interface, works naturally with AMUSE.

Address the following questions:

- What is the minimum distance and relative velocity you find for M67 and the solar system?
- Is this a realistic solution? Take into account here that the orbital speed of Neptune is $\sim$5.5 km s^{-1}, that the solar system probably already had the planet by the time it left the cluster, and Neptune still orbits the Sun.
- What difference does it make if you change the potential of the Galaxy, change the number of spiral arms using the `BarAndSpirals3D()` routine from `Galaxia` as discussed in Section 9.4.1.2, or introduce giant molecular clouds (see Section 9.4.1)?
- Instead of searching for the optimal position and velocity of the cluster, you could relax the assumptions about the Galaxy, such as the mass of the bulge and bar, the pitch angle of the spiral arms or their rotation speed. One advantage of the Metropolis–Hastings algorithm is that it is rather agnostic about the parameters you change so long as you define a proper phase-space distance on which to optimize.

9.5.3 Half-Treecode, Half-Direct

We want to study a large system of self-gravitating particles initially distributed in a virialized Plummer (1911) sphere, with a stellar mass function running from 0.1 $M_\odot$ to 100 $M_\odot$. We have insufficient computer resources to perform the entire calculation using a direct N^2 method, but we require a precision that exceeds that of a standard treecode. We therefore decide to create a hybrid *split method* in which we combine a treecode with a direct N-body code by interfacing the two codes using `Bridge` within the AMUSE framework.

We divide the stars among the two codes, where the treecode handles the low-mass stars (those with $m < m_{\rm cut}$) and the direct N-body code takes care of the high-mass stars (with $m \gtrsim m_{\rm cut}$). In order to optimize the simulation, we will study the behavior of the hybrid code as a function of the mass cut $m_{\rm cut}$, which determines which particles are treated by which code. Although we will perform the numerical experiments with units, we will not take stellar evolution into account because we are currently only interested in validating the proposed hybrid method.

1. Why do we split the particle set in mass rather than, for example, in distance from the center of mass or as a function of local density?
2. Estimate the bridge time step relative to the typical orbital timescale you will use for your calculation.
3. Write the hybrid script and give it an interface so that you can run it from the command line.

Initialize the system with $N = 10^4$ stars with a virial radius of $r = 3$ pc, and integrate it for 10 Myr. Perform the following three runs and store the half-mass–radius, the core radius, and the total kinetic and potential energies at a time resolution of 0.1 Myr:

- All stars in the treecode;
- All stars in the direct code;
- Split code, with $m_{\rm cut} = 1.0\,{\rm M}_\odot$.

Make the following two plots

- The relative energy error compared to the initial energy ($(E - E_0)/E_0$) as a function of time;
- The half-mass and core radii as functions of time.

Also make a table of the wall clock times of these three runs.

Discuss your findings, and say whether or not running low-mass stars with a tree code and high-mass stars with a direct code is an advantageous strategy.

9.5.4 The Accreting black hole in HLX-1

HLX-1 is a variable and bright ($\gtrsim 10^{42}$ erg/s) X-ray source in the direction of the spiral galaxy ESO 243-49 (Farrell et al. 2009). The reasons for its repeated outbursts and the nature of the X-ray source remain elusive (Webb et al. 2017). Figure 9.17 shows a portion of the X-ray counts reported by the SWIFT XRT instrument from Figure 3 of Titarchuk & Seifina (2016).

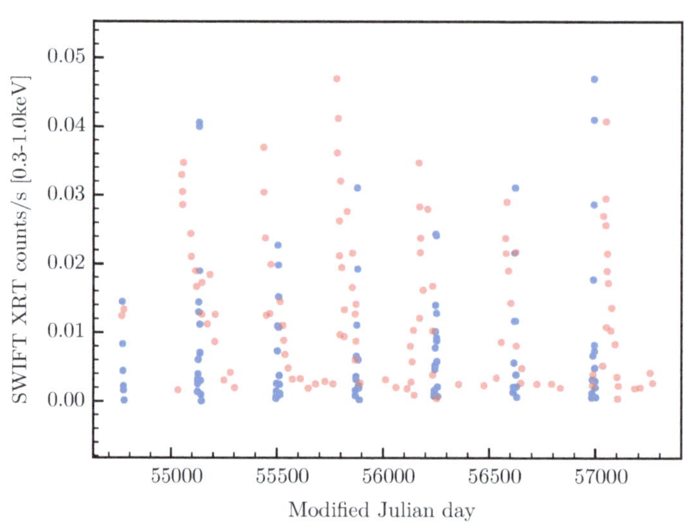

Figure 9.17. X-ray counts of the source HLX-1 as observed by SWIFT XRT in the 0.3–1.0 keV band (Titarchuk & Seifina 2016) (red bullets). Overplotted are the results of the simple model calculation described in the text (blue bullets).

Observations support the hypothesis that the X-rays are produced by a black hole of mass $10^3 - 10^5\,M_\odot$ that is orbited by a massive star (Webb et al. 2012). The companion star is thought to be evolved, so that it fills its Roche lobe near pericenter, but at apocenter no mass is transferred and therefore no X-rays are produced. We can make a very simple model to test this hypothesis by integrating the orbit of a $20\,M_\odot$ star around a black hole. In Figure 9.17, we present the results of such a simple calculation in which the giant donor orbits a $10^3\,M_\odot$ black hole in a 370 day orbit. To ensure that the donor overflows its Roche lobe near pericenter, we adopt an eccentricity of 0.95. The X-ray count rate is assumed to be proportional to the degree to which the $20\,M_\odot$ overfills its Roche lobe, which only happens near pericenter.

The counts calculated in the simple model do not align with the observed counts, although a number of features of the observations are shared by the model calculation. Tidal effects, as well as the viscous evolution of the donor star, probably contribute to the variations observed in the X-ray outburst period. In this assignment we will try to improve the simple model used to produce Figure 9.17. We adopt the `MESA` (Paxton et al. 2010, 2011) stellar evolution package with the Huayno N-body integrator (Pelupessy et al. 2012). We evolve a $20\,M_\odot$ star for 9.0 Myr, by which time it has ascended the giant branch. We start the simulation when the star has lost $1.54\,M_\odot$ as a stellar wind, and its radius is 4.19 au, sufficient to fill its Roche lobe at pericenter for the adopted binary period.

There are two basic approaches we can follow here—model the tidal evolution of the donor star in the black hole field, or use a smoothed-particle hydrodynamics code to resolve the donor. For tidal evolution, we recommend using the `tidal_interaction` routine from the `stellar_tidal_evolution` package in `amuse.ext`. This prescription follows the tidal evolution of planets and stars according to the model by Hansen (2010), which is based on the theory of equilibrium tidal models by Eggleton et al. (1998).

To use an SPH code to resolve the donor star in its orbit around the black hole, we convert the outer layers of the donor star to an SPH realization using at least 1000 particles, with the central portion of the star represented by a single massive particle. This can be realized by generating an SPH realization of a star in which a core mass is specified (see Section 8.5.4.1 and in particular Code example 8.11, where we discussed how to convert a stellar evolution model to an SPH realization). Code example 9.15 shows how a star can be evolved to the giant stage and subsequently converted to a SPH realization with a single particle representing its core. We used the same method in Section 9.4.3 to generate the giant star with a core in order to simulate the triple system Xi Tau. In Code example 9.15, we use the argument `with_core_partricle` to indicate that the stellar model has a core, which should be returned in a separate particle set.

A bridge can then be built with the gravity code to integrate the black hole and the donor's core, using the SPH code to resolve the stellar envelope. Allow the black hole to accrete using a sink particle and measure, as a function of time, the amount of material accreted by the black hole. Integrate the system for half a dozen orbits, as

```
stellar = Mesa()
stellar.particles.add_particle(Particle(mass=o.mass))
XiTau = stellar.particles[0]
while XiTau.radius <= o.radius:
    print(
        f"Time={stellar.model_time.in_(units.Myr)} "
        f"{stellar.particles.radius.in_(units.AU)} "
        f"{stellar.particles.stellar_type}"
    )
    stellar.evolve_model()

target_core_mass = XiTau.core_mass
print(XiTau)
print("Core mass:", XiTau.core_mass.in_(units.MSun))
sph_model = convert_stellar_model_to_sph(
    XiTau,
    o.Nsph,
    with_core_particle=True,
    target_core_mass=target_core_mass,
    do_store_composition=False,
    base_grid_options=dict(type="fcc")
)
core_particle = sph_model.core_particle.as_set()
gas_particles = sph_model.gas_particles

print(f"Ngas={len(gas_particles)} Ncore={core_particle}")
write_set_to_file(
    core_particle, "Hydro_PrimaryStar_XiTau.amuse", "amuse",
    append_to_file=False
)
write_set_to_file(gas_particles, "Hydro_PrimaryStar_XiTau.amuse", "amuse")
stellar.stop()
```

Code example 9.15: Evolving a star and turning it into a SPH realization with a single particle representing the stellar core.

demonstrated in Figure 9.17, and plot the accretion rate of the black hole together with the observational data.

Which of the two models, tidal evolution or detailed hydrodynamics, gives a better comparison with observations?

9.5.5 The Formation of the Widest Binary Stars

In 1915, Robert Innes discovered that the binary system α Centauri is orbited by a third single star (Innes 1915). The discovery was made by examining photographic plates of the region around α Centauri using a blink microscope to search for faint stars with large proper motion (Innes 1915). It was not until 1917 that Innes (1917) referred to this star as Proxima Centauri: "If this small star had a name it would be convenient—it is therefore suggested that it should be referred to as Proxima Centaurus."

Today the orbit of Proxima Centauri is well constrained to have semimajor axis $8.7^{+0.7}_{-0.4} \times 10^3$ au and eccentricity $0.50^{+0.08}_{-0.09}$ (Kervella et al. 2017). The inner α Centauri binary has a semimajor axis of about 23.7 au and an eccentricity of 0.524 ± 0.0011; the three stars have masses $1.105 \pm 0.0070\,M_\odot$, $0.934 \pm 0.0061\,M_\odot$ and $\sim 0.123\,M_\odot$ (Pourbaix et al. 2002). This very wide orbit is not explained from standard star (and binary) formation theories. An interesting alternative for the formation was proposed by Reipurth & Mikkola (2012). They argue that three stars embedded in a gas cloud can find each other and engage in a triple resonant interaction. The result of the interaction is the ejection of one of the stars, but due to the presence of the surrounding gas, the star does not escape but finds itself parked in a wide orbit around the binary system.

The calculations performed in Reipurth & Mikkola (2012) employed a direct N-body code that in AMUSE is called `Mikkola`. In order to perform the calculations, Reipurth & Mikkola (2012) added a smooth background potential representing the gas. They concluded that the existence of extremely hierarchical systems is a natural consequence of resonant interactions embedded in a primordial gas reservoir. Fundamental to this formation channel is the interaction with the background gas potential, accretion onto the stars, and stellar gravitational interactions. With AMUSE, we can relatively easily reproduce their results, but we can also include the actual background gas, rather than a static background potential.

In Section 9.5.3, we had an assignment in which we bridged two gravity codes: one for the low-mass "background" stars and the other for the more massive stellar population. Here, we will replace the background population with a gas distribution, simulated using an SPH code:

1. Adopt a mass of $0.1\,M_\odot$ for each of the three stars and distribute them in a Plummer sphere with a characteristic radius of 400 au. Reipurth & Mikkola (2012) adopted a mean separation between 40 au and 400 au.
2. Add an SPH particle representation to the system, centered on the center of mass of the three-body system, with a mass of $10\,M_\odot$ and characteristic radius 7500au.
3. Reipurth & Mikkola (2012) allowed the three stars to accrete from the ambient gas. We can take accretion onto the three stars into account by having a sink particle move along with the stars (see Section 6.3.2). Stop the simulation after 100 Myr and study the resulting triple system.

Reipurth & Mikkola (2012) argued that the resulting projected orbital separation s matches the power law (see their Figure S1)

$$f(s)ds \sim s^{-1.6}ds. \tag{9.18}$$

Perform a number of simulations (Reipurth & Mikkola 2012 performed a total of 180,000) with randomized initial realizations from the above initial conditions, and compare their projected separation with the distribution suggested by Reipurth & Mikkola (2012).

A key aspect of this model is that the three embryonic stars are initially identical, but are forced to compete for gas to accrete through dynamical interactions. The consequential change in mass subsequently feeds back into their orbital evolution, resulting in the power-law distribution in orbital separation. Note, however, that the obtained $s^{-1.6}$ power law is in reasonable agreement with observations (or maybe a bit steeper, see Reipurth & Mikkola 2012).

Repeat the experiment using four stars rather than three, each with a mass of $0.1\,\mathrm{M}_{\odot}$, adopting the same initial conditions as before, and including the sink particles. Answer the following questions:

- What is the fate of the fourth star?
- What are the distributions of the final masses of the four stars?
- Do you expect that the inner binary contains the two most massive stars, as in the observed Alpha-Proxima Centauri triple system?

9.5.6 Photoevaporating Disk in a Binary System

In Section 10.1.3, we discuss the possibility of coupling multiple codes into a self-scheduling organized fashion. The described method, `Venice`, allows multiple codes to be coupled on their own internal temporal resolved and spatially resolved scales. This leads AMUSE to address intricate multiscale and multiphysics problems.

One such problem is the photoevaporation and gravitational perturbation of a circumstellar disk around a low-mass star in an eccentric orbit around a massive companion star. Such a disk is hydrodynamically affected by the gravity of the companion, while being photoevaporated. In AMUSE, two methods can be effective in addressing this problem. It is possible to resolve the disk using a hydrodynamics solver together with the gravity using an *N*-body solver and the radiation with a radiative transfer code. The other way would be using a combination of `Vader` (see Section B.2.3), as a disk-evolution code, and a Kepler solver together with the FRIED grid (see Section B.2.4). Both methods allow you to study various aspects of such a problem. In this assignment we are going to try both methods.

The key aspect for the evolution of an irradiated and gravitationally influenced disk around a low-mass star is the truncation, evaporation, and structural changes in the disk due to the other star. The first method, in principle, incorporates all these effects, whereas the more parameterized method only informs us about the disk's photoevaporation.

Write a script to address the coupled problem by bridging an SPH code for the disk with a radiative transfer code, a stellar evolution code, and a direct *N*-body solver for the mutual gravity.

Write a second script, using `Venice`, where the gravity is addressed with a Kepler solver, the disk using `vader`, the stellar evolution with the appropriate solver, and the radiative interaction with the FRIED grid.

Answer the following questions:

- To what extent is the disk gravitationally affected by the companion star?
- To what extent is the disk gravitationally affected by the radiative transfer?
- What can you say about the difference is speed when comparing the two scripts?
- Which of the two approaches is most satisfactory for addressing the problem's key questions?

References

Adams, D. N. 1979, The Hitchhiker's Guide to the Galaxy (London: Pan Books)

Allen, C., & Santillan, A. 1991, RMxAA, 22, 255

Barnes, J., & Hut, P. 1986, Natur, 324, 446

Bottke, W. F., Vokrouhlický, D., Rubincam, D. P., & Broz, M. 2002, Asteroids III, ed. W. F. Bottke Jr., A. Cellino, P. Paolicchi, & R. P. Binzel (Tucson, AZ: Univ. Arizona Press) 395

Bovy, J. 2015, ApJS, 216, 29

Camarillo, T., Mathur, V., Mitchell, T., & Ratra, B. 2018, PASP, 130, 024101

Chambers, J. E., & Migliorini, F. 1997, AAS Meeting, 29, 27.06

Chambers, J. E. 2012, Mercury: A Software Package for Orbital Dynamics, Astrophysics Source Code Library, ascl:1201.008

Cooder, R. 1972, Crow Black Chicken, track 3 on Boomer's Story, Reprise Records

Cox, D. P., & Gómez, G. C. 2002, ApJS, 142, 261

Eggleton, P. P., Tout, C., Pols, O., et al. 2011, STARS: A Stellar Evolution Code, Astrophysics Source Code Library, ascl:1107.008

Eggleton, P. 2006, Evolutionary Processes in Binary and Multiple Stars (Cambridge: Cambridge Univ. Press)

Eggleton, P. P. 1971, MNRAS, 151, 351

Eggleton, P. P., Kiseleva, L. G., & Hut, P. 1998, ApJ, 499, 853

Farrell, S. A., Webb, N. A., Barret, D., Godet, O., & Rodrigues, J. M. 2009, Natur, 460, 73

Feroz, F., Hobson, M. P., & Bridges, M. 2009, MNRAS, 398, 1601

Foreman-Mackey, D., Hogg, D. W., Lang, D., & Goodman, J. 2013, PASP, 125, 306

Fujii, M., Iwasawa, M., Funato, Y., & Makino, J. 2007, PASJ, 59, 1095

Gaia CollaborationBrown, A. G. A., Vallenari, A., Prusti, T., et al. 2016, A&A, 595, A2

Gerritsen, J. P. E., & Icke, V. 1997, A&A, 325, 972

Gustafsson, B., Church, R. P., Davies, M. B., & Rickman, H. 2016, A&A, 593, A85

Hansen, B. M. S. 2010, ApJ, 723, 285

Harfst, S., Portegies Zwart, S., & Stolte, A. 2010, MNRAS, 409, 628

Hastings, W. K. 1970, Biometrika, 57, 97

Hernquist, L., & Katz, N. 1989, ApJS, 70, 419

Heyer, M., & Dame, T. M. 2015, ARA&A, 53, 583

Huang, R. Q., & Taam, R. E. 1990, A&A, 236, 107

Hut, P., Makino, J., & McMillan, S. 1995, ApJL, 443, L93

Innes, R. T. A. 1915, CiUO, 30, 235

Innes, R. T. A. 1917, CiUO, 40, 331

Ivanova, N., Justham, S., Chen, X., et al. 2013, A&ARv, 21, 59

Karachentsev, I. D., & Makarov, D. I. 1996, AstL, 22, 455
Kervella, P., Thévenin, F., & Lovis, C. 2017, A&A, 598, L7
King, I. R. 1966, AJ, 71, 64
Kokubo, E., & Ida, S. 2002, ApJ, 581, 666
Kozai, Y. 1962, AJ, 67, 591
Landau, L. D., & Lifshitz, E. M. 1969, Mechanics (2nd ed.; Oxford: Pergamon Press)
Lidov, M. L. 1962, Planet. Space Sci., 9, 719
Makino, J. 1991, ApJ, 369, 200
Martínez-Barbosa, C. A., Brown, A. G. A., Boekholt, T., et al. 2016, MNRAS, 457, 1062
McMillan, S. L. W., & Portegies Zwart, S. F. 2003, ApJ, 596, 314
Metropolis, N., Rosenbluth, A. W., Rosenbluth, M. N., Teller, A. H., & Teller, E. 1953, JChph, 21, 1087
Mezger, P. G., Duschl, W. J., & Zylka, R. 1996, A&ARv, 7, 289
Mishurov, Y. N., & Acharova, I. A. 2011, MNRAS, 412, 1771
Miyamoto, M., & Nagai, R. 1975, PASJ, 27, 533
Navarro, J. F., Frenk, C. S., & White, S. D. M. 1996, ApJ, 462, 563
Önehag, A., Korn, A., Gustafsson, B., Stempels, E., & Vandenberg, D. A. 2011, A&A, 528, A85
Paxton, B., Bildsten, L., Dotter, A., et al. 2010, MESA: Modules for Experiments in Stellar Astrophysics, Astrophysics Source Code Library, ascl:1010.083
Paxton, B., Bildsten, L., Dotter, A., et al. 2011, ApJS, 192, 3
Pelupessy, F. I., & Portegies Zwart, S. 2012, MNRAS, 420, 1503
Pelupessy, F. I., van der Werf, P. P., & Icke, V. 2004, A&A, 422, 55
Pelupessy, F. I., Jänes, J., & Portegies Zwart, S. 2012, NewA, 17, 711
Pichardo, B., Moreno, E., Allen, C., et al. 2012, AJ, 143, 73
Plummer, H. C. 1911, MNRAS, 71, 460
Portegies Zwart, S. F. 2009, ApJL, 696, L13
Pourbaix, D., Nidever, D., McCarthy, C., et al. 2002, A&A, 386, 280
Rein, H., & Spiegel, D. S. 2015, MNRAS, 446, 1424
Reipurth, B., & Mikkola, S. 2012, Natur, 492, 221
Romero-Gómez, M., Athanassoula, E., Masdemont, J. J., & García-Gómez, C. 2007, A&A, 472, 63
Romero-Gómez, M., Athanassoula, E., Antoja, T., & Figueras, F. 2011, MNRAS, 418, 1176
Salpeter, E. E. 1955, ApJ, 121, 161
Saz Ulibarrena, V., & Portegies Zwart, S. 2025, CNSNS, 145, 108723
Schönrich, R., Binney, J., & Dehnen, W. 2010, MNRAS, 403, 1829
Springel, V. 2005, MNRAS, 364, 1105
Stroustrup, B. 2013, The C++ Programming Language (4th ed.; Reading, MA: Addison-Wesley)
Taam, R. E., & Sandquist, E. L. 2000, ARA&A, 38, 113
Titarchuk, L., & Seifina, E. 2016, A&A, 595, A101
Tremaine, S. 2015, ApJ, 807, 157
Tricarico, P. 2012, ORSA: Orbit Reconstruction, Simulation and Analysis, Astrophysics Source Code Library, ascl:1204.013
Verlet, L. 1967, PhRv, 159, 98
Webb, N. A., Guérou, A., Ciambur, B., et al. 2017, A&A, 602, A103
Webb, N., Cseh, D., Lenc, E., et al. 2012, Sci., 337, 554
Webbink, R. F. 1984, ApJ, 277, 355

Whipple, F. L. 1950, ApJ, 111, 375
Winn, J. N., & Fabrycky, D. C. 2015, ARA&A, 53, 409
Xiang, M., Liu, X., Shi, J., et al. 2017, ApJS, 232, 2
Yoshida, H. 1990, PhLA, 150, 262
Zemp, M. 2014, Halogen: Multimass Spherical Structure Models for N-body Simulations, Astrophysics Source Code Library, ascl:1407.020
Zemp, M., Moore, B., Stadel, J., Carollo, C. M., & Madau, P. 2008, MNRAS, 386, 1543

Chapter 10

Expanding and Extending AMUSE

> No limit may be set to art,
> neither is there any craftsman that is fully master of his craft.
>
> —The Instruction of Ptahhotep

Congratulations on reaching the final chapter of this book! In the last several chapters, we have demonstrated how AMUSE can be used to run individual codes and, more importantly, how to couple codes to simulate more complex systems. In this final chapter, we present several examples of advanced strategies for coupling multiple codes trading performance for accuracy, and vice versa. In addition, we explore two tightly coupled physical problems in more depth. In the first example, we explore accretion onto a supermassive black hole from a population of massive stars in orbit around it. In the second, we study the effect of a nearby supernova on a young planetary system with a circumstellar disk. The problems in this chapter require considerable computer resources for high-resolution results, but the examples can be run at lower but acceptable resolution with some patience on a good workstation.

10.1 Intricate Multiscale Coupling Strategies

The stars in a cluster are often found in pairs, triples, or even higher-order hierarchies. Democratic hierarchies are notoriously unstable and fall apart into smaller components. In some rare cases, higher-order hierarchies can be stable for a longer time, but eventually most of them dissolve. The fundamental building blocks of a star cluster are ultimately single stars and binaries.

So far, we have only discussed binaries in our discussion on stellar evolution (Chapter 5, Sections 8.7.1 and 8.7.2, and Section 9.4.5). In this chapter, we will discuss some of the numerical complications with binaries. The terms binaries and

doi:10.1088/2514-3433/adfc3cch10

higher-order hierarchies should be read here in their broadest context because we will use these terms to address planetary systems, stars around a supermassive black hole, and ordinary stellar multiples.

Within the AMUSE framework, we have two fundamental quite different methods to deal with hierarchical systems as part of a large star cluster. We discuss them separately in Sections 10.1.1 and 10.1.2.

In each case, we make slightly different assumptions which help us to resolve the dynamics of the multiple systems in a larger stellar environment. As a rule of thumb, we could argue that the `encounters` module (see Section 10.1.1) is meant for simulating star clusters with lots of binaries, whereas the `Nemesis` module (see Section 10.1.2) is used for integrating planetary systems including moons and asteroids inside a star cluster.

Although we present here two methods in which hierarchical systems can be resolved in a large environment, other strategies are possible. One such possibility, effectively employed in several studies, is the use of Kepler's second law (Kepler 1609) to resolve close binaries in a star cluster. The interaction with other stars can then be incorporated as perturbations applied externally onto the binary system. With the various bridging strategies and the information from this chapter, it is quite easy to construct a hybrid integrator in which Kepler solutions replace some of the orbital integrations.

In the final two sections, we discuss various modules developed within the framework. These include the Encounters (Section 10.1.1), Nemesis (Section 10.1.2), Venice (Section 10.1.3), Torch (Section 10.2.1), Ekster (Section 10.2.2), and Fresco (Section 10.5.1) modules.

10.1.1 Encounters: Multiples in Stellar Clusters

The `encounters` module is based on the principle that encounters between a number (>2) of stars are short-lived compared to the global evolution timescale of the cluster. And that, if such an encounter is long-lived, it can be separated into a number of shorter encounters. The underlying assumption of the `encounters` module then is that any encounter is instantaneous with respect to the cluster's dynamical evolution.

The crossing time for a typical stellar cluster is on the order of a few million years (see Section 4.1.2), whereas a 100 au encounter between two solar-mass stars lasts a few thousand years. A democratic resonance encounter does not necessarily last much longer, unless it is associated with wide excursions to large distance in a hierarchical resonance. A star in such an excursion-orbit may then be perturbed again by another star in the cluster, and we argue that such encounters can be separated into individual encounters.

Binaries and multiples can form dynamically in the cluster; but in many cases, we want to start with a population of primordial binaries. In Code example 10.1, we demonstrate how binary stars are set up in the `multiples` module. It encompasses three particle sets, one for the single stars (not in binaries), one for the binary centers of mass, and one for the stars in binaries. Note that the latter are replicated (as

```
    for bi in binaries:
        mp = bi.mass
        ms = np.random.uniform(mmin.value_in(units.MSun),
                                mp.value_in(units.MSun)) | units.MSun

        nb = new_binary_orbit(mp, ms, bi.semi_major_axis, bi.eccentricity)
        nb.position += bi.position
        nb.velocity += bi.velocity
        nb = singles_in_binaries.add_particles(nb)
        nb.radius = ZAMS_radius(bi.mass)
        nb.type = "planet"
        nb.name = "planet"
        binary_particle = bi.copy()
        binary_particle.child1 = nb[0]
        binary_particle.child2 = nb[1]
        resulting_binaries.add_particle(binary_particle)

    single_stars = center_of_masses-binaries
    return single_stars, resulting_binaries, singles_in_binaries
```

Code example 10.1: Setting up binaries for the `multiples` module. The full code example is in `amuse.examples.multiples_minimal`.

pointers) in the binary center of mass particle set as `child1` and `child2`. The particle set provided (called `binaries`) contains three attributes which are relevant for setting up the binaries and for the operation of the `multiples` module, these are the `semimajor_axis`, `eccentricity`, and the `radius`. The first two are used to set up the binary, and could equally well be provided externally. The radius, however, is an important parameter that tells the `multiples` module within what distance a binary should be considered isolated. Here, we adopt three times the apocenter distance, but this parameter is problem-dependent and probably should evolve at runtime.

Once the stars and binaries are generated, we can set up the code structure. When two or more stars in an evolving star cluster approach each other within this `radius`, we stop the global cluster integration. A new N-body code is started within which the encountering stars are integrated until the system dissolves again. The resolved system is subsequently put back in the global N-body integrator and the interim integrator is deleted.

In Code example 10.2, we present the fundamental setup for the `multiples` module. This starts with generating the converter and starting the N-body code. We subsequently start the `encounters` handling module and the `multiples` module. The `encounters` module has two important arguments: the code used for resolving two-body encounters and the code for resolving multibody ($\gtrsim$3) encounters. For the former, we use the standard Kepler solver, and for the latter, we use the `SmallN` solver in AMUSE.

The `multiples` module also has two handling codes: one is the gravity code for integrating the global system, and the other is the `encounters` module, which we have just defined. For the gravity module, we use in `ph4` in this example, but it could be any N-body code.

```
    converter=nbody_system.nbody_to_si(stars.mass.sum(), t_end)
    gravity = Ph4(converter)
    gravity.parameters.force_sync = True

    encounter_code = encounters.HandleEncounter(
        kepler_code=new_kepler(converter),
        resolve_collision_code=new_smalln(converter),
        interaction_over_code=None,
        G=constants.G
    )
    multiples_code = encounters.Multiples(
        gravity_code=gravity,
        handle_encounter_code=encounter_code,
        G=constants.G
    )
```

Code example 10.2: Setting up the multiples module.

```
    multiples_code.particles.add_particles(stars.copy())
    multiples_code.singles_in_binaries.add_particles(planets_in_JuMBOs)
    multiples_code.binaries.add_particles(JuMBOs)
    multiples_code.commit_particles()

    stopping_condition \
        = multiples_code.stopping_conditions.binaries_change_detection
    stopping_condition.enable()
```

Code example 10.3: Providing the particles to the multiples module.

In Code example 10.3, we demonstrate how to add the various particles to the `multiples` module. Here, three types of particles are included: the `stars`, the `planets_in_JuMBOs`, and the `JuMBOs`. The `stars` are all the gravitational objects handed to the code. The `planets_in_JuMBOs` include the particles which are the member of a binary system, and the `JuMBOs`, in this example, represent the centers of mass of the binaries. The term `JuMBO` may be a bit arcane here, but is related to the recent discovery of Jupiter-mass binary objects in the direction of Orion's Trapezium cluster (Pearson & McCaughrean 2023). On the other hand the detection of `JuMBOs` were not confirmed, and the original paper was never published. The scripts to stimulate the dynamical evolution of these hypothetical `JuMBOs` can be used to demonstrate why these objects have no chance to survive in the violent dynamical environment of a young stellar cluster.

One advantage of the `multiples` module is the strict separation of binaries (and multiples) from the single stars. This allows us to invoke binary and triple evolution by treating those substructures with separate methods compared to single stellar evolution. An example of this is presented in the `kira` code in Code example 10.4 and Section 8.7.2.

Although the `encounters` module is quite powerful and speeds up N-body codes with a large number of binaries or multiples, it is best suited to the rather specific problem of relatively separated multiple encounters in a stellar cluster. The assumption that the rest of the cluster is not evolving while the subsystem experiences an interaction may break down when an encounter lasts for a substantial

```
resulting_binaries = Particles()
singles_in_binaries = Particles()
binaries = center_of_masses.random_sample(Nbin)
mmin = center_of_masses.mass.min()
for bi in binaries:
    mp = bi.mass
    ms = (
        np.random.uniform(mmin.value_in(units.MSun), mp.value_in(units.MSun))
        | units.MSun
    )
    a = random_semimajor_axis_PPE(mp, ms)
    e = np.sqrt(np.random.random())

    nb = new_binary_orbit(mp, ms, a, e)
    nb.position += bi.position
    nb.velocity += bi.velocity
    nb = singles_in_binaries.add_particles(nb)
    nb.radius = 0.01 * a
```

Code example 10.4: Example of how to use the encounter module.

amount of time, and when many encounters occur simultaneously. An example would be a star cluster in which the majority of stars are orbited by planets, maybe including some binary stars. To address such a complex multiscale but rather narrow range of physical processes, we developed the more general `Nemesis` module, which we discuss in the following section.

10.1.2 Nemesis: Planetary Systems in Stellar Clusters

What the `multiples` module does for binary stars or more hierarchical multiples, `Nemesis` does for systems with one or more dominating body and several minor bodies, such as a planetary system. The main advantage of `Nemesis` over `encounters` is the consistent time integration. `Encounters` assumes interactions between multiple subsystems to be instantaneous, whereas in `Nemesis` time progresses gradually and continuously, although objects can switch from child to parent system, and vice versa, on a local crossing timescale.

The realization that many stars are potentially born with multiple planets (and smaller objects) begs for a numerical treatment of such hierarchical systems. Stellar distributions around supermassive black holes are quite similar to planetary systems, except that the distributions of orbital eccentricity and inclinations of stars around a supermassive black hole tend to have a wider range compared to planetary systems.

The relatively short orbital periods of planetary systems with respect to the crossing time of a star cluster gives rise to a number of interesting numerical challenges. One of these is the time to solution, which, due to the short time steps for integrating planets, calls for multi-time-step algorithms. A disadvantage of these algorithms, however, is their lack of symplecticity.

In `Nemesis`, each planetary system is integrated with its own individual integrator. When several stars or planetary systems approach each other within a prespecified limit, both systems are merged and further integration is realized by a single integrator. If a multibody system dissolves, single objects are picked up by the global N-body integrator, whereas the multibody subsystems continue to be integrated with their individual solvers. Integrators in this method are dynamically created and deleted at runtime. `Nemesis` is fast, efficient, parallel, and effective in addressing multiplanetary systems in a star cluster (van Elteren et al. 2019). In addition, it is possible for each subsystem to be addressed with its own private integrator. This flexibility makes `Nemesis` quite unique, and applicable to a wide range of problems. One could even address one system with a hydrodisk, while around another star the disk has already coagulated into a planetary system. In principle, a child system could itself be a version of `venice` or `multiples`, or even a complete magneto radiative hydrodynamical simulation. Here, however, we focus the dynamical aspects of `Nemesis`.

In `Nemesis`, we integrate planetary systems and stars together, but not entirely self-consistently. Planets or planetesimals on wide orbits may move between their host child (the planetary system) and the parent system (star cluster). As a consequence, that particular particle will be integrated for parts of its orbit with different integrators. An example would be an Oort-cloud object orbiting a star which is a cluster member. The planetesimal then will be integrated in the child system when near pericenter, say within 1000au, but beyond a certain point it will be considered part of the parent system. The logic associated with this child-to-parent transition is, despite being parallelized, expensive in the current implementation of AMUSE. In addition, the child system will lose its symplectic qualities by regularly adding and removing particles. The errors associated with these changes are small, but noticeable for long timescale integrations.

Child systems are composed of small numbers of stars, planets, and other minor bodies. These stars and planets have a relative interaction time step smaller than some fraction η of the free-fall time. The parent system (including the collective of child system centers of mass) is typically integrated using a fourth-order Hermite code, or maybe a treecode. Each child system is integrated using a high-order N-body code for small-N systems, such as `rebound` (Rein & Liu 2012), `Mercury` (Chambers & Migliorini 1997), or `Huayno` (Pelupessy et al. 2012).

In calculating the force on the center of mass of a subsystem, we take the substructure of the other subsystems self-consistently into account. This procedure ignores the gravity of the child system's minor bodies on the internal evolution of each of the other child systems' minor bodies, and vice versa. While integrating the Alpha Centauri system, we would then, for example, ignore the gravitational effect of planetesimals in the solar system. The errors that result from this procedure can be estimated, and turn out to be negligible, as expected.

Nemesis is implemented in a number of bridge schemes (see Chapter 9). These bridges integrate the equations of motion of the individual components (collective and subsystems) self-consistently via a second-order kick-drift-kick (KDK) method. One advantage is that the entire method is independent of the integrators used for either the collective or of the subsystems. Individual bridges are needed between the cluster and the center-of-mass subsystems, and between all subsystems and the cluster. These bridges connect the split parts by employing a second-order KDK method. Kicks are calculated from the gravitational field of the kicking cluster, and drifts are the evolution of the models in the separate N-body solvers (Portegies Zwart et al. 2020).

The bridge between the cluster and each subsystem needs to take into account that the subsystem is represented in the cluster by a center-of-mass particle. Two adaptations to a vanilla KDK method are needed for this. First, to avoid an extra self-gravity force, the kick has to be calculated with all particles except the center-of-mass particle of the subsystem in question. Next, to avoid doubly calculating the gravitational field of the cluster on the ensemble of the particles, once during the evolution of the particles in the direct N-body code and once with this kick, the kick on the central particle from the cluster must be subtracted from all kicks on the particles in the subsystem.

The bridge between the subsystems and the cluster needs similar adaptations. Now everything is reversed; the particles in each subsystem kick the center-of-mass particles in the cluster, except the one representing the subsystem. The kick of the center-of-mass particle of each subsystem is subtracted from the sum of the kicks of each particle in the subsystem. Figure 10.1 presents a schema of the method. Of all the possible interactions between the particles, we ignore the forces between the particles of one subsystem and another.

Although we have accounted for the interactions between the cluster and the subsystems, we do not yet have a working method because each kick on a subsystem will result in a velocity component of the center of mass of that subsystem. The same is true of each drift phase: the N-body solvers should conserve the center-of-mass position and velocity but small errors will cumulatively result in an increasing position offset and velocity component. After the first kick and for each drift of a KDK step, we need to absorb the translation of the center-of-mass velocity and position into the center-of-mass particle in the cluster. This is not necessary for the

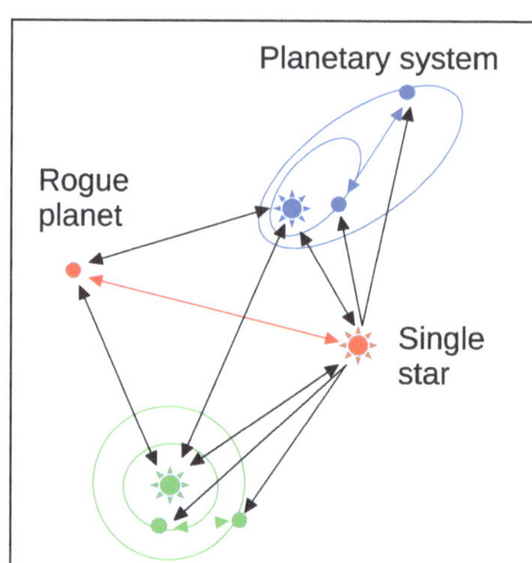

Figure 10.1. Strategy to reduce computing cost without loss of accuracy in `Nemesis`. The planets in one system (blue) interact with the center of mass of another system (green) but not with its planets. Single stars and free-floating planets are marked in red.

second kick because the kick is independent of the velocity component and this component will be resolved in the first kick of the next step.

Dividing the cluster in this way introduces a new problem: the gravitational evolution of each subsystem can disintegrate or subdivide the subsystem. The subsystem would grow in size and no longer evolve on a separate space scale as the cluster. A clustering algorithm is used to determine for each subsystem if it needs to be restructured in one or more new subsystems. This algorithm is applied after each KDK step.

Similar to subsystem disintegration, subsystems can merge to form a new subsystem. We detect these mergers by defining a radius on each center-of-mass particle and interrupting the drift phase of the center-of-mass system whenever the distance between two of these particles is smaller than the sum the individual radii. Before merging the subsystems, we need to bring the state of each system to the merge time by applying a kick and drift on each subsystem. The kick will be subtracted if the time is before half the step time and added otherwise. Next, each system is added to the new subsystem and the new center-of-mass particle is added to the cluster. The drift of the cluster can then continue as normal.

10.1.3 Venice: A Generalized Multiscale Bridge

Bridge is powerful, but limited by the rigidity of its implementation. These limitations include the order of the integration and the constant time step. Both can be relaxed. Here, we show how these limitations can be mitigated by using a generalization we call `Venice` (Wilhelm & Portegies Zwart 2024).

`Venice` is an operator-splitting algorithm designed to address the challenges of integrating numerical models across multiple timescales in multiphysics simulations. As in bridge, `Venice` is a multiphysics coupling strategy for combining a diverse collection of physical processes operating on vastly different timescales.

The traditional bridge approach becomes computationally expensive and complex due to the need to synchronize all processes on the shortest common time step. In `Venice`, we overcome these limitations by enabling adaptive coupling of independent components on their individually relevant timescales, thereby improving both performance and maintainability.

In `Venice`, we connect all available codes in a matrix of bridges. Each bridge coupling is associated with an interaction timescale, resulting in a matrix of timescales. The coupling is second-order, as in classic bridge, but in principle this is easily generalized. Dynamically changing the order of individual bridges, however, is currently not supported within `Venice`. The coupling strategy, however, is considerably more flexible than in the traditional bridge. In searching for a name for an elegant method representing a matrix of bridges, we settled on *La Serenissima*: the Italian city of canals.

The adaptive time-stepping allows all components to operate on an individual natural timescale. As in Equation (9.3), we split the general operator and solve each part separately. After solving the individual parts, we combine them into a unique, self-consistent solution. The mathematics is a bit daunting, for which we refer to

Wilhelm & Portegies Zwart (2024). With `Venice` in mind, one could perceive bridge as a simpler bidirectional strategy.

The main advantages of `Venice` include its temporal flexibility and the modular design, making it easier to maintain and extend simulation frameworks and optimize multiscale and multiphysics solvers. The algorithm supports an arbitrary number of operators and coupling strategies, making it suitable for a wide range of multi-physics problems.

To make `Venice` work properly in AMUSE, we use channels to copy the data from one code to another. These are defined directly between codes and allow any subset of data to be copied from one code to the other. They must be explicitly called, which is done between operations. In a more relaxed approach, the copying could be automatically realized after the call to the specific code, but this would result in an end state where the data of different codes is not synchronized, in which case it is nontrivial to determine which code contains the most recent state. The explicit approach, where we copy all data from a code after its evolution step, introduces some unnecessary copy operations, but is less prone to mistakes. Of course, this can be optimized, in particular if one takes the multiscale modeling paradigms (Borgdorff et al. 2011) properly into account.

As an example of how `Venice` operates, we dynamically evolve a stellar cluster with a few very massive stars. The fundamental setup can be written as follows

```
from venice import Venice
def make_gravity_stellar (converter, timestep, verbose=False):
    stellar = Seba()
    gravity = Ph4(converter)
    gravity.parameters.force_sync = True

    system = Venice()
    system.verbose = verbose
    system.add_code(stellar)
    system.add_code(gravity)

    system.add_channel(0, 1, from_attributes=["mass"], to_attributes=["mass"])
    system.timestep_matrix[0,1] = timestep
    return stellar, gravity, system
```

First, we initialize both codes for stellar evolution and dynamics. Note that we added the line

```
    gravity.parameters.force_sync = True
```

to guarantee that the N-body code properly synchronizes to the requested time (see also Section 8.4.4). Some N-body codes do not naturally synchronize to precisely the time requested because it introduces extra numerical errors, slows down the code, and is often not necessary. For `Venice`, however, we do require the codes to be synchronized at the end of a step.

The actual `Venice` system is initiated in much the same way we initialize a bridge in the lines

```
system = Venice()
system.add_code(stellar)
system.add_code(gravity)
```

Here, the script is started, and the (in this case) two codes are added. Running the code is not very different than a usual code bridge.

```
timesteps = [3., 1., 0.3, 0.1, 0.03, 0.01, 0.003, 0.001] | units.kyr
stellar_l, gravity_l, system_l = make_gravity_stellar(converter, timesteps[i])

stellar_l.particles.add_particles(cluster_l_copy)
gravity_l.particles.add_particles(cluster_l_copy)
channel_l = gravity_l.particles.new_channel_to(cluster_l_copy)

stellar_i, gravity_i, system_i = make_gravity_stellar(
    converter,
    timesteps[i]
)
system_i.interlaced_drift = True

stellar_i.particles.add_particles(cluster_i_copy)
gravity_i.particles.add_particles(cluster_i_copy)
channel_i = gravity_i.particles.new_channel_to(cluster_i_copy)

system_l.evolve_model(end_time)
channel_l.copy()
system_l.stop()
```

Subsequently, running the code is not much different than any other code in AMUSE.

With `Venice`, we can couple an arbitrary number of codes and run them on their own preferred timescales. If you encounter an astrophysical problem in which two codes are coupled, `Venice` is probably overkill, but you should consider using this strategy once more than two codes require coupling. In Figure 10.2, we present an example of the working of `Venice`, showing the evolution of the stellar time step and type as a function of time.

10.2 Advanced Coupling Projects

10.2.1 Torch: Forming Star Clusters from Molecular Clouds

One of the objectives of the MODEST community is to model the formation and evolution of a star cluster from the moment a giant molecular cloud starts to collapse. This has led to the `Torch` project,[1] in which AMUSE is used as coupling framework to model star cluster formation by integrating multiple simulation codes.

Installing `Torch` is nontrivial. After AMUSE is properly installed, the MHD code FLASH needs to be installed, after which the `Torch` package can be added. You can download an academic version of FLASH from http://flash.uchicago.edu/site/flashcode.

If not all packages are installed in a conda environment, you may have to set some environment variables in your shell

[1] See https://torch-sf.bitbucket.io/quickstart.pdf

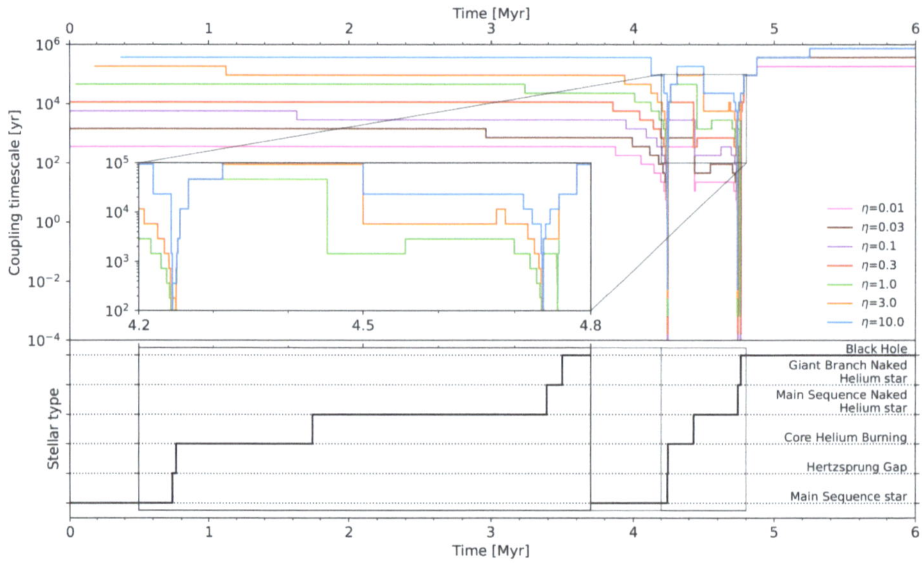

Figure 10.2. Illustration of the dynamic coupling timescale between gravitational dynamics and stellar evolution (top panel). The calculation is performed for a range of coupling timescales η from 0.01 to 10. The bottom panel illustrates the type of the most massive star in the cluster (of $50\,M_\odot$) star in time. We illustrate how different coupling timescales lead to slightly different moments the most massive star in the cluster is updated (see Wilhelm & Portegies Zwart 2024). `amuse.examples.venice_minimal`. It should take about 10 minutes to run on a laptop.

```
export AMUSE_DIR=/path/to/amuse
export FLASH_DIR=/path/to/FLASH4.5
export TORCH_DIR=/path/to/torch
export PYTHONPATH=$PYTHONPATH:$TORCH_DIR/src
```

Then, run the installation run script:

```
./install.sh
```

You will be prompted to confirm your environment variables, and continue with the installation. The installer will then automatically copy interface code from `TORCH_DIR/src` to their correct locations in your AMUSE and FLASH repositories. It will also ensure that the FLASH4.5-a patch for HDF5 >1.10.x compatibility is applied in your FLASH repository. The main script is `torch_user.py`, but we encourage you to make a copy in some local directory and change it to your liking.

In Figure 10.3, we show the analytic energy injection of a pressure-dominated snowplow phase of a supernova explosion in a gaseous medium using the `Torch` simulation package. The various vertical lines give the Sedov–Taylor (t_{ST}), thermal (t_{TR}), and snowplow (t_{PDS}) transition times. Figure 10.3 demonstrates how in `Torch` these high-energy phenomena are satisfactorily represented.

The Torch module uses three codes coupled using the AMUSE framework. These codes are the MHD-radiative-transfer hydrodynamics code FLASH (Fryxell et al.

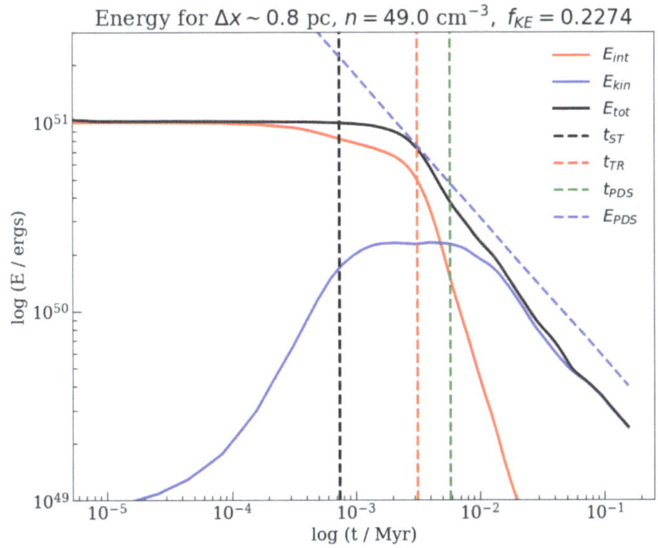

Figure 10.3. The Sedov–Taylor (t_{ST}), thermal (t_{TR}), and snowplow (t_{PDS}) transition times (Draine 2011) are shown as vertical dashed lines, and the analytic form of the energy during the pressure-dominated snowplow phase is shown as a blue dashed line above the energy during this phase. Reproduced from Wall et al. (2020). The Author(s). CC BY 4.0.

2000), the direct N-body code `PeTar` (Wang et al. 2020), and the binary evolution code `SeBa` (Portegies Zwart & Verbunt 1996). Although each of these codes could in principle be replaced by any other in AMUSE, the `Torch` script has become quite intricate, and by now it may not be trivial to replace, e.g., FLASH for Ramses (Teyssier 2002). On the other hand, it is somewhat easier to replace `PeTar` for the `multiples` module, and `SeBa` for `Mesa`, for example.

The performance and accuracy of `PeTar` are tuned using the parameter `r_out` which is a boundary radius between the tree- and direct N-body gravity. The default is tuned for simulating star clusters, but by all means test this and modify it for your particular use case. `Torch` can also use `VorAMR` (Lewis et al. 2023), a utility developed within the `Torch` framework that allows for the conversion of output data from `AREPO` to be converted to an AMR grid. An example is provided in Figure 2.1 of Section 2.2.2.

A `Torch` calculation starts with a giant molecular cloud that collapses under its own gravity. Star formation is mediated through sink particles. We use AMUSE's sink-particles module to handling the sinks (see Section 6.3.2). Stars are spawned from these sinks according to a predetermined mass function. Stellar feedback is facilitated by the transfer of stellar parameters (luminosity, mass-loss rates) from `SeBa` to FLASH, where these parameters are used to calculate the effects of radiation, winds, and supernovae on the surrounding gas.

Recently, we added anisotropic winds, or jets to simulate the formation of low-mass clusters (Appel et al. 2025). Protostellar jets and outflows are critical phenomena associated with star formation, launched by the interaction of magnetic fields and accretion disks. In Figure 10.4, we present a test result of a simulation using jet outflows from a young stellar object in `Torch`.

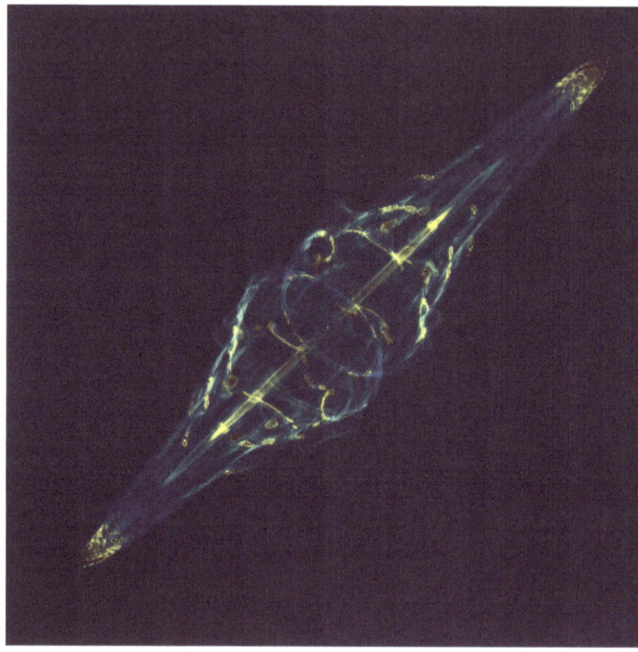

Figure 10.4. A volume rendering of a protostellar jet produced by a single star particle in a uniform density box. This volume rendering was generated from a simulation run using the jets branch of the Torch code. Note that although the injection weighting is smooth, as described in Appel et al. (2025), hydrodynamical instabilities develop as the injected material propagates into the denser background material.

10.2.2 `Ekster`: Forming Clusters from Gas to Star

The `Ekster`[2] module in AMUSE is somewhat similar to the Torch endeavor (Rieder 2019; Rieder & Liow 2021). `Ekster` has a similar philosophy as `Torch`, except that it uses an SPH method for the hydrodynamics. `Ekster` was developed with the SPH code `Phantom` (Price et al. 2018) as its primary hydrodynamics code, although following the AMUSE philosophy it is possible to use any other SPH code, provided it supports the high-density stopping condition, see Section A.5.2. Like `Torch`, `Ekster` combines hydrodynamics with gravity and stellar evolution, by default using the same codes as in `Torch` (`PeTar` and `SeBa`).

The application domain for `Ekster` is quite similar to that for `Torch`, as was illustrated in Sills et al. (2023). `Ekster` was developed with relatively low mass resolution in mind ($\sim 10^5\ M_\odot$ per gas particle), allowing it to run on modest-sized machines (anything from a laptop to a small cluster). When gas reaches a sufficiently high density, star formation starts through the formation of sink particles, each of which acts as a star-forming region. The mass in the sinks is converted into stars, which are spatially distributed within the sink's accretion radius using a Gaussian distribution. The mass of the stars is sampled from a predetermined mass function.

[2] Ekster is the Dutch name for the European magpie. Also, it is a homophone for Exeter, where the module was primarily developed. `Ekster` is available from: https://github.com/amusecode/amuse-ekster.

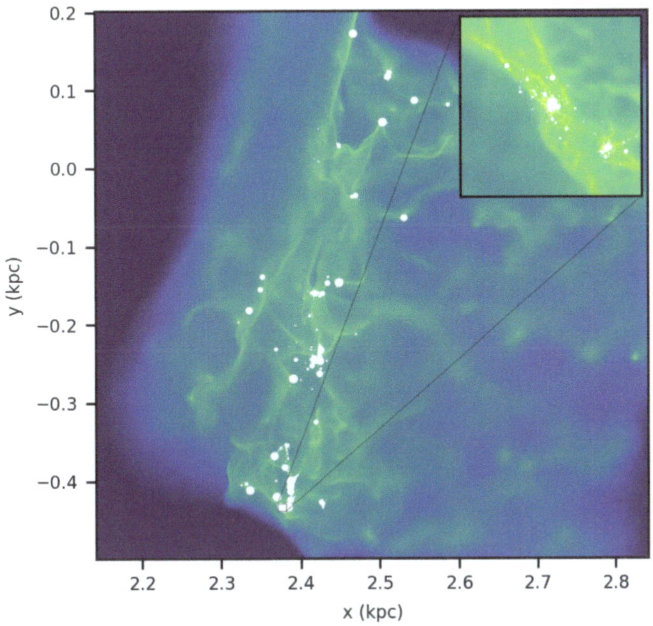

Figure 10.5. One of the regions CRC0060 at an age of 3 Myr (Rieder et al. 2022). The inset shows a 10 parsec zoom of the largest star cluster. Symbol sizes are proportional to the stellar radius. Adapted with permission from Rieder et al. (2022). © 2022 The Authors. Published by Oxford University Press on behalf of the Royal Astronomical Society.

After star formation, the sink's accretion radius shrinks so that its "density" remains constant. This prevents the sink moving towards and accreting low-density gas regions.

As an alternative for simulations with a slightly higher mass resolution, `Ekster` also has a mode where multiple sinks can be grouped according to their positions, velocities, and ages. These groups can then "borrow" mass from each other for star formation. This allows massive stars to form even when individual sinks are unable to accrete enough mass for such stars (Liow et al. 2022).

In Figure 10.5, we show a small section of a simulated galaxy (Dobbs & Pringle 2013) for which the simulation was continued using `Ekster` in AMUSE (Rieder et al. 2022). It shows one of the active star-forming regions in the simulated galaxy at an age of 3 Myr after the collapse of the giant molecular cloud.

10.3 Supernova Impact on the Early Solar System

Meteoritic evidence suggests that short-lived radionuclides—specifically ^{26}Al and ^{60}Fe, with half lives of 0.717 and 1.49 Myr, respectively (Schiller et al. 2015)—were incorporated into grains in the primitive solar nebula about 2 Myr after those grains formed (Luu et al. 2015; Scott 2007). As discussed in Section 8.6.1, the origin of these unstable elements is uncertain, but one hypothesis is that a nearby supernova might have enriched the solar circumstellar disk with these radionuclides shortly after the Sun formed. The process would have required high temperatures (1500–2000 K) to melt the grain material. In principle, the same supernova could have produced the ^{26}Al and ^{60}Fe radionuclides, delivered them to the disk, and heated the disk enough to melt the hardest grains. For this to work, the supernova must have exploded within about 0.5 pc and within the first 1.5 Myr of the budding solar system. The proximity is necessary to guarantee a sufficient amount of short-lived radionuclides, and the timescale stems from the half-life of the volatile elements.

It might seem extraordinary to posit a type II supernova so close to the Sun just 2 Myr after it formed. However, considering that the Sun was probably born in a star cluster, the probability of a such an event is actually quite high. Looney et al. (2006) argue that a 23 $M_{\odot}$ star could deliver a sufficient amount of short-lived radionuclides when it explodes at an age of about 8.4 Myr. The time difference can plausibly be explained by an age spread in the star formation process, in which the Sun formed somewhat later than the most massive stars (Stahler & Palla 2005).

The possibility of a nearby supernova was first suggested by Chevalier (2000). Subsequent hydrodynamical calculations have been performed to study the amount of supernova remnant material that could be accreted by the protoplanetary disk (Ouellette et al. 2007). These simulations suggest that the supernova might not be able to deposit sufficient mass on the disk to match the observations, but made some simplifying assumptions and were not conclusive. One of those assumptions was homogeneous density throughout the slab of supernova material hitting the disk. Here, we take the opportunity to revisit the problem using AMUSE, combining gravitational dynamics, hydrodynamics, and radiative transfer codes to explore this scenario. Unlike previous studies, we investigate both the initial irradiation of the disk by the supernova flash and the direct effects of the subsequent blast wave, both of which could have contributed to the heating of the disk. Most of the ingredients of the scripts described below have already been discussed in previous chapters.

We start by generating a protoplanetary system with a star, some gas-giant planets, a gas disk, and debris. This system is evolved using a bridged hydrodynamics–gravity solver (see Section 9.3.2) in which we integrate the debris disk as test particles in the combined potentials of the planets and the gas disk. We ignite a supernova at a distance of $d_{\rm sn} = 0.1$ pc and assume that the normal to the disk is inclined at an angle of 60° to the direction to the supernova, so the radiation and the blast wave impinge on the disk almost edge-on. A head-on blast wave would be less effective in tilting the disk, and would damage the disk more severely.

The supernova explosion is modeled in two stages. First, a radiative transfer code is started to irradiate the protoplanetary disk roughly $d_{\rm sn}/c \sim 100$ days after the explosion. Then, approximately 10 years later, the blast wave arrives at a speed of $\sim$10,000 km s^{-1}. At the adopted small distance, the supernova shell is still in its free-expansion phase and the hot gas has not been significantly slowed by the interstellar medium. The short distance, and therefore short travel, time guarantees that the short-lived radionuclides are still abundant in the blast wave. We study the effect of the supernova irradiation and the subsequent blast wave on the disk, and determine if the disk survives and if a sufficient amount of short-lived radionuclides can be deposited.

10.3.1 Initial Conditions and Model Parameters

We start the simulation by generating a young solar system with four giant planets using the method discussed in Section 4.4.1. The underlying assumption is that the rocky planets have not yet emerged, so instead we incorporate a debris disk. The solar system was generated at an arbitrary Julian date of 2474649.5 days[3] After initialization, we recenter the system to place the center of mass at the origin.

The gas disk around the star is generated using $n_{\rm disk} = 10000$ equal-mass SPH particles between $r_{\rm min} = 1$ au and $r_{max} = 100$ au from the Sun, with total mass of $0.01\,M_\odot$. We adopt a flat density profile with Safronov–Toomre Q-parameter $q_{out} = 25$ (see Section 6.4.1). You might wish to experiment with higher resolution, but note that the calculation is quite expensive in terms of computer time, and parallelization is limited to the OpenMP implementation of the adopted hydrodynamics solver (see below).

```
converter = nbody_system.nbody_to_si(mass_star, radius_min)
disk = ProtoPlanetaryDisk(n_disk, converter, radius_min, radius_max, q_out=25.0,
        discfraction=0.01).result
n_planets = int(planets.mass.sum() / disk[0].mass)
disk = disk.random_sample(n_disk - n_planets)
```

Since we generate the particle set `planets` independently of the gas, we randomly remove gas particles from the disk until the total mass is $0.01\,M_\odot$. We then generate a debris disk composed of $n_{\rm debris} = 1000$ planetesimals orbiting the star. This is a small number, but sufficient to trace the potential. The debris disk density distribution is generated in much the same way as the gas disk, except that we do not correct for the planetesimal masses, which are assumed to be negligible.

As in Section 8.6.1, we model the supernova light curve using the multiple power-law fit of Sanders et al. (2015) to supernova PS1-10a. We use parameters from their Table 3, with a peak luminosity of 2.3×10^{42} erg/s ($10^{8.8}\,L_\odot$) a few days after the explosion.

10.3.2 The Combined Solver

The first part of the calculation combines a hydrodynamics solver for the circumstellar disk, a radiative transfer code, a gravity solver for the giant planets, and a test-particle solver for the massless debris. The radiative transfer and the hydrodynamics are coupled as described in Section 8.6.1. In fact, the script for the combined hydro and radiative transfer is quite similar to the one presented in that section. It differs mainly in that the Sun is no longer a dark-matter particle in the hydro code; now it is integrated together with the giant planets using the N-body code.

[3] Of course, the precise Julian date is irrelevant, but this corresponds to First Contact: http://www.startrek.com/article/origin-of-first-contact-day-explained.

The debris disk is resolved using the Kepler solver in the Huayno integrator:

```
debris_gravity = Huayno(converter, number_of_workers=12)
debris_gravity.inttype = KEPLER
debris_gravity.particles.add_particles(debris)
debris_gravity.timestep = 1 | units.hour
```

Here, we start Huayno with a Kepler solver (KEPLER = 14) and run it on 12 cores concurrently (see Section A.3.3 for some of the other options available in Huayno). The debris disk of massless particles is subsequently sent to the N-body code, with a fixed time step of 1 hour. This is very short, but the time step in this experiment is set not by the dynamics but by the radiative transfer. Later, we could increase the time step again.

For the N-body code to integrate the planets and the Sun, we adopt ph4, which is initialized with

```
planet_gravity = Ph4(converter)
planet_gravity.particles.add_particles(star_and_planets)
```

After declaring the various solvers, we build the bridge, in this case composed of two gravity solvers and the radiation-hydrodynamics solver discussed in Section 8.4.2:

```
gravity_hydro = bridge.Bridge(use_threading=True)
gravity_hydro.add_system(debris_gravity, (planet_gravity, radhydro,))
gravity_hydro.add_system(planet_gravity, (radhydro,))
gravity_hydro.add_system(radhydro, (planet_gravity,))
gravity_hydro.timestep = min(1 | units.hour, time_step / 4.)
```

In the first line, we start the bridge, setting threading to True to speed up the calculation. In the second line, we add the debris_gravity code to the bridge, specifying that it is affected by planet_gravity and the radiative-hydrodynamics solver radhydro. In the third and fourth lines, we add the gravity code for the planets and the radiative-hydrodynamics solver. These are affected by each other but not by the debris disk. In the last line, we set the bridge time step to 1 hour (or even smaller), similar to the internal time step for the Kepler solver.

10.3.3 Radiative Hydrodynamics with Cooling and Heating

The simulation starts when the bright irradiation of the supernova reaches the young solar system. During this period, the heating and cooling of the disk is calculated self-consistently with the radiative transfer code. Figure 10.6 presents side and top views of the solar system 6 years after the ignition of the supernova. In the radiative-hydro calculation we assume that the gas is isothermal and free of dust. The lack of dust reduces the cooling efficiency of the disk material. With a dust mass fraction of 0.1, the equilibrium temperature in the radiation field drops by about 20%.

Once the luminosity of the supernova drops below $10^6\,L_\odot$ (after about a year of irradiation), we switch off the radiative transfer code, because it is rather expensive to keep using it just for the cooling of the gas, and turn on the external cooling functions for the gas (see Section 6.3.4). The remaining codes continue to run as before.

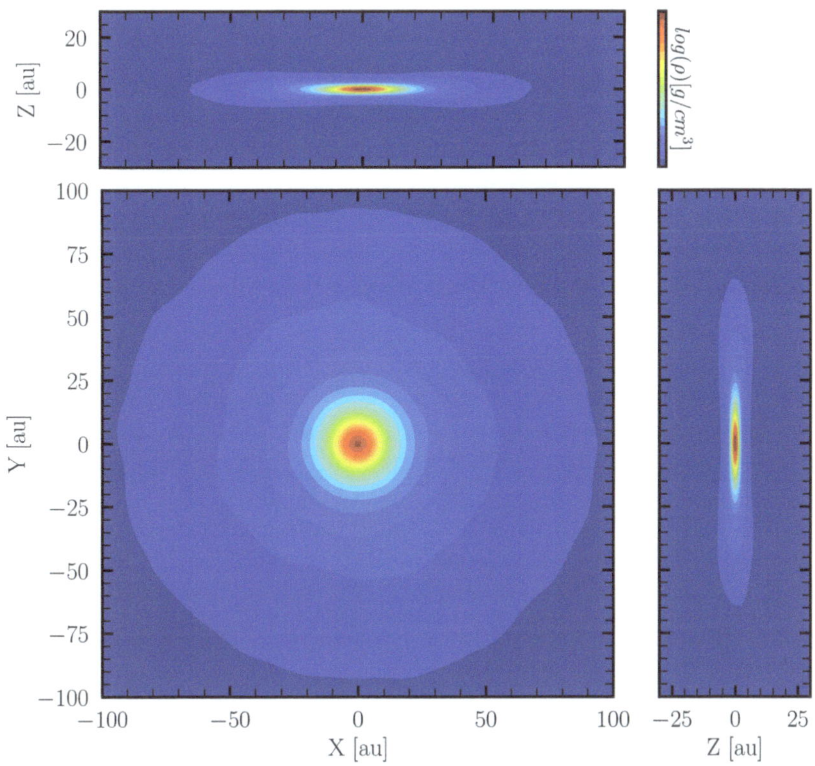

Figure 10.6. View of the solar system 6.4 years after the supernova explosion. By this time, the supernova has dimmed to such a low luminosity that there is no reason to continue the radiative transfer, but the blast wave has not yet arrived. The color bar at the top right of the figure presents a scaling of ρ ranging from $\sim 10^{-13}$ to $\sim 10^{-10}$ g cm^{-3}. The code to reproduce this result is `amuse.examples.supernova_irradiation`, which took a few days to run on a fast workstation. The figure was made using `amuse.examples.plot_projected_disk_density_from_top`.

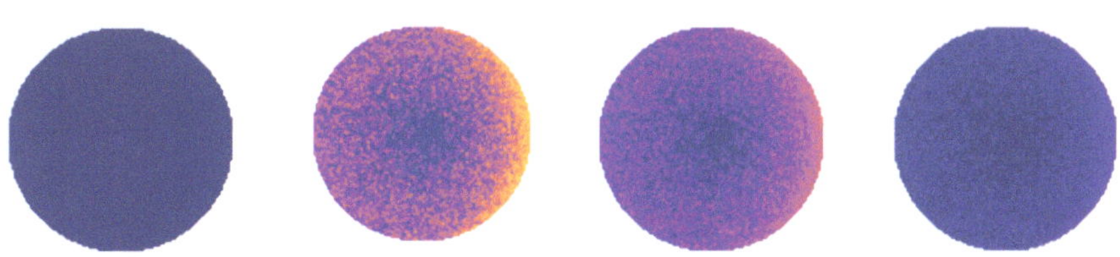

Figure 10.7. Top view of the midplane temperature of the circumstellar disk at four epochs: the start of the simulation (leftmost image); 26 days after the onset of supernova PS1-11aof; about 100 days after the supernova; and about 2 years later, when the disk has almost completely cooled down again (rightmost image). Each image was taken from the simulation where the supernova exploded at a distance of 0.2 pc and at an angle of $\theta = 6°$. The figure was made using `amuse.examples.plot_projected_disk_temp_from_top`.

Figure 10.7 shows top view images of the surface temperature of the circumstellar disk at four moments in time. Although the flash of radiation does briefly raise the temperature of the disk to well over 2000 K (second image from the left in

Figure 10.7), the subsequent cooling is so efficient that by the time the supernova shock wave reaches the young solar system 10 years later, the disk has cooled back to its “normal” (presupernova) temperature of 20–200 K. In addition, the heating and cooling are so sudden that the disk hardly has time to respond, and as a result its morphology is only slightly affected by the supernova flash. The upshot is that, by the time the blast wave arrives, the disk structure is essentially the same as its initial state.

10.3.4 Injection of the Supernova Blast Wave

The hot radioactive outflow from the supernova reaches the solar system about 9.8 years after the explosion. We inject the hot gas at a distance of 10 au from the solar system. The hot shell is initialized in a cylinder directed at the solar system. This cylinder has a radius of $r_{\rm shell} = 150$ au, to ensure that the entire solar system is engulfed, and the gas moves parallel to its axis. The density is uniform across the cylinder, but along the axis the density varies according to

$$\rho \propto \frac{1}{z + 10\,\mathrm{au}}, \tag{10.1}$$

where z is distance along the axis, with $z = 0$ at the shock front. The total mass in the cylinder is

$$m_{\rm shell} = \frac{1}{4} m_{\rm eject} \left(\frac{r_{\rm shell}}{d_{\rm sn}} \right)^2. \tag{10.2}$$

For SN PS1-10a, the estimated total mass of the ejecta is $m_{\rm eject} = 6\,\mathrm{M}_\odot$.

The initial temperature of the expanding supernova shell is about $T_{\rm eject} = 9 \times 10^7$ K at a shell radius of 10 au, but by the time the gas reaches the solar system it has cooled to (Spicer et al. 1990)

$$T_{\rm shell} = T_{\rm eject} \left(\frac{d_{\rm sn}}{10\,\mathrm{au}} \right)^{-3(\gamma-1)}, \tag{10.3}$$

so $T_{\rm shell} \simeq 9500$ K. Here, we have adopted $\gamma = 7/5$. The sound speed in the shell is about $c_s \simeq 5.8\,\mathrm{km\,s^{-1}}$, which is negligible compared to the speed of the shell at impact.

10.3.5 The Supernova Blast Wave Hits the Disk

In Figure 10.8, we present the surface density profile of the disk at three moments in time: the moment the blast wave hits the disk, 55 days later, and 3 years after the impact. The last curve (dark purple) shows the solar system’s debris disk, if we spread out all planets and planetesimals over a narrow ring. As we have just seen, the disk structure just before the blast wave hits is almost indistinguishable from the initial disk profile because the disk has cooled back to its preirradiation state. Shortly after the blast wave hits, the disk is most strongly affected near the edges by the ram pressure of the shock front, but the disk has little time to respond—the

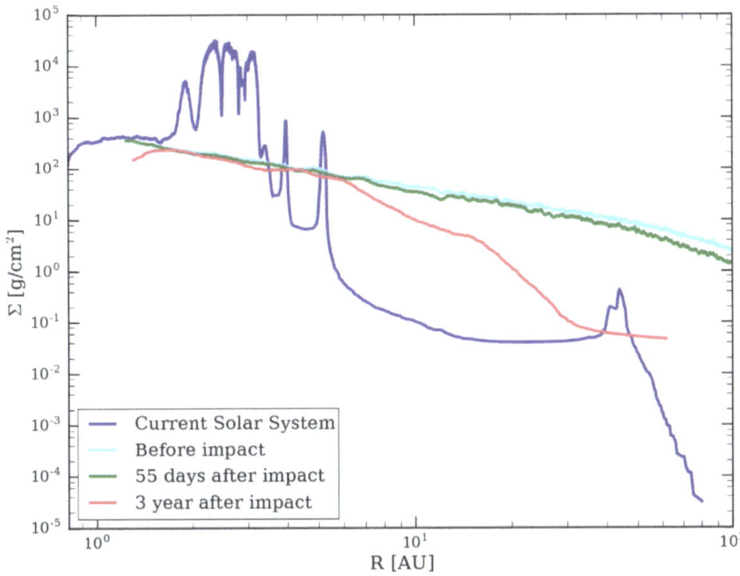

Figure 10.8. Surface density profile of the protoplanetary disk at three moments in time: just before the blast wave hits the disk (blue), 55 days later (green), and 3 years later (red). After about a year the disk has a mass of $\lesssim 0.001\,M_\odot$. Also shown (dark purple) is the current mass distribution of planetesimals in the solar system.

shock front takes only a few weeks to cross the disk. Afterwards, the disk erodes more rapidly. Fifty-five days after the supernova blast wave hit the disk, the disk mass has dropped to half its initial value ($\sim 0.0052\,M_\odot$, from an initial $0.01\,M_\odot$). Most of the mass is lost from the outermost parts of the disk. After 1 year, the disk mass has dropped by an order of magnitude (to less than $0.001\,M_\odot$).

From that point on, the disk structure remains largely unchanged, mainly due to the low density of the supernova material at late times. The final state of the disk is shown as the red curve in Figure 10.8, $\sim$3 yr after first contact. We note that the planets and planetesimals are barely affected by the supernova, remaining in basically unchanged orbits around the central star. For comparison, Figure 10.8 also presents an estimate of the density profile of the current solar system's planetesimal disk. Here, we simply assume that each planetesimal contributes equally to the local density and that the solar system disk has a total mass of $0.01\,M_\odot$, equivalent to the initially adopted disk mass. If the planetesimals had already formed by the time of the supernova shell impact, they would hardly be affected by the blast wave. It is interesting to note in Figure 10.8 that the disk becomes considerably depleted beyond 6–7 au, where we also find a huge drop in the current planetesimal distribution. One might wonder if the current depletion beyond this range can in part be explained by such a supernova blast wave.

Figure 10.9 presents the temperature evolution of the protoplanetary disk. The temperature rises about 100 days after the supernova explosion due to the supernova flash. During this first irradiation phase, the disk becomes sufficiently hot to melt the

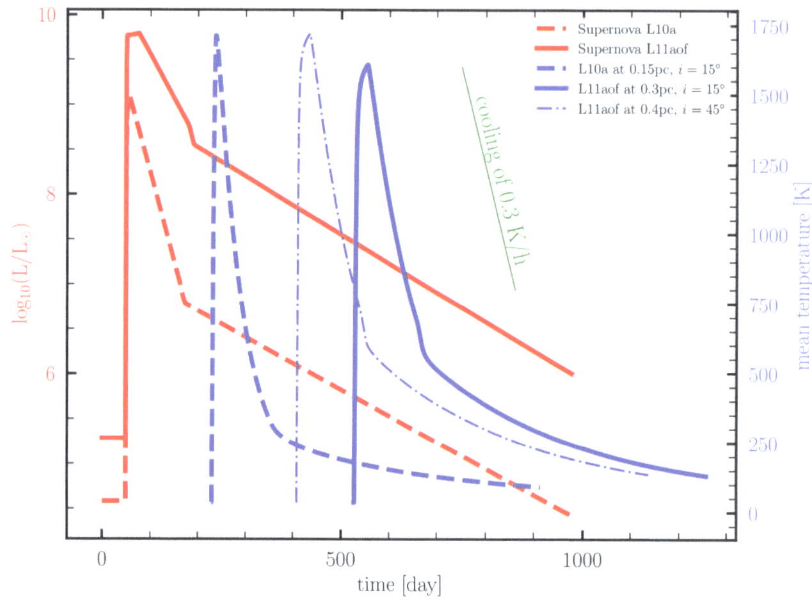

Figure 10.9. Evolution of the mean temperature of the disk due to the supernova explosion. The supernova radiation starts to heat the disk some 100 days after the explosion, and a decade later the hot nuclear blast wave starts to heat the disk. The two horizontal lines indicates temperatures of 1500 and 2000 K, which are sufficient to melt hard chondrules.

hardest chondrules. One can wonder, however, if this initial melting of the disk has any long-term consequences for the evolution of the solar system. The problem as initially posed sought to explain the enrichment by radioactive elements, which are delivered with the blast wave, but clearly the time between the irradiation and the blast wave is sufficiently long for the disk to cool down again before the shock arrives.

The nuclear blast wave hits the disk about 10 years later, heating it again to high temperatures. The blast wave does deposit some material in the Sun's protoplanetary disk, although the total amount of mass is small, less than 1% of the current disk mass. This is probably too small to explain the measured enrichment of radioactive elements in the early solar system, although these melting periods might have been important for other aspects of the formation and evolution of the solar system, such as the formation of molybdenum isotopic elements in carbonaceous and noncarbonaceous meteorites (Budde et al. 2016). A supernova explosion, therefore, could have had important consequences for the formation and evolution of the early solar system, but it is debatable whether it could have deposited a sufficient amount of short-lived radionuclides to explain the abundances of their decay products in primitive chondrules.

Finally, we note that the continuous flow of hot gas causes the gas in the surviving disk to become eccentric, and the disk starts to tilt (see also Wijnen et al. 2017) to an average inclination of about $5° \pm 3$. In Figure 10.10, we show the evolution of the

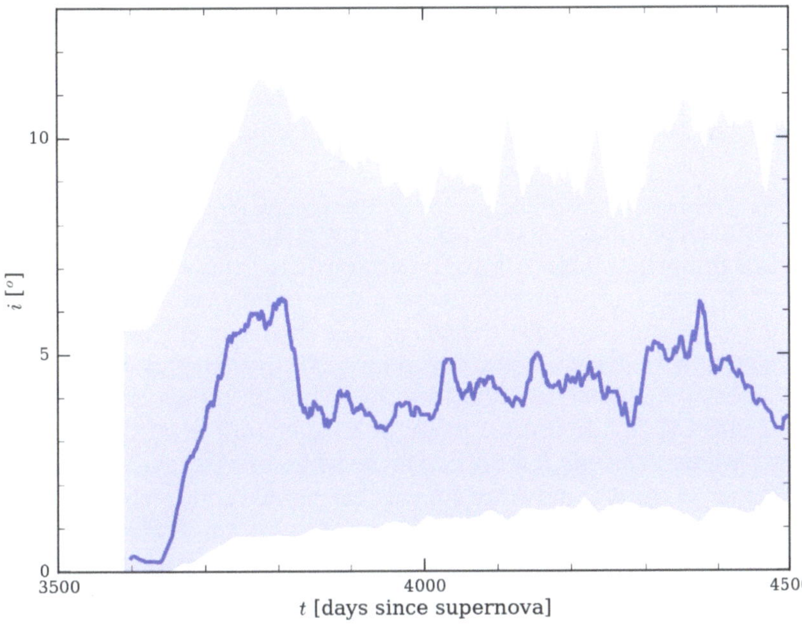

Figure 10.10. Evolution of the inclination of the planetary disk as a function of time since the disk was hit by the hot gas of the supernova blast wave. The median of the disk material inclination distribution is presented as the blue curve, with the shaded region giving the 5th and 95th percent percentiles.

inclination of the protoplanetary disk (relative to its initial value). This was measured by calculating the orbital elements for each of the gas particles and binning them in inclination using $10\sqrt{n}$ bins between zero and the maximum measured inclination. We subsequently adopted the median for display in the figure.

During the first 100 days, the median inclination of the disk grows to a maximum of about 6° before settling down to about 4° throughout the remainder of the period studied. The bulk of the disk gas has inclination between 1° and 9°. The inclination of the Sun with respect to the ecliptic plane in the solar system is $5.6° \pm 1.2°$, which is not inconsistent with the results from these simulations, illustrating that external events such as supernovae can plausibly account for the Sun's current inclination with respect to the ecliptic. As noted earlier, the planets were unaffected by the blast wave. If the calculation had been continued for a longer time, both the gaseous disk and the planets would probably have realigned to form a new combined disk of gas, debris, and planets. This would have led to an averaging of the angular-momentum axis. However, we did not follow this process but stopped the calculations after the impact shell traverses the disk.

The model provides a plausible explanation for the tilt of the ecliptic with respect to the Sun's equator, and also of the truncation of the outer edge of the solar system. Solving two problems with a single model is very satisfactory for researchers, particularly when complex models have trouble explaining even one of them. In this way, simulations can be used to guide the researcher's intuition, or to quantify a qualitative description of a physical process.

10.4 Accretion in the Galactic Center from S-Star Winds

The supermassive black hole in the Galactic center, Sgr A*, is orbited by a few dozen young (7–11 Myr), massive (19–25 $M_\odot$) stars (Schödel et al. 2002; Ghez et al. 2005), often called the "S stars." Because their orbits are dominated by the mass of the central black hole, these stars allow an accurate measurement of the black hole's mass (Ghez et al. 2003; Genzel et al. 2010) and give us unique insights into resonant and relativistic effects in its vicinity (see Merritt et al. 2010). The evolution of the S stars also has important implications for the local environment. With relatively high masses and advanced ages, these stars have strong stellar winds. Some of this wind material may find its way into the black hole, possibly giving rise to the X-ray emission observed by the Chandra X-ray observatory (Baganoff et al. 2003). The Chandra observations constrain the emitting region in the vicinity of Sgr A* to lie within ~12,000 AU of the central black hole (Baganoff et al. 2003). We can ask whether the winds from the S stars can account for the X-ray luminosity, or if the accreting material must come from some other population farther out.

The relatively small amount of accretion (just $\sim 10^{-7}$ $M_\odot$ yr^{-1}) onto the black hole during the pericenter passage of the nearest star, S2, led Nayakshin (2005) to argue that the S stars cannot be responsible for the general accretion onto the black hole. However, the accretion process is probably more complicated than just the wind of a single star, as was demonstrated in Cuadra et al. (2006), where the winds of the dozen S stars nearest to the black hole were resolved by hydrodynamical simulations. Recently, Lützgendorf et al. (2016) repeated this calculation with 27 S stars using the AMUSE framework. This calculation combined SPH simulations of the gas dynamics with direct N-body integration of the orbits of the S stars, including the effects of the stellar winds and accretion onto both the black hole and the S stars themselves. In this section, we show how the script used to perform these calculations was constructed.

We have encountered all of these ingredients in earlier chapters (specifically, Chapters 4 to 6). The coupling strategy was discussed in detail in Chapter 9. We repeat the Lützgendorf et al. (2016) calculation here without radiative transfer because this would put rather severe constraints on available computer resources. However, following the discussion on code-coupling strategies between hydrodynamics and radiative transfer in Chapter 7, it would be straightforward to expand the discussion to include radiative transfer if desired.

10.4.1 Initial Conditions

We start the simulation by constructing a realization of the 27 innermost S stars. The S stars are relatively massive and have most likely lost significant mass to stellar winds since their formation. Their current mass-loss rates are observed to be on the order of 10^{-5} $M_\odot$ yr^{-1}. In principle, knowing the masses and ages of these stars, we could simply model them using a stellar evolution code, feeding the mass loss directly into the hydro calculation. However, in practice, the masses and ages are quite poorly known. In this experiment, we adopt a simpler description that illustrates the general approach.

Figure 10.11 presents the results of a series of stellar evolution calculations using the parameterized stellar evolution code `SSE` (see Section 5.2.2). Reconstructing this figure should by now be relatively straightforward, given the examples in Chapter 5, and we do not provide the source code here. Note that the observed mass-loss rate for the S stars (gray bar) is consistent with winds from either 7–11 Myr old giants with initial masses between 19 $M_{\odot}$ and 25 $M_{\odot}$, or 18 $M_{\odot}$ supergiants at an age of about 14 Myr. Accordingly, we simply assume that our model S stars were all born at the same time, each with a mass of 20 $M_{\odot}$ and solar metallicity ($z = 0.02$), and are now 9 Myr old. These masses, ages, and metallicities are broadly consistent with the observed stellar parameters in the Galactic center. We evolve our model S stars to this age before integrating the equations of motion. At the start of the simulation, their masses are 18.1 $M_{\odot}$, radii 890 $R_{\odot}$, effective temperatures 3630 K, and luminosities $1.25 \times 10^5 L_{\odot}$. Using the theoretical wind model of Kudritzki et al. (1989)

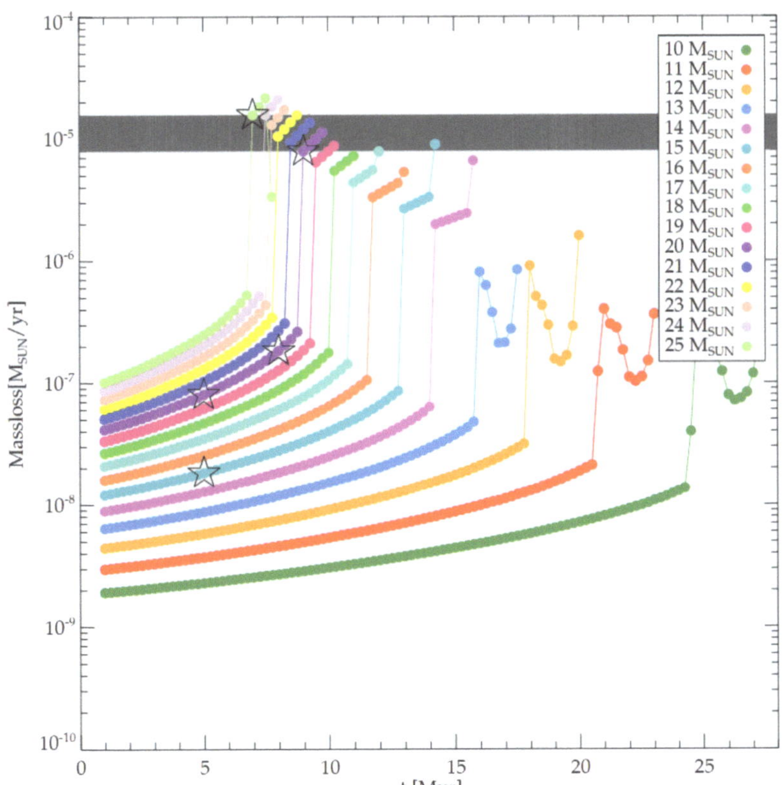

Figure 10.11. Mass-loss rates of high-mass stars, as functions of age and mass. The observed mass-loss range for the S stars, between 8×10^{-6} and 2×10^{-5} $M_{\odot}$ yr^{-1} is indicated by the gray horizontal bar. To make the mass-loss rates of the stars in our simulations consistent with the observations, we opted to use 20 $M_{\odot}$ stars at an age of 9 Myr (marked with an open star). Adapted with permission from Lutzgendorf et al. (2016). © 2016 The Authors. Published by Oxford University Press on behalf of the Royal Astronomical Society.

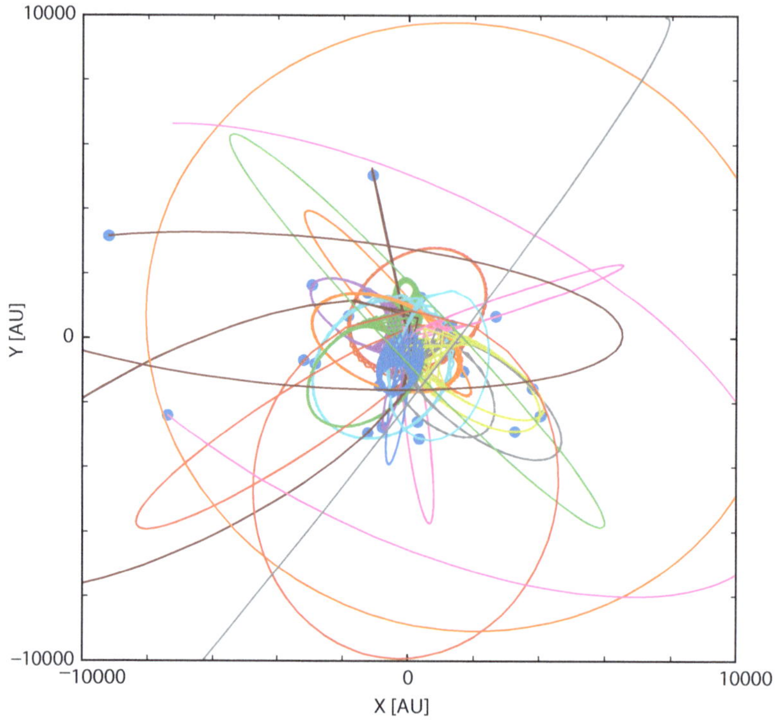

Figure 10.12. Orbital integration of the S stars over an interval of 500 yr, corresponding to the time frame of the simulation. The integration was performed using the symplectic N-body integrator `Mercury` (see Section 4.3). This figure was made using the script `amuse.examples.plot_sstar_orbits`.

(see Kudritzki & Puls 2000 for a review), we find a mass-loss rate of $\sim 7.9 \times 10^{-6}\,M_\odot\,yr^{-1}$ with a relatively low wind velocity of about $80\,km\,s^{-1}$.

Once their physical properties are established, the stars are placed in orbit around the central supermassive black hole using the orbital parameters provided in Table 7 of Gillessen et al. (2009b), calculated using the methods discussed in Section 4.4.1. Our simulation starts with the positions and velocities of the S stars on January 1, 2001. We adopt a black hole of mass $4.45 \times 10^6\,M_\odot$ located at the origin of the coordinate system, with zero velocity. The initial coordinates of the S stars are generated with the script `amuse.examples.initialize_sstars` (see also Figure 10.12).

10.4.2 The Combined Solver

We start the simulation with no interstellar gas yet present; we assume that all such gas will be created by the winds of the S stars as they orbit the central supermassive black hole. We follow the hydrodynamics of the stellar winds while integrating the orbits of the S stars in the black hole potential. Although we do not expect the stellar parameters to change much during the integration, we run a parameterized stellar evolution code in tandem with the other codes to determine the stellar wind parameters. This makes the script more flexible, homogeneous, and self-contained.

10.4.2.1 Bridging Gravity with Hydrodynamics

For the gravitational dynamics, we use the direct N-body integrator `Mercury`. Since we are integrating only a few dozen stars, we use the single-core non-GPU enabled version of this code (see Chapter 4 for details). We adopted the initial conditions from Gillessen et al. (2009a) and converted them to Cartesian coordinates as described in Section 4.4.1. The result of integrating the initial conditions for 500 yr is presented in Figure 10.12.

The complicated boundary conditions, the winds and accretion processes, and the lack of intrinsic symmetries make a grid-based code impractical for this application. With the SPH formalism, we can relatively easily tune the stellar winds, incorporate cooling, and monitor the accretion processes, we therefore use the SPH code `Fi` to resolve the hydrodynamics in the system. Shocks (see Section 6.1.2) are an issue in these calculations, and internal temperatures can rise high enough to emit X-rays. We deal with these rising temperatures by incorporating the cooling mechanism introduced in Section 6.3.4.

The combined solver for the hydrodynamics and the gravity is constructed using `bridge` (see Chapter 9). In this case, the gravity is affected by the hydrodynamics and the hydrodynamics by the gravity, much as discussed in Code example 9.3:

```
gravity_hydro = bridge.Bridge()
gravity_hydro.add_system(gravity, (hydro,))
gravity_hydro.add_system(hydro, (gravity,))
gravity_hydro.timestep = 2.0 * hydro.parameters.timestep
```

The final line initializes the bridge time step to a conservative value.

10.4.2.2 Stellar Winds

New SPH particles are generated at runtime when stars lose sufficient mass in their stellar winds. Instead of using the simple wind description discussed in Section 6.4.3, we adopt the built-in AMUSE wind module, which is imported via

```
from amuse.ext.stellar_wind import new_stellar_wind
```

This wind module is a class, similar to but more sophisticated than the method presented in Code example 6.2 in Section 6.3.1. This built-in wind module can be started by initializing the mass of the SPH particles modeling the wind (`mass_gas`), the target particle set to which this gas should be added (here called `gas`), and the time step (`dt`). We also set an additional parameter called `derive_-from_evolution`, which indicates that the winds of the stars should be derived from the running stellar evolution module,

```
wind = new_stellar_wind(
    mass_gas, target_gas=gas, timestep=time_step / 2., derive_from_evolution=True
)
```

After declaring the stellar wind class, we add the particles, in this case the particle set of 20 $M_\odot$ stars at an age of 9 Myr (see `amuse.examples.initialize_s-stars.py`). The channel from the stars to the wind particles ensures that the wind parameters calculated in the stellar evolution code are properly copied to the particles in the wind module.

```
wind.particles.add_particles(stars)
channel_stars_to_wind = stars.new_channel_to(wind.particles)
```

In the eventual script, we update the wind module as for any other code:

```
wind.evolve_model(time)
```

10.4.2.3 The Accretion Process

During the integration of the equations of motion, we allow the black hole and the S stars to accrete from the ambient gas. As with the wind module, we use the internal sink particle solution in AMUSE (see Section 6.3.2) which can be importing by:

```
from amuse.ext.sink import new_sink_particles
```

after which we initiate the sink particles for the S stars and have them accrete:

```
sink_stars = new_sink_particles(stars)
sink_stars.accrete(gas)
```

The stellar parameters are provided by the stellar evolution code, and the stellar radius is used as the sink radius.

For the black hole, the sink is initialized with an accretion radius smaller than the pericenter distance of the innermost S star.

```
sink_BH = new_sink_particles(BH, sink_radius=accretion_radius)
sink_BH.accrete(gas)
```

Choosing the proper value for this accretion radius is nontrivial and affects the accretion rate onto the black hole. A small accretion radius is inefficient in terms of computer time, whereas a large radius causes too much mass to accrete onto the black hole. First-order estimates of the accretion radius are given by Bondi & Hoyle (1944) and Hoyle & Lyttleton (1939):

$$r_{\rm BHL} = \frac{2GM_{\rm BH}}{(v_{\rm wind}^2 + c_s^2)}. \tag{10.4}$$

Here, c_s is the local sound speed, which depends on the temperature of the gas (see Equation (6.17)). Although the temperature near the black hole is very high (it is an

X-ray emitting source), the sound speed is still small compared to the speeds of the S star winds. For a stellar wind velocity of $\sim$1800 km s^{-1}, we estimate the accretion radius to be $\sim$2400 au. For our calculation, we adopt 100 au as the radius at which accretion actually occurred—considerably smaller than the Bondi–Hoyle–Lyttleton radius, but large enough to avoid a dramatic breakdown or inefficiency of the hydrodynamics solver.

10.4.2.4 Removing Escaping Gas

Some of the gas liberated by the stars will escape the computational domain. If a gas particle drifts too far from the Galactic center, it will find that the Galactic center is isolated instead of embedded in a larger environment. Thus, if we want to resolve the hydrodynamics on the scale of the S stars, we cannot also resolve the motion of the gas on kiloparsec scales. To err on the safe side, we strip gas as soon as it passes a distance of twice the apocenter distance of the farthest S star, if its speed exceeds the local escape speed from the black hole. We calculate these two parameters once for the system:

```
system_size = (2 * stars.a * (1 + stars.ecc)).amax()
system_escape_velocity = (
    2 * constants.G * (stars.mass.sum() + BH.mass) / system_size
).sqrt()
```

For our system, these values are $\sim$0.7 parsec and $\sim$760 km s^{-1}, respectively. Gas particles meeting these criteria are identified and removed from the simulation:

```
out_of_box = gas[gas.position.lengths() > system_size]
unbound = out_of_box[escaping.velocity.lengths() > system_escape_velocity]
escaping = unbound.select(
    lambda pos, vel:pos.dot(vel) > quantities.zero, ["position", "velocity"]
)
if escaping.is_empty():
    return
escaped_gas.add_particles(escaping)
gas.remove_particles(escaping)
```

Every time the amount of gas in the simulation changes (increases due to stellar winds, decreases by accretion or escape), the hydrodynamics solver is updated. We realize this by synchronizing the local gas particles with the particles in the hydrodynamics solver:

```
gas.synchronize_to(hydro.gas_particles)
```

10.4.2.5 The Event Loop

After declaring and initializing all the separate ingredients, we can write the event loop for this simulation. It is not complicated, but rather long due to the accumulation of the separate pieces. We have removed the diagnostics, validation, and data-writing operations for brevity and to clarify the basic functionality. But of course, these checks are of the utmost importance for a production simulation.

```
while time < time_end:
    time = time + time_step / 2
    stellar.evolve_model(pretime+time)
    time = time + time_step / 2
    gravity_hydro.evolve_model(time)
    stellar.evolve_model(pretime+time)

    channel_stars_to_wind.copy()
    wind.evolve_model(time)
    gas.synchronize_to(hydro.gas_particles)

    sink_BH.accrete(gas)
    gas.synchronize_to(hydro.gas_particles)
    sink_stars = new_sink_particles(stars)

    sink_stars.accrete(gas)
    gas.synchronize_to(hydro.gas_particles)
    remove_escaping_gas(gas, system_size, system_escape_velocity,
    escaped_gas)
    gas.synchronize_to(hydro.gas_particles)
```

The sinks keep track of the amount of angular momentum accreted by the black hole particle in order to estimate the size of the accretion disk.

```
    m_accreted = BH_sink.particles.mass.sum()
    diskrad = (
        sink_BH.angular_momentum.length()**2
        / (constants.G * m_accreted**2 * BH.mass[0])
    )
```

Running this script for any length of time and at a reasonable resolution of at least $5 \times 10^{-8}\,M_\odot$ per gas particle requires a substantial amount of computer time, really requiring the use of a large computing cluster or supercomputer rather than a laptop or workstation. Figure 10.13 shows the simulation results as published in Lützgendorf et al. (2016).

10.4.3 Results of the Simulation

The highest resolution run performed by Lützgendorf et al. (2016) was for a gas-particle mass of $4.8 \times 10^{-8}\,M_\odot$, with new SPH particles generated during each time step of 1 yr. It lasted for 500 yr, which is about 3 times the orbital period of the star S8. Figure 10.13 presents several snapshots of this simulation. As discussed previously, the initial snapshot (upper left) contains no gas; it takes a few hundred years for the system to reach a richer gas distribution. The simulation

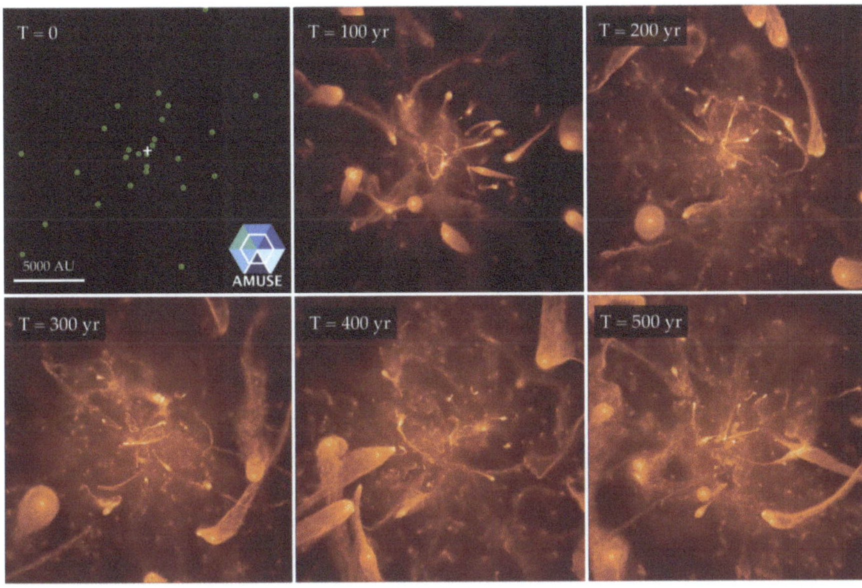

Figure 10.13. Evolution of the S stars in the Galactic center. This is Figure 3 from Lützgendorf et al. (2016); a spectacular animation of this simulation can be found at https://youtu.be/soGFgzRso3c. Adapted with permission from Lutzgendorf et al. (2016). © 2016 The Authors. Published by Oxford University Press on behalf of the Royal Astronomical Society.

was stopped at 500 yr because, although the gas distribution continues to change, the accretion onto the central black hole stabilizes at $\sim 1.4 \times 10^{-7}\,M_\odot\,yr^{-1}$. This slightly exceeds the observed accretion rate of 10^{-8}–$10^{-7}\,M_\odot\,yr^{-1}$ (Baganoff et al. 2003; Xu et al. 2006).

Figure 10.14 presents the total amount of mass accreted by the black hole during several simulations. The various curves show the result of different calculations in which we varied the ages of the S stars. As discussed earlier, the ages are important for determining the S star winds, and therefore accretion onto the black hole. If the S stars are young ($\lesssim$8 Gyr), they are unable to account for the observed accretion onto the black hole, but if they are giants with ages $\gtrsim$9 Myr, the observed accretion rate onto the black hole could be explained by the S star winds.

In every simulation, the contribution of star S2 provides at most ~2% of the total accretion rate onto the black hole. The idea that S2, being the nearest star, contributes most to the accretion onto the back hole is clearly not valid. During our 500 yr simulation, the star S14 contributed most mass to the black hole—almost an order of magnitude more than S2—even though S2 has the smallest pericenter distance. Of course, our assumption that all of the S stars have the same mass and age is artificial and somewhat contrived. When improved observations of the S stars provide better constraints on their masses and evolutionary states, we can easily improve our treatment of accretion onto the black hole.

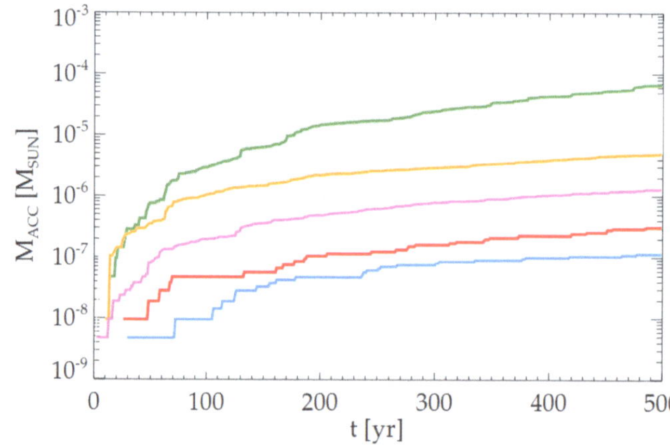

Figure 10.14. Total amount of mass accreted by the central supermassive black hole, as a function of time. The gray shaded area is an estimate of the total amount of mass accreted by the black hole, using observed mass accretion rates (Baganoff et al. 2003; Xu et al. 2006). The curves represent calculations for 20 $M_\odot$ stars at ages of 5 Myr (red curve), 8 Myr, (purple curve) and 9 Myr (green curve). The other two curves are for 15 $M_\odot$ stars at 5 Myr (blue) and 25 $M_\odot$ at 7 Myr (yellow). Adapted with permission from Lutzgendorf et al. (2016). © 2016 The Authors. Published by Oxford University Press on behalf of the Royal Astronomical Society.

10.5 Visualization

10.5.1 Fresco: Visualizing Gas-Rich Dusty Star Clusters

`Fresco` is a visualization package specifically for making "observational" images of simulations, focused on simulations in which stellar evolution, stellar dynamics, and possibly gas dynamics are incorporated. It takes the luminosities and temperatures from stars, and uses these to calculate their brightness in a series of color bands (by default the Bessell *UBVRI bands*). Then, using the positions of the stars, it calculates a base image for each color band, which is then convolved with a point-spread function (PSF). Finally, the color bands are combined to create an RGB image, by default mimicking the colors as the human eye would see them.

An additional feature of `Fresco` is the ability to include extinction from interstellar dust, using gas particles as a proxy. In this case, it creates a three-dimensional extinction map using the `FiMap` module (part of `Fi`), by default using a fixed gas-to-dust ratio, though this can be customized. After the brightness of each star is calculated in each color band, the extinction is then applied before convolving with the PSF. Additionally, the dust can reflect a fraction of the nearby starlight, using a tweakable albedo parameter.

`Fresco` is not in the main `amuse GitHub` repository, but can be downloaded from https://github.com/amusecode/amuse-fresco. An example of an image made using `Fresco` is presented in Figure 10.15, which was generated with exactly the same data as were used to generate Figure 6.10.

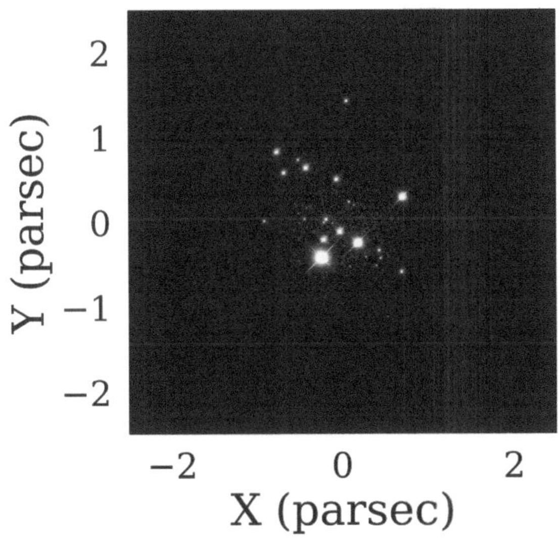

Figure 10.15. Projection of a collapsing molecular cloud at an age of 4.5 Myr. The stars are represented with the PSF of the Hubble Space Telescope, and the underlying gas distribution is given in contours. The data from which this figure is made is identical to those used to generate Figure 6.10.

10.6 Closing

You have reached the end of the final chapter of this book. Congratulations! We have discussed several examples on the use of AMUSE, and we gave a number of pointers to other projects in which the framework is used. Several of these examples are published in peer-reviewed journals. They generally address somewhat more complex problems, but the fundamental working remains unchanged. What is left now, are the appendices and some philosophical concluding remarks in Part III.

References

Appel, S. M., Andersson, E. P., Burkhart, B., et al. 2025, AAS Meeting, 245, 343.08

Baganoff, F. K., Maeda, Y., Morris, M., et al. 2003, ApJ, 591, 891

Bondi, H., & Hoyle, F. 1944, MNRAS, 104, 273

Borgdorff, J., Lorenz, E., Hoekstra, A. G., Falcone, J. L., & Chopard, B. 2011, 7th IEEE International Conference on e-Science (Los Alamitos, CA: IEEE Computer Society Press) 97

Budde, G., Burkhardt, C., Brennecka, G. A., et al. 2016, E&PSL, 454, 293

Chambers, J. E., & Migliorini, F. 1997, AAS/Division for Planetary Sciences Meeting Abstracts, 29, 27.06

Chevalier, R. A. 2000, ApJL, 538, L151

Cuadra, J., Nayakshin, S., Springel, V., & Di Matteo, T. 2006, MNRAS, 366, 358

Dobbs, C. L., & Pringle, J. E. 2013, MNRAS, 432, 653

Draine, B. T. 2011, ApJ, 732, 100

Fryxell, B., Olson, K., Ricker, P., et al. 2000, ApJS, 131, 273

Genzel, R., Eisenhauer, F., & Gillessen, S. 2010, RvMP, 82, 3121

Ghez, A. M., Duchêne, G., Matthews, K., et al. 2003, ApJL, 586, L127
Ghez, A. M., Salim, S., Hornstein, S. D., et al. 2005, ApJ, 620, 744
Gillessen, S., Eisenhauer, F., Trippe, S., et al. 2009a, ApJ, 692, 1075
Gillessen, S., Eisenhauer, F., Fritz, T. K., et al. 2009b, ApJL, 707, L114
Hoyle, F., & Lyttleton, R. A. 1939, PCPS, 35, 405
Kepler, J. 1609, Astronomia Nova (Heidelberg: G. Voegelinus), 1
Kudritzki, R. P., Pauldrach, A., Puls, J., & Abbott, D. C. 1989, A&A, 219, 205
Kudritzki, R.-P., & Puls, J. 2000, ARA&A, 38, 613
Lewis, S., McMillan, S., Mac Low, M. M., et al. 2023, AAS Meeting, 241, 09.03D
Liow, K. Y., Rieder, S., Dobbs, C. L., & Jaffa, S. E. 2022, MNRAS, 510, 2657
Looney, L. W., Tobin, J. J., & Fields, B. D. 2006, ApJ, 652, 1755
Lützgendorf, N., Helm, E., van der, Pelupessy, F. I., & Portegies Zwart, S. 2016, MNRAS, 456, 3645
Luu, T.-H., Young, E. D., Gounelle, M., & Chaussidon, M. 2015, PNAS, 112, 1298
Merritt, D., Alexander, T., Mikkola, S., & Will, C. M. 2010, PhRvD, 81, 062002
Nayakshin, S. 2005, A&A, 429, L33
Ouellette, N., Desch, S. J., & Hester, J. J. 2007, ApJ, 662, 1268
Pearson, S. G., & McCaughrean, M. J. 2023, arXiv:2310.01231
Pelupessy, F. I., Jänes, J., & Portegies Zwart, S. 2012, NewA, 17, 711
Portegies Zwart, S. F., & Verbunt, F. 1996, A&A, 309, 179
Portegies Zwart, S., Pelupessy, I., Martínez-Barbosa, C., van Elteren, A., & McMillan, S. 2020, CNSNS, 85, 105240
Price, D. J., Wurster, J., Tricco, T. S., et al. 2018, PASA, 35, e031
Rein, H., & Liu, S. F. 2012, A&A, 537, A128
Rieder, S. 2019, rieder/ekster: IAUS 351, v1.0.1a, Zenodo, doi:10.5281/zenodo.3356893
Rieder, S., & Liow, K. Y. 2021, Zenodo, doi:10.5281/zenodo.5520944, v1.01
Rieder, S., Dobbs, C., Bending, T., Liow, K. Y., & Wurster, J. 2022, MNRAS, 509, 6155
Sanders, N. E., Soderberg, A. M., Gezari, S., et al. 2015, ApJ, 799, 208
Schiller, M., Paton, C., & Bizzarro, M. 2015, GeCoA, 149, 88
Schödel, R., Ott, T., Genzel, R., et al. 2002, Natur, 419, 694
Scott, E. R. D. 2007, AREPS, 35, 577
Sills, A., Rieder, S., Buckner, A. S. M., et al. 2023, MNRAS, 519, 4142
Spicer, D. S., Clark, R. W., & Maran, S. P. 1990, ApJ, 356, 549
Stahler, S. W., & Palla, F. 2005, The Formation of Stars (New York: Wiley)
Teyssier, R. 2002, A&A, 385, 337
van Elteren, A., Portegies Zwart, S., Pelupessy, I., Cai, M. X., & McMillan, S. L. W. 2019, A&A, 624, A120
Wall, J. E., Mac Low, M.-M., McMillan, S. L. W., et al. 2020, ApJ, 904, 192
Wang, L., Iwasawa, M., Nitadori, K., & Makino, J. 2020, MNRAS, 497, 536
Wijnen, T. P. G., Pelupessy, F. I., Pols, O. R., & Portegies Zwart, S. 2017, A&A, 604, A88
Wilhelm, M. J. C., & Portegies Zwart, S. 2024, A&A, 691, A71
Xu, Y.-D., Narayan, R., Quataert, E., Yuan, F., & Baganoff, F. K. 2006, ApJ, 640, 319

AAS | IOP Astronomy

Astrophysical Recipes (Second Edition)
The art of AMUSE
Simon Portegies Zwart, Steve McMillan and Steven Rieder

Chapter 11

Epilogue

There is no dark side of the Moon really.

—Gerry O'Driscoll

In the last 400 or so pages we have presented the Astrophysical Multipurpose Software Environment (AMUSE) and motivated its development and use. We have glossed over many underlying physical processes, and tried to stay away from the many mathematical details. Equally, we have barely discussed the technical challenges and algorithms needed to make all this work. Our objectives have been to present an interesting and novel way in which domain applications can be combined at a meta-level in a homogeneous environment. The methods presented allow developers to preserve their software in a stable and homogeneous habitat, yet also provide tools that allow them to couple these codes. In this coupled environment, it is possible to both test the individual parts and address complex multiscale and multiphysical processes.

The methods discussed and explained in this book are not limited to astronomy. We have worked on expanding the methods to long-term weather prediction and oceanography in a project we call `OMUSE` (Pelupessy et al. 2016a, 2016b), which stands for (as you might guess) the Oceanographic Multipurpose Software Environment. Further extrapolations on the concept make us wonder if similar environments could be developed for simulations in biology (`BMUSE`) or chemistry (`CMUSE`). For the latter, we have worked on the implementation of the chemical evolution codes `KROME` (Grassi et al. 2014) and `UCLChem` (Holdship et al. 2017). Both are operational, but so far we have not published any results using them in a peer-reviewed scientific journal.

Furthermore, we have been working on developing a general `*MUSE` with modules for various individual domains, such as `A` for Astronomy, `B` for biology,

doi:10.1088/2514-3433/adfc3cch11

etc. The basic functionality of AMUSE could provide a template for such an endeavor. The development of AMUSE makes for a very interesting first push in that direction, although we are still far from a global simulation environment that might be utilized for a wide range of physical applications.

Maybe the biggest challenge to the development of such a meta-framework is not technical but psycho-sociological. In our experience, senior researchers tend to stick to the methods and techniques they developed when they finished their PhDs. Graduate students are directed by their supervisors, which usually confines them to familiar methodologies. Postdoctoral researchers are naturally preoccupied with securing a solid foundation for their future careers. One can then wonder if there is an avenue leading to broad acceptance of the ideas presented here. Maybe there is. We have found on a daily basis that AMUSE has enriched our academic repertoire at all levels, from freshman to postdoc. Most importantly, using AMUSE turns out to be a lot of fun, and working with the package is enormously rewarding.

We hope you have enjoyed this book, using the codes, and hopefully developing more scripts and even publishing the results. We hope you will acknowledge AMUSE (and the constituent code references thrust upon you by our software) in any such work. As a final comment, we emphasize that we see AMUSE as a community effort. In that respect, we hope that any scripts you develop and any software you write will be made freely available to everyone in the field. The AMUSE team will be happy to work with you to incorporate your work as part of the project.

References

Grassi, T., Bovino, S., Schleicher, D. R., et al. 2014, MNRAS, 439, 2386

Holdship, J., Viti, S., Jiménez-Serra, I., Makrymallis, A., Priestley, F., et al. 2017, AJ, 154, 38

Pelupessy, I., van Werkhoven, B., van Elteren, A., et al. 2016a, The 5th Int. Workshop on Multi-scale (Un)-structured Mesh Numerical Modeling for Coastal, Shelf, and Global Ocean Dynamic (Delft: TU Delft) 1

Pelupessy, I., van Werkhoven, B., van Elteren, A., et al. 2016b, 12th International Conf. on e-Science (e-Science) (Piscataway, NJ: IEEE) 399

Part IV

Appendices

AAS | IOP Astronomy

Astrophysical Recipes (Second Edition)
The art of AMUSE
Simon Portegies Zwart, Steve McMillan and Steven Rieder

Appendix A

AMUSE Fundamentals

> You take the blue pill — the story ends, you wake up in your bed and believe whatever you want to believe. You take the red pill — you stay in Wonderland, and I show you how deep the rabbit hole goes.
>
> —Morpheus

We present implementation details of the AMUSE framework. AMUSE contains more than 70 highly optimized codes and more than 5000 handy routines to manipulate data. We would love to discuss them all, but this would result in a horribly boring text to write, and read. We limit ourselves to describing the most important and most used methods, such as file I/O, data manipulation, how to tune the handling of units for diagnostics, and we list the most important parameters used in several of the domain-specific modules. We further list the most important packages currently interfaced with AMUSE.

A.1 Internal Parameters, Units, and Types

Most of the community modules in AMUSE have internal parameters that control the performance of the code. Some tune the accuracy or the details of the physics, while others inform the user of the state of the running module. Most are preset to reasonable values, often the values initially set by the original authors of the package, and for which the module has been validated. Under certain circumstances, you may want to change these parameters.

As discussed in Section 2.3.3, community codes can be queried for information on parameter settings, but even then some parameters stay hidden deep within the source code. Here, we present a brief overview of some of the parameters that appear in this book, and for which a reference may be needed.

doi:10.1088/2514-3433/adfc3cch12

Table A.1. Stellar Classifications in AMUSE

`Stellar_type`	Physical Description
0	Fully convective low mass main-sequence star
1	Main-sequence star
2	Hertzsprung gap
3	First giant branch
4	Core-helium burning
5	First asymptotic giant branch
6	Second asymptotic giant branch
7	Main-sequence naked helium star
8	Hertzsprung gap naked helium star
9	Giant branch naked helium star
10	Helium white dwarf
11	Carbon/oxygen white dwarf
12	Oxygen/neon white dwarf
13	Neutron star
14	Black hole
15	Massless post-supernova object

A.1.1 Nontunable Parameters

There are a number of nontunable parameters in the AMUSE community codes. These often inform the user of the state of the underlying code. An example is the `stellar_type` parameter in the stellar evolution codes, as discussed in Section 5.3.12. The standard AMUSE stellar types are listed in Table A.1. In Section 5.2.2 we demonstrated how to query the stellar evolution type from a running stellar evolution code.

We have adopted the designation of stellar types used in `SSE` (Single-Star Evolution; Hurley et al. 2000). These designations are used to indicate the astrophysical state of a star, even though they are not 100% accurate—these types should be homogeneous across codes, but it is hard to guarantee consistency. While implementing the other stellar evolution codes, we attempted to port the `SSE` designations, but this did not always lead to a perfect classification. One example can be seen in Figure 5.3, where an asymptotic giant branch star is misidentified as a stellar remnant. Another is in `MESA` (Paxton et al. 2010, 2011, 2013, 2015, 2018; Gaburov et al. 2018), where the interface fails to report the core-helium burning stage even though this phase is actually properly modeled in the code.

A.1.2 Tunable Parameters

Many of the code-specific parameters can be tuned. Examples include the integration-controlling time-step parameter in gravity (see Section 4.3.6), metallicity of the star in stellar evolution (see Section 5.2.3), the adiabatic index (gamma) in hydrodynamics (see Section 6.2.3), or the number of photons in radiative transfer (see Chapter 7).

The following tables, Tables A.2 to A.6, present the tunable and nontunable parameters for the core models in gravity, stellar evolution, hydrodynamics, and radiative transfer.

In Table A.4, for example, we present an overview of the tunable parameters in the SPH code `Fi`. Some of them are quite similar to those used in other codes, such as `time step`, `verbosity`, and `epsilon_squared`, which have similar meaning across disciplines. Others may sound somewhat obscure, but if you are an SPH expert, they actually make a lot of sense. Now, it is not our intention to turn the reader into an SPH expert, and we will explain a few of the more obscure names. We refer to Springel (2010b) for an excellent overview of the free parameters in SPH. Some physical parameters tune the behavior of the gas, rather than the numerics. Setting the viscosity of a gas (`alpha`) allows one to model various density domains, or gas compositions. Although it is a numerical parameter, its interpretation 234567890 is physical, but regrettably not calculable from first principles.

Let us just provide an example of the sort of discussion one of these parameters leads to. For example, the parameter `alpha` (a common name for a parameter in astrophysics) is an artificial viscosity parameter, with a typical value of 0.5. The parameter `beta`, between 0 and 1, refers to another artificial viscosity, often simply called the "beta parameter", the value of which is 2**alpha*, with a typical value of 1. Similarly, one might wonder why two artificial viscosity parameters, (`alpha` and `beta`) are needed, but numerically this seems to make sense.

Table A.2. Overview of the Most Common Tunable and Nontunable Gravity-Specific Parameters

Tunable Parameters			Gravitational Dynamics
Parameter Name	Value	Unit	Explanation
`begin_time`	0	Time	Star time for integrator
`epsilon_squared`	0	Length^2	Gravitational softening
`force_sync`	0	None	Force synchronization
`gpu_id`	−1	None	Identity of the GPU
`sync_time`	0	Time	...
`timestep_parameter`	0.14	None	Time-step parameter
`use_gpu`	0	Bool	Use the GPU
`zero_step_mode`	0	None	Do a zero time step
Nontunable Parameters			
Parameter Name	Value	Unit	Explanation
`block_steps`	0	None	Number of block-time steps
`gpu_id`	0	None	Identity of the GPU
`total_steps`	0	None	Number of integration steps

Table A.3. Overview of the Most Common Tunable and Nontunable Stellar Evolution Codes

Tunable Parameters			Stellar Evolution
Parameter Name	Value	Unit	Explanation
supernova_kick_velocity	600	$km\,s^{-1}$	Supernova kick velocity
is_logging_of_evolve_enabled	False	Bool	Log stellar evolution
OB_wind_setting	1	Bool	O-star wind parameter
RGB_wind_setting	1	Bool	RGB wind parameter
AGB_wind_setting	1	Bool	AGB wind parameter
convective_overshoot_parameter	0.12	None	Convective overshoot length
import_model_entropy_accuracy	0.0001	None	Entropy solver accuracy
import_model_entropy_force	20	None	Entropy forcing efficiency
max_age_stop_condition	$2 \cdot 10^{12}$	yr	Stopping after a time
maximum_number_of_stars	10	None	Number of stars
metallicity	0.02	None	Stellar metalicity
min_timestep_stop_condition	10^6	s	Minimal time step
mixing_length_ratio	2	None	Mixing length parameter
number_of_ionization_elements	5	None	Number of ionizing elements
path_to_data	/data/input		Output directory path
semi_convection_efficiency	0.04	None	Convection efficiecy
thermohaline_efficiency	1	None	Efficiency of the thermohalinety
verbosity	True	Bool	Maximal output

Table A.4. Overview of the Most Common Tunable and Nontunable Hydrodynamics Codes

Tunable Parameters	Hydrodynamics	
Parameter Name	Value	Unit
acc_timestep_crit_constant	0.25	None
artificial_viscosity_alpha	0.5	None
beta	1	None
code_length_unit	1	au
code_mass_unit	1	$M_\odot$
cool_par	1	None
courant	0.3	None
epsilon_squared	0	m^2
feedback	fuv	None
fi_data_directory	data/	input/
<code>_cell_opening_constant	0.01	None

gamma	5/3	None
gas_epsilon	0.005	au
max_density	100	$M_\odot$ au^{-3}
n_smooth	64	None
n_smooth_tol	0.1	None
nn_tol	0.1	None
opening_angle	0.5	None
optical_depth	0	None
periodic_box_size	1	au
sph_artificial_viscosity_eps	0.01	None
sph_h_const	0.2	au
sph_viscosity	sph	None
time step	58	Day
verbosity	0	None

We omit explaining the details of the parameter because most of them are rather detailed, require considerable explanation, and are probably hardly ever changed. The best way to learn what these parameters do is to look in the source code, and try them out. Just run a simple example script, such as `amuse.examples.stellar_minimal`, change one of the parameters and see what it does in the output. Some of the more obscure names include `beta`, `gamma`, `courant`, `feedback`, `heat_par1`, `heat_par2`.

Table A.5. Overview of the Boolean Parameters in SPH Codes

Boolean Value	Boolean Parameter Name
True	acc_timestep_flag, conservative_sph_flag, eps_is_h_flag, gadget_cell_opening_flag, integrate_entropy_flag, sph_dens_init_flag, use_hydro_flag, self_gravity_flag.
False	adaptive_smoothing_flag, balsara_flag, direct_sum_flag, enforce_min_sph_grav_softening_flag, fixed_halo_flag, freeform_timestep_flag, freeform_timestep_flag, isothermal_flag, periodic_boundaries_flag, quadrupole_moments_flag, radiation_flag, smooth_input_flag, square_root_timestep_flag, star_formation_flag.

Table A.6. Overview of the Most Common Tunable and Nontunable Radiative Transfer Specific Parameters

Tunable Parameters			Radiative Transfer
Parameter Name	Value	Unit	Explanation
`boundary_condition`	0	None	See below
`box_size`	13	kpc	Box size, if bound
`default_spectral_type`	−1	None	Radiation spectrum
`helium_case_A_flag`	True	Bool	Helium recombination
`hydrogen_case_A_flag`	True	Bool	Hydrogen recombination
`isothermal_flag`	False	book	Evolve temperature or not
`momentum_kicks_flag`	False	Book	Use the momentum kick
`number_of_rays`	1000	yr^{-1}	Number of rays
`spectral_file`	`./spectra/thermal30k.cdf`		

Another example is (as you might guess) `gamma`. This parameter has a completely different meaning from `alpha` and `beta`, and its effect on the computational results is profound. The parameter `gamma` is the adiabatic index for the equation of state of the gas,

$$P = K\rho^{\gamma}. \tag{A.1}$$

Here, K is a proportionality constant and γ typically has a value of $5/3$. For historical reasons, Poisson's equation for a gas of adiabatic index γ is often written in dimensionless form as the *Lane–Emden equation* (Lane 1870; Emden 1907),

$$\frac{1}{\xi^2}\frac{d}{d\xi}\left(\xi^2\frac{d\theta}{d\xi}\right) + \theta^n = 0, \tag{A.2}$$

where ξ is dimensionless radius, $\theta = \phi/\phi_0$ is the dimensionless potential, and the polytropic index $n = 1/(\gamma - 1)$. Solutions to the Lane–Emden equation result in finite-radius stars for $0 < n < 5$, while stars with $n \geqslant 5$ have infinite radii and stars with $n > 5$ have infinite mass. The Eddington model for radiative stars, such as the Sun, corresponds to $n = 3$. The adopted value for n in Fi in AMUSE in $n = 3/2$, or $\gamma = 5/2$, which ensures energy conservation in systems without shocks or dissipative processes. We note that hydro codes, such as `Fi`, contain many such parameters, each of which might be worth an extensive discussion.

Apart from the adiabatic indices, the most important free parameters in SPH include `n_smooth`, the number of neighbors over which the results are smoothed, `sph_h_const`, specifying a fixed smoothing length, if desired, and `sph_artificial_viscosity_eps`, and `sph_viscosity`, controlling the viscosity of the simulated gas. The first, `n_smooth`, is a tunable parameter, which is set to a relatively small number (64 by default). For modern high-resolution simulations, one might wish to set this to a higher value.

Some hydro codes have, in addition to the already mentioned long list of free parameters, specific time-step parameters to control drifts in velocity and acceleration. These parameters are named `free_timestep_crit_constant_a, ... _aexp`, and similarly for the velocity with a `..._v` instead of `..._a`, for the velocity and acceleration.

Some of the flags may sound unfamiliar, such as `balsara_flag`. This parameter in the SPH code Fi refers to the activation of the Balsara switch (or Balsara limiter), a numerical technique used to reduce the effect of artificial viscosity in regions of effective shear, while still allowing artificial viscosity to be effective. Enabling the Balsara flag is important in simulations of differentially rotating disks, galaxies, or accretion flows, where shear is important and artificial viscosity otherwise dampens real structures.

In AMUSE, the Balsara limiter is turned off by default (see Table A.5), but it is one of these parameters that make the dissipation and viscosity in a disk simulation more realistic. We emphasize that default values are what they are for a reason, and changing them is risky, but we recognize that not changing them in certain circumstances may be risky too.

A.1.2.1 Tunable Radiative-transfer Parameters

A parameter not mentioned in Table A.6 is `ionization_temperature_solver`, which is really a flag for various options in solvers. The default value of 2 here indicates the backward differentiation formula (or bdf) method for solving the equations of motion (Gear 1967). It is an implicit time integration scheme, which solves the time-dependent radiative-transfer equations on SPH data, the bdf solver is used to advance the solution of the radiative-transfer equations in time, particularly when dealing with stiff equations that can arise in radiative processes. An alternative to the bdf solver would be to set `ionization_temperature_solver` to 1, in which case an explicit Euler solver is used.

The bdf solver allows for larger time steps and improved numerical stability compared to explicit Euler methods, especially when the radiative-transfer equations involve rapid changes or processes with widely varying timescales. This makes it suitable for accurately and efficiently modeling the evolution of ionization and temperature in simulations where radiative transfer is coupled to hydrodynamics.

Another common parameter in radiative transfer is `global_helium_mass_fraction`. By default, this fraction is set to 0, because in astrophysics we often deal with Hydrogen, rather than Helium-rich interstellar gas. It is one of these parameters that is important to set properly depending on your specific application.

The `isothermal_flag` controls whether to evolve temperature (0) or keep the gas isothermal (1). The `boundary_condition` flag can either be reflective (use a value of -1), vacuum (0), periodic (-1). In a more modern code, one would probably opt for strings, rather than integers for specifying such options.

Table A.7. Recognized Read/Write File Formats

Keyword	Brief Description
csv	Comma-separated values file
dyn	Starlab binary structured file
starlab	Extensive Starlab data format
gadget	Gadget binary data file
hdf5	HDF5 file
nemo	NEMO binary structured file
tsf	NEMO binary structured file
txt	Text file containing a table of values separated by a space
vts	Text file containing a table of values separated by a predefined character
vtu	Text file of unstructured XMLVTK-formatted data
amuse	Default selection (uses HDF5)

For a full appreciation of this routine, we refer to the source codes in amuse.io, in particular in base for the base class of the storage module, and store for the specifics.

A.1.3 Input/Output File Types

In Section 2.2.4 we discussed how to write and read particle sets from permanent storage using write_set_to_file and read_set_from_file. Normally we adopt hdf5 binary formatted files for input and output, but there are circumstances in which another format might be desirable. Table A.7 presents an overview of the file types recognized in AMUSE. There are restrictions to some file formats: ASCII files, for example, will not be able to represent pointers other than by showing the address which is not very useful when reading the data from permanent storage. Of course, we could add other file formats, but we think that with this repertoire covers some of the most common ones. The default AMUSE file format is hdf5, which is widely accepted within the astronomical community.

Each of these storage format settings has its own derived class and associated Python file. The module amuse.io.starlab specifies the starlab data format, whereas the module amuse.io.gadget does the same for gadget. If you want to make your own derived data format, the best way to start is to copy one of the existing derived class files and edit it to your needs, filling in the specific requirements of your module.

A.2 Initial Conditions

A.2.1 Generalized Initial Condition Generating Routines

In Chapter 3, we discussed a few examples of initial conditions. Of course, any output from any high-fidelity simulation can be considered as input for another simulation. Table A.8 presents a list of the most common initial conditions generating functions and routines (as of the publication of this book). Some of these create new data, others append or manipulate data. Many of these generator functions are part of the standard AMUSE sandbox, but a large number of

Table A.8. Overview of the Routines for Generating and Manipulating Initial Conditions

Particle-set Generators	
Stars	make_single_star.py, make_binary_stars.py, make_stars_in_mass_function.py, make_cluster_of_stars.py, make_cube_of_stars.py, make_plummer_sphere.py, make_gas_plummer.py
Planets	make_solar_system_planets.py, make_exoplanetary_system.py, make_nice_like_system.py, make_planets_oligarch.py, new_solar_system_with_sun_and_planets.py.
Complexes	make_hierarchical_cluster_with_planets.py.
Append to Particle Set	
Planets	add_asteroids_to_solar_system.py, add_comets.py, add_debris_disk.py, add_moons_to_solar_system_planets.py, add_planets_oligarch.py, add_proto_oort_cloud.py, add_scattered_asteroids.py.
Stars	add_binary_companion.py, add_star_in_orbit.py.
Complexes	add_perturbation_to_model.py.
Manipulators	
Manipulators	move_all_particles.py, move_bodies_to_stellar_position.py, rotate.py.
Diagnostics	determine_orbital_elements.py, orbital_elements_to_cartesian.py.
Converters	conv_amuse_to_brutus.py, name_stars.py, write_set_as_ascii.py.
Plotting	plot_bodies.py, print_snapshot.py.

additional routines can be found on GitHub at https://github.com/amusecode/initial-conditions-generator.

These routines are used as follows (see also Figure A.1):

```
python make_plummer_sphere.py --nstars 16 --radius 0.01 -Q 0.3 -F Pl16.amuse
python name_stars.py -f Pl16.amuse --nstars 12 --name system -F PlN.amuse
python add_planets_oligarch.py -f PlN.amuse --fplanets 1 --name system -F PlNO.amuse
python add_debris_disk.py -f PlNO.amuse --name system --ndisk 300 -F PlNOA.amuse
python rotate.py -f PlNOA.amuse -F Pl_N16NO_A1000R.amuse --type star
```

This daunting series of calls results in a Plummer distribution of 16 stars, 12 of which have a Oligarchic planetary system, totaling 44 planets, including a debris

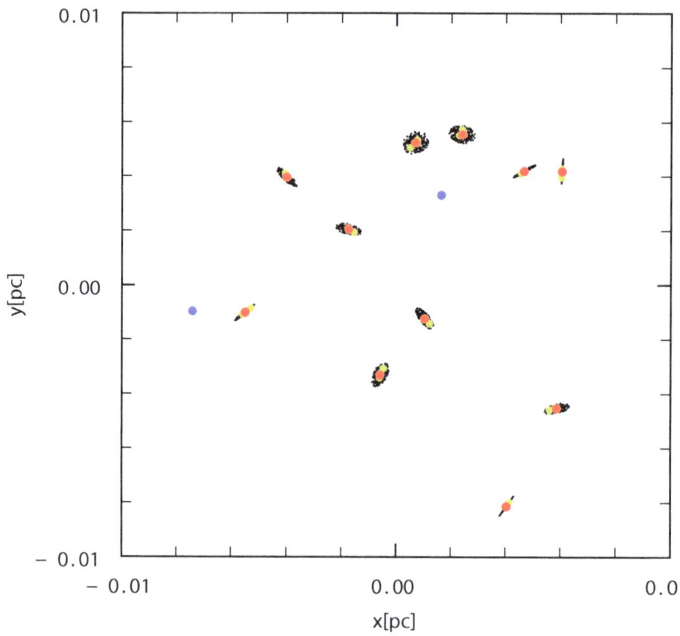

Figure A.1. Representation of a simple 16 star cluster with planetary systems and disks. Single stars are blue bullets, the host star for a planetary system is red. Planets are yellow (hardly visible), but they can be perceived since the stars are surrounded by tiny black dots (the debris). The script to generate these initial conditions is the shell script `make_cluster_with_planets.sh` (which we copied below), and the code to plot the results is `amuse.examples.plot_plummer_stars_and_planets`.

disk of 3816 test particles each in random orientations in space. The results are stored in a series of intermediate files `Pl16.amuse`, `PlN.amuse`, `PlNO.amuse`, `PlNOA.amuse`, and eventually the final file `Pl_N16NO_A1000R.amuse` containing the initial realization of the simulation. Figure A.1 shows this cluster.

The shell script `make_cluster_with_planets.sh` generates a cluster of stars of which some stars have planets and a disk of planetesimals. The resulting data file (`Pl_N16NO_A1000R.amuse` can be plotted using the script `amuse.examples.plot_plummer_stars_and_planets`, resulting in Figure A.1.

A.2.1.1 Uniform Gas Distributions for SPH

Some codes for generating initial conditions are quite versatile, and can take numerous arguments to tune their function. The classes provided by `spherical_model` in `amuse/ext` are examples of such rich functionality. The `Python` class `Uniform`SphericalDistribution in `amuse.ext.spherical_model` forms the basis for a number of associated routines, including `new_uniform_spherical_particle_distribution`, `new_spherical_particle_distribution`, and `new_plummer_spatial_distribution`. The underlying class generates a uniform spherical grid of particles, which can then be used for more

Table A.9. Additional Options Used in the Class `UniformSphericalDistribution` to Tune the Generation of Initial Gas-Density Distributions

Keyword	Brief Description
`cubic`	Cubic grid with particles at each corner
`bcc`	Body-centered cubic, with additional particles at the center of each cube
`fcc`	Face-centered cubic, but with additional particles on the face of each cube
`random`	Particles distributed randomly using `numpy.random.uniform`
`glass`	As for `random`, but stabilized using hydro pressure without gravity
`sobol`	3D sobol sequence (low discrepancy, quasirandom) (see Sobol 1967)

For the sobol method, low discrepancy indicates that the points are on average close to equally distributed.

useful initial distribution functions. Some of the most important parameters used in the class are listed in Table A.9.

To generate a spherical quasirandom distribution of `N = 10000` particles with radius `R = 1 | units.km` and mass `M = 1 | units.kg`, we write

```
from amuse.ext.spherical_model import new_uniform_spherical_particle_distribution
particles = new_uniform_spherical_particle_distribution(N, R, M, type="sobol")
```

The class has two more options: `offset` and `target_rms`. Option `offset` is only used for the regular grids ("`cubic`", "`bcc`", "`fcc`) and should be specified using a tuple of three numbers in the range [0, 1), which define the offset between the origin of the grid and the corner of the unit cell, normalized to the unit cell size. Option `target_rms` is only used for `glass` as the rms density criterion for convergence (see Section 9.3.2.1).

A.2.2 Writing Composite Datafiles

Simulations such as those presented in Section 9.4.4 and Chapter 10 often require writing composite files, which in Section 9.4.4 include both gravitating and gaseous particles. There is no best or optimal way to write composite data files, but we recommend keeping the format as simple as possible. In the cases just referenced, we accomplish this by creating a common particle set containing all particles and writing this to file. This particle set contains a variety of attributes, depending on the types of particles resulting from several independently operating codes, which can later be disentangled to recover the individual components.

Such a common particle set is generated by the `ParticlesSuperset` function:

```
moving_bodies = ParticlesSuperset([gravity.code.particles, hydro.gas_particles])
```

It can later be saved by

```
write_set_to_file(moving_bodies, filename)
```

or subsequently read from a file:

```
moving_bodies = read_set_from_file(filename)
```

Here, `filename` is the name of the input or output file. In this example, we write the data in the default `amuse` format (which uses `hdf5`), but several of the alternatives listed in Table A.7 could be used.

Another way to write composite data files is by simply storing the various particles sets successively, subsequently recovering them in the same order. In this example, we would create two files (using the default, we omitted specifying the data format):

```
write_set_to_file(gravity_particles, gravity_filename)
write_set_to_file(hydro_particles, hydro_filename)
```

and in reverse read both files:

```
gravity_particles = read_set_from_file(gravity_filename)
hydro_particles = read_set_write_from_file(hydro_filename)
```

A.3 Notes on Specific Codes

A.3.1 Test Particles in *N*-body Codes

Some codes can handle test masses, objects with zero mass. They can be quite handy for following the gravitational potential. However, if you have to pay the N^2-price for integrating them, they do not make a lot of sense. Some of the N-body codes in AMUSE are therefore equipped to treat zero-mass objects as test masses. This feature is enabled in `Huayno` and `ph4`, and maybe more codes will be enabled with this handy feature as this book goes to print. To enable test-mass integration, simply set the mass of a particular particle to 0 (or to some negative value).

```
particle.mass = 0 | nbody_system.mass
```

Once set to zero, other nonzero-mass particles accelerate the test mass, but the test mass does not affect any other particles.

A.3.2 Post-Newtonian Integration with Relativistic Cross Terms

Between 1907 and 1915, Einstein developed general relativity (see Weinstein 2012; Poisson & Will 2014; for an interesting read on the history of this development) and viewed gravity as the result of the curvature of spacetime (Einstein & Grossmann 1914; Einstein 1915b, a; Hilbert 1915). The Einstein field equations describe the gravitational interaction between particles, but these equations are nonlinear and notoriously hard to solve. Schwarzschild (1916) found the first nontrivial solution to the Einstein field equations. The Schwarzschild metric describes a point-like particle. A rotating black hole was first described analytically as a solution to the field equations by Kerr (1963). Today, there are standard software implementations to solve general relativistic dynamical systems (Mewes et al. 2018), even including magnetic fields (Mewes et al. 2020). However, simulating multiple black holes in a relativistic context is very expensive in terms of computer time.

The N-body code Hermite-GRX includes relativistic effects in the post-Newtonian approximation. It was introduced in Portegies Zwart et al. (2022). Apart from the usual parameters of any direct N-body code, `Hermite-GRX` has four extra settings, listed in Table A.10.

The parameters `integrator` and `perturbation` control the integration method and the perturbation. Integrators include various forms of symmetrization and regularization. Both help in speeding up the code and reducing energy errors. The perturbation is mainly used to include various post-Newtonian effects, the first post-Newtonian and 2.5 post-Newtonian orders of the Einstein Infeld Hoffman equation, and the first post-Newtonian equations of motions that include the relativistic cross terms (useful for multiple black holes).

Table A.10. Overview of the Most Common Tunable and Nontunable Parameters for General Relativity Calculations in the Gravitational N-body Problem

Tunable Parameters			Gravitational Dynamics
Parameter Name	Value	Unit	Explanation
`integrator`	Hermite		Integrator type (see Table A.11)
`light_speed`	1	Length/time	Speed of light
`num_threads`	1		Number of threads on the processing core
`perturbation`	None		Perturbation mode (see Table A.11)

Table A.11. Integrators Available in `Hermite-GRX`

Parameter Name	Keyword
`integrator`	`Hermite` (default) `SymmetrizedHermiteRegularizedHermite` `SymmetrizedRegularizedHermite`
`perturbation`	`None` (default), `1PN_Pairwise`, `1PN_EIH2.5PN_EIH`

These integrators can be selected by providing the appropriate term in the `mode` keyword as an argument for `Hermite-GRX`.

In addition, Hermite-GRX has two additional particle types, large_particles and small_particles. The former is used for particles that are relativistically affected, whereas the others remain unaffected by general relativity. Normally, assigning new particles to the general particle set (particles) will cause all objects to be affected by general relativistic effects. But for performance considerations, sometimes one might decide to assign certain minor bodies to the small_particles set, excluding them from relativistic effects.

A.3.3 The Huayno Multicomponent Symplectic *N*-body Code

Most codes use specific parameters to control their internal solvers. An example of a code with a quite extensive parameter set is Huayno (Pelupessy et al. 2012), which uses a recursive Hamiltonian splitting strategy (much like bridge, see Chapter 9) to generate multiple-time step integrators that conserve momentum to machine precision. A wide variety of different integrators is available within Huayno. They are listed in Table A.12, but for more details we refer to Pelupessy et al. (2012). The code was developed within the AMUSE environment, and can be compiled with GPU support using OpenCL (for which the mode "OpenCL" should be specified in the run-time environment). Another gravity code with many options is Rebound (see Section A.8).

Huayno has two types of shared but variable time step codes: with and without a treatment of collisions. Both types come in all even orders, from second to 10th. Such an integrator is initialized as follows

```
gravity = Huayno(mode="SHARED10_COLLISION")
```

or equivalently by

```
gravity = Huayno()
gravity.set_integrator("SHARED10_COLLISION")
```

Table A.12. Integrators Available in Huayno

Brief Description	Keyword
Shared time step	SHARED2, …, SHARED10,
Collision-enabled	SHARED2_COLLISIONS, …, SHARED10_COLLISION
Kepler solvers	KEPLER, CC_KEPLER, CCC_KEPLER, OK
Connected components	CC, CCC_BS, CC_BSA, CCC_BSA
Bulirsch–Stoer	SHAREDBS, BS_CC, BS_CC_KEPLER
KDK	PASS_KDK, HOLD_KDK, BRIDGE_KDK
Drift–kick–drift	PASS_DKD, HOLD_DKD, PPASS_DKD, BRIDGE_DKD

These integrators can be selected by providing the appropriate term in the mode keyword as an argument for Huayno,

The Kepler solvers evolve the two-body problem, with a variety of alternative approaches. The `OK` solver (see Pelupessy et al. 2012), for example, adopts the optimal-kick Hamiltonian splitting method to evaluate every interaction in the system at the optimal-split time step, thereby using the least number of force evaluations to evolve the system.

The designation `CC` in Table A.12 stands for connected components. In the `CC` method, we split the Hamiltonian to find isolated subsystems which are integrated separately for better accuracy compared to the rest system (see also Section 10.1.2). The `CC` method is applicable not only to Kepler solvers but also to the Bulirsch–Stoer integrator (Bulirsch & Stoer 1964). This latter integrator can adopt shared time steps and run with connected components or with a Kepler solver. Each of these leads to better accuracy, particularly for hierarchical systems with complex configurations.

Finally, `Huayno` supports various types of KDK and drift–kick–drift schemes, much as in `bridge` (see Section 9.1).

A.3.4 Symple: The Simplest Possible Sympectic N-body Integrator

An example of a code that was written specifically for AMUSE is the direct symplectic N-body code `Symple`. Since there is no documentation on this code, we use it here as an example of a code that was written as an AMUSE module. `Symple` is simple, and symplectic, hence its name. The code can operate as a symplectic N-body integration module of orders 1, 2, 3, 4, 5, 6, 8, and 10, with fixed, shared time steps. Alternatively, setting the time step parameter $\eta > 0$ enables variable and shared time steps, as in, say `Hermite` (but the integrators then lose their symplectic quality). The default is $\eta = 0.05$.

Initially, we developed `Symple` to be able to effectively integrate the orbit of the triple θ Muscae (see Section 8.4.5). As we saw, this triple comprises three Wolf–Rayet stars, each with a hefty stellar wind. The combination of tight orbits, massive stars, and copious stellar winds makes this system extremely challenging to integrate. It turns out that the linear coupling strategy adopted in Section 8.4.5 is slow and prone to numerical errors. Using a bridge for this system makes the integration more robust, but at an enormous cost in computing time. This made us decide to write `Symple`, a simple symplectic N-body integrator that included a mass-loss term for each star.

Although symplectic integrators do conserve energy on the long run much better than the usually adopted explicit Hermite-type integrators, they can still lead to precession in an elliptic two-body orbit. Orbital phase and the argument of pericenter are not conserved quantities, and therefore are very hard to preserve in a numerical simulation. Although generally this hardly has only minor consequences for the integration, in some cases, such as when integrating von Zeipel–Lidov–Kozai (Lidov 1962; Kozai 1962) affected triples, it is of crucial importance.

Even in symplectic integrators, noncommensurate steps lead to long-period modulations in the energy error, without a long-term drift. The amplitude of the modulation can be reduced by making the step an integral fraction of the period, and reduced more by making it an integral fraction of the actual (numerical) period.

Table A.13. Symple Integration Order and Number of Evaluation Steps

Integrator/Order	Function Evals/Step	Evals/Period (Const. dE)
1	1	–
2	1	32k
3	3	3k
4	3	3k
5	7	1792
6	7	1792
8	15	3840
10	33	4224

`Symple` is not as simple as the name suggests: it facilitates an implicit Euler (Euler 1768), second-order predictor-corrector, and higher-order symplectic integrators. The choice of integrators is made in your script through the code-related parameter. In Table A.13 we present the various integration possibilities. But let us first discuss the advantages and disadvantages of each of the integrators. It is important to bear in mind that a simple symplectic integrator, like `Symple`, works best if the time step is a power of 2 times the orbital period. Not obeying this requirement leads to a rather large drift in orbital phase. For higher-order integrators (eighth- and tenth-order schemes) this time step requirement is not necessary.

The first order scheme is really an implicit Euler integrator. It has no appreciable long-term energy drift. Instead, it will exhibit large orbital variations. For longer time steps, the orbit will modulate with small amplitude. The second-order predictor-corrector scheme in `Symple`, however, is probably your best choice for most problems. Long-term orbital integration hardly shows any net drift of energy, exhibits long-term periodic modulation, and the integration is not appreciably affected by commensurate/noncommensurate steps. We recommend making the step commensurate with the numerical period, which reduces the modulation amplitude.

For the order of integrator in `Symple` to 10th order, see Table A.13. The first column gives the order of the integrator, followed by the number of evaluations per step, which quickly grows beyond sixth order. The last column gives the number of actual evolutions of the integrator per orbital period of a standard binary system (in N-body units $m1 = m2 = 0.5$, $a = 0.57$, $e = 0.75$, resulting in an orbital period of $P = 2.71$) for constant error in the energy. Considering these values, it may be hard to see why anybody would ever use a sixth-order symplectic integrator for solving a regular N-body problem, but for multiobject integrations these numbers will be different and there may be a sweet spot in parameter space for optimal use of a 10th-order symplectic integrator.

A.3.4.1 Radiative Transfer in SPH codes with `SPHRay`

SPHray (Altay et al. 2008) is an SPH ray tracer designed to solve the three-dimensional, time-dependent, radiative-transfer equation. The method relies on a Monte Carlo ray-tracing scheme that does not interpolate the SPH particles on to a grid but instead integrates directly through the SPH kernels. An Axis-Aligned Bounding Box (AABB) (Zhao et al. 2021) test taken from computer graphics applications allows for the acceleration of the ray-tracing component. The AABB is defined as a rectangular box (in three dimensions) with edges parallel to the coordinate axes.

SPHRay is capable of handling multiple ionizing sources and complex geometries, making it suitable for studying complex astronomical structures such as giant molecular clouds. Its treatments for time-dependent photoionization, recombination, heating processes, and temperatures makes is excellently suited for coupling with other applications to study time-dependent processes.

The atomic rates are stored in a number of text files for various chemical species, including Hydrogen, Helium, and Deuterium, in various degrees of ionization. SPHRay also supports a wide range of ionization spectra, which can be incorporated using

```
radiative = Sphray(convert=converter, mode="openmp")
radiative.set_spectra_file("spectra/thermal1e5.cdf")
```

alternative spectra include `spectra/thermal6e4.cdf`, `spectra/power2.0.cdf`, and `spectra/power4.0.cdf`.

A.3.5 External Galactic Potential Models

In many cases, we want to simulate a self-gravitating system in some background Galactic potential. In order to accommodate such studies, we have incorporated several potential models for the Milky Way and other point-symmetric potentials.

The `galpy` (Bovy 2015, 2014) package includes a bulge modeled as a power-law density profile with an exponential cut-off, a Miyamoto & Nagai (1975) disk, and a Navarro–Frenk–White (NFW; Navarro et al. 1996) dark-matter halo. The parameters of each of the individual components are based on fits to observational data, with the additional overall constraint that the Sun's distance to the Galactic center is $R = 8.0$ kpc and the solar circular velocity is $v = 220\ \mathrm{km\ s^{-1}}$ Bovy (2014, 2015). We presented an example of how to initialize such a model in Section 9.4.1.1.

The asymmetric disk of Miyamoto & Nagai (1975) can also be run separately, in which case a different bulge and halo model can be added. For example, it can be complemented using the spherically symmetric Plummer (1911) profile. These profiles can be found in `amuse.ext.galactic_potentials`. The Galactic disk model is initialized using the following lines:

```
disk_mass = 1.0e+10 | units.MSun
disk_size = 10 | units.kpc
disk_height = 100 | units.parsec
from amuse.ext.galactic_potentials import MiyamotoNagai_profile
galactic_disk_model = MiyamotoNagai_profile(disk_mass, disk_size, disk_height)
```

A point-symmetric Plummer potential model can be initiated in a similar fashion using

```
total_mass = 1.0e+6 | units.MSun characteristic_size = 3 | units.parsec
from amuse.ext.galactic_potentials import Plummer_profile
globular_cluster_model = Plummer_profile(total_mass, characteristic_size)
```

In Table A.14, we present a brief overview of the semianalytic potentials that are directly available within AMUSE.

The most versatile and complete semianalytic potential model for the Milky Way is `Galaxia` (Allen & Santillan 1991; Cox & Gómez 2002) This parameterized galaxy-evolution model includes multiple potentials to describe the Milky Way or another galaxy. To make a representation of the current Milky Way, a number of parameters are set by default. They are listed in Table A.15.

An alternative would be to use an external package for either generating complex initial conditions based on some predefined distribution function, or use a specific background potential. The AGAMA package, discussed in Section B.2.2 introduced in Vasiliev (2019), will prove to be quite useful for such studies. It would require the separate compilation of AGAMA, though, which is explained in Section B.2.2.

Table A.14. Short Description of the Potential Models Available in AMUSE

Model	Brief Description
`NFW_profile`	Navarro, Frenk, & White (Navarro et al. 1996) dark matter profile. Free parameters include the scale radius and central density.
`MiyamotoNagai_profile`	Semianalytic disk model (Miyamoto & Nagai 1975). Free parameters include the total mass and the disk scale size and height.
`Plummer_profile`	Plummer density profile (Plummer 1911). Free parameters include the total mass and the scale radius.
`PowerLawCutoff_profile`	Spherically symmetric power-law density profile with exponential cut-off. Free parameters include the central density, power-law scaling radius, the power-law index and the cut-off radius.
`MWpotentialBovy2015`	Jo Bovy's `galpy` density profile model (Bovy 2015).
`Galaxia`	Galaxy potential with bulge, bar, disk, spiral arms and halo (Sharma et al. 2011).

Table A.15. Default Parameters for the Milky Way in `galaxia` (See `amuse.examples.solar_cluster_in_galaxy_potential` discussed in in Section 9.4.1.1)

Parameter Name	Value	Unit
phase	-0.34906	...
a_disk	5.3178	kpc
a_halo	12.0	kpc
aaxis_bar	3.12	kpc
amplitude	1300	$\mathrm{kms^2\,kpc^{-1}}$
amplitude2	680	$\mathrm{kms^2\,kpc^{-1}}$
axis_ratio_bar	0.37	...
b_bulge	0.3873	kpc
b_disk	0.25	kpc
bar_phase	-0.349	...
caxis_bar	1.0×10^{-6}	kpc
fiducial_radius	8.0	kpc
initial_phase	-0.34906	...
mass_bar	1.4000×10^{10}	$M_\odot$
mass_bulge	1.4059×10^{10}	$M_\odot$
mass_disk	8.5608×10^{10}	$M_\odot$
mass_halo	10.7068×10^{10}	$M_\odot$
nbt	0.0	...
omega_bar	40	$\mathrm{km\,(s\,kpc)^{-1}}$
omega_spiral	20	$\mathrm{kms\,(kpc)^{-1}}$
omega_spiral2	20	$\mathrm{km\,(s\,kpc)^{-1}}$
omega_system	40	$\mathrm{km\,(s\,kpc)^{-1}}$
phi21_spiral	-3.49	...
r_sigma	2.5	kpc
rsp	3.12	kpc
scale_height	0.18	kpc
sigma_s	100	Myr
spiral_density_amplitude	5.80×10^{14}	$M_\odot\,\mathrm{(kpc)^{-3}}$
spiral_model	0.0	...
spiral_phase	-0.349	...
t_sim	-34.132	Myr
tan_pitch_angle	0.227	...

These are many free parameters in this model. Here, we have rounded off some of the numbers; in the package, the pitch angle is given to 9 decimal places and the scale height of the disk to 12 digits.

You can call `Galaxia` and set the mass of the disk to $1.0 \times 10^{11}\,M_\odot$ as follows:

```
from amuse.community.galaxia.interface importBarAndSpirals3D
galaxy = BarAndSpirals3D()
galaxy.parameters.mass_disk = 1.0e+11 | units.MSun
```

All the parameters listed in Table A.15 can be set separately to construct original background-potential models.

A.4 Higher-order Bridges

The standard bridge is second order, and here we derive the fundamental method. As a result, the numerical error (in particular in energy) scales with the square of the time step dt^2. Sometimes, higher-order integration methods are desirable. For this reason, we present some higher-order bridge schemes.

A.4.1 Fundamentals of High-order Bridges

In some cases, the Bridge scheme may encounter numerical difficulties because the method is only second-order accurate in the time step `dt`. This can become a limiting factor when coupling two higher-order methods (e.g., sixth- and fourth-order Hermite schemes), or when coupling a fourth-order scheme with a background potential expressed as an analytic function or look-up table. For these cases, a higher-order bridging method is desirable, especially if the symplectic quality can be preserved. Here, we show how Bridge can be extended to an arbitrarily high (even) order. There exist symplectic integrators of odd order, for example the implicit Euler, but they are not very suitable for constructing bridges (Jiao et al. 2007).

We can derive higher-order bridges formally as follows. This starts by assuming a system of particles composed of some number Q of subsystems k. In this case, the total Hamiltonian of the system,

$$H = \sum_{i\in S}\frac{\mathbf{p}_i^2}{2m_i} - \sum_{\substack{i,j\in S\\ i\neq j}}\frac{Gm_im_j}{r_{ij}}, \tag{A.3}$$

can be split to obtain

$$H = \sum_{k=1}^{Q} H_k + \sum_{\substack{k,l=1\\ k\neq l}}^{Q} H_{kl}^{int}. \tag{A.4}$$

The terms in Equation A.4 are given by

$$\begin{aligned} H_k &= \sum_{i\in k}\frac{\mathbf{p}_i^2}{2m_i} - \sum_{\substack{i,j\in k\\ i\neq j}}\frac{Gm_im_j}{r_{ij}} \\ H_{kl}^{int} &= -\sum_{\substack{i\in k\\ j\in l}}\frac{Gm_im_j}{r_{ij}}. \end{aligned} \tag{A.5}$$

Note that this is a slightly different notation than in the original paper by Fujii et al. (2007), but the end result is the same.

Based on this splitting, a general, multisubsystem, second-order time evolution operator can be constructed as follows:

$$\texttt{Bridge}_2(\tau) = \prod_{k=1}^{Q} e^{\frac{1}{2}\tau D_{H_k}} \prod_{\substack{k,l=1\\k\neq l}}^{Q} e^{\tau D_{H_{kl}^{int}}} \prod_{k=1}^{Q} e^{\frac{1}{2}\tau D_{H_k}}. \tag{A.6}$$

As in the classic Bridge, the operators $e^{\tau D_{H_k}}$ independently evolve each of the subsystems k in isolation. The operators $e^{\tau D_{H_{kl}^{int}}}$ represent pure momentum kicks due to interactions between subsystems k and l. In the case of the Hamiltonian in Equation A.4, the interaction terms do not depend on velocity, so the operators $e^{\tau D_{H_{kl}^{int}}}$ commute. We note, however, that this is not the case for velocity-dependent forces, so special treatment is required in that case. Because each of the operators $e^{\tau D_{H_k}}$ and $e^{\tau D_{H_{kl}^{int}}}$ can, in principle, be associated with different solvers running concurrently, a highly efficient parallel implementation can be achieved for the generalized Bridge scheme.

A higher-order symplectic Bridge scheme can be constructed in a similar fashion. By defining the drift and kick operators

$$\mathcal{D}(\tau) = \prod_{k=1}^{Q} e^{\tau D_{H_k}} \tag{A.7}$$

and

$$\mathcal{K}(\tau) = \prod_{\substack{k,l=1\\k\neq l}}^{Q} e^{\tau D_{H_{kl}^{int}}}, \tag{A.8}$$

Equation A.6 can be extended to higher order by a suitable combination of $D(\tau)$ and $K(\tau)$ operators (Hairer et al. 2006). For a fourth-order symmetric composition with four stages, the high-order Bridge takes the form:

$$\begin{aligned}\texttt{Bridge}_4(\tau) = {} & \mathcal{D}(u_0\tau)\mathcal{K}(u_0\tau)\mathcal{D}(u_1\tau)\mathcal{K}(v_1\tau)\\ & \mathcal{D}(u_2\tau)\mathcal{K}(v_1\tau)\mathcal{D}(u_1\tau)\mathcal{K}(v_0\tau)\mathcal{D}(u_0\tau)\end{aligned} \tag{A.9}$$

(see Yoshida 1990; Hairer et al. 2006; for the coefficients u_i and v_i). The sixth-order method becomes

$$\begin{aligned}\texttt{Bridge}_6(\tau) = {} & \mathcal{D}\left[\frac{1}{2}w_0\tau\right]\mathcal{K}\\ & [w_0\tau]\mathcal{D}\left[\frac{1}{2}(w_0 + w_1)\tau\right] \quad \cdots \quad \mathcal{D}\left[\frac{1}{2}(w_{s-1} + w_s)\tau\right]\mathcal{K}[w_s\tau]\mathcal{D}\left[\frac{1}{2}w_s\tau\right],\end{aligned} \tag{A.10}$$

where the coefficients w_i are given by Sofroniou & Spaletta (2005) and Hairer et al. (2006). The higher-order bridges are identified by order: Leapfrog, fourth, sixth, etc., and by the number of force evaluations: $m = 5$ or $m = 6$ for the fourth order, $m = 11$ or $m = 13$ for the sixth order. AMUSE includes implementations up to and including 10th order (with $m = 35$), which you will probably never need. A complete list of methods for bridge can be found in Section A.4.2 (see Table A.16).

Table A.16. Bridge Methods Identified by the Order of Their Integrator

Name	Order	S	M
LEAPFROG	...	...	...
SPLIT_4TH_S_M4	4	S	4
SPLIT_4TH_S_M5	4	S	5
SPLIT_4TH_S_M6	4	S	6
SPLIT_6TH_SS_M11	6	SS	11
SPLIT_6TH_SS_M13	6	SS	13
SPLIT_8TH_SS_M21	8	SS	21
SPLIT_10TH_SS_M35	10	SS	35

The columns give the order of the integrator, the splitting strategy (`S` for symmetric composition and `SS` for symmetric composition of symmetric maps), and the number of evaluations per bridge step (M).

The above formulation provides a fully symplectic time evolution if the codes being bridged are themselves symplectic. Using this scheme, codes/integrators of different orders can be coupled to construct a formally high-order scheme by matching the order of the bridge used during the coupling. For example, when coupling a sixth order to a fourth-order method, it is appropriate to choose $\mathtt{Bridge}_4(\tau)$ to create a fourth-order compound method. If we chose $\mathtt{Bridge}_6(\tau)$, the method would still be only fourth-order because of the fourth-order subintegrator. In what follows, we refer to the second-order (original) bridge as the classic bridge, and use the term higher-order bridge for the generic case.

The logic of selecting an appropriate bridge order may not always be straightforward. For example, if second-, fourth-, and sixth-order schemes were coupled using the classic bridge [$\mathtt{Bridge}_2(\tau)$], the result would be a second-order compound method. However, we could improve this by adopting a *hierarchical coupling strategy*: bridge the fourth- and sixth-order codes using $\mathtt{Bridge}_4(\tau)$, then bridge this compound method with the remaining second-order code using $\mathtt{Bridge}_2(\tau)$. Although the overall compound solver is still formally second order, the subsystems evolved with the more precise codes are coupled at a higher order than the rest of the system. This can be advantageous if such a subsystem dominates the overall error or requires higher precision—for example, because of the requirement to resolve close encounters. Finally, we note that, while the methods presented here are very flexible, the proper choice of integrators is highly problem-dependent.

A.4.2 Tuning the Higher-Order Bridge

In Chapter 9 (see Section 9.3.1) we discussed coupling codes using `bridge`, and in Section A.4 we explained how higher-order bridges can be constructed by repeated steps with a bridge operator. Several higher-order bridge methods are incorporated into AMUSE; some of them worked out in `composition_methods`, which can be incorporated in your script via

```
from amuse.ext.composition_methods import {method_name}
```

where method_name is the name of the method you want to use. Table A.16 presents a brief overview of the available methods.

An example of the working of a higher-order bridge can be found in the code example amuse.examples.three_body_bridge_order4m4, which we discuss in Section 9.3.1. There we declare a fourth-order bridge with four evaluations per bridge step, using

```
gravity = bridge.Bridge(use_threading=False, method=SPLIT_4TH_S_M4)
```

A.5 Stopping and Boundary Conditions

Any code has a domain of operation. This limitation can be controlled in terms of stopping conditions (at what point should a code hand back control to the parent script) and boundary conditions (when should a code hand back data that does not belong to its domain).

A.5.1 Hydrodynamical Boundary Conditions

In many circumstances, smoothed-particle hydrodynamics solvers do not have to worry much about boundary conditions. Simulations of star formation or stellar collisions, for example, simply remove SPH particles that stray too far from the region of interest. In other circumstances, however, boundary conditions are important, and these are implemented and accessible through the AMUSE interface (see Table 6.2). For example, periodic boundary conditions in an SPH code can be declared by setting the parameter periodic_boundaries_flag to True and giving a dimension to periodic_box_size.

On the other hand, grid-based hydrodynamics codes, as discussed in Section 6.2.3, always require some sort of boundary condition, and the number of options is more extensive. For these codes, the boundary conditions are an attribute of the code that can be set at run time. For example, to apply periodic boundary conditions along the Cartesian coordinates, you could write:

```
hydro_code.parameters.x_boundary_conditions = ("periodic", "periodic")
hydro_code.parameters.y_boundary_conditions = ("periodic", "periodic")
hydro_code.parameters.z_boundary_conditions = ("periodic", "periodic")
```

Alternatives to periodic boundary conditions are "cont," "inflow," "interface," "outflow," and "reflective," although not all of these may be defined for each of the available hydrodynamics solvers. These terms may sound a bit arcane, but if you perform hydrodynamical simulations, they are probably quite familiar.

A.5.2 Stopping Conditions

Throughout this book we have stressed the importance of being able to stop a code cleanly when certain conditions are met. These conditions are called stopping conditions. We currently maintain a finite set of such conditions; but if more are needed, we encourage you to implement them because they make the interaction between the modules in AMUSE more flexible and versatile.

Table A.17 lists the most commonly used stopping conditions in the AMUSE community modules. Not all codes support all of the stopping conditions for that particular domain. The main reasons that some codes do not support some or all possible stopping conditions are either that the stopping conditions were never implemented (or tested, and therefore not advertised) or that a particular code does not behave well under externally initiated interrupts. Adding a new stopping condition is not hard for most codes, but it does require modifying the source code and it tends to adversely affect performance.

Table A.17. Stopping Condition Keywords for Gravitational Dynamics, Stellar Evolution, and Hydrodynamics

Stopping Condition Keywords	
Gravitational Dynamics	
`collision_detection`	Detect collision between two particles
`stopping_conditions_timeout`	Detect if the predetermined end time is reached
`out_of_box_detection`	Check if particles have left some predefined box size
`number_of_steps_detection`	Stop after predetermined number of integration steps are realized
`escape_detection`	Check if a particles has escaped
`pair_detection`	Check if two particles have become mutually bound
Stellar Evolution	
`supernova_detection`	Check if a supernova has occurred
`max_age_stop_condition`	Check of some maximum age has been reached
`max_iter_stop_condition`	Check for maximum number of internal code iteration-steps
`min_timestep_stop_condition`	Guarantee the code to stop when the time step becomes too small
Hydrodynamics	
`minimum_density_parameter`	Check for a minimun density threshold
`maximum_density_parameter`	Check for a maximum density threshold

`minimum_internal_energy_parameter`	Check for the minimun energy of a particle
`maximum_internal_energy_parameter`	Check for the maximum energy of a particle
`number_of_steps_parameter`	Parameter to set the maximum number of steps
`out_of_box_parameter`	Parameter to set the size of the box
`out_of_box_use_center_of_mass_parameter`	Assure that the out-of-box detection is done from the center of mass
`timeout_parameter`	Set the maximum simulation time

There are no dedicated stopping conditions for radiative-transfer codes, simply because up to now they have not been needed. The lack of stopping conditions for radiative-transfer codes does not mean that these are not needed, but rather that these codes have not been used as frequently yet in production simulations.

A.6 Other Handy Routines

There are quite a few handy routines in AMUSE. If ever you find yourself in a position to write a new routine because you want to manipulate a specific particle set, chances are that we already wrote such a routine. This book, however, is too short to list all possibilities and refer explicitly to each of the handy routines we have written and tested. If it exists, it is likely that our routine also turns out to be effectively optimized. On the other hand, there are many ways in which one can manipulate specific data. The best place to look for our own implementation, and hopefully in the near future also including your contributions, is in the actual source code. Particularly handy routines to handle data can be found in the following directories: `coupledata`, `datamodel`, `ext`, `ic`, `io`, `plot`, `rfi`, `support`, `test`, and `units`.

An example of such a handy routine is the method to find a subset of particles that belong to a larger particle set:

```
L4_trojans = planetesimals.get_intersecting_subset_in(L4_trojans_rot)
```

here the particle set `planetesimals` is a set that supposedly contains members of another particle set, called `asteroids`. This routine returns the particular particles in the particle set `planetesimals` that are identical to those in the other particle set `asteroids`.

It may seem that this is a rather contrived operation, but in practice we need this more often than one might naively expect. There is a total of 5000 routines in the collective AMUSE directories. Some are more useful than others, but usually there a good reasons why we have written them. Before you start writing your own data-handling routine, we strongly recommend that you first search for an existing routine that does the job.

A.6.1 Particles and Particle Sets

Particles form a fundamental part of the AMUSE framework. They are a form of containers with attributes, member functions, manipulators, and communicators. They form the heart of the AMUSE data representation.

A.6.2 Particle Attributes

Each domain has a standard selection of attributes. Some codes have additional sets. The main attributes are listed in `amuse.datamodel.attributes`.

In Tables A.18 to A.21, we list the most fundamental attributes and their default values for the various particles sets for that specific domain.

Table A.18. Overview of the Minimum Set of Attributes for the Gravity Domain in AMUSE

Attribute Name			Gravitational Dynamics
Attribute	Value	Unit	Explanation
Mass	0.0	kg	Particle mass
Radius	0.0	m	Particle radius
Position	[0.0, 0.0, 0.0]	m	Position vector of a particle
Velocity	[0.0, 0.0, 0.0]	$\mathrm{m\ s^{-1}}$	Velocity vector of a particle
Acceleration	[0.0, 0.0, 0.0]	$\mathrm{m\,s^{-2}}$	Acceleration vector

Table A.19. Overview of the Minimum Set of Attributes for the Stellar Evolution Domain in AMUSE

Attribute Name			Stellar Evolution
Attribute	Value	Unit	Explanation
Mass	0.0	kg	Particle mass
Radius	0.0	m	Particle radius
Age	0.0	s	Particle age

Table A.20. Overview of the Minimum Set of Attributes for the Hydrodynamics Domain in AMUSE

Attribute Name			Hydrodynamics
Attribute	Value	Unit	Explanation
pressure	0.0	Pa	The pressure in a region of space
density	0.0	$\mathrm{kg\,m^{-3}}$	The density of molecules or solid matter

temperature	0.0	K	The temperature of the gas
magnetic_field	0.0	T	Magnetic field created by gas and stars
velocity_field	0.0	$m\,s^{-1}$	Velocity of the gas
gravitational_potential	0.0	$m^2\,s^{-2}$	Gravity forces from stars and gas
viscosity	0.0	$kg\,m^{-1}\,s^{-1}$	Viscosity of the gas cloud

Table A.21. Overview of the Minimum Set of Attributes for the Radiative-transfer Domain in AMUSE

Attribute Name			Radiative Transfer
Attribute	Value	Unit	Explanation
temperature_gas	0.0	K	The temperature of the gas
temperature_dust	0.0	K	The temperature of the dust
temperature_background	0.0	K	The temperature of the background
density	0.0	$mol\,m^{-3}$	Molecular density
magnetic_field	0.0	tesla	Magnetic field created by gas and stars
velocity_field	0.0	$m\,s^{-1}$	Velocity of the gas

A.6.3 Particle Queries and Functions

Apart from a simple storage container, there are numerous functions associated with particles and particle sets. Here, we list the most useful ones. Many of these have already been presented in various forms, but we will give a brief overview of a few of the most useful. Note that we have both singular (`Particle()`) and plural `Particles()`. The latter, we call a set. The difference is similar to an integer and an array of integers. Some functions will work on a single particle, others on a set of particles. Converting a single particle to a set can be realized using `as_set()` (see below):

- `get_timeline_of_attribute_as_vector()` returns an array of a specific attribute from the time-line of a particle set.
- `copy_to_new_particles` copies all of a particle's attribute data to a new particle.
- `new_channel_to(self, other, attributes=None, target_names =None)` makes a channel between two particle sets.
- `copy_values_of_attributes_to(self, attribute_names, particles)` copies the attribute values of one particle to another particle.
- `add_particles(self, particles)` adds a particle to a particle set.

- `ensure_presence_of(self, x)` ensures that a particle has a specific attribute.
- `remove_particles(self, particles)` removes an existing particle.
- `synchronize_to(self, other_particles)` synchronizes one particle set to another. This routine does not copy data, but makes sure that particles missing in one set will be synchronized to the other, or that particles not present in the one set are removed from the other.
- `as_set(self)` converts a single particle to a particle set.
- `select_array(self, selection_function, attributes=())` returns a subset view of a particle set, containing all particles for which the selection function returns True.
- `difference(self, other)` returns a new subset containing the difference between two particle sets.

```
particles = Particles(3)
particles.x = [1.0, 2.0, 3.0] | units.pc
subset = particles.select(lambda pos : pos < 2.5 | units.pc,
["x"])
inner_stars = particles.difference(subset)
print(inner_stars)
```

- `has_duplicates(self)` returns True when a set contains a particle with the same key more than once. Particles with the same key are interpreted as the same particle. This small example:

```
particles = Particles()
p1 = particles.add_particle(Particle(1))
p2 = particles.add_particle(Particle(2))
particles.has_duplicates()
```

returns `False`.

- `reversed(self)` returns a subset with the same particles, but with reversed sequential order (the first particle will become last)
- `sorted_by_attributes(self, *attributes)` sorts particles by a specific attribute. The last attribute name in the call is used for the primary sort order, the second-to-last attribute name for the secondary sort order, and so forth.

```
particles = Particles(4)
particles.mass = [2.0, 3.0, 1.0, 4.0] | units.kg
particles.radius = [3.0, 2.0, 1.0, 2.0] | units.m
sorted = particles.sorted_by_attributes('mass', 'radius')
print(sorted.radius)
```

- `random_sample(self, number_of_particles)` randomly selects a subset of `number_of_particles` particles from a parent set.

- `remove_attribute_from_store(self, name)` removes a specific attribute from a particle.

```
particles = Particles(3)
particles.temporary_attribute = 42
particles.remove_attribute_from_store("temporary_attribute")
```

Particles can be organized in supersets using the `ParticlesSuperset` particle set, which is a derived class from the `AbstractParticleSet`. In a superset of particles, attribute values are not stored by the superset, but kept locally in the subset. The superset provides a view of two or more sets of particles.

Superset objects are not supposed to be created directly, but are always derived from two or more particle sets. A simple example is

```
p1 = Particles(3)
p1.mass = [10.0, 20.0, 30.0] | units.kg
p2 = Particles(3)
p2.mass = [40.0, 50.0, 60.0] | units.kg
p = ParticlesSuperset([p1, p2])
print(len(p))
6
print(p.mass)
[10.0, 20.0, 30.0, 40.0, 50.0, 60.0] kg
p[4].mass = 70 | units.kg
print(p.mass)
[10.0, 20.0, 30.0, 40.0, 70.0, 60.0] kg
p2[1].mass
quantity<70.0 kg>
cp = p.copy()
print(len(cp))
6
print(cp.mass)
[10.0, 20.0, 30.0, 40.0, 70.0, 60.0] kg
```

A.6.4 Grid-based Data

Similarly to `Particle` and `Particles` sets, AMUSE recognizes grids. The grid class is based on the `AbstractGrid` class. This abstract class as an endless range of member functions for grid manipulation. It is best to look in the module `amuse.datamodel.grids`. Derived classes from the `AbstractGrid` include `UnstructuredGrid`, `StructuredGrid`, `RectilinearGrid`, `RegularGrid`, and `CartesianGrid`. The functions to generate those grids include `new_cartesian_grid()`, `new_regular_grid()`, `new_rectilinear_grid()`, `new_structured_grid()`, and `new_unstructured_grid()`. And then we did not even mention the entire repository of `SubGrid` structures.

Apart from those grid-generating functions, there are numerous grid manipulation routines, query functions, attribute setters, and methods to sample the grid, similarly to the enormous range of functions associated for particles and particle sets. The number of routines is so large that even for this appendix it would go too far to mention them all. The best advice we have is to read the source code in the `amuse.datamodel` module.

A.7 Plotting

Making figures and plotting in general is an art. A good figure has a message, and it is important to have the right tools to communicate it. Here, we discuss a few `Python` tools that we like and have used to make the figures in this book.

A.7.1 Basic Plotting in `Python`

The most basic plotting can be done with `gnuplot` (Williams & Kelley 2013; Vaught 1996), which can be downloaded from `http://www.gnuplot.info`. If you have installed `gnuplot` on your computer, you can use the `gnuplot-py` package to run it directly from your script. To install `gnuplot-py`, open a shell and type

```
>% easy_install gnuplot-py
```

An extensive manual for `gnuplot` is available at: http://www.gnuplot.info/docs_5.0/gnuplot.pdf.

Sometimes more control is needed to create a plot exactly in accord with your demanding co-author's wishes. For more elaborate plotting, we have opted for `pyplot`, which is part of the `matplotlib` package (Hunter 2007). `Matplotlib` is a `Python` plotting library which can produce publication-quality figures. Many of the AMUSE example scripts use this library to produce graphical output, and we refer you to `amuse.examples` for many examples, from simple to complex. If you would like to take advantage of `matplotlib`, get it from https://matplotlib.org and install it as a `Python` package. For your own work, it is of course also possible to print the required data to a file and use your favorite plotting tool to make the figures.

A.7.2 Plotting with Units

Most of the figures in this book are made using `matplotlib`. In most cases, we used the standard `matplotlib` and set the units by hand. However, this is not really necessary in most cases because we have overloaded the most important functions to allow them to handle arrays with units. The AMUSE enabled `matplotlib` functions can be imported via

```
from amuse.plot import scatter, plot, hist
```

The majority of the matplotlib routines can be imported in this way, but we have also created a few new ones that are specific to particular codes used in AMUSE. They are listed below in Section A.7.3.

An interesting tool for plotting that supports AMUSE particle sets is vaex at https://vaex.io/ (Breddels 2017). This package is particularly interesting if the particle sets become very large and hard to handle because vaex can process a billion objects per second on a regular workstation. This enables direct exploration of large data sets, making slices and selections using Boolean expressions. Vaex is easily installed from the git server using

```
%> pip install git+https://github.com/vaexio/vaex/
```

It is also possible to use the jupyter (Kluyver et al. 2016) version of vaex, which is called ipyvolume. This enables you to construct a portable html compliant version of your plotted data which can easily be incorporated in a web interface. This package can be downloaded from the github repository.[1]

A.7.3 Additional Plotting Utilities

We have created supporting routines for a variety of plotting applications. Some of these routines have found their way into AMUSE. Particle-based plots are generally rather easy to make, including animations, as we demonstrated in Chapter 4. Plotting results from radiative-transfer and hydrodynamical simulations are more complicated, and for these purposes we provide a few additional plotting routines. Most can be found in the amuse.plot module.

- The results from SPH simulations are often easiest to interpret by simply plotting the individual particles. For this purpose, we provide the function

```
sph_particles_plot()
```

which is a very simple and fast way to make a plot of the hydrodynamical state of a set of SPH particles. It takes a particle set as argument, with several additional arguments to control the energy range, the size of the viewport, and the transparency of the particles. The particles must have position, internal energy e, and a smoothing length h_smooth, otherwise the routine will not know how to calculate the density maps. Figure A.2 shows the 2010 Leiden Observatory Christmas card that we made using this routine.

- For more flexible plotting, we can convert a particle set in AMUSE to a pynbody data set by calling[2]

[1] See https://github.com/widgetti/ipyvolume

[2] A concise manual for pynbody is available online at https://pynbody.github.io/pynbody/

Figure A.2. False color image of the HII region NGC 2264 in the constellation Monoceros (the Unicorn). Overplotted are snapshots of a hydrodynamical simulation (using `Fi`) of two stars of masses 0.3 and $3.0\,M_\odot$ that collide head-on with a relative velocity of $3000\,\mathrm{km\,s^{-1}}$ at an age of 310 Myr, when the primary is a giant. The Christmas card was made with AMUSE and used by Leiden Observatory in 2010. The figure was made using the script `amuse.examples.christmas_card_2010`; it takes a couple of hours to run on a laptop.

```
convert_particles_to_pynbody_data()
```

Here, in addition to providing a particle set, the arguments `length_unit` and `pynbody_unit` should be set to the desired units, for example `length_-unit=units.kpc` and `pynbody_unit="kpc"`.

- Since `pynbody` is a convenient package, we also provide a column density plotting routing for it:

```
pynbody_column_density_plot()
```

There are a number of limitations to these plotting packages, and they are not as well developed as the rest of AMUSE, but they may at least serve as a basis for more complete routines.

- Finally, for plotting equipotential surfaces we provide

```
effective_iso_potential_plot(gravity_code, ...)
```

which has a large number of possible arguments. This routine creates a contour plot of the effective potential of particles in a gravity code. The code must support `get_potential_at_point`, which is used to map the iso-contour. The routine is practical because it is almost completely independent of the underlying code, and, as a result, it is possible to make iso-density plots or equipotential-surface plots from any potential, even if the underlying potential is provided by a complicated bridging of multiple codes. It requests the gravity code from which the data are read, and some additional parameters, each of which is listed in the original source code, which can be found in the `amuse.plot` module.

A.7.4 Multiprocessing Codes

We can implement multiprocessing in this example by slightly rewriting the main evolve function in Code example 5.4. In Code example A.1, we adopt a similar approach as in Code example 5.4, but we can enable multiprocessing by introducing the specific stellar evolution code as an argument. This routine takes `code` as an argument and uses it to evolve a 10 $M_\odot$ star to an age of 1 Myr. We could expand its functionality by introducing more arguments, such as mass, end time, and metallicity, but that would make the code example longer and not necessarily easier to understand.

The driving sequence for the main evolution routine in Code example A.1 is presented in Code example A.2. Here, we use the `Python` packages `queue` and `threading` for synchronizing queues and enabling threading, respectively, along with the process-based threading interface `multiprocessing`. We recognize that this example uses several fairly sophisticated multiprocessing concepts that the novice reader may find intimidating. If that is the case, you can safely skip to the next subsection.

We construct the task queue as a global variable code_queue, which in our case contains the list of codes used in our experiment. The actual work is done in the function remote_worker_code, which reads the next task from the queue, calls the evolution function evolve_single_star, and notifies the queue when the task has finished. The routine evolve_with_different_stellar_model forms the heart of the multiprocessing task. It fills the queue with the various codes, checks how many

```
def evolve_single_star(code):
    stars = Particles(mass=10 | units.MSun)
    stellar = code()
    stellar.particles.add_particles(stars)
    channel = stellar.particles.new_channel_to(stars)

    stellar.evolve_model(1 | units.Myr)
    channel.copy()
    print(
        f"Star evolved to time= {stellar.model_time} "
        f"M= {stars.mass} R= {stars.radius}"
    )
    stellar.stop()
```

Code example A.1:. Rewritten routine (from Code example 5.4) for evolving a single 10 $M_\odot$ star to an age of 1 Myr with the stellar evolution code as an argument. Code fragment from `amuse.examples.multiple_stellar_threaded`.

```
code_queue = queue.Queue()


def remote_worker_code():
    code = code_queue.get()
    evolve_single_star(code)
    code_queue.task_done()


def evolve_with_different_stellar_model(codes):
    for code in codes:
        if code is not None:
            code_queue.put(code)
    n_cpu = multiprocessing.cpu_count()
    for i in range(n_cpu):
        thread = threading.Thread(target=remote_worker_code)
        thread.daemon = True
        thread.start()
    code_queue.join()  # block until all tasks are done
```

Code example A.2:. Multiprocess version of the main loop for evolving stars, again from `amuse.examples.multiple_stellar_threaded`.

cores are available (to prevent overloading your system), and starts the threads. It ends by blocking each thread until all items in the queue have finished. The entire routine to launch multiple stellar evolution codes is given in `amuse.examples.stellar.multiple`_stellar_threaded.

Now, we can evolve the same (10 $M_{\odot}$) star to the same age (1 Myr), using four different stellar evolution codes in parallel. These codes are listed in an array and passed as an argument to the driving routine:

```
codes = [SeBa, SSE, MESA, EVTwin]
evolve_with_different_stellar_model(codes)
```

Note that simulating the evolution of a large number of stars may run into fundamental limitations of the operating system. Modern operating systems are limited, by their standard settings, to $2^{15} = 32{,}768$ concurrent jobs (even on large supercomputers), and it often requires root access to change this.

A.8 AMUSE Community Modules

AMUSE contains a collection of some 50 different community modules, each of which can be used separately, or in conjunction with one or more other codes. Not all combinations make sense, even though technically the codes can be combined. Here, we give a brief overview of all current community modules. This list is bound to change with time.[3]

[3] For updates, see `AMUSE_updates.pdf`.

We recognize three classes of modules, the core selection, envelope, and the rest. Originally, we designed AMUSE following a "Noah's Ark" philosophy (Portegies Zwart 2018). According to this paradigm, each domain should have two working implementations. Table A.22 presents an overview of the core selection of codes. In practice, we sometimes deviate from this, but two modules are the minimum necessary to be able to develop a generalized interface.

A.8.1 Modules by Domain and Functionality

Table A.23 lists the modules in AMUSE by domain and functionality. Note that, in the gravitational domain, we have expanded the code type category to include "dedicated" codes, which simply solve Kepler's equations, and "limited" codes, which make such rigorous assumptions about the gravitational force that they should be considered potential solvers, rather than *N*-body codes.

Because the efficiency of a code depends on the algorithm, time-stepping, data transport, initial conditions, available hardware, parameter settings, etc., it is hard to give an absolute ranking, and in addition speed and efficiency are not always the same thing. Generally, but not always, slower codes tend to produce more accurate results, and therefore should not necessarily be avoided; the difference in speed may translate to higher accuracy or a better precision (see Portegies Zwart & Boekholt 2018 for a discussion of the difference between accuracy and precision).

A.8.2 Modules by Name

Finally, we list all current modules alphabetically, for reference. For each code, we (1) provide an indication of its main working domain, (2) point to a script in which the code's usage is demonstrated, (3) list extra requirements or features regarding software or hardware, and give a short list of (4) equivalent and (5) comparable codes. Equivalent codes use the same or a similar algorithm to solve the same underlying equations, while comparable codes solve the same physics but using a different numerical method. Here, *the same* and *different* are used rather loosely; the terms are meant to indicate what alternative code might be used to address a particular problem. Specifically, it should be possible to substitute an equivalent code into a script designed for any given code, with minimal changes, so if a code has no script listed, try one of its equivalents.

Core selection: These are codes we support and continue to support. We either update the interface, or incorporate bug fixes or even new version of the code if they become available.

`Athena` (core)

Grid-based astrophysical hydrodynamics solver with support for magnetic fields (Stone et al. 2010). It was originally developed for studying the interstellar medium, star formation, and accretion flows, but can address many other hydrodynamics problems (Gardiner & Stone 2005, 2008; Stone et al. 2008).

Work domain: Hydrodynamics/radiative transfer
Working example: `amuse.examples.test_jeans_instability`
Extra requirements/features: MPI parallel
Equivalent codes: `Flash`, `Capreole`

Table A.22. Overview of the Core Selection of Codes in AMUSE

Module	**Gravitational Dynamics**	
Code Name	**Application Domain**	
`ph4`	Direct *N*-body	
`SmallN`	Direct *N*-body	
`BHTree`	Tree-code	
`Petar`	Hybrid direct-tree *N*-body	
`Mercury`	Planetary systems	
`SecularMultiple`	Hierarchical multiples	
Module		**Stellar Evolution**
Code Name	**Application Domain**	
`MESA`	Henyey stellar evolution	
`SeBa`	Parametrized stellar evolution	
Module		**Hydrodynamics**
Code Name	**Application Domain**	
`Fi`	Smoothed-particles Hydrodynamics	
`Athena`	Grid-based hydrodynamics	
Module		**Radiative-transfer Dynamics**
Code Name	**Application Domain**	
`Mocassin`	Monte Carlo radiative transfer	
`SPHray`	Monte Carlo radiative transfer	
Module		**Chemical Dynamics**
Code Name	**Application Domain**	
`Krome`	Equilibrium chemistry	
`UCLChem`	Nonequilibirum chemistry	
Module		**Coupling and Initial Conditions**
Code Name	**Application Domain**	
`Kepler`	Two-body solver	
`Kepler_orbiters`	*N*-body planetary solvers	
`mmams`	Stellar collision routines	
`mameclot`	Initial conditions generator	
`Fastkick`	GPU-kick routine for bridge	
`FractalCluster`	Initial conditions generator	
`GalactICs`	Initial conditions generator	
`HaloGen`	Initial conditions generator	
`hop`	Clump finder	

Table A.23. Community Modules in AMUSE, Listed by Domain and Algorithm

GRAVITATIONAL DYNAMICS	
Pure	AarsethZare, Brutus, Hermite, ph4, SmallN.
Direct	Nbody6++, Mercury, HiGPU, Huayno, Rebound,
Approximate	BHTree, Bonsai, Petar.
Dedicated	Kepler, kepler_orbiters, Sakura, TwoBody. SecularMultiple, TidyMess.
Limited	Etics.
STELLAR EVOLUTION	
Henyey	EVTwin, MESA.
Parameterized	BSE, SeBa, SSE, CashedSE.
HYDRODYNAMICS	
SPH	Fi, Gadget2
Grid	Athena, Capreole, Flash, MPI-AMRVAC.
Moving mesh	Arepo.
RADIATIVE TRANSFER	
Monte Carlo	Mocassin, SPHray.
Photon transport	SimpleX2.
INTERSTELLAR CHEMISTRY	
Chemistry	Krome. UCLChem.
MISCELLANEOUS	
Analysis	Hop.
Initial conditions	GalactICs, FractalCluster, mameclot.
Potentials	Galactics, HaloGen, Galpy.
Visualization	Vaex, Fresco.
Stellar evolution	Evtwin2Sse.
Disk evolution	Vader.
Planet formation	PACE.
Specialized	ImpactCrater, PPD.
Coupling	Bridge, Venice

Comparable codes: `Fi`, `Gadget2`

`BHTree` (core)

Second-order softened gravitational N-body code with shared time steps and approximate forces. This is a parallel implementation of the original Barnes–Hut tree code (Barnes & Hut 1986).

Work domain: gravitational dynamics, in particular galaxy dynamics
Working example: `amuse.examples.particles_and_gas_in_cluster`
Extra requirements/features: MPI parallel, GPU
Equivalent codes: `Bonsai`, `Octgrav`
Comparable codes: `MI6`

`EVTwin2SSE` (core)
Version of `EVTwin` modified to handle exceptions and hand them to `SSE` when `EVTwin` crashes. The handover algorithm performs an rms search on the relative differences in mass, radius, and luminosity between the last known EVTwin state and the entire stellar history of SSE until the first state of a stellar remnant type. A brief description was presented in Section 8.4.3.

Note that although the code supports running several stars simultaneously, the point where EVTwin crashes is strongly dependent on the initial mass of the stars. It is advisable to run the EVTwin2SSE code for one initial mass at a time.
Work domain: Stellar evolution
Working example: `amuse.examples.hr_diagram_cluster`
Extra requirements/features: `EVTwin`, `SSE`
Equivalent codes: None
Comparable codes: None.

`Fi` (core)
MPI-parallel SPH code (Hernquist & Katz 1989; Pelupessy et al. 2004; Gerritsen & Icke 1997).
Work domain: gravitational dynamics/hydrodynamics
Working example: `amuse.examples.gas_in_cluster`
Extra requirements/features: OpenMP parallel
Equivalent codes: `Gadget2`
Comparable codes: `Flash`, `Athena`

`Flash` (core)
Grid-based radiative magnetohydrodynamics solver originally written to simulate nova explosions on white dwarfs (Fryxell et al. 2010). Flash is modular at the compile level, with fundamental structure not dissimilar to AMUSE, using modular codes and providing an environment to combine them.
Work domain: Hydrodynamics, radiative transport
Working example: None
Extra requirements/features: OpenMP parallel
Equivalent codes: `Athena`
Comparable codes: `Fi`, `GadGet2`

`Hermite_GRX` (core)
MPI-parallel shared block-time step regularized fourth-order Hermite direct N-body code with post-Newtonian expansions.

Work domain: Collisional relativistic gravitational dynamics, in particular for Galactic nuclei.
Working example: `amuse.examples.montgomery`
Equivalent codes: `hacs64`, `AarsethZare`, `ph4`, `Hermite`
Comparable codes: `Bonsai`, `MI6`, `BHTree`

`Huayno` (core)
Class containing a large variety of N-body codes based on KDK algorithms using coupled second-order Hamiltonian splitting strategies of tunable order (Pelupessy et al. 2012). Developed for integrating planetary systems using shared or individual time steps.
Work domain: Collisional gravitational dynamics, in particular for planetary systems and star clusters
Working example: `amuse.examples.pythagorean`
Extra requirements/features: MPI parallel, GPU, GRAPE
Equivalent codes: `Hermite`, `Kepler`
Comparable codes: `HiGPUs`, `ph4`, `Hermite`

`Kepler` (core)
Kepler orbit manipulation functions, imported from Starlab (Portegies Zwart et al. 1999; Hut et al. 2010).
Work domain: Gravitational dynamics/data processing and analysis
Working example: `amuse.examples.kira`
Extra requirements/features: MPI parallel, GPU, GRAPE
Equivalent codes: `Hermite`, `Huayno`, `Kepler`
Comparable codes: `HiGPUs`, `ph4`, `Hermite`

`kepler_orbiters` (core)
Solve Kepler's equations for some number of two-body problems with one shared body, using universal variables (Bate et al. 1971), with a choice between `Sakura` (Gonçalves Ferrari et al. 2014) or `Huayno` (Pelupessy et al. 2012).
Work domain: Gravitational dynamics/data processing and analysis
Working example: amuse.examples.kepler_orbiters_planets_around_sun
Extra requirements/features: MPI parallel, GPU, GRAPE
Equivalent codes: `Hermite`, `Kepler`
Comparable codes: `HiGPUs`, `ph4`, `Hermite`

`MESA` (core)
Henyey stellar evolution solver with adaptive mesh refinement (AMR; Paxton et al. 2010, 2011).
Work domain: Stellar evolution
Working example: `amuse.examples.helium_star` see Code example 6.7
Extra requirements/features: OpenMP parallel
Equivalent codes: `EVTwin`
Comparable codes: `SeBa`, `SSE`

`MWpotentialBovy2015` (aka `galpy`) (core)

Static galactic potential model with characteristics similar to the Milky Way (Bovy 2015, 2014). The model includes semianalytic prescriptions for the bulge, disk (Miyamoto & Nagai 1975), and a dark-matter halo (Navarro et al. 1996).

Work domain: Gravitational dynamics, for galactic background potentials
Working example: None
Extra requirements/features: None
Equivalent codes: `Galaxia`
Comparable codes: `GalactICs`, `FractalCluster`, `HaloGen`

`ph4` (core)

MPI-parallel individual block-time step fourth-order Hermite direct *N*-body code with GPU support.

Work domain: Collisional gravitational dynamics, in particular star clusters
Extra requirements/features: MPI parallel, GPU
Working example: `amuse.examples.evolve_orbit_in_potential`
Equivalent codes: `hacs64`, `phiGRAPE`
Comparable codes: `Bonsai`, MI6

`Phantom` (core)

Smoothed particles hydrodynamics solver

MPI-parallel SPH code written to perform cosmological structure-formation simulations (Price et al. 2018).

Work domain: Gravitational dynamics/hydrodynamics
Working example: see `Ekster`. Extra requirements/features: MPI parallel
Equivalent codes: `Fi`, `Gadget2`.
Comparable codes: `Athena`, `Flash`

`SeBa` (core)

Parameterized stellar evolution code, with heuristic recipes for the evolution of binary stars of arbitrary metallicity (Portegies Zwart & Verbunt 2012). The code includes treatments of wind accretion, mass transfer through Roche-lobe overflow, and the effect of supernova explosions on binary orbital parameters. The code was developed for studying the formation and evolution of high-mass X-ray binaries and compact object mergers (Portegies Zwart & Verbunt 1996; Portegies Zwart & Yungelson 1998), but later expanded to include the prescription of the evolution of white dwarf binaries (Toonen et al. 2012). In a new version, it also provides support for triple evolution (Toonen et al. 2016).

Work domain: Stellar, binary, and triple evolution
Working example: `amuse.examples.evolve_binary`

Extra requirements/features: None
Equivalent codes: `BSE`
Comparable codes: `MESA`, `EVTwin`

`SecularMultiple` (core)
Secular (orbit-averaged) gravitational dynamics of hierarchical multiple systems composed of nested binary orbits (Hamers & Portegies Zwart 2016); see also Section 4.4.4. Tidal interactions (companion as a single body) and relativistic corrections are incorporated.
Work domain: Gravitational dynamics, in particular planetary systems
Working example: `amuse.examples.hierarchical_quadruple`, and see Section 4.4.4
Extra requirements/features: None
Equivalent codes: `Sakura`
Comparable codes: `Mikkola`

`SmallN` (core)
Shared time step direct N body integrator using a fourth-order time-symmetrized Hermite scheme, and including a modified unperturbed treatment of close approaches. `SmallN` was first introduced in Starlab (Portegies Zwart et al. 1999; Hut et al. 2010) to handle close encounters and resonant multi-particle interactions.
Work domain: Few-body gravitational dynamics, in particular democratic multiple systems
Working example: `amuse.examples.particles_and_multiples`
Extra requirements/features: `amuse.examples.kira`, `amuse.examples.basic_multiples_si` (see Section 8.7.1)
Equivalent codes: `Kepler`. `Kepler_orbiter`, `Sakura`
Comparable codes: `Hermite`, `HiGPUs`, `ph4`, `Hermite`

`SPHRay` (core)
Time-dependent Monte Carlo radiative-transfer code (Altay et al. 2011). The code was developed to couple an SPH code with radiative transfer to study the nonequilibrium ionization state (HI, HII, HeI, HeII, HeIII, e) and temperature (internal energy/entropy) of arbitrary density fields (Altay et al. 2008). Table 7.1 presents an overview of the code's application domain.
Work domain: Radiative transfer
Working example: `amuse.examples.iliev_test5`
Extra requirements/features: OpenMP parallel
Equivalent codes: `Mocassin`, `SimpleX`
Comparable codes: `MPI-AMRVAC`, `Flash`.

Tres (core)

The Triple-star Evolution Code (Toonen et al. 2012, 2016). More a coupled AMUSE script than a separate code, combining SeBa and Secularmultiples by including tidal effects, spin evolution, and mass transfer in triples.

Work domain: Triple-star evolution

Extra requirements/features: None

Equivalent codes: None

Comparable codes: None

Envelope selection: These are codes we support, but not explicitly. It is largely up to the user community to help maintain these codes. If you need to use one of these codes, it may be easiest to first find an equivalent alternative from the core selection of codes. If the demand for some of these codes grows, we will consider making them part of the core selection. At this point, few research projects have requested or used these codes, and therefore we cannot spare the effort to maintain them. Our recommendation is to test these codes before usage.

AarsethZare (envelope)

Regularized direct N-body code with a Bulirsch–Stoer (Bulirsch & Stoer 1964) integrator with support for post-Newtonian relativistic corrections to 2.5th order (Aarseth & Zare 1974a, b).

Work domain: Gravitational dynamics, in particular few-body interactions

Working example: None

Extra requirements/features: None

Equivalent codes: Brutus, Hermite_GRX, Huayno

Comparable codes: MI6, SmallN

Bonsai (envelope)

Second-order softened gravitational N-body code with shared time steps and approximate tree-code forces controlled by an opening angle θ. This code is massively parallel and designed to run concurrently on a large number (tested to up to 18,600) of graphical processing units (Bédorf et al. 2014).

Work domain: Gravitational dynamics, in particular galaxy dynamics

Working example: None

Extra requirements/features: MPI parallel, GPU, parallel-GPU

Equivalent codes from core: BHTree, Octgrav

Comparable codes: MI6

BSE (envelope)

Parameterized stellar evolution code expanded with heuristic recipes for the evolution of binary stars (Hurley et al. 2013a). The code includes treatments for wind accretion, mass transfer through Roche-lobe overflow, and the effect of supernova explosions on binary orbital parameters.

Work domain: Stellar and binary evolution
Working example: None
Extra requirements/features: None
Equivalent codes: `SeBa`
Comparable codes: `MESA`, `EVTwin`

`Brutus` (envelope)
Direct N-body code using Bulirsch–Stoer integration (Bulirsch & Stoer 1964), with support for arbitrary-precision arithmetic to calculate the orbits of multiple self-gravitating objects (Boekholt & Portegies Zwart 2015). An older version was called `adaptb`.
Work domain: Collisionless gravitational dynamics, in particular few-body systems to high precision and accuracy
Working example: None
Extra requirements/features: MPI parallel, GPU
Equivalent codes: None
Comparable codes: `AarsethZare`

`EVTwin` (envelope)
Henyey stellar evolution code for all ranges of mass and composition (Eggleton et al. 2011).
Work domain: Stellar and binary evolution
Working example: `amuse.examples.hr_diagram_cluster`
Extra requirements/features: None
Equivalent codes: `MESA`
Comparable codes: `SeBa`, `SSE`, `BSE`

`GadGet2` (envelope)
MPI-parallel SPH code written to perform cosmological structure-formation simulations, but implemented for more general applications (Springel 2000).
Work domain: Gravitational dynamics/hydrodynamics, in particular cosmology
Working example: `amuse.examples.gas_in_cluster`
Extra requirements/features: MPI parallel
Equivalent codes: `Fi`
Comparable codes: `Athena`, `Flash`

`Hermite` (envelope)
MPI-parallel shared block-time step fourth-order Hermite direct N-body code with GPU support (McMillan 2014).
Work domain: Collisional gravitational dynamics, in particular for star clusters
Working example: `amuse.examples.pythagorean`
Extra requirements/features: MPI parallel, GPU
Equivalent codes: `hacs64`, `phiGRAPE`, `ph4`
Comparable codes: `Bonsai`, `MI6`, `BHTree`

`Krome` (envelope)

Chemical network for solving the chemodynamical kinetics of a system modeled by coupled ordinary differential equations (Grassi et al. 2014). Incorporated processes include photochemistry, cooling, heating, dust treatment, and reverse kinetics. It was designed to be used for studying primordial chemistry in low-metallicity environments, molecular cloud chemistry, and planetary atmospheres and dust grain populations.

Work domain: Chemical evolution/hydrodynamics

Working example: `amuse.examples.molecular_cloud_chemistry`

Extra requirements/features: None

Equivalent codes: None

Comparable codes: `UCLChem`

`MI6` (envelope)

MPI-parallel block-time step sixth-order Hermite direct N-body code with support for GPU and GRAPE (Iwasawa et al. 2015). It was designed for integrating stellar orbits around a supermassive black hole, and has special treatment for this. It includes post-Newtonian expansions to general relativity up to 3.5th order (gravitational wave kick).

Work domain: Collisional gravitational dynamics, in particular for galactic nuclei

Working example: `amuse.examples.plot_nbody_performance`

Extra requirements/features: MPI parallel, GPU, GRAPE

Equivalent codes: `Mikkola`, `Pikachu` (Fujii et al. 2012)

Comparable codes: `HiGPUs`, `ph4`, `Hermite`

`Mikkola` (envelope)

Direct N-body code for self-gravitating systems, using algorithmic regularization and incorporating post-Newtonian effects (Mikkola 1983; Mikkola & Merritt 2008). The code was designed for integrating stellar orbits around a supermassive black hole. It includes post-Newtonian expansions to general relativity to 2.5th order (gravitational wave radiation).

Work domain: Collisional gravitational dynamics, in particular galactic nuclei

Extra requirements/features: None

Equivalent codes: `MI6`

Comparable codes: `HiGPUs`, `ph4`, `Hermite`, `SmallN`

`Mercury` (envelope)

Symplectic N-body solver dedicated to studying the long-term dynamical evolution of planetary systems (Chambers & Migliorini 1997; Chambers 2012). The equations of motion are solved using a variety of integrators, including mixed-variable symplectic methods (which are in fact quite similar to the Bridge method discussed in Chapter 9), the Bulirsch–Stoer method, and a hybridization for studying planetary coagulation.

Work domain: Gravitational dynamics, in particular planetary systems
Working example: `amuse.examples.mercury_example`
Extra requirements/features: None
Equivalent codes: `Huayno`, `Sakura`
Comparable codes: `HiGPUs`, `ph4`, `Hermite`, `Kepler`

`Mocassin` (envelope)

Three-dimensional Monte Carlo photoionization code with radiative transfer through arbitrary geometry and density distributions (Ercolano 2011). The code was originally developed to the model photoionized regions such as HII regions and planetary nebulae (Ritzerveld & Icke 2006). Table 7.1 presents an overview of the code's application domain.

Work domain: Radiative transfer
Working example: `amuse.examples.h2region`
Extra requirements/features: None
Equivalent codes: `SimpleX2`, `SPHray`
Comparable codes: `MPI-AMRVAC`, `Flash`

`Nbody6++` (envelope)

Direct N-body code using a fourth-order Hermite integrator with individual block-time steps and regularized treatment of multiple interactions (Aarseth 2011).

Work domain: Collisional gravitational dynamics, in particular star clusters
Working example: None
Extra requirements/features: MPI parallel, GPU, GRAPE
Equivalent codes: `hacs64`, `MI6`
Comparable codes: `HiGPUs`, `ph4`, `Hermite`

`Octgrav` (envelope)

MPI-parallel and GPU-enabled tree code up to octupole moments (Gaburov et al. 2010b).

Work domain: Collisionless gravitational dynamics, in particular star clusters
Working example: None
Extra requirements/features: MPI parallel, GPU
Equivalent codes: `BHTree`, `Bonsai`
Comparable codes: `HiGPUs`, `ph4`, `Hermite`

`phiGRAPE` (envelope)

Block time step fourth-order Hermite direct N-body code with support for GRAPE and/or GPU (Harfst et al. 2007).

Work domain: Collisional gravitational dynamics, in particular star clusters
Extra requirements/features: MPI parallel, GRAPE-4, GRAPE-6, GPU

Equivalent codes: `hacs64`, `ph4`
Comparable codes: `Bonsai`, MI6

`PPD` (envelope)
The FRIED interpolator object. Performs linear interpolation on the FRIED grid (Haworth et al. 2020).
Work domain: Proto-planetary disks in stellar clusters

Extra requirements/features: None
Equivalent codes: None
Comparable codes: `Mocassin`, `SPHRay`

`Rebound` (envelope)
Symplectic direct N-body integrator designed to study planetary rings, but of more general applicability (Rein & Liu 2012; Rein & Tamayo 2018; Rein & Liu 2011).
Work domain: Gravitational dynamics, in particular planetary systems
Working example: None
Extra requirements/features: MPI parallel
Equivalent codes: `Huayno`, `Mercury`, `Sakura`
Comparable codes: `ph4`, `Hermite`

`Sakura` (envelope)
Kepler-based Hamiltonian splitting method for solving the gravitational N-body problem (Gonçalves Ferrari et al. 2014). This is a general N-body code, but due to the imposed data structure it is ideally suited for integrating systems with large binary fractions.
Work domain: Collisional gravitational dynamics, in particular planetary systems and star clusters
Extra requirements/features: MPI parallel
Working example: None
Equivalent codes: `SecularMultiple`, `Huayno`, `Kepler_orbiter`
Comparable codes: `hacs64`, `phiGRAPE`

`SimpleX` (envelope)
Radiative-transfer solver for an optically thin medium using unstructured Delaunay grid (Paardekooper et al. 2010). Simplex was originally developed for cosmological radiative transfer (Altay et al. 2008). Table 7.1 presents an overview of the code's application domain.
Work domain: Radiative transfer
Working example: `amuse.examples.iliev_test5`
Extra requirements/features: OpenMP parallel
Equivalent codes: `Mocassin`, `SPHRay`
Comparable codes: `MPI-AMRVAC`, `Flash`

SSE (envelope)
Parameterized stellar evolution code with heuristic recipes for the evolution of single stars with arbitrary metallicity (Hurley et al. 2013b). The code was originally used for studying stellar populations in star clusters (Hurley et al. 2000).
Work domain: single stellar evolution
Working example: None
Extra requirements/features: None
Equivalent codes: BSE, SeBa
Comparable codes: MESA, EVTwin

TPI (envelope)
CUDA code for test-particle integration.
Work domain: Gravitational dynamics, in particular planetesimal or test-particle dynamics
Working example: none. Extra requirements/features: CUDA
Equivalent codes: SmallN, Kepler
Comparable codes: Huayno, Sakura

twobody (envelope)
Python module to solve Kepler's equations for two-body problems using universal variables (Bate et al. 1971).
Work domain: Gravitational dynamics, in particular two-body dynamics
Working example: None
Extra requirements/features: None
Equivalent codes: SmallN, Kepler
Comparable codes: Huayno, Sakura

Vader (envelope)
Viscous disk evolution code (Krumholz & Forbes 2015). We extended the code to include internal photoevaporation with mass-loss profiles. It is built to work with the FRIED grid through the PPD class in AMUSE.
Work domain: Proto-planetary disks in stellar clusters

Extra requirements/features: None
Equivalent codes: None
Comparable codes: Mocassin, SPHRay.

Incomplete interface: These codes have been incorporated at some time into AMUSE, but are not longer (or not yet) explicitly supported. We consider them interesting, but they are not often used and we do not have the resources to maintain them. If you plan on using one of these, contact us to see if other users are interested in joining in to improve and complete the interface.

- `Arepo`

  Vorenoi mesh hydrodynamical solver for simulating interstellar gas and solids (Springel 2010a).

  Work domain: Collisional gravitational dynamics with tides
  Working example: None
  Extra requirements/features: MPI parallel
  Equivalent codes: None.
  Comparable codes: `Fi`, `Gadget2`, `Athena`, `Flash`

- `Capreole`

  MPI-parallel grid-based astrophysical hydrodynamics code (Mellema et al. 1991). Works in one, two, or three spatial dimensions. For the hydrodynamics, it relies on the Roe–Eulderink–Mellema solver, which is an approximate Riemann solver for arbitrary metrics. It can solve different hydrodynamical problems. Capreole has run not only on single processors but also on massively parallel systems (up to 512 processors on BlueGene/L).

  Work domain: Hydrodynamics, in particular cosmology
  Working example: `amuse.examples.linear_wave`
  Extra requirements/features: OpenMP parallel
  Equivalent codes: `Athena`, `Flash`,
  Comparable codes: `Fi`, `GadGet2`

- `Etics`

  Gravity solver for collisionless stellar dynamical simulations using two fast force calculation methods based on series expansions of the Poisson equation (Meiron et al. 2014). One method is the self-consistent field model, a Fourier-like expansion of the density field in some basis set. The other method is a multipole expansion method, a Taylor-like expansion of the Green's function.

  Work domain: Collisionless gravitational dynamics, in particular elliptical galaxies
  Working example: None
  Extra requirements/features: MPI parallel, GPU
  Equivalent codes: None
  Comparable codes: `Bonsai`, `Octgrav`

- `HiGPUs`

  Block-time step sixth-order Hermite direct gravitational N-body code (Spera et al. 2012).

  Work domain: Collisional gravitational dynamics, in particular star clusters or planetary systems
  Working example: None
  Extra requirements/features: MPI parallel, GPU
  Equivalent codes: `MI6`
  Comparable codes: `ph4`, `Hermite`

- `hacs64`

  Block time step Hermite direct N-body code with Ahmad–Cohen neighbor scheme and GPU support. The code uses a sixth-order Hermite integrator for the irregular step and a fourth-order Hermites for the regular step.

Work domain: Collisional gravitational dynamics
Working example: None
Extra requirements/features: GPU
Equivalent codes: `MI6`
Comparable codes: `HiGPUs`, `ph4`, `Hermite`

- `ImpactCrater`
  Impact crater size calculator. Simulates the effect of an impact of a minor body on a rocky major body.
  Work domain: Collisional dynamics
  Working example: None
  Extra requirements/features: None
  Equivalent codes: none
  Comparable codes: `Fi`, `Gadget2`, `Phantom`, `Arepo`
- `MPI-AMRVAC`
  Three-dimensional magnetohydrodynamics AMR code with radiative transfer (van der Holst et al. 2012; Xia et al. 2018),[4] using the versatile advection method (Tóth & Keppens 2012). The code was developed to simulate shock-dominated problems (Meliani et al. 2007; Keppens & Meliani 2012), but handles smooth flows as well (Porth et al. 2014).
  Work domain: Hydrodynamics
  Working example: `amuse.examples.linear_wave_boundaries`
  Extra requirements/features: MPI parallel
  Equivalent codes: `Flash`
  Comparable codes: `Athena`
- `PACE`
  Planet population synthesis code for simulating the formation of planetary systems from photoevaporated (PPD) viscous (vader) disks (Huang et al. 2024).
  Work domain: Planet formation
  Working example: None
  Extra requirements/features: `Venice`, `Vader`, `PPD`
  Equivalent codes: None
  Comparable codes: None

- `Ramses` (deprecated)
  AMR hydrodynamics solver with a tree-based data structure allowing recursive grid refinements on a cell-by-cell basis (Teyssier 2010). Based on the method of Kravtsov et al. (1997) and designed to study structure formation in the Universe (Teyssier 2002).
  Work domain: Hydrodynamics/radiative transfer, in particular cosmology
  Working example: `amuse.examples.linear_wave_boundaries`
  Extra requirements/features: None

[4] See also http://amrvac.org/ or the `github` version at https://github.com/amrvac/amrvac.

Equivalent codes: `Flash`, `Athena`, `Capreole`, `MPI-AMRVAC`
Comparable codes: `Fi`, `Gadget2`

- `Tupan`
  Direct N-body code for collisional stellar dynamics. Like `Huayno`, it contains a wide range of different integrators. The code was originally written by Guilherme Goncalves, but regrettably never published.
  Work domain: Collisional gravitational dynamics
  Working example: None
  Extra requirements/features: MPI parallel, GPU
  Equivalent codes: `Huayno`, `Sakura`, `Sei`
  Comparable codes: `Kepler`
- `TidyMess`
  Direct N-body code with tidal evolution solver. Excellently suited for chaotic N-body problems for which tides are relevant (Boekholt & Correia 2023).
  Work domain: Collisional gravitational dynamics with tides
  Working example: None
  Extra requirements/features: MPI parallel
  Equivalent codes: `Huayno`, `Sakura`, `Sei`
  Comparable codes: `Kepler`
- `Venice`
  General solver for matrices of differential equations, primarily used for bridging multiple codes (Wilhelm & Portegies Zwart 2024)
  Work domain: General multicode bridging strategy
  Working example: see Section 10.1.3
  Extra requirements/features: None
  Equivalent codes: None
  Comparable codes: `PACE`

Support codes: These are codes for

- `FractalCluster`
  Generator for producing fractal stellar distributions by defining a cube of side 2 which is subsequently divided into subcubes, on which the same star addition algorithm is used. Eventually, this procedure leads to a three-dimensional, scale-free distribution of equal-mass particles (Goodwin & Whitworth 2004).
  Work domain: Initial conditions
  Working example: `amuse.examples.plot_fractal_clumpy_cluster`
  Extra requirements/features: None
  Equivalent codes: None
  Comparable codes: `GalactICsHaloGen`

- `Fresco`
  Visualization package specific for making "observational" images of simulations, focused on simulations in which stellar evolution, stellar dynamics, and possibly gas dynamics are incorporated. `Fresco` is not on the `amuse` `github` page, but can be downloaded from `github` at https://github.com/rieder/Fresco.
  Work domain: Visualization
  Working example: None
  Extra requirements/features: None
  Equivalent codes: None
  Comparable codes: `vaex`
- `GalactICs`
  Generator for an N-body particle realization of an axisymmetric galaxy that consists of a disk, bulge, and halo (Kuijken & Dubinski 2011), based on the original work by Kuijken & Dubinski (1995).
  Work domain: Initial conditions
  Working example: None
  Extra requirements/features: None
  Equivalent codes: `FractalCluster`, `HaloGen`
  Comparable codes: `Galaxia`
- `Galaxia`
  Dynamic potential of a galactic model with characteristics similar to the Milky Way (Sharma et al. 2011). The model includes semianalytic prescriptions for the axisymmetric parts (bulge, disk, halo; Allen & Santillan 1991), a parameterization for the central (Ferrers) bar (Romero-Gómez et al. 2007, 2011) and transient two-dimensional and three-dimensional spiral arms (Cox & Gómez 2002).
  Work domain: Gravitational dynamics, in particular for background potentials
  Working example: None
  Extra requirements/features: None
  Equivalent codes: `MWpotentialBovy2015`
  Comparable codes: `GalactICs`, `FractalCluster`, `HaloGen`
- `HaloGen`
  Generator for an N-body particle realization of the Galactic halo (Zemp et al. 2008; Zemp 2014; Avila Perez & Murray 2015) by generating cosmological tracer distributions (halos) for a range of different cosmologies.
  Work domain: Initial conditions, in particular for galaxies
  Working example: `amuse.examples.solar_cluster_in_live_galaxy`
  Extra requirements/features: Parallel
  Equivalent codes: `GalactICs`, `FractalCluster`.
  Comparable codes: `Galaxia`

- `hop`
  Coordinate-free group-finding algorithm for particle distributions in three dimensions (Eisenstein & Hut 2011). Original use was finding groups of particles in N-body realizations in Cartesian coordinates.
  Work domain: Data processing and analysis
  Working example: `amuse.examples.clump_finding`
  Extra requirements/features: None
  Equivalent codes: None
  Comparable codes: None
- `mameclot`
  Generator for an N-body particle realization of one or two star clusters in orbit around each other (see https://github.com/mgieles/mameclot). The velocities can be isotropic or radially anisotropic (Osipkov 1979; Merritt 1985).
  Work domain: Initial conditions
  Working example: None
  Extra requirements/features: None
  Equivalent codes: None
  Comparable codes: `FractalCluster`, `GalactICs`
- `mmams`
  Fast stellar collision method using Archimedes's principle to merge stars, including rudimentary treatments for shock heating, hydrodynamic mixing, mass ejection, and angular momentum transfer (Gaburov et al. 2008; Lombardi & Sawyer Warren 2014).
  Work domain: Resolving stellar collisions
  Working example: See Code example 8.10
  Extra requirements/features: None
  Equivalent codes: None
  Comparable codes: `Gadget2`, `Fi`
- `MMC`
  Monte Carlo star cluster dynamics simulation code (Askar et al. 2017). The code was developed to compare simulation results with observed globular clusters (Giersz 1998, 2001).
  Work domain: Gravitational dynamics
  Working example: None
  Extra requirements/features: None
  Equivalent codes: None
  Comparable codes: None
- `UCLChem`
  Chemical network for solving the chemodynamical kinetics of a system modeled by coupled ordinary differential equations (Holdship et al. 2017; Keil et al. 2022). Incorporated processes include photochemistry, cooling, heating, dust treatment, and reverse kinetics. It was designed to be used for studying primordial chemistry in low-metallicity environments, molecular cloud chemistry, and planetary atmospheres and dust grain populations.

Work domain: Chemical evolution/hydrodynamics
Working example: None
Extra requirements/features: None
Equivalent codes: None
Comparable codes: `Krome`

- `Vaex`

  Visualization package written in `Python` for analyzing large data volumes. It was written by Maarten Breddels as a project at Groningen University (Breddels 2017), see also https://vaex.io.
  Work domain: Visualization and analysis
  Working example: None
  Extra requirements/features: Does not require but can use the GPU
  Equivalent codes: None
  Comparable codes: `pyplot`, `gnuplot`

We are sorry if your code is not among the listed modules, and we will gladly work with you to include it. Writing a custom interface to a new code is relatively easy, but, regrettably, this is not the right place to explain how to develop a new AMUSE interface. However, we can teach you how to do this, or you may want to visit the amusecode.org website for a brief tutorial.

A.8.3 Packages Removed Since the 2018 Version of the Book

Regrettably, we cannot muster the support we would love to provide for all packages to which AMUSE has an interface. Some of these packages have similar functionality as others, and some are simply not used a lot. The result is that bit-rot sets in, compilers change, and it turns out that modules are not always written in a sustainable way. Most FORTRAN code remains compilable, but for many C and C ++ packages this is not the case. In addition, some of these codes have become obsolete, as new versions or better implementations of the same processes have become available. So long as no new bugs are found, or if the original authors remain accessible, this does not necessarily pose a problem. However, not using a code for a while does not stop compilers and computers from changing, and therefore the codes may stop working eventually.

An example is the code `Adaptb`, which basically is an early version of `Brutus` (Boekholt & Portegies Zwart 2015). The code `mmc` (Leveque et al. 2022) was never properly incorporated into the AMUSE framework. We will miss the GPU-supported treecode in `Octgrav` (Gaburov et al. 2010b, a), but this was simply replaced by `Bonsai` (Bédorf et al. 2014). The `pikachu` code (Iwasawa et al. 2015) was replaced by `PeTar` (Wang et al. 2015). Although `hacs64` has a sixth-order predictor-corrector Hermite integrator for direct N-body with an Ahmad–Cohen scheme (Ahmad & Cohen 1973) running on GPU, a combination of methods nowhere else available in AMUSE, there are high-order integrators in `Symple` and we can relatively easily expand `ph4` to include higher orders. Eventually, `sei` (Rein

& Tremaine 2011) also had to go, even though it is a nice and effective 183-line direct N-body code; it was replaced with `Rebound` (Rein & Spiegel 2015), which utilizes the faster IAS15 integrator.

References

Aarseth, S. J., & Zare, K. 1974a, CeMec, 10, 185

Aarseth, S. J., & Zare, K. 1974b, CeMec, 10, 516

Aarseth, S. J. 2011, NBODY Codes: Numerical Simulations of Many-body (N-body) Gravitational Interactions, Astrophysics Source Code Library, ascl:1102.006

Ahmad, A., & Cohen, L. 1973, ApJ, 179, 885

Allen, C., & Santillan, A. 1991, RMxAA, 22, 255

Altay, G., Croft, R. A. C., & Pelupessy, I. 2008, MNRAS, 386, 1931

Altay, G., Croft, R. A. C., & Pelupessy, I. 2011, SPHRAY: A Smoothed Particle Hydrodynamics Ray Tracer for Radiative Transfer, Astrophysics Source Code Library, ascl:1103.009

Askar, A., Giersz, M., Pych, W., & Dalessandro, E. 2017, COCOA: Simulating Observations of Star Cluster Simulations, Astrophysics Source Code Library, ascl:1703.002

Avila Perez, S., & Murray, S. 2015, HALOGEN: Approximate synthetic halo catalog generator, Astrophysics Source Code Library, ascl:1505.017

Barnes, J., & Hut, P. 1986, Nature, 324, 446

Bate, R. R., Mueller, D. D., & White, J. E. 1971, Fundamentals of Astrodynamics (New York: Dover)

Bédorf, J., Gaburov, E., Fujii, M. S., et al. 2014, Proc. of the International Conference for High Performance Computing, Networking, Storage and Analysis, SC '14 (Piscataway, NJ: IEEE) 54

Boekholt, T., & Portegies Zwart, S. 2015, ComAC, 2, 2

Boekholt, T. C. N., & Correia, A. C. M. 2023, MNRAS, 522, 2885

Bovy, J. 2014, galpy: Galactic Dynamics Package, Astrophysics Source Code Library, ascl:1411.008

Bovy, J. 2015, ApJS, 216, 29

Breddels, M. A. 2017, IAU Symp 325, Astroinformatics ed. M. Brescia, S. G. Djorgovski, E. D. Feigelson, G. Longo, & S. Cavuoti (Cambridge: Cambridge Univ. Press) 299

Bulirsch, R., & Stoer, J. 1964, NuMat, 6, 413

Chambers, J. E., & Migliorini, F. 1997, AAS/DPS Meeting, 29, 27.06

Chambers, J. E. 2012, Mercury: A Software Package for Orbital Dynamics, Astrophysics Source Code Library, ascl:1201.008

Cox, D. P., & Gómez, G. C. 2002, ApJS, 142, 261

Eggleton, P. P., Tout, C., Pols, O., et al. 2011, STARS: A Stellar Evolution Code, Astrophysics Source Code Library, ascl:1107.008

Einstein, A. 1915a, SPAW, 778,

Einstein, A. 1915b, SPAW 799,

Einstein, A., & Grossmann, M. 1914, ZMP, 63, 215

Eisenstein, D., & Hut, P. 2011, HOP: A Group-finding Algorithm for N-body Simulations, Astrophysics Source Code Library, ascl:1102.019

Emden, R. 1907, Gaskugeln: Anwendungen der mechanischen Wärmetheorie auf kosmologische und meteorologische Probleme (Leipzig: B. G. Teubner)

Ercolano, B. 2011, MOCASSIN: MOnte CArlo SimulationS of Ionized Nebulae, Astrophysics Source Code Library, ascl:1110.010
Euler, L. 1768, Institutionum Calculi Integralis (1-3,; St. Petersburg: Academy of Sciences)
Fryxell, B., Olson, K., Ricker, P., et al. 2010, FLASH: Adaptive Mesh Hydrodynamics Code for Modeling Astrophysical Thermonuclear Flashes, Astrophysics Source Code Library, ascl:1010.082
Fujii, M. S., Iwasawa, M., Saitoh, T. R., & Portegies Zwart, S. 2012, ASP Conf. Ser. 453, Advances in Computational Astrophysics: Methods, Tools, and Outcome ed. R. Capuzzo-Dolcetta, M. Limongi, & A. Tornambè (San Francisco, CA: ASP)
Fujii, M., Iwasawa, M., Funato, Y., & Makino, J. 2007, PASJ, 59, 1095
Gaburov, E., Lombardi, J. C., & Portegies Zwart, S. 2008, MNRAS, 383, L5
Gaburov, E., Bédorf, J., & Portegies Zwart, S. 2010a, Procedia Computer Science, 1, 1119
Gaburov, E., Bédorf, J., & Portegies Zwart, S. 2010b, OCTGRAV: Sparse Octree Gravitational N-body Code on Graphics Processing Units, Astrophysics Source Code Library, ascl:1010.048
Gaburov, E., Lombardi, J. C., Portegies Zwart, S., & Rasio, F. A. 2018, StarSmasher: Smoothed Particle Hydrodynamics Code for Smashing Stars and Planets, Astrophysics Source Code Library, ascl:1805.010
Gardiner, T. A., & Stone, J. M. 2005, JCoPh, 205, 509
Gardiner, T. A., & Stone, J. M. 2008, JCoPh, 227, 4123
Gear, C. W. 1967, MaCom, 21, 146
Gerritsen, J. P. E., & Icke, V. 1997, A&A, 325, 972
Giersz, M. 1998, MNRAS, 298, 1239
Giersz, M. 2001, MNRAS, 324, 218
Gonçalves Ferrari, G., Boekholt, T., & Portegies Zwart, S. F. 2014, MNRAS, 440, 719
Goodwin, S. P., & Whitworth, A. P. 2004, A&A, 413, 929
Grassi, T., Bovino, S., Prieto, J., et al. 2014, KROME: Chemistry package for astrophysical simulations, Astrophysics Source Code Library, ascl:1402.011
Hairer, E., Lubich, C., & Wanner, G. 2006, Geometric Numerical Integration: Structure-Preserving Algorithms for Ordinary Differential Equations (2nd ed.; Dordrecht: Springer)
Hamers, A. S., & Portegies Zwart, S. F. 2016, MNRAS, 459, 2827
Harfst, S., Gualandris, A., Merritt, D., et al. 2007, NewA, 12, 357
Haworth, T. J., Cadman, J., Meru, F., et al. 2020, MNRAS, 494, 4130
Hernquist, L., & Katz, N. 1989, ApJS, 70, 419
Hilbert, D. 1915, Gott. Nachr., 27, 395
Holdship, J., Viti, S., Jiménez-Serra, I., Makrymallis, A., & Priestley, F. 2017, AJ, 154, 38
Huang, S., Portegies Zwart, S., & Wilhelm, M. J. C. 2024, A&A, 689, A338
Hunter, J. D. 2007, CSE, 9, 90
Hurley, J. R., Pols, O. R., & Tout, C. A. 2000, MNRAS, 315, 543
Hurley, J. R., Tout, C. A., & Pols, O. R. 2013a, BSE: Binary Star Evolution, Astrophysics Source Code Library, ascl:1303.014
Hurley, J. R., Pols, O. R., & Tout, C. A. 2013b, SSE: Single Star Evolution, Astrophysics Source Code Library, ascl:1303.015
Hut, P., McMillan, S., Makino, J., & Portegies Zwart, S. 2010, Starlab: A Software Environment for Collisional Stellar Dynamics, Astrophysics Source Code Library, ascl:1010.076
Iwasawa, M., Portegies Zwart, S., & Makino, J. 2015, ComAC, 2, 6

Jiao, Y., Dai, G., Feng, Q., & Tang, Y. 2007, JCM, 25, 690
Keil, M., Viti, S., & Holdship, J. 2022, ApJ, 927, 203
Keppens, R., Meliani, Z., van Marle, A. J., et al. 2012, JCoPh, 231, 718
Kerr, R. P. 1963, PhRvL, 11, 237
Kluyver, T., Ragan-Kelley, B., Perez, F., et al. 2016, Positioning and Power in Academic Publishing: Players, Agents and Agendas, ed. F. Loizides, & B. Schmidt (Amsterdam: IOS Press) 87
Kozai, Y. 1962, AJ, 67, 591
Kravtsov, A. V., Klypin, A. A., & Khokhlov, A. M. 1997, ApJS, 111, 73
Krumholz, M. R., & Forbes, J. C. 2015, A&C, 11, 1
Kuijken, K., & Dubinski, J. 1995, MNRAS, 277, 1341
Kuijken, K., & Dubinski, J. 2011, GalactICS: Galaxy Model Building Package, Astrophysics Source Code Library, ascl:1109.011
Lane, J. H. 1870, AmJSA, 50, 57
Leveque, A., Giersz, M., Banerjee, S., et al. 2022, MNRAS, 514, 5739
Lidov, M. L. 1962, Planet. Space Sci., 9, 719
Lombardi, J., & Sawyer Warren, J. 2014, MMAS: Make Me A Star, Astrophysics Source Code Library, ascl:1412.010
McMillan, S. L. W. 2014, AAS/DDA Meeting, 45, 303.01
Meiron, Y., Li, B., Holley-Bockelmann, K., & Spurzem, R. 2014, ApJ, 792, 98
Meliani, Z., Keppens, R., Casse, F., & Giannios, D. 2007, MNRAS, 376, 1189
Mellema, G., Eulderink, F., & Icke, V. 1991, A&A, 252, 718
Merritt, D. 1985, AJ, 90, 1027
Mewes, V., Zlochower, Y., Campanelli, M., et al. 2018, PhRvD, 97, 084059
Mewes, V., Zlochower, Y., Campanelli, M., et al. 2020, PhRvD, 101, 104007
Mikkola, S. 1983, MNRAS, 203, 1107
Mikkola, S., & Merritt, D. 2008, AJ, 135, 2398
Miyamoto, M., & Nagai, R. 1975, PASJ, 27, 533
Navarro, J. F., Frenk, C. S., & White, S. D. M. 1996, ApJ, 462, 563
Osipkov, L. P. 1979, PAZh, 5, 77
Paardekooper, J. P., Kruip, C. J. H., & Icke, V. 2010, A&A, 515, A79
Paxton, B., Bildsten, L., Dotter, A., et al. 2010, MESA: Modules for Experiments in Stellar Astrophysics, Astrophysics Source Code Library, ascl:1010.083
Paxton, B., Bildsten, L., Dotter, A., et al. 2011, ApJS, 192, 3
Paxton, B., Cantiello, M., Arras, P., et al. 2013, ApJS, 208, 4
Paxton, B., Marchant, P., Schwab, J., et al. 2015, ApJS, 220, 15
Paxton, B., Schwab, J., Bauer, E. B., et al. 2018, ApJS, 234, 34
Pelupessy, F. I., van der Werf, P. P., & Icke, V. 2004, A&A, 422, 55
Pelupessy, F. I., Jänes, J., & Portegies Zwart, S. 2012, NewA, 17, 711
Plummer, H. C. 1911, MNRAS, 71, 460
Poisson, E., & Will, C. M. 2014, Gravity (Cambridge: Cambridge Univ. Press)
Portegies Zwart, S. F., & Verbunt, F. 1996, A&A, 309, 179
Portegies Zwart, S. F., & Verbunt, F. 2012, SeBa: Stellar and Binary Evolution, Astrophysics Source Code Library, ascl:1201.003
Portegies Zwart, S. F., & Yungelson, L. R. 1998, A&A, 332, 173
Portegies Zwart, S. F., Makino, J., McMillan, S. L. W., & Hut, P. 1999, A&A, 348, 117

Portegies Zwart, S. F., Boekholt, T. C. N., Por, E. H., Hamers, A. S., & McMillan, S. L. W. 2022, A&A, 659, A86
Portegies Zwart, S. 2018, Sci., 361, 979
Portegies Zwart, S. F., & Boekholt, T. C. N. 2018, CNSNS, 61, 160
Porth, O., Xia, C., Hendrix, T., Moschou, S. P., & Keppens, R. 2014, ApJS, 214, 4
Price, D. J., Wurster, J., Tricco, T. S., et al. 2018, PASA, 35, e031
Rein, H., & Liu, S. F. 2012, A&A, 537, A128
Rein, H., & Liu, S.-F. 2011, REBOUND: Multi-purpose N-body Code for Collisional Dynamics, Astrophysics Source Code Library, ascl:1110.016
Rein, H., & Spiegel, D. S. 2015, MNRAS, 446, 1424
Rein, H., & Tamayo, D. 2018, MNRAS, 473, 3351
Rein, H., & Tremaine, S. 2011, MNRAS, 415, 3168
Ritzerveld, J., & Icke, V. 2006, PhRvE, 74, 026704
Romero-Gómez, M., Athanassoula, E., Masdemont, J. J., & García-Gómez, C. 2007, A&A, 472, 63
Romero-Gómez, M., Athanassoula, E., Antoja, T., & Figueras, F. 2011, MNRAS, 418, 1176
Schwarzschild, K. 1916, Sitzungsber. Königl. Preuss. Akad. Wissenschaften, 7, 189
Sharma, S., Bland-Hawthorn, J., Johnston, K. V., & Binney, J. 2011, Galaxia: A Code to Generate a Synthetic Survey of the Milky Way, Astrophysics Source Code Library, ascl:1101.007
Sobol, I. M. 1967, ZVMMF, 7, 784
Sofroniou, M., & Spaletta, G. 2005, Optimization Methods and Software, 20, 597
Spera, M., Capuzzo Dolcetta, R., & Punzo, D. 2012, HiGPUs: Hermite's N-body Integrator Running on Graphic Processing Units, Astrophysics Source Code Library, ascl:1207.002
Springel, V. 2000, GADGET-2: A Code for Cosmological Simulations of Structure Formation, Astrophysics Source Code Library, ascl:0003.001
Springel, V. 2010a, MNRAS, 401, 791
Springel, V. 2010b, ARA&A, 48, 391
Stone, J. M., Gardiner, T. A., Teuben, P., Hawley, J. F., & Simon, J. B. 2008, ApJS, 178, 137
Stone, J. M., Gardiner, T. A., Teuben, P., Hawley, J. F., & Simon, J. B. 2010, Athena: Grid-based Code for Astrophysical Magnetohydrodynamics (MHD), Astrophysics Source Code Library, ascl:1010.014
Teyssier, R. 2002, A&A, 385, 337
Teyssier, R. 2010, RAMSES: A New N-body and Hydrodynamical Code, Astrophysics Source Code Library, ascl:1011.007
Toonen, S., Nelemans, G., & Portegies Zwart, S. 2012, A&A, 546, A70
Toonen, S., Hamers, A., & Portegies Zwart, S. 2016, ComAC, 3, 6
Tóth, G., & Keppens, R. 2012, VAC: Versatile Advection Code, Astrophysics Source Code Library, ascl:1207.003
van der Holst, B., Keppens, R., Meliani, Z., et al. 2012, MPI-AMRVAC: MPI-Adaptive Mesh Refinement-Versatile Advection Code, Astrophysics Source Code Library, record ascl:1208.014
Vasiliev, E. 2019, MNRAS, 482, 1525
Vaught, A. 1996, Linux J., 1996, 7–es
Wang, L., Spurzem, R., Aarseth, S., et al. 2015, MNRAS, 450, 4070
Weinstein, G. 2012, arXiv:1202.2791

Wilhelm, M. J. C., & Portegies Zwart, S. 2024, A&A, 691, A71

Williams, T., Kelley, C., et al. 2013, Gnuplot 4.6: An Interactive Plotting Program, http://gnuplot.sourceforge.net/.

Xia, C., Teunissen, J., El Mellah, I., Chané, E., & Keppens, R. 2018, ApJS, 234, 30

Yoshida, H. 1990, PhLA, 150, 262

Zemp, M. 2014, Halogen: Multimass Spherical Structure Models for N-body Simulations, Astrophysics Source Code Library, ascl:1407.020

Zemp, M., Moore, B., Stadel, J., Carollo, C. M., & Madau, P. 2008, MNRAS, 386, 1543

Zhao, Y., Zhang, Y., & Wang, J. 2021, International Journal of Advanced Network, Monitoring and Controls, 6, 20

Appendix B

How to Add a Code to AMUSE

Jag har aldrig provat det förut, så jag tror absolut att jag borde kunna göra det.

I have never tried that before, so I think I should definitely be able to do that.

—Pippi Långstrump (apocryphal)

The main part of this manuscript presents an extended description of how to use the AMUSE framework. Here, we focus on the basic structure and design of AMUSE.

B.1 The AMUSE Framework

At first sight, the approach of assimilating existing software into a larger framework would appear to be a difficult undertaking, particularly since community software is written in a wide variety of languages, such as FORTRAN, C, and C++, and exhibits an enormous diversity of internal units and data structures. On the other hand, such a framework, if properly designed, could be relatively easy to use, since learning one simulation package is much easier than mastering the idiosyncrasies of many separate programs. The use of standard calling procedures can enable even novice researchers to quickly become acquainted with the environment and to perform complicated simulations using it.

Within AMUSE, a user writes a relatively simple script with a standardized calling sequence to each of a set of underlying *community codes*. Instructions are communicated through a message-passing interface to any number of spawned community modules, which respond by performing operations and transferring data back through the interface to the user script. This appendix is intended to help you write production-quality (meaning working and clearly written) user scripts. There

doi:10.1088/2514-3433/adfc3cch13

are numerous examples in the amuse/examples directory, and here we discuss several of them (see for example the code examples in Chapters 4 and 5).

As illustrated in Figure B.1, the two top-level components of AMUSE are:

- The *user script*, which implements a specific physical problem or set of problems, in the form of system-provided or user-written scripts that serve as the user interface to the AMUSE framework. Coupling between community codes is implemented at this level only, with the help of support classes provided by the *manager*.
- The *community module*, comprising three key elements:
  1. The *manager*, which provides an object-oriented interface onto the *communication* layer via a suite of system-provided utility functions.

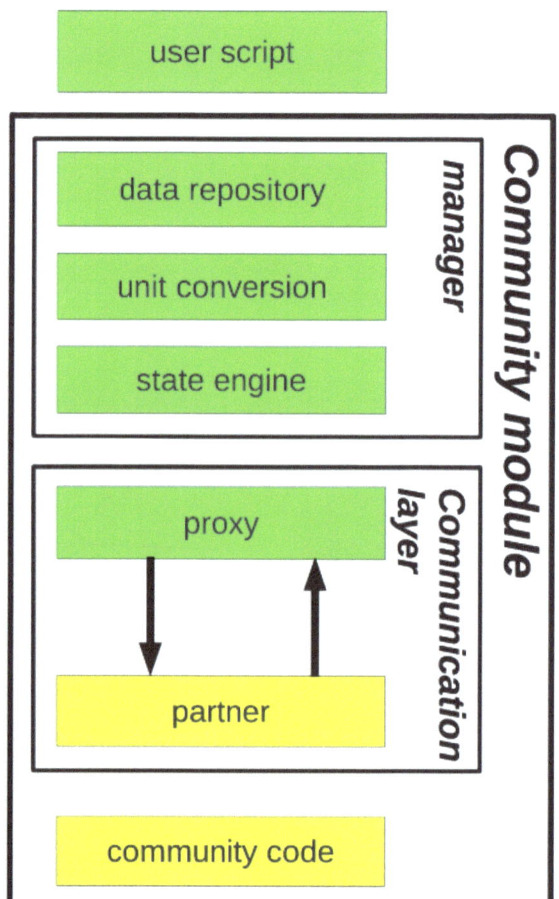

Figure B.1. The AMUSE framework architecture is based on a three-layered design. The top layer consists of a user script written by the end user in `Python` (top box). The *manager* is part of the *community module* and consists of a data repository, unit conversion, and a state engine. The *community module* also includes the partner–proxy pair in the communication layer—a bidirectional message-passing framework—and a community code (bottom box).

This layer handles unit conversion, as discussed below, and also contains the state engine and the associated data repository, both of which are needed to guarantee consistency of data across modules. The role of this layer is generic—it is not specific to any particular problem or to any single physical domain. In the rest of this manuscript, we discuss an actual implementation of the *manager* for astrophysical problems and give some examples of coupling codes.

2. The *communication* layer, which realizes the bidirectional communication between the *manager* and the community code layer. This is implemented via a proxy and an associated partner, which together provide the connection to the community code.
3. The *community code* layer, which contains the actual community codes and implements control and data management operations on them. Each piece of code in this layer is domain-specific, although the code may be designed to be very general within its particular physical domain.

B.1.1 The Community Module

The community module consists of the actual community code and the bidirectional communication layer enabling low-level MPI (Gropp et al. 1996) communication between the proxy and the partner. Each community module contains a `Python` class of functions dedicated to communication with the community code. Direct use of this low-level class from within a user script is possible but discouraged because it requires considerable replication of code dealing with data models, unit conversion, and exception handling. Instead, the AMUSE framework provides the manager layer on top of the community module layer to manage the bookkeeping needed to maintain the interface with the low-level class.

B.1.1.1 The Community Code

Community applications may be programmed in any computer language that supports smart sockets or bindings to the MPI protocol. In practice, these codes exhibit a wide variety of structural properties, model diverse physical domains, and span a broad range of temporal and spatial scales. In AMUSE this diversity is exemplified by a suite of community programs ranging from toy codes to production-quality, high-precision dedicated solvers. A user script can be developed quickly using toy codes until the production and data analysis pipelines are fully developed, and can then easily be switched to production-quality implementations of the physical solvers for actual large-scale simulations.

In principle, each of the community modules can be used standalone, but the main strength of AMUSE is in the coupling between them. Many of the codes incorporated in our practical implementation were not written by us, but are publicly available and are amply described in the literature.

B.1.1.2 The Communication Layer

Bidirectional communication between a running community process and the manager could in principle be realized using any protocol for remote interprocess communication. Our main reasons for choosing MPI are similar to those for adopting `Python`—widespread acceptance in the computational science community and broad support by many programming languages and on most computing platforms.

The communication layer consists of two main parts, a `Python` proxy on the manager side of the framework and a partner on the community code side. The proxy converts `Python` commands into MPI messages, transmits them to the partner, and then waits for a response. The partner waits for and decodes MPI messages from the proxy into community code commands, executes them, and then sends a reply. The MPI interface allows the user script and community modules to operate asynchronously and independently. As a practical matter, the detailed coding, decoding, and communications operations in the proxy and the partner are never hand-coded. Rather, they are automatically generated as part of the AMUSE build process, from a high-level `Python` description of the community module interface.

We note that other solutions for linking high-level languages within `Python` exist —e.g. `swig` (see http://www.swig.org) and `f2py` (Peterson 2009), and in fact both were used in earlier implementations of the AMUSE framework (Portegies Zwart et al. 2009). However, despite their generality, these solutions cannot maintain namespace independence between community modules, and in addition are incapable of accommodating high-performance parallel community codes. For these reasons, we abandoned the standard solutions in favor of our custom, high-performance MPI alternative.

The use of MPI may seem like overkill in the case of serial operation on a single computing node (and this might well be the case), but it imposes negligible overhead, and even here it offers significant practical advantages. It rigorously separates all community processes in memory, guaranteeing that multiple instances of a community module are independent, and explicitly avoids namespace conflicts. It also ensures that the framework remains thread-safe when using older community codes that may not have been written with that concept in mind.

Despite the initial threshold for its implementation and use, the relative simplicity of the message-passing framework allows us to easily couple multiple independent community codes as community modules, or multiple copies of the same community module, if desired (see Figure B.1). An example of the latter might be to simulate simultaneously two separate objects of the same type, or the same physics on different scales, with different boundary conditions, without having to modify the data structures in the community code. Specifically, we might want to evolve two stars of the same mass but with different metallicities.

As a bonus, the framework naturally accommodates inherently parallel community modules and allows simultaneous execution of independent modules from the user script. Individual community modules can be offloaded to other processors, in a cluster or on the grid, and can be run concurrently without requiring any changes in the user script, which still looks like a simple serial code. This makes the AMUSE framework well suited for distributed computing with a wide diversity of hardware architectures.

B.1.1.3 The Manager

Between the user script and the communication layer, we introduce the manager, which as part of the community module. This part of the code is visible to the script-writing user and guarantees that all accessible data are up to date, have the proper format, and are expressed in the correct units.

The manager layer constructs data models (e.g., particle sets or tessellated grids) on top of all functions in the low-level communication class and checks the error codes they return. To allow different modules to work in their preferred systems of units, the manager also incorporates a unit-conversion protocol (see Section B.1.3). It also includes a state engine to ensure that codes are not in the middle of other activities when they hand control back to the framework, since many community codes are written with specific assumptions about calling procedures. The state engine is important, for example in adding particle to a tree code: for performance, it is much better to delay rebuilding the tree until all particles have been added rather than rebuilding the tree every time a particle is added. This repository can be structured in one or more native formats, such as particles, grids, tessellation, etc., depending on the topology of the data adopted in the community code.

Each community code has its own internal data, needed for modeling operations but not routinely exposed to the user script. However, some of these data may be needed by other parts of the framework. To share data effectively and to minimize the bookkeeping for data coherency, the most fundamental parts of the community data are replicated as a structured repository in the manager. The internal data structures differ, but the repository imposes a standard format. It allows the user to access community module data from the user script without having to make individual (and often idiosyncratic) calls to individual community modules. The repository is updated from the community module on demand and is considered to be authoritative within the script.

The separation of the manager from the community module realizes a flexibility that allows the user to swap modules and recompute the same problem using different physics solvers, providing an independent check of the implementation, validation of the model, and verification of the results, simply by rerunning the same realization of the initial conditions using another community module. Potentially even more interesting is the possibility of solving some parts of a problem with one solver and others with a different solver of the same type, and then combining the results at a later stage to complete the solution. While studying the general behavior of a simulation, perhaps to test a hypothesis or guide our physical intuition, we may elect to use computationally cheaper physics solvers, and then switch to more robust, but computationally more expensive, modules in a production run. Alternatively, we might adopt less accurate solvers for physical domains that are not deemed crucial for capturing the correct overall behavior.

A further advantage of separating the manager from the community module is the possibility of combining existing community modules into new community modules which can themselves be coupled via the manager, building in this way a hierarchical environment for simulating complex physical systems with multiple components.

B.1.2 The User Script

The AMUSE framework is controlled by the user via `Python` scripts. The main tasks of these scripts are to identify and spawn community modules, control their calling sequences, and manage the data flow among them. The user script and the spawned community code are connected by a communication channel embedded in the community module and maintained by the manager. Since the community modules do not communicate directly with one another, the only communication occurs between modules and the user script. The number of community modules controlled by a user script is effectively unlimited, and multiple instances of identical modules can be spawned.

The main objective of the user script is to read in or generate an input model, process these initial conditions through one or more simulation modules, and subsequently mine and analyze the resultant data. In practice, the user script will itself be composed of a series of scripts, each performing one specific functionality.

B.1.3 Intermodule Data Transfer and Unit Conversion

Computer scientists tend to attribute types to parameters and variables, and generally impose strict rules for the conversion from one type to another. In physics, data types are almost never important (everything is real or integer), but units are crucial for validating and checking the consistency of a calculation or simulation. The lack of coherent unit handling in programming environments is notorious, and can cause significant problems in multiphysics simulations, particularly for inexperienced researchers.[1]

In many cases, a researcher can (and arguably should) define a set of dimensionless variables applicable to a specific simulation, within the confines of a set of rather strict assumptions. For a monophysics simulation, the consistent use of such variables is generally clear to most users. However, as soon as an expert from another field attempts to interpret the results, or when the output of one simulation is used by another program, units must be restored and the data converted to another physical system for further processing and/or interpretation.

The absence of units in scientific software raises two important problems. The first is the loss of an independent consistency check for theoretical calculations. The second is the expert knowledge and intuition required to manage dimensionless variables or otherwise unfamiliar units. Few would recognize 4.303×10^{43} as the

[1] The Mars Climate Orbiter crashed on 1998 December 11 on the red planet due to a miscalculation during Trajectory Correction Maneuver 4 in the conversion of imperial unit pound-seconds adopted by NASA to the metric units of Newton-seconds used by Lockheed.

radius of the Sun in units of the Planck length, even though it is not completely inconceivable that an astronomer might use such units in a cosmological simulation. To address these issues, AMUSE incorporates a unit-conversion module as part of the manager (see Section B.1.1.3) to guarantee that all communications within the top-level user script are performed in the proper units. In order to prevent unit checking and conversion from becoming a performance bottleneck, we adopt "lazy" evaluation, performing unit conversion only when explicitly required.

B.1.4 Installing Prerequisites and Debugging

The non-`Python` packages on which AMUSE depends are listed in Table B.1. These dependencies may differ in later releases; so to be sure that you install the correct packages, it is best to consult the AMUSE web site https://www.amusecode.org. Many of these dependencies may already be installed in recent `Linux` and `macOS` releases, but some fine-tuning may be needed to complete the dependency installation.

The Message Passing Interface (or MPI for short) is an important package, although the same functionality can be realized using sockets (Gay 2000). We recommend using either sockets, `OpenMPI` (Dagum & Menon 1998) or `MPICH` (Gropp 2002). The former has the advantage that the communication does not require a process when running, which saves computer time for other tasks, but the latter is more standard, usually faster, and you will then have an MPI running for other parallel tasks.

When running on a supercomputer, it is probably best to use the native `numpy` (van der Walt et al. 2011) and the MPI interface specific to the architecture rather

Table B.1. Dependencies Needed to Compile and Run AMUSE

Package Name	Minimum Version Number	Note
`Python`	3.7	Scripting programming language
`HDF5`	1.6.5	Little/big endian neutral binary formatted I/O
`MPI`	N/A	Message-passing interface
`FFTW`	3.0	Subroutine library for discrete Fourier transforms
`GSL`	2.0	Scientific Library for C and C++ programmers
`CMake`	2.4	Cross-platform, open-source build system
`GMP`	4.2.1	Library for arbitrary precision arithmetic
`MPFR`	2.3.1	C Library for multiple-precision floating-point computations
`Numpy`	1.18.0	Numerical toolbox for `Python`
`h5py`	3.0	HDF5 interface for `Python`
`mpi4py`	3.0	`Python` interface for MPI
`pytest`	3.0	Extension of the `Python` testing framework
`docutils`	0.18	Documentation support for `Python` packages

than installing your own, but for all other packages this should not make a difference. When installing AMUSE on multiple machines in order to use distributed operation (using `ibis`–see Section B.1.6; Bal et al. 2010; Drost et al. 2012), it is important to ensure that the each of the machines runs the same version of `Java` (Gosling et al. 1996, 2014); otherwise, the distributed interface will become confused and your scripts will fail with unpleasant distributed error messages, which are hard to debug.

B.1.5 Reproducibility with `git`

A nice feature of `git` is that older versions are guaranteed to exist, and can be downloaded and installed if needed. In this way, we can, for example, ensure that the version of AMUSE we are using is identical to the version used to write this book. To check the version number, you can use the command

```
>% git rev-parse HEAD
```

which, on the day we submitted the manuscript to the publisher, gave

```
8d81c91a9fbd55b7510b3949700d3a594e5cc179
```

or version v2025.5.0.rc1 from 2025 May 30. This is the unique key for the particular version of AMUSE under which we tested many of the scripts in the book. If you wish to use exactly the same version of `amuse`, you can download and install it using the bash script presented in Code example B.1. The script automatically downloads and starts the script that will guide you through installing the AMUSE package.

Note, by the way, that we test AMUSE frequently to ensure that our scripts and test suites run as expected, so there is no actual reason to use only this specific version. However, for purposes of code verification, it is often important to be able to specify precisely the version of the software that was used to obtain a specific result, so this feature of `git` is key to writing reproducible code.

```
#! /bin/bash

GITHASH=450d20c0b542be53ff6fdf82bb6c2cce1c80c7bf
temp=amuse-temp-$$
mkdir $temp; cd $temp

git clone https://github.com/amusecode/amuse.git
cd amuse; git reset --hard $GITHASH; cd ../..

export AMUSE_DIR="$PWD/amuse-$GITHASH"
mv $temp/amuse $AMUSE_DIR
rm -rf $temp

cd $AMUSE_DIR
./setup
```

Code example B.1: Automated script to download a specific version of AMUSE. The script is available via `wget https://amusecode.org/get_amuse_and_install.sh`.

B.1.6 Tuning Your AMUSE Installation

The operation of AMUSE can be controlled and tuned using environment variables. These are generally stored in your home directory in the file `.amuserc`. These settings are superseded by the same named file in the directory from which you launch your AMUSE script.

The file `.amuserc` can contain a number of directives. These include the choice between `MPI`, `sockets`, or `ibis`. `MPI` is the default; to use sockets as your main communication channel with the following lines in your `.amuserc` file:

```
[channel]

channel_type=sockets
```

The `ibis` channel allows the user to run `amuse` in distributed mode using the `ibis` package.

Another option is to redirect the output of any of the AMUSE modules you may be running. The native output of the community codes is usually discarded, this would be reflected by

```
redirection=null
```

but by writing

```
redirection=none
```

it is redirected to `stdout`. An alternative would be to write `file` instead of `null` or `none`. This latter option may produce a number of files, depending on the codes you run. The two lines

```
redirect_stdout_file=code.out
redirect_stderr_file=code.out
```

will redirect the standard output and error channels `stdout` and `stderr`.

A debugger, in this example the GNU debugger `gdb` (see Robbins 2005), may be enabled by adding the line

```
debugger=gdb
```

to the `.amuserc` file. The standard setting is `debugger=none`, indicating that no debugger is to be used. Alternatives are `xterm` for debugging with an `xterm` screen, `ddd` for a `ddd` session using a graphical debugger, or `valgrind` (Nethercote & Seward 2007) to run the community codes under `valgrind` to check for memory leaks. We present more detail on debugging options in Section B.1.7.

These options can also be provided as part of the argument list for the various modules at runtime. For example, one could start a N-body code `ph4` with the following arguments

```
gravity = Ph4(redirection="file", number_of_workers=16, mode="gpu",
              debugger="gdb")
```

to redirect the code output to the standard files identified in the original source code, run the code concurrently on 16 cores, each equipped with a GPU, and run the debugger in order to track numerical problems.

Several more options can be specified in the `.amuserc` file, but these are less relevant for the day-to-day operation of the package. An example of an `.amuserc` file is provided in the installation directory in your AMUSE repository.

B.1.7 Debugging in AMUSE

Debugging a monophysics solver is sometimes difficult, but debugging an AMUSE script when the problem arises in one of the coupled codes can be much more challenging. With any luck, you will never have to do this because we hope that the authors of the individual community codes will continue to maintain their packages, but you never know.

There are several ways to debug a community code. Most basic is to set the code's internal "verbosity" level to a high value. Almost all codes have something like this, but there is absolutely no consensus on what value that should be because every author has a private method to do this. Since the community codes in AMUSE are (almost) virgin, meaning that we have touched them as little as possible while integrating them into the framework, the private verbosity settings are unchanged. However, apparently due to some unwritten understanding, setting the verbosity of most codes to a large positive integer value leads to an immense flow of information from the running code. This is illustrated in the example below:

```
self.hydro = Fi(converter, redirection="none")
self.hydro.parameters.verbosity = 99
```

Combining high verbosity with the `redirection` argument (indicating that the code's standard output should be written to your terminal) and running the code again generally produces enough (probably too much) information to debug obvious bugs.

A more systematic way is to start an online debugger. although this generally requires significant knowledge of the community code. In the following snippet, we start the GNU debugger `gdb` by providing it as an argument in the code (see also Section B.1.6):

```
hydro = Fi(converter, debugger="gdb")
```

Table B.2. Debugging Directives

Method	Explanation
None	No debugger, normal run
gdb	Start a gdb debugger in a separate xterm
xterm	Start the code under xterm, ideal for console logging
ddd	Start a ddd session (a graphical debugger)
valgrind	Check for memory leaks at runtime (Nethercote & Seward 2007)
gdb-remote	Run gdb as a server that listens on a port
	For remote debugging specify a port to listen to
debugger	None
debugger	gdb-remote
debugger_port	4343

You can also specify these arguments in the `.amuserc` file in your root directory, or the directory from which you run the AMUSE script (see Section B.1.6). There are several options, as listed in Table B.2.

B.2 Other Associated Codes and Packages

Many people contributed to AMUSE. They helped with development, maintenance, expansion, bug fixes, and provided interesting examples. Without the community, AMUSE would only be a shadow of its current form. Here, we discuss a few of the more prominent packages, not as big as `PeTar`, `Torch`, `Ekster`, or `Venice`, but still important. Many of these packages are still in development—honestly, all software is continuously in development. But they have been used in scientific production.

B.2.1 TRES: Triple-Star Evolution

The Triple-star Evolution Code (`Tres`) (Toonen et al. 2012, 2016, 2020) is a numerical simulation framework designed to model the evolution of hierarchical triple systems, including both stellar and planetary components. `Tres` combines stellar evolution (e.g., changes in mass, radius, and luminosity) using the `SeBa` package with secular three-body orbital dynamics (Hamers & Portegies Zwart 2016), allowing for mutual influences between the stars' evolution and their orbital interactions.

In `Tres`, three stars with orbits are integrated synchronously. The stellar evolution is realized with `SeBa`, as well as the mass-transfer recipes between two stars. The tertiary orbit, and its effect on the inner binary, is integrated using the secular dynamics code. In addition, `Tres` accounts for the tidal evolution of the three stars, which is important because of von Zeipel–Lidov–Kozai cycles (von Zeipel 1910; Lidov 1962; Kozai 1962) can render the triple highly eccentric.

`Tres` is particularly useful for population synthesis studies, enabling researchers to explore the evolutionary pathways and interaction rates in large samples of triple systems, as presented in Toonen et al. (2020).

```
from tres import run_tres, initialize_triple_class
from tres.setup import make_particle_sets


inner_primary_mass = 10.0 | units.MSun
inner_secondary_mass = 8.0 | units.MSun
outer_mass = 5 | units.MSun
inner_semimajor_axis = 1.0 | units.au
outer_semimajor_axis = 12.0 | units.au
inner_eccentricity = 0.5
outer_eccentricity = 0.5
relative_inclination = np.pi / 3
inner_argument_of_pericenter = 0.0
outer_argument_of_pericenter = 0.0
inner_longitude_of_ascending_node = 0.0

stars, bins, correct_params = make_particle_sets(
    inner_primary_mass,
    inner_secondary_mass,
    outer_mass,
    inner_semimajor_axis,
    outer_semimajor_axis,
    inner_eccentricity,
    outer_eccentricity,
    relative_inclination,
    inner_argument_of_pericenter,
    outer_argument_of_pericenter,
    inner_longitude_of_ascending_node,
)

stellar_code = Seba()
secular_code = Seculartriple()


triple = initialize_triple_class(
    stars, bins, correct_params, stellar_code, secular_code
)

triple.stellar_code.parameters.metallicity = 0.02
triple.evolve_model(10 | units.Myr)
triple.print_stellar_system()
```

Code example B.2: Minimal example to set up a simulation using `Tres` (see `amuse.examples.tres_minimal`).

Code example B.2 shows a small example script on how to set-up a simulation using `Tres`. This is considerably more elaborate, mainly because now we have three stars and two orbits to initialize.

B.2.2 AGAMA: Action-based Galaxy Modelling Architecture

AGAMA is an open-source toolkit for generating realistic *N*-body representation of galaxies (Vasiliev 2019); see also Vasiliev (2018). It offers a suite of computational methods for galactic modeling and analysis. It was developed since 2015 and is documented in over 130 pages of technical specifications.

AGAMA is a comprehensive software library for stellar dynamics and galactic modeling. Its core functionality centers on the computation of gravitational potentials and forces, which can be derived from analytic density profiles or N-body simulations using spherical-harmonic expansions and adaptive solvers. AGAMA supports the numerical integration of stellar orbits in both static and time-dependent potentials, accommodating composite systems such as interacting galaxies.

The library provides robust tools for coordinate transformations, enabling conversion between position-velocity and action-angle variables via the Stäckel approximation or torus mapping techniques. AGAMA implements a range of action-based distribution functions for disk and spheroidal systems, facilitating predictions of velocity distributions and the construction of self-consistent, multi-component galaxy models.

Scientifically, AGAMA is suited for galactic structure reconstruction using Schwarzschild orbit-superposition methods, dark matter halo profiling from stellar kinematics, equilibrium initial condition generation for N-body simulations, and the analysis of tidal streams and dynamical populations. The framework also supports parameter optimization against observational data.

Technically, AGAMA is written in C++ with performance optimizations using OpenMP parallelization. It offers a native `Python` API for interactive analysis and is interoperable with other astrophysical software ecosystems such as AMUSE, GALPY, and NEMO. The package is modular, allowing users to extend its capabilities with custom potential models and distribution functions. Included are advanced mathematical utilities for spline interpolation and multidimensional sampling, as well as extensive test suites and example workflows.

B.2.3 Vader: Viscous Accretion Disk Evolution Resource

We implemented an interface to the VADER viscous evolution code (Krumholz & Forbes 2015) within the AMUSE framework. Our interface supports EPE using mass loss rates from the FRIED grid (Section B.2.4), and we extended the code to include IPE with mass loss profiles from Picogna et al. (2019), scaled by stellar mass using the relations in Owen et al. (2012). Stellar X-ray luminosities, which drive IPE rates, are based on Flaccomio et al. (2012) and decrease over time, as given by Equation B.4.

To model accretion onto the central star, we apply a time-dependent accretion rate following viscous disc theory (Lynden-Bell & Pringle 1974):

$$\dot{\mathrm{M}}_{\mathrm{accr}}(t) = \frac{\mathrm{M_d}}{2t_s}\left(1 + \frac{t}{t_s}\right)^{-3/2}, \tag{B.1}$$

with viscous timescale

$$t_s = \frac{R_d^2}{3}\frac{\Omega}{\alpha c_s^2}, \tag{B.2}$$

which we approximate using the model's temperature profile and ideal gas law:

$$t_s = 11\ \text{Myr}\left(\frac{\alpha}{10^{-3}}\right)^{-1}\left(\frac{R_d}{200\ \text{au}}\right)\left(\frac{M_*}{M_\odot}\right)^{1/4}. \tag{B.3}$$

Rather than explicitly evaluating Equation B.1, we scale the input accretion rate by $(1 + t/t_s)^{-3/2}$.

Boundary conditions are handled dynamically: when the inner mass flux is insufficient to sustain the prescribed accretion, we first attempt to reduce the accretion rate to empty the innermost cell; if this fails, we switch to a vanishing-torque condition.

EPE rates are fixed per model and scaled linearly with disc radius, with optional interpolation from the FRIED grid (Haworth et al. 2018) under a fixed 10 G_0 radiation field. Dust is evolved as a passive scalar, initially 1% of the gas mass, and subject to EPE following Haworth et al. (2018) with an exponential decay representing grain growth.

The implementation uses a logarithmic grid of 330 radial zones from 0.01 to 3000 au and assumes a turbulent viscosity parameter of $\alpha = 10^{-3}$.

B.2.4 FRIED: Far-ultraviolet Radiation Induced Evaporation of Discs

EPE occurs when nearby stellar radiation heats disc material until it becomes unbound, driving a thermal wind. Far-ultraviolet radiation usually dominates, with extreme ultraviolet important near massive stars.

The FRIED grid (Haworth et al. 2018) provides EPE mass loss rates across host star mass, radiation field, disc radius, and mass. Coleman & Haworth (2020) used a simplified scaling due to the high minimum radiation field in FRIED, which yields mass loss rates too large for Peter Pan discs around M dwarfs.

IPE, driven by the host star's radiation, shows mass loss nearly independent of stellar mass but strongly dependent on X-ray luminosity (Owen et al. 2012; Picogna et al. 2019). X-ray luminosity scales with stellar mass (Flaccomio et al. 2012), giving a mass loss scaling exponent $\beta \approx 1.87$. X-ray luminosity decreases by an order of magnitude from 1 to 110 Myr, approximately as $L_X \propto t^{-2/5}$ after radiative core formation (Johnstone et al. 2021). We adopt:

$$L_X(t) = \begin{cases} L_{X,0} & t < 1\text{Myr} \\ L_{X,0}\left(\dfrac{t}{1\text{Myr}}\right)^{-2/5} & t > 1\text{Myr}. \end{cases} \tag{B.4}$$

Here, $L_{X,0}$ was adopted from Flaccomio et al. (2012).

In Figures B.2 and B.3, we present examples of how the FRIED grid can be used in combination with `SeBa` and `VADER`. In the former figure we show the mean circumstellar disk mass as a function of the projected local stellar density. For higher density environment-disk masses are typically smaller. The thick curves in Figure B.3 for a 16 $M_\odot$ star irradiating a 0.01 $M_\odot$ disk around a 1 $M_\odot$ star at a distance of 0.1 pc. The thin curves are calculated for the same star but then at a distance of 1 pc.

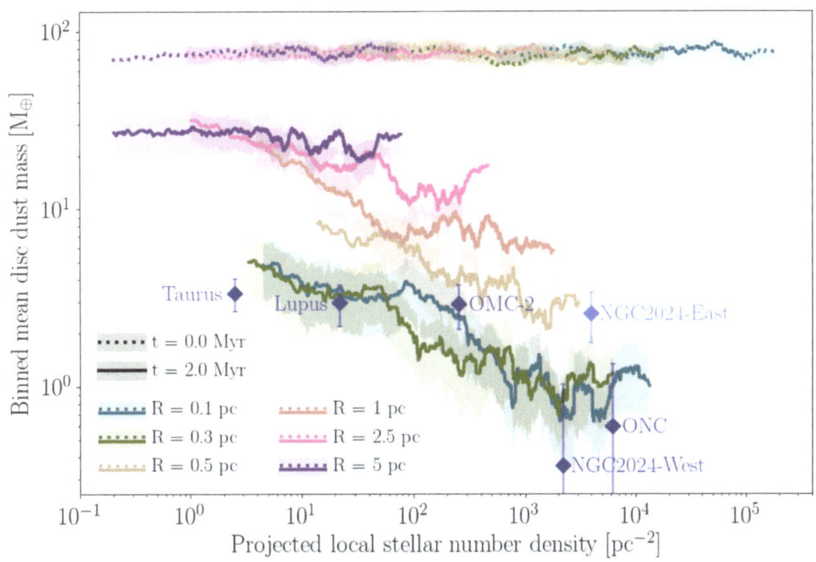

Figure B.2. Binned mean disc dust mass versus local stellar density, projected in two dimensions. Mean values use bins of 100 stars. Observational comparisons are included from Lupus, ONC, OMC-2, Taurus, and NGC 2024. Adapted with permission from Concha-Ramírez et al. (2021). © 2020 The Authors. Published by Oxford University Press on behalf of the Royal Astronomical Society

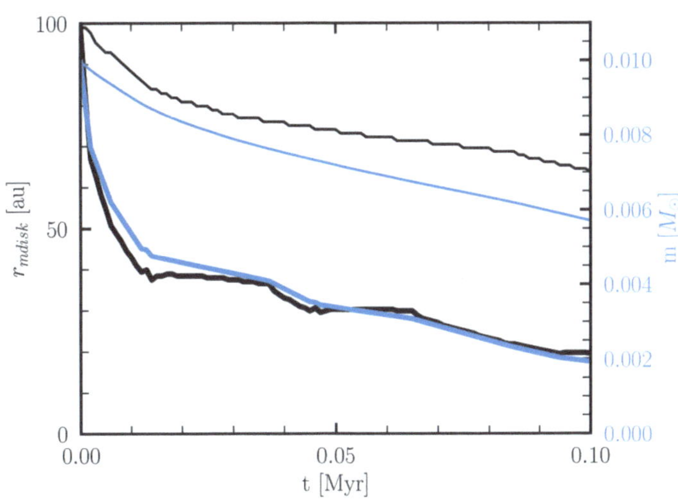

Figure B.3. Demonstration of how SeBa, VADER and the Fried grid can lead to interesting results in terms of the EPE of a circumstellar disk due to a nearby massive star. The blue curves adopt the disk mass as a function of time (right-hand axis scale), and the black curve give the disk radius (left). A script to generate this figure can be found at `amuse.examples.ppd_minimal`.

B.2.5 PACE: Planet Accretion and Evolution

Planet growth in our model proceeds via core accretion (Perri & Cameron 1974), beginning with the formation of a solid core that grows by accreting pebbles. Pebble accretion (Ormel et al. 2015) is efficient for Moon- to Ceres-sized embryos and is described by

$$\dot{M}_{\rm peb} = \varepsilon 2\pi r u_{\rm peb} \Sigma_{\rm peb}, \tag{B.5}$$

where ε is the pebble accretion efficiency and $u_{\rm peb}$ the pebble drift velocity, both of which depend on the Stokes number St. These are calculated in a mass-weighted fashion from the two-population dust model:

$$\mathrm{St} = \left(1 - f_{\rm m}\right)\mathrm{St}_0 + f_{\rm m}\,\mathrm{St}_1, \quad u_{\rm peb} = \left(1 - f_{\rm m}\right)v_0 + f_{\rm m}\,v_1. \tag{B.6}$$

Accretion ceases when the planet reaches the pebble isolation mass (Bitsch et al. 2018), at which point it creates a pressure bump that halts pebble inflow:

$$\begin{aligned} M_{\rm iso} = 25 M_{\oplus}\left(\frac{h}{0.05}\right)^3 &\left[0.34\left(\frac{-3}{\log_{10}\alpha_t}\right)^4 + 0.66\right] \\ \times &\left[1 - \frac{\partial \ln P/\partial \ln r + 2.5}{6}\right]\frac{M_\star}{M_\odot}. \end{aligned} \tag{B.7}$$

Gas accretion onto the core then follows, initially regulated by Kelvin–Helmholtz contraction (Ikoma et al. 2000):

$$\dot{M}_{\rm KH} = 10^{-5}\left(\frac{M_{\rm p}}{M_\oplus}\right)^4\left(\frac{\kappa}{0.1\ {\rm m^2 kg^{-1}}}\right)^{-1} M_\oplus\ {\rm yr}^{-1}, \tag{B.8}$$

but may later be limited by gas inflow across the planetary Hill sphere (Tanigawa & Tanaka 2016):

$$\dot{M}_{\rm Hill} = \frac{0.29}{3\pi h^4}\left(\frac{M_{\rm p}}{M_\star}\right)^{4/3}\frac{\dot{M}_{\rm g}}{\alpha_{\rm acc}}\frac{\Sigma_{\rm gap}}{\Sigma_{\rm g}}, \tag{B.9}$$

with gap depth factor

$$K = \left(\frac{M_{\rm p}}{M_\star}\right)^2 h^{-5}\alpha_{\rm t}^{-1}. \tag{B.10}$$

The final gas accretion rate is taken as the minimum of all limiting rates:

$$\dot{M}_{\rm GA} = \min\left[\dot{M}_{\rm KH}, \dot{M}_{\rm Hill}, \dot{M}_{\rm g}\right]. \tag{B.11}$$

Figure B.4 illustrates the diversity observed in planetary systems from two runs using PACE, one without dynamics, and one in which we took the dynamical evolution of the stars and planetary systems into account. In the latter run, we also incorporated the FRIED grid (Section B.2.4) to affect the evolution of circumstellar disks using VADER (Section B.2.3). In Figure B.4, we include the solar system for comparison: the orbital parameters range over more than four orders of magnitude, provoking many questions about their formation, evolution, the role of birth

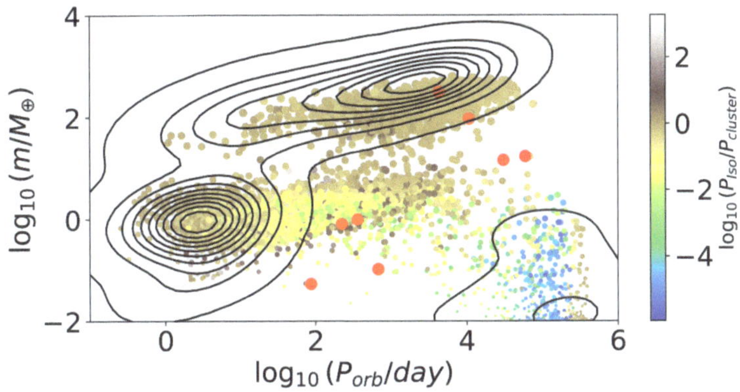

Figure B.4. Simulated planet population for isolated stars (contours) and for stars in a cluster (bullets) (Huang et al. 2024). The initial conditions are identical in both cases, but the planetary systems in a clustered environment are subject to EPE onto the planet-forming disk. The colors indicate the logarithmic change in the planetary system's orbital period from isolation to perturbed.

environment and later isolated secular evolution. So far, research has focused on the isolated formation and evolution of planetary systems, but it has become clear that stars form clustered and that this clustering has a long-term effect on the evolution of planetary systems.

B.2.6 Packages Not Yet Used in Scientific Production

Several packages are being developed as this book goes to print. Some of them have an operational interface, but are not yet tested in production simulations. Experience has taught us that software not used for producing scientific results is buggy, at best. We therefore do not necessarily recommend using these packages, but by the time this book is available some of them may have a working operational interface and will have been tested in practice.

They may also inspire the reader to start another project, and contribute to the framework. The more people contribute, the better and more useful the software becomes.

B.2.7 Solar System Initial Conditions

As part of a Master's project, Nadia Kempers developed a collection of routines to generate a realization of the solar system as we know it, based on available ephemerides from NASA and JPL. The script generates a current solar system particle set, allowing us to integrate it in a larger environment, or to study the orbital evolution of specific objects. It enables to selection of specific objects, planets, moons, planetesimals, comets, etc. Although, there are a number of hidden assumptions, such as the mass of objects, or their sizes, where observational data is unavailable, it is, so far, the most complete solar system orbital generator we have seen.

B.2.8 The Interface to Interstellar Nonequilibrium Chemistry

As part of a Master's project, Jelmer Stroo developed an interface to the UCLCHEM. UCLCHEM (Holdship et al. 2017; Keil et al. 2022) is a gas–grain chemical code designed for astrochemical modeling across diverse environments, such as molecular clouds, protostellar cores, and shocks. Unlike many codes that focus solely on gas-phase reactions, UCLCHEM incorporates comprehensive gas–grain processes, including freeze-out, nonthermal and thermal desorption, and grain-surface reactions

UCLCHEM models both gas-phase and grain-surface reactions, treating molecules on grains as distinct species from their gas-phase counterparts. Nonthermal desorption processes include molecule release due to H_2 formation, cosmic ray interactions, and UV irradiation. Thermal desorption is uniquely modeled as multiple events, reflecting laboratory findings, rather than a single peak as in other codes. The code propagates species abundances through a user-defined network of reactions, adapting to the physical conditions of the modeled environment.

Written in Fortran 90, UCLCHEM utilizes the DVODE algorithm to solve the resulting system of rate equations. The code is optimized for high-density, low-temperature environments (e.g., protostellar cores and shocks), and is less suitable for hot, low-density conditions where gas–grain interactions are minimal.

When used within AMUSE, it is essential to ensure that simulations remain within UCLCHEM's intended parameter space to maintain accuracy.

Although not yet used in production, we envision the use of UCLCHEM for protostellar core chemistry, shock and jet-driven desorption, chemistry in Seyfert galaxies, prebiotic molecule formation, effects of cosmic rays in dense regions, and machine learning emulators for astrochemical networks.

While this proof-of-concept simulation has notable limitations, it demonstrates the significant potential of integrating UCLCHEM with AMUSE for advanced scientific studies. The current implementation offers several features that can support more sophisticated simulations.

In hydrodynamical simulations, sink particles are employed to represent high-density regions that cannot be resolved at the simulation's spatial resolution (Bate et al. 1995). These are modeled as point masses that interact gravitationally but not hydrodynamically, and can accrete surrounding material. AMUSE facilitates the creation, management, and tracking of sink particles, including the chemical abundances of accreted material, which is valuable for core-scale modeling.

The choice of hydrodynamics code within AMUSE influences the physical processes included in a simulation. For example, the SPH code `Fi` is straightforward to use but lacks magnetic fields, turbulence, and molecular cooling. In contrast, codes such as `Athena2` (AMR: Stone et al. 2008, 2020) and `Phantom` (SPH: Price et al. 2018) support magnetohydrodynamics, with `Phantom` also incorporating a basic chemical network for molecular cooling—an important process in molecular clouds.

AMUSE's multiphysics framework allows for the integration of additional physical processes. This includes *N*-body codes for simulating planetary embryos, coupling chemical codes with stellar evolution models to study stellar winds, or combining stellar evolution, hydrodynamics, and radiative transfer to investigate radiative feedback in the interstellar medium. While increasing the complexity and computational cost, AMUSE's modular design streamlines the coupling of diverse codes, now including astrochemistry, enabling comprehensive and flexible astrophysical simulations.

In Figure B.5 we show the result of a simulated molecular cloud on the verge of collapsing and making the first stars. The setup for the on-the-fly simulation is similar to that in Priestley et al. (2023a): two spherical molecular clouds, each with mass $M = 10^4\,\mathrm{M}_\odot$, radius of $r = 19$ pc, and a uniform number density of $n_H = 10\,\mathrm{cm}^{-3}$, are initialized on either side of the x-axis with their centers at $\pm r$, so that the clouds' edges touch at $x = 0$. The clouds are then each given a bulk motion of $v = \pm 7\ \mathrm{km}^{-1}$ to cause them to collide. The initial gas temperature is set at $T = 300$ K. The SPH code `fi` (Pelupessy et al., 2004) is used to simulate the hydrodynamics of the molecular gas. Each molecular cloud consists of 100,000 particles, for a mass resolution of 0.1 $\mathrm{M}_\odot$ per particle. Time steps of 4.4×10^3 yr are used between steps of the hydrodynamical and chemical codes, and the simulation is stopped at a time of $t_{end} = 5$ Myr.

As in Priestley et al. (2023a), the UV radiation field is set at 1.7 times the Habing (1968) field, and the cosmic ray ionization rate at $\xi = 10^{-16}\ \mathrm{s}^{-1}$. While the assumption of a constant cosmic ray ionization rate is common in similar works (Priestley et al. 2024, 2023b), recent work has shown that this rate varies drastically based on the environment (see Luo et al. 2023) for diffuse clouds, and for different locations within star formation regions (Sabatini et al. 2023). Though the AMUSE framework is capable of simulating a variable cosmic ray field, in this proof-of-concept simulation the cosmic ray ionization rate is kept constant. This is done to make comparison with earlier work easier, and also to reduce the complexity of this simulation. Due to this same argument of simplicity, turbulence and magnetic fields are not included either.

In the center of the cloud, substructures with higher densities are visible, while near the outside scattered particles that were likely expelled from the clouds during

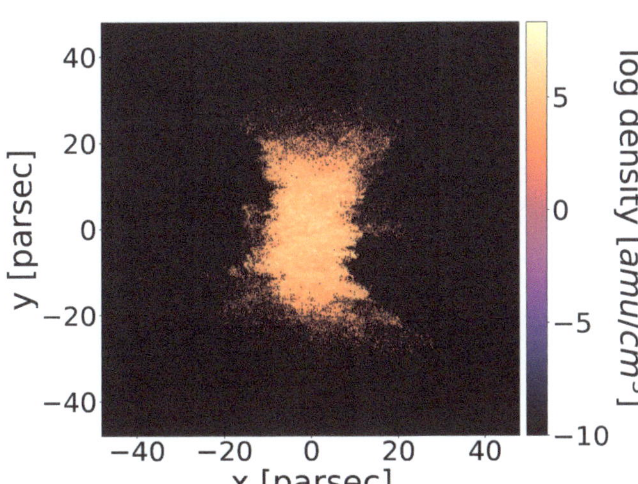

Figure B.5. Projected view of a simulated molecular cloud.

the collision can be seen. This distinction is also visible in Figure B.5, which shows the distribution of the gas temperature as a function of the density for the chemical particles at the end of the simulation. As expected, the bulk of the particles are at relatively high density and low temperature, while lower-density particles, near the outside of the cloud, have higher temperatures. The relatively low number of particles prevents a more comprehensive analysis of this distribution. The temperatures of the full set of hydrodynamical particles would allow such an analysis, but (due to an error) they are unfortunately not accessible. Regardless, this distribution is qualitatively in line with expectations.

Unexpectedly, the simulation shows relatively low values for the number density. Priestley et al. (2023a) reached number densities up to 10^6 cm^{-3} with a very similar setup, which is comparable to observed densities in dense cores. The fact that this simulation does not reproduce these values indicates it is missing some crucial components needed to form these dense structures. Unlike Priestley et al. (2023a), the simulation in this work does not contain turbulence, magnetic fields, or sink particles. Turbulence and magnetic field effects together cause an effect named reconnection diffusion, which is able to mediate cloud collapse. This simulation also does not contain sink particles to model the areas with densities beyond the simulation's resolution. The mass resolution is also higher than in Priestley et al. (2023a): they used a mass resolution of $0.005\,M_\odot$ per particle.

The hydrodynamics of the simulation have a clear effect on the chemistry. The jump in density caused by the collision of the clouds translates to higher abundances for a wide variety of gas-phase molecules. The specific molecules in this figure were chosen to allow for comparison with Priestley et al. (2023b), who present their results for the same selection of molecules, all in the gas phase. It should be noted, however, that the abundance on the grain surface for each of these molecules is also available in this simulation, as well as every other molecule in the network. These two are chosen because they are commonly observed in the gas phase. This shows the successful coupling between the hydrodynamics and chemistry in this simulation. The abundances show a similar spatial distribution to the hydrogen density shown in Figure B.6.

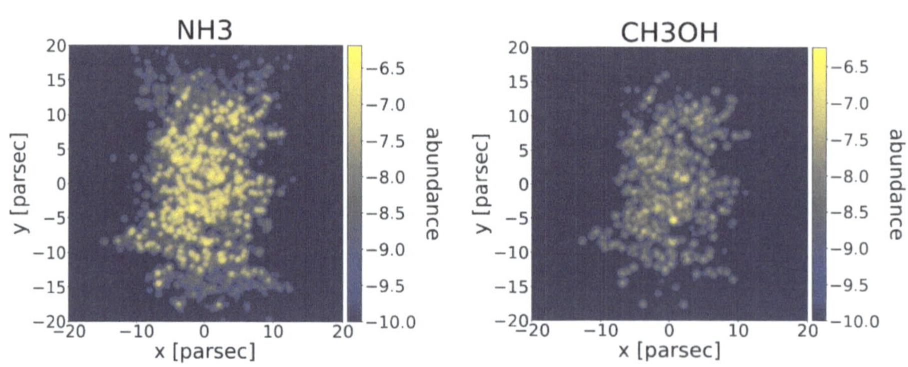

Figure B.6. Projected distribution of abundances for NH_3 and CH_3OH.

B.2.9 `Tidymess`: Tidal Dynamics of Multibody Extrasolar Systems

As part of a Master's project, Tünde Meyer developed an interface to the `Tidymess` code. `Tidymess` (Tidal Dynamics of Multibody Extrasolar Systems; Boekholt & Correia 2023a, b) is an open-source *N*-body code designed to model the complex dynamical evolution of systems where tidal forces and deformations play a critical role, such as in planetary systems, moons, stars, and compact objects. `Tidymess` enables detailed, efficient, and flexible modeling of tidal dynamics in multibody astrophysical systems, filling a unique niche for studying chaotic and nonlinear tidal evolution.

`Tidymess` implements the Creep deformation law (Norton 1929; Bailey 1935), allowing each body's tidal response to be characterized by its fluid Love number and relaxation time. This approach is more realistic for long-term evolution and can handle both linear and nonlinear tidal regimes. The code models bodies as ellipsoids, accounting for both tidal and centrifugal deformations. The gravitational field is calculated up to quadrupole order, enabling accurate computation of accelerations and torques. Orbits, spins, and deformations are integrated self-consistently using a fourth-order symplectic method. A novel aspect is that deformation integration uses a time step based on orbital timescales, independent of spin or relaxation times, which improves efficiency and consistency across different tidal regimes `Tidymess` is particularly suited for systems with many bodies, chaotic orbital evolution, and strong, nonlinear tidal interactions

References

Bailey, R. W. 1935, PIME, 131, 1

Bal, H. E., Maassen, J., van Nieuwpoort, R., et al. 2010, Compr, 43, 54

Bate, , et al. 1995, MNRAS, 277, 362

Bitsch, B., Morbidelli, A., Johansen, A., et al. 2018, A&A, 612, A30

Boekholt, T. C. N., & Correia, A. C. M. 2023a, MNRAS, 522, 2885

Boekholt, T. C. N., & Correia, A. C. M. 2023b, TIDYMESS: TIdal DYnamics of Multi-body ExtraSolar Systems, Astrophysics Source Code Library, ascl:2306.004

Coleman, G. A. L., & Haworth, T. J. 2020, MNRAS, 496, L111

Concha-Ramírez, F., Wilhelm, M. J. C., Portegies Zwart, S., van Terwisga, S. E., & Hacar, A. 2021, MNRAS, 501, 1782

Dagum, L., & Menon, R. 1998, ICSEn, 5, 46

Drost, N., Maassen, J., van Meersbergen, M., et al. 2012, IPDPS 2012: 26th {IEEE} International Parallel and Distributed Processing Symposium Workshops (Piscataway, NJ: IEEE) 150

Flaccomio, E., Micela, G., & Sciortino, S. 2012, A&A, 548, A85

Gay, W. W. 2000, Linux Socket Programming: By Example (Indianapolis, IN: Que Corp.)

Gosling, J., Joy, B., & Steele, G. L. 1996, The Java Language Specification (1st ed.; Boston, MA: Addison-Wesley)

Gosling, J., Joy, B., Steele, G. L., Bracha, G., & Buckley, A. 2014, The Java Language Specification, Java SE 8 Edition (1st ed.; Boston, MA: Addison-Wesley)

Gropp, W. 2002, MPICH2: A New Start for MPI Implementations 7 (Berlin: Springer)

Gropp, W., Lusk, E., Doss, N., & Skjellum, A. 1996, ParC, 22, 789

Habing, H. 1968, BAN, 19, 421

Hamers, A. S., & Portegies Zwart, S. F. 2016, MNRAS, 459, 2827
Haworth, T. J., Clarke, C. J., Rahman, W., Winter, A. J., & Facchini, S. 2018, MNRAS, 481, 452
Holdship, J., Viti, S., Jiménez-Serra, I., Makrymallis, A., & Priestley, F. 2017, AJ, 154, 38
Huang, S., Portegies Zwart, S., & Wilhelm, M. J. C. 2024, A&A, 689, A338
Ikoma, M., Nakazawa, K., & Emori, H. 2000, ApJ, 537, 1013
Johnstone, C. P., Bartel, M., & Güdel, M. 2021, A&A, 649, A96
Keil, M., Viti, S., & Holdship, J. 2022, ApJ, 927, 203
Kozai, Y. 1962, AJ, 67, 591
Krumholz, M. R., & Forbes, J. C. 2015, A&C, 11, 1
Lidov, M. L. 1962, Planet. Space Sci., 9, 719
Luo, G., Zhang, Z.-Y., Bisbas, T. G., et al. 2023, ApJ, 946, 91
Lynden-Bell, D., & Pringle, J. E. 1974, MNRAS, 168, 603
Nethercote, N., & Seward, J. 2007, PLDI '07: Proc. 28th SIGPLAN Conference on Programming Language Design and Implementation ed. K. S. McKinley, & J. Ferrante (New York: ACM) 89
Norton, F. H. 1929, The Creep of Steel at High Temperatures (New York: McGraw-Hill)
Ormel, C. W., Shi, J.-M., & Kuiper, R. 2015, MNRAS, 447, 3512
Owen, J. E., Clarke, C. J., & Ercolano, B. 2012, MNRAS, 422, 1880
Peterson, P. 2009, IJCSE, 4, 296
Pelupessy, F. I., van der Werf, P. P., & Icke, V. 2004, A&A, 422, 55
Perri, F., & Cameron, A. G. W. 1974, Icar, 22, 416
Picogna, G., Ercolano, B., Owen, J. E., & Weber, M. L. 2019, MNRAS, 487, 691
Portegies Zwart, S., McMillan, S., Harfst, S., et al. 2009, NewA, 14, 369
Price, D. J., Wurster, J., Tricco, T. S., et al. 2018, PASA, 35, e031
Priestley, F. D., Clark, P. C., Glover, S. C. O., et al. 2023a, MNRAS, 526, 4952
Priestley, F. D., Clark, P. C., Glover, S. C. O., et al. 2023b, MNRAS, 524, 5971
Priestley, F. D., Clark, P. C., Glover, S. C. O., et al. 2024, MNRAS, 531, 4408
Robbins, A. 2005, GDB - Pocket Reference: Debugging Quickly and Painlessly (Sebastopol, CA: O'Reilly & Associates)
Sabatini, G., Bovino, S., & Redaelli, E. 2023, ApJL, 947, L18
Stone, J. M., Gardiner, T. A., Teuben, P., Hawley, J. F., & Simon, J. B. 2008, ApJS, 178, 137
Stone, J. M., Tomida, K., White, C. J., & Felker, K. G. 2020, ApJS, 249, 4
Tanigawa, T., & Tanaka, H. 2016, ApJ, 823, 48
Toonen, S., Nelemans, G., & Portegies Zwart, S. 2012, A&A, 546, A70
Toonen, S., Portegies Zwart, S., Hamers, A. S., & Bandopadhyay, D. 2020, A&A, 640, A16
Toonen, S., Hamers, A., & Portegies Zwart, S. 2016, ComAC, 3, 6
van der Walt, S., Colbert, S. C., & Varoquaux, G. 2011, CISE, 13, 22
Vasiliev, E. 2018, AGAMA: Action-based Galaxy Modeling Framework, Astrophysics Source Code Library, ascl:1805.008
Vasiliev, E. 2019, MNRAS, 482, 1525
von Zeipel, H. 1910, AN, 183, 345

Appendix C

Programming Primer

> To be a programmer is to develop a carefully managed relationship with error.
>
> —Ellen Ullman

In this book, we assume that you are familiar with the Linux operating system, with the Python programming language, and with the use of git. If this is not the case, we present here a very basic primer about them. Linux is a UNIX-based operating system designed in 1991 by Linus Torvalds. Python is a third-generation computer language designed in 1991 by Guido van Rossum. Git is a source code management system that began in 2005 as a development system for Linus Torvalds's Linux kernel.

C.1 Linux

The Linux operating system was originally designed for microcomputers.[1] It is composed of a number of functional components that together form the operating system. These components include (see http://linux.com):

The **bootloader**, the software that manages the startup process of your computer.

The **kernel**, the core of the system that manages the CPU, RAM memory, and peripheral devices.

Daemons, background services that manage printing, sound, scheduling, etc., that either startup during boot, or when you log in to the desktop.

The **shell**, which manages commands to control the computer. Commands are typically typed into a terminal with a text interface, or contained in scripts.

[1] "Microcomputer" refers to the type of machine we nowadays call a workstation. Minicomputers used to be the midlevel machines that we now call cluster computers or Beowulfs (Sterling 2002), not to be confused with the Raspberry Pi processor board, which is sometimes also referred to as a minicomputer.

doi:10.1088/2514-3433/adfc3cch14

The **graphical server** (or X server), the subsystem that displays graphics on your monitor.

The desktop environment, the customizable user interface. There are many choices, such as Gnome, KDE, Cinnamon, Enlightenment, etc.

Applications (apps), the programs that run on your computer. In this book, you write your own applications using AMUSE, but you can download applications or write them in other languages too. One application you will be using a lot is a text editor. This can be emacs (Ciccarelli 1978), Pico, vim, or any of a number of others. Choice of editor is squarely in the category of religion and politics, and we make no recommendation. Another application you will be using a lot is the Python interpreter (van Rossum 1995a, b, c, d).

You generally communicate with your computer via the desktop environment. Most modern operating systems have desktop environments, but the sophisticated shell and the programming style it enables make Linux (and its cousin UNIX) rather unique. Some other operating systems have command-line interfaces, but they are often undeveloped and almost completely useless. Note that there is not a single Linux, but many flavors, including Fedora, Ubuntu, RedHat, SuSE, Debian, Gentoo, and many more (see http://distrowatch.com/). The practical differences are generally not so large, although true OS geeks swear by their preferred flavor. MacOS is not Linux, but is based on FreeBSD, a version of UNIX, and presents a programming environment essentially equivalent to that found in Linux.

The shell allows you to send commands directly to your computer, or launch scripts to automate this process. Some typical Linux commands, in no particular order, include:

%> man -t ascii | ps2pdf - > ascii.pdf makes a PDF file of a manual page.

%> apropos whatis combines two commands: apropos, which lists alternatives on a specific command, and whatis, a rudimentary help function.

%> which python shows the full path name of the command Python.

%> mkdir amuse creates a directory with the name amuse.

%> ls amuse lists the contents of directory amuse.

%> cd amuse go to directory named amuse.

%> find . -name *.py finds all files with the extension .py in all subdirectories.

%> locate sun_venus_earth.py finds the file sun_venus_earth.py.

%> tar -x dir | gzip make a zip-compressed archive of the directory dir.

%> sort -u file1 file2 make a union of unsorted files.

%> gzip < /dev/cdrom > cdrom.iso.gz saves a copy of data from a CD-ROM.

%> ps -aux shows a list of all processes running on your computer.

%> top shows an interactive list of your current processes.

%> grep MemTotal /proc/meminfo shows the total random access memory in the system. (Note that macOS does not have a /proc filesystem.)

`%> cat /proc/cpuinfo` displays the contents of the file named `/proc/cpuinfo`.
`%> tail -lines=10 filename` shows the last 10 lines of the file named `filename`.
`%> less filename` displays the content of the file `filename`, page by page.

Note the use of the operators "<" (read from a file or the keyboard), ">" (write to a file or the computer screen), and "|" (connect the output of one command to the input of another).

C.1.1 A Little `awk`ing and `grep`ping

The `Linux` shell includes a large number of handy utilities, which can help you work more efficiently with the operating system. The commands `awk` (see http://www.gnu.org) and `grep` (Kernighan 1984) are particularly useful for rudimentary data processing.

- `awk` is a scripting language designed to manipulate ASCII text documents and automate the extraction of data. It was developed in the 1970s by Alfred Aho, Peter Weinberger, and Brian Kernighan (Aho et al. 1987).
- `grep` is a utility run from the `Linux` command line that searches for regular expressions in plain-text data sets.

Typically, `grep` is used to search and catch certain phrases, whereas `awk` is used to manipulate them. Knowing some basic `awk` and `grep` commands will save you valuable time, and we strongly recommend that you learn how to work with them. You might begin with

```
%> man awk
```

and

```
%> man grep
```

both of which will provide more information than you need right now. These utilities make for a powerful experience in the command shell.

Another handy command-line tool is the stream editor `sed`, which is very efficient at manipulating a data stream (from a file or another program) on the fly.

C.2 `Git`

While working through this book, you are likely to write source code for several numerical experiments. `Git` is an excellent source code management system that lets you keep track of and manage the different versions of your code. Tutorials and complete documentation are readily available online at many locations, but we recommend `https://www.atlassian.com/git/tutorials/`.

Probably the simplest way to start using `git` is to create an account on one of the online `git` brokers, such as `bitbucket.org`, `github.com`, or `sourceforge.`

net. These sites are widely used and free. There are many more sites, some providing better services and enhanced security, but at a cost. The details vary, but after you create an account, you will generally be guided through the steps needed to create a new minimal project, which you can then clone to your own computer and populate with your own files, as described below.

Alternatively, you can make and manage your own git repository. First create a directory on the computer where you want the repository to reside. The following commands create a directory containing an empty git repository.

```
%> mkdir project-home
%> cd project-home
%> git init -bare
%> git update-server-info
```

Now you can update the repository with your project source code, copied from another computer or another directory on the same computer. We assume you have placed it in the local subdirectory project. You can add the code to the repository as follows.

```
%> cd project
%> git init
%> git add *
%> git commit -m "First version of my code"
%> git remote add origin spz@eucalypta.strw.leidenuniv.
nl:project-home


%> git push -u origin master
```

Here, spz@eucalypta.strw.leidenuniv.nl:project-home refers to your git server's URL and the particular directory on that server where the project resides. Of course, you do not want to store your data on the first author's computer at the Leiden Observatory in the Netherlands, so you had better create your own!

Now, suppose you want to access the code on another computer, maybe your laptop, make changes, and upload them to the master repository. To do this, you first have to make a local directory for your code and clone the master copy of the code:

```
%> git clone spz@eucalypta.strw.leidenuniv.nl:project-
home project
```

Now, you can make changes to the source code using your favorite editor and upload them:

```
%> cd project
%> ...(edit files)...
```

```
%> git add filename.cpp
%> git commit -m "Small change made in filename.cpp"
%> git push origin master
```

Your changes are now added to the master repository, but that does not mean that they are have been propagated to other clones of the code. Suppose your friend's computer also has a working copy of the repository. She is still working with the version from before the changes you made, but can update her local version with the commands

```
%> cd project
%> git pull
```

Now she also has the latest version.

This all sounds very nice, although maybe a bit arcane, but believe us, it is a great improvement over exchanging source code by e-mail or using other version control systems. More information on `git` can be found at http://rogerdudler.github.io/git-guide/.

C.3 `Python`

If you learned to program computers in the dark ages, like two of the authors, `Python` (van Rossum 1995a, b, c, d) is a breath of fresh air. Almost like air travel in the 1980s: few restrictions, and nothing to declare! `Python` is easy to learn and it is almost impossible to write obfuscated code in (in contrast to C: see http://www.ioccc.org/). In this section, we review a few basics you should know about `Python` in order to use this book. This section will be as short as possible because there are a great many excellent books written about the language. We recommend especially the many books about the language by Hans Petter Langtangen, and specifically `Python` *Scripting for Computational Science* (Langtangen 2009).

`Python` is a procedural language, which means that it works with lists of instructions that tell the computer what to do with input and data. Other examples of procedural languages include `C` (Kernighan & Ritchie 1988) and `FORTRAN` (American National Standards Institute 1966). Some procedural languages incorporate abstract data types that allow objects can be created and manipulated. Examples of these object-oriented languages include `C++` (Stroustrup 2013) and `Java` (Gosling et al. 1996, 2014). `Python` is also an object-oriented language, and includes some other interesting features drawn from `C++` and `Lisp`.

`Python` is a typed language—variables have specific types that can only be combined in certain ways—but unlike the "scientific" programming languages listed above, types are not declared at compile time, but instead are inferred from context as needed. This makes `Python` both very flexible and much slower than `C++` or `FORTRAN`. Indeed, the whole point of AMUSE is to combine the flexibility of `Python` with the speed of compiled procedural languages. `Python` also supports reflection—the program structure can be modified during execution.

You generally start a Python session on the command line by invoking the interpreter

```
%> python
```

or, for the somewhat more practical interactive shell interpreter,

```
%> ipython
```

Both are basically the same language (Perez & Granger 2007). If you wish, you can try the Jupyter notebook environment (Kluyver et al. 2016), which can be found at http://jupyter.org/. This is a browser-based support structure on top of the usual Python interpreter, and has a number of powerful and interesting features.

C.3.1 Python Types and Printing

As a procedural language, a Python script is composed of a list of instructions. One of the simplest is a scalar assignment:

```
>>> star = 5
```

(here and below >>> represents the interactive Python prompt), which here assigns the integer 5 to the variable star. If we had wanted to use a floating point number, we would have said

```
>>> star = 5.0
```

The instructions can contain arithmetic operations

```
>>> other_star = 4*star - 2
>>> print(other_star)
18
```

Note that the Python interpreter decides a variable's type at run time, and the type can change from line to line, e.g.,

```
>>> star = 5
>>> a = star
>>> star = 5.0
>>> print(a, star)
5. 5.0
```

One "feature" of `Python` 2 was that dividing an integer by an integer returns an integer. For example, if you divided two integers and printed the result, you would get

```
>>> print 1/2
0
```

This is very familiar to `C` and `FORTRAN` programmers, and is even very useful once you understand that it happens, but it is a never-ending source of confusion for beginners. `Python` 3 has eliminated this feature (to the chagrin of the old-timers), and the same code would produce the following:

```
>>> print(1/2)
0.5
```

while the former behavior can be reproduced with the // operator:

```
>>> print(1//2)
0
```

In our opinion, it is generally better (meaning that it produces more predictable code) to write the decimal point explicitly:

```
>>> print(1./2.)
0.5
```

To be sure about the type of variable, you can always query its type explicitly:

```
>>> type(star)
<type 'int'>
```

We have used the function `print` several times above. In `Python` 2.x, we could have eliminated the parentheses:

```
>>> print 1./2.
0.5
```

but in `Python` 3 they are required. Integer division and the `print` statement are the two most noticeable differences between `Python` 2 and `Python` 3. Because AMUSE was originally designed and implemented using `Python` 2.7, we usually assumed that integer division yields an integer and used print statements without parentheses in the AMUSE source for the first edition of this book. In this edition, we have switched fully to Python 3.

C.3.2 Lists

Variables of different types can be combined into lists, e.g.

```
>>> stellar_list = [star, other_star, 1/2, 1./2.]
```

which will give

```
>>> print (stellar_list)
[5, 18, 0, 0.5]
```

Lists are nice. They can contain anything you like. Individual elements can be accessed by using their position in the list. Like C, but unlike FORTRAN, Python counts from zero, so the list here runs from element 0 to element 3.

The length of a list can be queried by the function len() with the list as an argument:

```
>>> len(stellar_list)
4
```

The last element of a list can also be accessed using by the index -1:

```
>>> print(stellar_list[-1])
0.5
```

You can check this by comparing the last element of a list referenced in two different ways:

```
>>> stellar_list[-1] == stellar_list[len(stellar_list)-1]
True
```

The value True indicates that the two expressions are indeed identical.

A list is really a class (see below). This means that a number of operations are defined to work with or on lists. For example, to add a new element to a list

```
>>> stellar_list.append(100)
```

or sort a list

```
>>> stellar_list.sort()
>>> print(stellar_list)
[0, 0.5, 4, 14, 100]
```

or remove an element from a list

```
>>> stellar_list.pop(0)
>>> print(stellar_list)
[0.5, 4, 14, 100]
```

Lists also have operators, for example the + operator that enables you to concatenate several lists

```
>>> stellar_list + [3, 2, 1]
[0.5, 4, 14, 100, 3, 2, 1]
```

You can do a lot more with lists, but by now you probably get the idea.

C.3.3 Dictionaries

Making a hash table in `Python` is easily accomplished by using a dictionary. This works as follows

```
>>> stellar_mass = {'Betlegeuse': 8, 'Rigel': 21, 'Spica': 10.25}
>>> for name, mass in stellar_mass.iteritems():
...:     print('star: %s has mass: %s' % (name, mass))
...:
star: Rigel has mass: 21
star: Spica has mass: 10.25
star: Betlegeuse has mass: 8
```

Each of the entries `Betlegeuse`, `Spica`, and `Rigel` are now associated with a value. Just as in a list, this value can be anything: a parameter, a function, or even a class.

C.3.4 Flow control

A simple `Python` program is usually composed of conditionals and loops, just as in any other procedural language. The main flow control elements in `Python` are `if`, `for`, and `while`.

A typical `if` statement chooses between two or more alternatives:

```
>>> if star == 6:
...:     print(star)
...:else:
...:     print(other_star)
...:
18
```

Other comparison operators (e.g. <, <=, >, >=, etc.) can be used in place of ==. One slight peculiarity of `Python` is that it also includes the operators `is` and `is not`. You might think that == and `is` should mean more or less the same, but actually == compares the value of two expressions, whereas `is` tests if the two objects are the same:

```
>>> 1 == True
True
>>> 1 is True
False
```

The `for` loop has a running parameter (here called `i`) that increases from a value 0 up to some value:

```
>>> for i in range(3):
...:     print(i)
...:
0
1
2
```

In this case, we simply print the variable `i`. The loop stops automatically when the objective is reached. Note that, in consistent `Python` usage, `range(3)` actually returns the list `[0, 1, 2]`. `Python` recognizes which instructions belong to the loop by indentation. In most AMUSE scripts, we adopt four spaces of indentation, although sometimes, if the lines become long and the number of inner loops large, we use two. Other languages use special symbols to indicate the start and the end of a loop, such as braces in `C` or the words `DO` and `END DO` in `FORTRAN`. (`C` and `FORTRAN` programmers generally hate `Python`'s strict reliance on indentation to indicate structure.)

A `while` loop continues until some Boolean expression is no longer true. The previous loop could have been written

```
>>> i = 0
>>> while (i < 3):
...:     print(i)
...:     i += 1
...:
0
1
2
```

To be honest (as is also the case in `C` and `FORTRAN`), we do not really need both `for` and `while` loop constructs—one can always be written in terms of the other, but tradition is a powerful thing. Unlike `C` or `C++`, `Python` has no `case` or `switch` statements. This is not a problem because again these functionalities can be constructed as needed using the available flow control operators.

C.3.5 Functions

`Python` allows the user to define functions. They can take and return any argument (including other functions). The purpose of a function is to prevent code duplication and to make your code as modular as possible. A simple function in `Python` might look like

```
def factorial(n):
   if n < 1:
       return 1
   else:
       return n*factorial(n-1)
```

(Note: you should never write the factorial function this way in a real script!)

A more useful function that calculates the radius of a star in solar radii, given its mass in solar masses, might be:

```
def stellar_radius(mass=1.0):
    radius = mass**0.8
    return radius
```

Note the use of a default value for the argument. You might want to save this handy function in a file called `calculate_stellar_radius.py` for later use. Then, from another script (or interactive `Python` session) you could load this function and use it:

```
from calculate_stellar_radius import stellar_radius
r = stellar_radius(2)
print(r)
```

We recommend trying to keep functions small, less than a page in length if possible, and focused on a single operation.

C.3.6 Importing packages

In an earlier example (in Section C.3.5), we imported a function from a file we had just created. Many functions and entire packages are already written, and you should always try to avoid rewriting anything that has already been written, except perhaps for pedagogical purposes. Just as we import the `amuse.lab` package, you can import other packages. Important packages include `numpy` (van der Walt et al. 2011), `scipy` (see for example Rojas et al. 2015), `math` for numerical, scientific, and mathematical extensions of the language. There are also packages for plotting, such as `pyplot`, machine learning `pyML`, MCMC (e.g., `emcee`; Foreman-Mackey et al. 2013), `AstroPy` (The Astropy Collaboration et al. 2013), and many more.

C.3.7 Classes

A class is a data container with associated operators and functions to manipulate the data it contains. An example of a widely used class in AMUSE is the `Particle` set. Here, we illustrate the use of classes by creating a simple `Body` class that can form the basis for more complex data structures. Code example C.1 illustrates how to declare a `Body` base class.

This class is not complicated—it requires only one initialization parameter, the mass of the object, contains two class variables, called `type` and `mass`, and two class methods (functions), one to initialize the class itself and one to return its contents as a string. A more elaborate example of the use of classes in AMUSE can be found in Section 9.3.2.

A derived class is a class that inherits the properties of a base class and extends its functionality. Code example C.2 presents two examples, for a planet (called `Planet`) and for a star (called `Star`), both derived from `Body`. (Class declaration names in `Python` conventionally start with a capital letter.) They share the `Body` class variable `mass` and the `Body` member functions. Both, however, extend the class, `Planet` by adding the orbital parameters semimajor axis and eccentricity, and `Star` by adding a radius. In addition, they both override the `type` class variable, to "`planet`" and "`star`", respectively, and modify the string returned by `__str__`. We can now manipulate these classes, make declarations, put them in arrays, give them to functions as arguments, etc.

A useful aspect of classes is the ability to make compound classes that themselves are classes. This is illustrated in Code example C.3. We could further develop this class structure by making it more object-oriented, but that would be beyond the scope of this book, particularly because there are many books that do an excellent job in explaining classes and inheritance.

Classes are nice and versatile. They make programming a lot more fun, and we strongly encourage you to explore their potential. Despite these recommendations,

```
class Body:
    def __init__(self, mass):
        self.type = "body"
        self.mass = mass

    def __str__(self):
        return f"{self.type} mass = {self.mass}"
```

Code example C.1:. A base class.

```
MassEarth = 3.0e-6
RadiusEarth = 0.01


class Planet(Body):
    def __init__(self, mass, semimajor_axis, eccentricity=0.0):
        Body.__init__(self, mass)
        self.type = "planet"
        self.semi = semimajor_axis
        self.ecc = eccentricity
        self.radius = RadiusEarth * (mass / MassEarth) ** 0.55

    def __str__(self):
        return f"{self.type} mass = {self.mass}, a = {self.semi}"


class Star(Body):
    def __init__(self, mass):
        Body.__init__(self, mass)
        self.type = "star"
        self.calculate_radius()

    def __str__(self):
        return f"{self.type} mass = {self.mass}, radius = {self.radius}"

    def calculate_radius(self):
        self.radius = self.mass**1.2
```

Code example C.2:. Derived class fragments.

in this book we have largely refrained from using classes, mainly to avoid going too deeply into unfamiliar territory.

C.3.8 Operator Overloading

The redefinition in the previous section of member functions with double leading and trailing underscores is a form of overloading. The member functions

```
def __init__(self):
```

```
class PlanetarySystem(Body):
    def __init__(self, star, planets):
        self.type = "planetary system"
        self.star = star
        self.planets = planets

    def print(self):
        print(self.star)
        for pi in self.planets:
            print(pi)
```

Code example C.3:. A compound class.

and

```
def __str__(self):
```

defined in Code example C.1 overload the previously existing member functions of the same names. Aside from these predefined functions, many other functions can be overloaded, including arithmetic and other operators. You can even define your own operators and give them specific functionality. In AMUSE we have overloaded the pipe command, |, to add a unit to a quantity. The `AstroPy` package overloads the * operator for much the same purpose.

For example, in the earlier example of the `Body` base class, we could overload the + sign for adding a planet to a star. This could be realized by adding the following method to the `Star` class

```
def __add__(self, planet):
    self.mass += planet.mass
    self.calculate_radius()
    return self
```

In an executable script, we could then write

```
s = Star(mass=1)
p = Planet(mass=0.01)
t = s + t
print(t)
```

Which would result in

```
 star mass = 1.01, radius= 1.01201196814
```

This simply returns a `Star` with a larger mass. Obviously, we would want to do more, perhaps adding functionality to compute and/pr evolve the planet's orbit, but the basic procedure should be clear. Quite a number of operators can be overloaded in this fashion.

C.3.9 Properties

You will find the line

```
@property
```

in a few places in the text, and in numerous class definitions in the AMUSE source code. This construction is unfamiliar to many casual `Python` users, so we present a brief indication here of what it means and how it works.

The `@property` line preceding a class method definition in `Python` is a concise way to build a data descriptor that triggers a function call when an attribute is accessed. What does that mean? For example, in the `Star` class, we might add the method

```
@property
def core_mass(self):
    return 0.3*self.mass
```

Without the `@property` line, we would access the core mass with

```
s = Star(1)
mc = s.core_mass()
```

With the `@property` line, the second line above would become

```
mc = s.core_mass
```

Why is it important to turn a function call into an attribute? In AMUSE, the principal reason for doing this is to make access to class data more homogeneous. For example, in the `Particles` class, calculating the `kinetic_energy` and `potential_energy` are done internally using functions, but it is convenient to use the `@properties` syntax to make these quantities accessible as class variables.

This may still seem like a minor syntactic change, but in `Python` this approach is part of a broader class of operations that control access to internal class data that should not be accessed directly. In other object-oriented languages, such as `C++`, setting and getting private data in a class is commonly realized using a `set` or a `get` function. This allows the designer of a class to control how the class data are modified and used. In `Python`, which has "best-practice" conventions for limiting data access but does not actually have private data, this is best done via the @property paradigm and related constructs.

The syntax allows the result of a complex calculation to appear as a class attribute, as we have just seen, but it can also be used to prevent users from directly setting variables with "side effects." For example, changing the mass of a star with a planetary system should also adjust the orbital elements of the planets orbiting it. Hence, we should not allow the mass to be modified directly, but instead intercept that attempt with a function that both changes the mass and modifies the orbits. However, since AMUSE does not use this extended functionality, we must stop the discussion here and refer the interested reader to the extensive Python documentation on the subject.

C.3.10 Testing and Debugging Python Scripts

Every script that is intended for anything other than testing should be tested, and the test scripts should be tested too. From personal experience, we know that any program we write is likely to have some serious shortcomings (aka bugs) if it is not thoroughly tested. Tests can be simple, but they should test fundamental functionalities of every script. It is not uncommon to spend more time writing tests than writing production code. In the end, writing extensive test functions can speed up the development of the production code enormously, and, once finished, the code is much more likely to work as intended. Time devoted to writing test scripts is always time well spent.

A test code starts by importing the relevant Python module

```
import unittest
```

Within the script to test, test_get_posvel_from_binary_elements.py, we add a test case, for example something simple like:

```
def test_get_posvel_from_binary_elements():
    a = 1 | units.au
    try:
        r, v = get_posvel_from_binary_elements(a, 0.0, 1)
        assert False
    except:
        pass
    r, v = get_posvel_from_binary_elements(a, 0.0, 1 |units.MSun)
    assert r[0] == a
```

If the assert statement fails, the test fails, and you should start debugging the relevant routine. You can test the script by calling:

```
%> pytest test_get_posvel_from_binary_elements.py
```

You should create tests for every routine you write. Give the test routine the same name as the routine that performs the particular test, and run `pytest` to perform the test.

Making and running test scripts will help you to write correct and (almost) bug-free code. However, the time will probably come when you have to start a debug session (see Appendix B.1.7). Debugging is a necessity at every level of programming. It is no fun to hunt for the source of a `segmentation fault`[2]. A run-time debugging tool may provide much needed relief. In `Python`, this can be done by importing the `logging` library, after which it is possible to run a script in the debugger. This is done as follows:

```
import logging
logging.basicConfig(level=logging.DEBUG)
logging.getLogger("code").setLevel(logging.DEBUG)
```

There are few hard and fast practical rules on debugging. Pin down the bug using lots of `print` statements, wade through the data they produce to augment the debugger data, and good luck!

References

Aho, A. V., Kernighan, B. W., & Weinberger, P. J. 1987, The AWK Programming Language (Boston, MA: Addison-Wesley)

American National Standards Institute 1966, ANSI Fortran X3.9-1966 (New York: American National Standards Institute)

Ciccarelli, E. C. 1978, An Introduction to the EMACS Editor, AI Memo 447, *MIT*

Foreman-Mackey, D., Hogg, D. W., Lang, D., & Goodman, J. 2013, PASP, 125, 306

Gosling, J., Joy, B., & Steele, G. L. 1996, The Java Language Specification (1st ed.; Boston, MA: Addison-Wesley)

Gosling, J., Joy, B., Steele, G. L., Bracha, G., & Buckley, A. 2014, The Java Language Specification, Java SE 8 Edition 1st ed. (Addison-Wesley)

Kernighan, B. W. 1984, The UNIX Programming Environment (Englewood Cliffs, NJ: Prentice-Hall)

Kernighan, B. W., & Ritchie, D. M. 1988, The C Programming Language (2nd ed; Englewood Cliffs, NJ: Prentice-Hall)

Kluyver, T., Ragan-Kelley, B., Perez, F., et al. 2016, Jupyter Notebooks: A Publishing Format for Reproducible Computational Workflows (Amsterdam: IOS Press)

Langtangen, H. P. 2009, Python Scripting for Computational Science (3rd ed.; Berlin: Springer)

Perez, F., & Granger, B. E. 2007, CSE, 9, 21

Press, W. H., Teukolsky,, Saul, A., Vetterling,, William, T., & Flannery, B. P. 1992, Numerical Recipes in C. The Art of Scientific Computing (Cambridge: Cambridge Univ. Press)

[2] If you do not know what a segmentation fault is, make a point of never finding out (after Press et al. 1992 in relation to the bubble-sort algorithm).

Rojas, S. J. G., Christensen, E. A., & Blanco-Silva, F. J. 2015, Learning SciPy for Numerical and Scientific Computing Second Edition (2nd ed.; Birmingham: Packt Publishing)
Sterling, T. L. 2002, Beowulf Cluster Computing with Linux (Cambridge, MA: MIT Press)
Stroustrup, B. 2013, The C++ Programming Language (4th ed.; Reading, MA: Addison-Wesley)
The Astropy CollaborationRobitaille, T. P., Tollerud, E. J., et al. 2013, A&A, 558, A33
van der Walt, S., Colbert, S. C., & Varoquaux, G. 2011, CSE, 13, 22
van Rossum, G. 1995a, Extending and Embedding the Python Interpreter CS-R9527, *CWI*
van Rossum, G. 1995b, Python Library Reference CS-R9524, *CWI*
van Rossum, G. 1995c, Python Reference Manual CS-R9525, *CWI*
van Rossum, G. 1995d, Python Tutorial CS-R9526, *CWI*

www.ingramcontent.com/pod-product-compliance
Ingram Content Group UK Ltd.
Pitfield, Milton Keynes, MK11 3LW, UK
UKHW061829100926
8556IPUK00011B/927
9780750353298